The Finite Element Method in Heat Transfer and Fluid Dynamics

J.N. Reddy
Department of Mechanical Engineering
Texas A&M University
College Station, Texas

D.K. Gartling
Computational Fluid Dynamics Department
Sandia National Laboratories
Albuquerque, New Mexico

CRC Press
Boca Raton Ann Arbor London Tokyo

Library of Congress Cataloging-in-Publication Data

Reddy, J. N. (Junuthula Narasimha), 1945–
The finite element method in heat transfer and fluid dynamics/
J.N. Reddy and D.K. Gartling.
p. cm.
Includes bibliographical references and index.
ISBN 0-8493-9410-4
1. Fluid dynamics—Mathematical models. 2. Head—Transmission-
-Mathematical models. 3. Finite element method. I. Gartling,
David K. II. Title.
TA357.R43 1994
620.1′06—dc20 94-8277
CIP

International Standard Book Number 0-8493-9410-4
Library of Congress Card Number 94-8277
Printed in the United States of America 2 3 4 5 6 7 8 9 0
Printed on acid-free paper

ABOUT THE AUTHORS

Professor J. N. Reddy is the inaugural holder of the *Oscar S. Wyatt Endowed Chair* in Mechanical Engineering at Texas A&M University, College Station, Texas. Prior to the current position, he was the *Clifton C. Garvin Professor* of Engineering Science and Mechanics at Virginia Polytechnic Institute and State University, Blacksburg, Virginia. He earned a Ph.D. in Engineering Mechanics, and worked as a Postdoctoral Fellow at the University of Texas at Austin, Research Scientist for Lockheed Missiles and Space Company during 1974–75, and taught at the University of Oklahoma from 1975 to 1980 and Virginia Polytechnic Institute & State University from 1980 to 1992. Dr. Reddy authored numerous papers on the theory and finite element analysis of problems in continuum mechanics, solid and structural mechanics, laminated composite plates and shells, computational fluid mechanics, numerical heat transfer, and applied mathematics. Dr. Reddy is the author of seven other books, including: *An Introduction to the Finite Element Method*, McGraw–Hill, 1984 (second edition, 1993); *Energy and Variational Methods in Applied Mechanics*, John Wiley & Sons, 1984; *Applied Functional Analysis and Variational Methods in Engineering*, McGraw–Hill, 1986. Dr. Reddy is the recipient of the *Walter L. Huber Civil Engineering Research Prize* of the American Society of Civil Engineers (ASCE) and the *Worcester Reed Warner Medal* of the American Society of Mechanical Engineers (ASME). Dr. Reddy is a *fellow* of the American Academy of Mechanics (AAM), the American Society of Civil Engineers (ASCE), the American Society of Mechanical Engineers (ASME), and the Aeronautical Society of India (ASI). Dr. Reddy serves on the editorial boards of numerous journals, including *International Journal for Numerical Methods in Engineering* and *International Journal for Numerical Methods in Fluids.*

Dr. D. K. Gartling is a Distinguished Member of the Technical Staff in the Computational Fluid Dynamics Department at Sandia National Laboratories, Albuquerque, New Mexico. He earned his B.S. and M.S. in Aerospace Engineering at the University of Texas at Austin and completed the diploma course at the von Kármán Institute for Fluid Dynamics in Brussels, Belgium. After completion of his Ph.D. in Aerospace Engineering at the University of Texas at Austin, he joined the technical staff at Sandia National Laboratories. Dr. Gartling was a Visiting Associate Professor in the Mechanical Engineering Department at the University of Sydney, Australia under a Fulbright Fellowship, and later he was a Supervisor in the Fluid and Thermal Sciences Department at Sandia National Laboratories. Dr. Gartling published numerous papers dealing with finite element model development and finite element analysis of heat transfer and fluid dynamics problems of practical importance. He is presently a member of several professional societies, and serves on the editorial boards of several archival journals, including *International Journal for Numerical Methods in Fluids* and *Computational Fluid Dynamics.*

PREFACE

The numerical simulation of fluid mechanics and heat transfer problems has become a routine part of engineering practice as well as a focus for fundamental and applied research. Though there are still various topical areas where our physical understanding and/or ineffective numerical algorithms limit the investigation, a large number of complex phenomena can now be confidently studied via numerical simulation. Though finite difference methods have been and will continue to play a major role in computational fluid dynamics (CFD) and heat transfer, finite element techniques have spurred the explosive development of "general purpose" methods and the growth of commercial software. The inherent strengths of the finite element method such as unstructured meshes, element–by–element formulation and processing, and the simplicity and rigor of boundary condition application are being coupled with modern developments in automatic mesh generation, adaptive meshing, and improved solution techniques to produce accurate and reliable simulation packages that are widely accessible. Improvements in computer hardware and system software (*e.g.*, powerful workstations and window environments) have contributed significantly to streamlining the numerical simulation process. The finite element method in fluid mechanics and heat transfer has rapidly caught up with the well–established solid mechanics community in simulation capabilities.

As in any rapidly developing field, the education of the nonexpert user community is of primary importance. The present text is an attempt to fill a need for those interested in using the finite element method in the study of fluid mechanics and heat transfer. It is a pragmatic book that views numerical computation as a means to an end—we do not dwell on theory or proof. Other fundamental and theoretical text books that cover these aspects are available or anticipated. The emphasis here is on presenting a useful methodology for a limited but significant class of problems dealing with heat conduction, incompressible viscous flows, and convection heat transfer.

The text has been developed out of our experience and course notes used in teaching graduate courses and continuing education courses to a wide spectrum of students. To gain the most from the text the student should have a reasonable background in fluid mechanics and heat transfer as would normally be found in most mechanical, aerospace, chemical or engineering mechanics curriculums. An introductory knowledge of finite element techniques would be very helpful but not essential; some familiarity with basic numerical analysis, linear algebra, and numerical integration would also be of assistance.

Our approach to the finite element method for fluid mechanics and heat transfer has been designed as a series of incremental steps of increasing complexity. In Chapter 1, the continuum boundary value problems that form the central focus of the book are described in some detail. We have tried to be as general as possible in describing the varied physical phenomena that may be encountered within the limits of nonisothermal, incompressible, viscous flows. Chapter 2 introduces the finite element method by application to a simplified, two–dimensional heat conduction problem. All of the necessary machinery for constructing weak forms of a partial differential equation and building a finite element model are introduced here and demonstrated by application. Chapter 3 recaps parts of Chapter 2 and extends the finite element method to three dimensions, time dependence, and practical applications in conduction heat transfer. Isothermal viscous fluid mechanics formulations are described in Chapter 4 along with a significant section on the solution of nonlinear equations developed from the flow problem. Chapter 5 extends the viscous flow problem to consider convective heat transfer formulations and applications. Inelastic non–Newtonian flows and free surface problems are covered in Chapter 6. The complex topic of viscoelastic flow simulation is surveyed in Chapter 7. The last chapter concludes the text with a survey of several advanced topics, including turbulence modeling. The coverage of each topic is sufficient to allow the reader to understand the basic methodology, use existing simulation software with confidence, and allow development of some simpler, special purpose computer codes. Example problems ranging from simple benchmarks to practical engineering solutions are included with each topical area. Adequate references to the relevant literature have also been included for those desiring a more encyclopedic coverage of a specific topic.

The text is organized into major sections within each chapter. Equations are numbered consecutively within each major section. Within a section, reference to an equation is by its sequential number; references to equations outside the current section have a full section, equation number citation. Vectors, tensors and matrices are denoted by boldface letters. The vectors of interpolation (shape) functions in this book are denoted by greek symbols ($\mathbf{\Psi}, \mathbf{\Theta}, \mathbf{\Phi}$). Bibliographic information for literature cited in the text is numbered sequentially within each chapter and collected at the end of the chapter.

The first author would like to thank Mr. M. S. Ravishankar for his help with Chapter 8 topics and Mr. Praveen Gramma for reading the manuscript. The second author would like to thank his numerous present and former colleagues at Sandia National Laboratories who have provided a seemingly endless stream of challenging problems and taught him much about the practice of computational mechanics. Specific acknowledgments must go to Drs. Charles Hichox, Rick Givler, Roy Hogan, Phil Sackinger, Randy Schunk, Rekha Rao, and Steve Rottler for their suggestions and comments on early versions of the text. Portions of this book are adapted from work performed at Sandia National Laboratories under Contract No. DE–AC04–76DP00789 awarded by the U.S. Department of Energy and are used with permission.

J. N. Reddy
College Station, Texas

D. K. Gartling
Albuquerque, New Mexico

CONTENTS

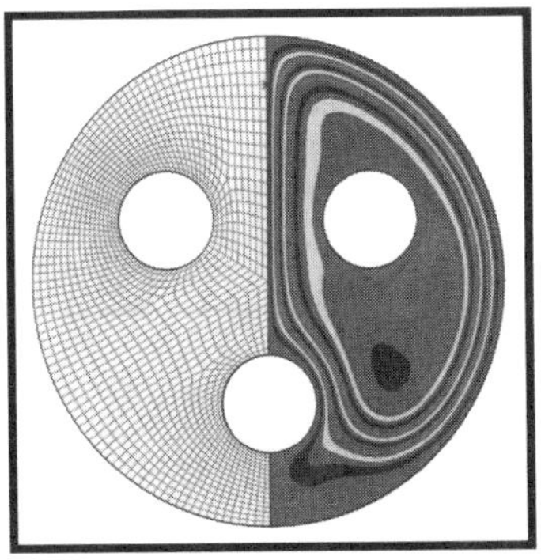

The Finite Element Method in Heat Transfer and Fluid Dynamics

Chapter One

EQUATIONS OF HEAT TRANSFER AND FLUID MECHANICS

1.1 INTRODUCTION

The disciplines of heat transfer and fluid dynamics are subjects that are closely coupled and form the core of many engineering and scientific studies. Areas as diverse as aerodynamics, meteorology, geology and geophysics, biology, material science and manufacturing all rely heavily on the ability to predict, understand and control complex fluid and thermal systems. In order to properly outline the scope of the present text, it is important to first provide brief definitions of the general areas of heat transfer and fluid dynamics.

1.1–1 Heat Transfer

Heat transfer is a branch of engineering that deals with the transfer of thermal energy from one point to another within a medium or from one medium to another due to the occurrence of a temperature difference. Heat transfer may take place in one or more of its three basic forms: conduction, convection, and radiation. The Fourier heat conduction law states that the heat flow is proportional to the temperature gradient. The constant of proportionality depends, among other things, on a material parameter known as the *thermal conductivity* of the material.

Convection heat transfer is usually defined as energy transport effected by the motion of a fluid. The convection heat transfer between two dissimilar media is governed by Newton's law of cooling. It states that the heat flow is proportional to the difference of the temperatures of the two media. The proportionality constant is called the *convection heat transfer coefficient* or *film conductance.*

Thermal radiation is defined as radiant (electromagnetic) energy emitted by a medium and is due solely to the temperature of the medium. Radiant energy exchange between surfaces or between a region and its surroundings is described by the Stefan–Boltzmann law, which states that the radiant energy transmitted is proportional to the difference of the fourth power of the temperatures of the surfaces. The proportionality parameter is known as the *Stefan–Boltzmann constant.*

1.1–2 Fluid Mechanics

Fluid mechanics is one of the oldest branches of physics, and is concerned with the motion of gases and liquids and their interaction with the surroundings. For example, the flight of birds in the air and the motion of fish in the water can be understood by the principles of fluid mechanics. Such understanding helps us design airplanes and ships. The formation of tornadoes, hurricanes, and thunderstorms can also be explained with the help of the equations of fluid mechanics.

A fluid state of matter, as opposed to the solid state of matter, is characterized by the relative mobility of its molecules. Very strong intermolecular attractive forces exist in solids which are responsible for the property of relative rigidity (or stiffness) in solids. The intermolecular forces are weaker in liquids and extremely small in gases. The stress in a solid body is proportional to the strain (*i.e.* deformation per unit length), while the stress in a fluid is proportional to the time rate of strain (*i.e.*, rate of deformation). The proportionality parameter in the case of fluids is known as the ***viscosity***. It is a measure of the intermolecular forces exerted as layers of fluid attempt to slide past one another. The viscosity of a fluid, in general, is a function of the thermodynamic state of the fluid and in some cases the strain rate.

Fluid mechanics is a very broad area and is traditionally divided into smaller topical areas based on characteristics of the fluid properties or the basic nature of the flow. An *inviscid fluid* is one where the viscosity is assumed to be zero. An *incompressible fluid* is one with constant density or an *incompressible flow* is one in which density variations (compared to a reference density) are negligible. An inviscid and incompressible fluid is termed an *ideal* or a *perfect* fluid. A *real fluid* is one with finite viscosity, and it may or may not be incompressible. When the viscosity of a fluid depends only on thermodynamic properties, and the stress is linearly related to the strain rate, the fluid is said to be *Newtonian*. A *non-Newtonian fluid* is one which does not obey the Newtonian (*i.e.*, linear) stress–strain rate relation.

The flow of viscous fluids can also be classified into two major types: a smooth, orderly motion is called *laminar flow*, and a random, fluctuating motion is called *turbulent flow*. The nature of a viscous flow can be characterized by a nondimensional parameter, the *Reynolds number*, $Re = \rho UL/\mu$, which is defined as the ratio of inertial forces ρU^2 to viscous forces $\mu U/L$. Here ρ denotes the density of the fluid, U the characteristic flow velocity, μ is the fluid viscosity, and L is a characteristic dimension of the flow. High viscosity fluids and/or small velocities produce relatively small Reynolds numbers and a laminar flow. The flow of less viscous fluids and/or higher velocities lead to higher Reynolds numbers and a turbulent flow. Transitional flows contain regions of both laminar and turbulent flow.

1.2 PRESENT STUDY

The purpose of the present book is to describe the use of the finite element method for the solution of engineering problems in well–defined areas of heat transfer and fluid mechanics. In particular, we will concentrate on computational techniques for the areas of conduction and convection heat transfer. Since convection relies heavily an fluid dynamics, the study of isothermal flows will play a prominent role in developing the overall subject. To focus the presentation, all fluid mechanics applications here will be limited to the area of viscous, incompressible flows. Radiation heat transfer will be discussed to the extent that it affects thermal boundary conditions and heat transfer in enclosures. In addition, nonisothermal flows in porous media will be considered as an extension of the viscous flow problem. Complexities such as change of phase, free surface boundaries, and non–Newtonian constitutive behavior will also be considered.

In the remainder of this chapter, we will provide a more definitive outline of the heat transfer problems of interest. These sections will provide a résumé of the partial differential equations that describe heat transfer and fluid mechanics in the continuum. The boundary conditions for typical problems are also presented, and the special conditions associated with change of phase problems, free surface flows, and enclosure radiation are discussed.

1.3 FUNDAMENTAL EQUATIONS OF A CONTINUOUS MEDIUM

1.3–1 Introduction

In the present study we are interested in the combined convective and conductive transport of thermal energy in a material region, Ω. In the general case, the region Ω is made up of subregions containing a moving fluid in Ω_f and a solid body in Ω_s, where $\Omega = \Omega_f \cup \Omega_s$. The motion or equilibrium of a continuous medium is governed by global conservation principles.

There are two alternative descriptions used to express the conservation laws in analytical form. In the first, one considers the motion of all matter passing through a *fixed spatial location.* Here one is interested in various properties (*e.g.*, velocity, pressure, temperature, density, *etc.*) of the matter that instantly occupies the fixed spatial location. This description is called the *Eulerian Description* or *Spatial Description.* In the second, one focuses attention on a *set of fixed material particles*, irrespective of their spatial locations. The relative displacements of these particles and the stress caused by external forces and temperature are of interest in this case. This description is known as the *Lagrangian description* or *material description.* The Eulerian description is most commonly used to study fluid flows, while the Lagrangian description is used to study the stress and deformation of solid bodies. Here we present the governing equations of a continuous medium based on the Eulerian description. For a derivation of the equations, the reader may consult books on continuum mechanics, heat transfer, or fluid mechanics (see [1–8]).

1.3–2 Conservation of Mass

The law of conservation of mass states that the total time rate of change of mass in a fixed region is zero. The mathematical statement of the law results in the following equation, known as the *equation of continuity*,

$$\frac{D\rho}{Dt} + \rho \nabla \cdot \mathbf{u} = 0 \tag{1.3 – 1}$$

where ρ is the density of the fluid and D/Dt denotes the *material derivative* or *Eulerian derivative* operator

$$\frac{D}{Dt} = \frac{\partial}{\partial t} + \mathbf{u} \cdot \nabla \tag{1.3 – 2}$$

and $\mathbf{u}$ is the velocity vector, and ∇ is the vector differential operator. When the density changes following a fluid particle are negligible, then the fluid (or flow) is termed *incompressible* and we have $D\rho/Dt = 0$. The continuity equation (1) then becomes

$$\nabla \cdot \mathbf{u} = 0 \tag{1.3 – 3}$$

which is often referred to as the incompressibility constraint.

1.3–3 Conservation of Momentum

The law of conservation of linear momentum (or Newton's second law of motion) states that the total time rate of change of linear momentum is equal to the sum of external forces acting on the region. Newton's second law can be written as

$$\rho \frac{D\mathbf{u}}{Dt} = \nabla \cdot \sigma + \rho \mathbf{f} \qquad (1.3-4)$$

where σ is the Cauchy stress tensor and $\mathbf{f}$ is the body force vector, measured per unit mass. Equations (4) describe the equations of motion of a continuous medium, and in fluid mechanics they are also known as the *Navier equations.* The conservation of angular momentum, in the absence of any distributed moments, leads to the symmetry of the stress tensor: $\sigma = (\sigma)^T$.

1.3–4 Conservation of Energy

The law of conservation of energy (or the First Law of Thermodynamics) states that the time rate of change of the total energy is equal to the sum of the rate of work done by applied forces and the change of heat content per unit time. The first law of thermodynamics can be expressed, for incompressible fluids, as

$$\rho C_v \frac{DT}{Dt} = -\nabla \cdot \mathbf{q} + Q + \Phi \qquad (1.3-5)$$

where T is the temperature, $\mathbf{q}$ is the heat flux vector, Q is the internal heat generation (measured per unit volume), Φ is the viscous dissipation function

$$\Phi = \tau : \mathbf{D} \qquad (1.3-6)$$

and C_v is the specific heat at constant volume. In equation (6), τ is the viscous part of the stress tensor σ, and $\mathbf{D}$ is the strain rate tensor. These quantities are defined more completely in the next section.

1.3–5 Constitutive Equations

The fluids of interest, for the moment, are assumed to be Newtonian (*i.e.*, the constitutive relations are linear). Non–Newtonian fluids will be considered in a later chapter. Further, the fluids are assumed to be incompressible, and the flow is laminar. For flows involving buoyancy forces, an extended form of the Boussinesq approximation [9,10] is invoked, which allows the fluid properties to be expressed as functions of the thermodynamic state (*e.g.*, pressure and temperature) and the density ρ to vary with temperature T according to the relation

$$\rho = \rho_0[1 - \beta(T - T_0)] \qquad (1.3-7)$$

where β is the coefficient of thermal expansion and the subscript zero indicates a reference condition. The variation of density as given in (7) is permitted only in the description of the body force; the density in all other situations is assumed to be that of the reference state, ρ_0 .

For viscous incompressible fluids the total stress σ can be decomposed into hydrostatic and viscous parts:

$$\sigma = -P\mathbf{I} + \tau \qquad (1.3-8)$$

where P is the hydrostatic pressure, $\mathbf{I}$ is the unit tensor and τ is the viscous stress tensor. For Newtonian fluids, the viscous stress tensor is related to the strain rate tensor $\mathbf{D}$ by

$$\tau = \mathbf{C} : \mathbf{D} \qquad (1.3-9)$$

where $\mathbf{C}$ is the fourth–order tensor of fluid properties and $\mathbf{D}$ is the strain rate tensor

$$\mathbf{D} = \frac{1}{2}\left[(\nabla\mathbf{u}) + (\nabla\mathbf{u})^T\right] \qquad (1.3-10)$$

For an isotropic fluid (*i.e.*, whose material properties are independent of direction), the fourth–order tensor $\mathbf{C}$ can be expressed in terms of two constants, λ and μ (the so–called Lamé constants), and equation (9) takes the form

$$\tau = \lambda(\text{tr } \mathbf{D})\mathbf{I} + 2\mu\mathbf{D} \qquad (1.3-11a)$$

where (tr $\mathbf{D}$) denotes the *trace* (or sum of the diagonal elements) of the matrix $\mathbf{D}$. For an incompressibe fluid, we have tr $\mathbf{D} = 0$ and equation (11a) becomes

$$\tau = 2\mu\mathbf{D} \qquad (1.3-11b)$$

The Fourier heat conduction law states that

$$\mathbf{q} = -\mathbf{k} \cdot \nabla T \qquad (1.3-12a)$$

where $\mathbf{k}$ denotes the conductivity tensor of order two. For an isotropic medium, $\mathbf{k}$ is of the form

$$\mathbf{k} = k\mathbf{I} \qquad (1.3-12b)$$

where k denotes the thermal conductivity of the medium.

The material coefficients, μ, C_v, β, and $\mathbf{k}$ are generally functions of the fluid temperature; property variations as a function of a second thermodynamic variable (*e.g.*, pressure) are possible but usually are unimportant for an incompressible fluid. Solid materials may also have specific heats and conductivities that vary with temperature. In addition, the solid conductivity may vary with spatial position and in general remains a symmetric, second-order tensor (*i.e.*, $\mathbf{k}^T = \mathbf{k}$) for anisotropic materials. The volumetric heat source for the fluid and/or solid may be a function of temperature, time, and spatial location. In developing the finite element models in the coming chapters, the dependence of the material properties on the spatial location is assumed. The dependence of the viscosity and conductivity on the strain rates and/or temperature are also discussed in the sequel.

1.4 GOVERNING EQUATIONS IN TERMS OF PRIMITIVE VARIABLES

1.4–1 Vector Form

Equations (1.3–1), (1.3–4), and (1.3–5) can be expressed in terms of the primitive variables $(\mathbf{u}, P, T)$ by means of equations (1.3–6)–(1.3–12). The results are summarized below for isotropic, Newtonian, viscous, incompressible fluids in the presence of buoyancy forces:

$$\nabla \cdot \mathbf{u} = 0 \qquad (1.4-1)$$

$$\rho_0 \left(\frac{\partial \mathbf{u}}{\partial t} + \mathbf{u} \cdot \nabla \mathbf{u} \right) = -\nabla P + \mu \nabla \cdot \left[(\nabla \mathbf{u}) + (\nabla \mathbf{u})^T \right] + \rho_0 \mathbf{f} + \rho_0 \mathbf{g} \beta (T - T_0) \qquad (1.4-2)$$

$$\rho_0 C_v \left(\frac{\partial T}{\partial t} + \mathbf{u} \cdot \nabla T \right) = \nabla \cdot (k \nabla T) + Q + \Phi \qquad (1.4-3)$$

where $\mathbf{u}$ represents the velocity vector, ρ_0 the density, $\mathbf{g}$ the gravity force vector per unit mass, T the temperature, C_v the specific heat of the fluid at constant volume, and Q the rate of heat generation.

The above equations are valid for the fluid region Ω_f. In the solid region Ω_s, the fluid velocity is zero, $u = 0$, and the only relevent equation is (3). The energy equation (3) for the solid region is given by

$$\rho_s C_s \frac{\partial T}{\partial t} = \nabla \cdot (k_s \nabla T) + Q_s \qquad (1.4-4)$$

In writing equation (4) it is assumed that the solid is stationary with respect to the coordinate frame such that the advective transport of energy [*i.e.*, the nonlinear part of (3)] need not be considered.

1.4–2 Cartesian Component Form

The vector form of the equations in (1.4–1)–(1.4–3) allows us to express them in any coordinate system. In the Cartesian coordinate system (x_1, x_2, x_3), the kinematic and constitutive relations become

$$D_{ij} = \frac{1}{2}\left(\frac{\partial u_i}{\partial x_j} + \frac{\partial u_j}{\partial x_i}\right) \tag{1.4-5}$$

$$\sigma_{ij} = -P\delta_{ij} + \tau_{ij} \;;\quad \tau_{ij} = 2\mu D_{ij} \tag{1.4-6}$$

The conservation equations can be expressed as,

$$\frac{\partial u_i}{\partial x_i} = 0 \tag{1.4-7}$$

$$\rho_0\left(\frac{\partial u_i}{\partial t} + u_j\frac{\partial u_i}{\partial x_j}\right) = \frac{\partial}{\partial x_j}\left[-P\delta_{ij} + \mu\left(\frac{\partial u_i}{\partial x_j} + \frac{\partial u_j}{\partial x_i}\right)\right] + \rho_0 f_i - \rho_0 g_i\beta(T - T_0) \tag{1.4-8}$$

$$\rho_0 C\left(\frac{\partial T}{\partial t} + u_j\frac{\partial T}{\partial x_j}\right) = \frac{\partial}{\partial x_i}\left(k\frac{\partial T}{\partial x_i}\right) + Q + 2\mu D_{ij}D_{ij} \tag{1.4-9}$$

for the fluid region Ω_f and

$$\rho_s C_s\frac{\partial T}{\partial t} = \frac{\partial}{\partial x_i}\left(k_s\frac{\partial T}{\partial x_i}\right) + Q_s \tag{1.4-10}$$

for the solid region Ω_s. Equations (5)–(10) are written for a Cartesian geometry in an Eulerian reference frame, with the indices $i, j = 1, 2, 3$ (or 1 and 2 for two–dimensional problems); the Einstein summation convention on repeated indices is used (see [6], pp. 18–20).

1.4–3 Cylindrical Component Form

Equations (1.4–1)–(1.4–3) can also be expressed in a cylindrical coordinate system, (r, θ, z), by writing all vectors and tensors, including the del operator, in terms of components in a cylindrical coordinate system. For example, the del operator and the material time derivative operators in the cylindrical coordinate system are given by (see Reddy and Rasmussen [6], p. 94 and p.128)

$$\nabla = \hat{\mathbf{e}}_r\frac{\partial}{\partial r} + \hat{\mathbf{e}}_\theta\frac{1}{r}\frac{\partial}{\partial \theta} + \hat{\mathbf{e}}_z\frac{\partial}{\partial z} \tag{1.4-11}$$

$$\frac{D}{Dt} = \frac{\partial}{\partial t} + u_r\frac{\partial}{\partial r} + \frac{u_\theta}{r}\frac{\partial}{\partial \theta} + u_z\frac{\partial}{\partial z} \qquad (1.4-12)$$

where $(\hat{\mathbf{e}}_r, \hat{\mathbf{e}}_\theta, \hat{\mathbf{e}}_z)$ are the unit basis vectors and (u_r, u_θ, u_z) are the velocity components in the r, θ, and z directions, respectively. Note that the basis vector $\hat{\mathbf{e}}_z$ is constant while the vectors $\hat{\mathbf{e}}_r$ and $\hat{\mathbf{e}}_\theta$ depend on the angular coordinate θ. Thus the derivatives of the basis vectors $\hat{\mathbf{e}}_r$ and $\hat{\mathbf{e}}_\theta$ with respect to the coordinates r and z are zero, and the derivatives with respect to θ are given by

$$\frac{\partial \hat{\mathbf{e}}_r}{\partial \theta} = \hat{\mathbf{e}}_\theta; \quad \frac{\partial \hat{\mathbf{e}}_\theta}{\partial \theta} = -\hat{\mathbf{e}}_r \qquad (1.4-13)$$

The kinematic and constitutive relations are

$$\begin{aligned} D_{rr} &= \frac{\partial u_r}{\partial r}; \quad D_{\theta\theta} = \frac{1}{r}\frac{\partial u_\theta}{\partial \theta} + \frac{u_r}{r} \\ D_{zz} &= \frac{\partial u_z}{\partial z}\ ; \quad 2D_{r\theta} = \frac{\partial u_\theta}{\partial r} - \frac{u_\theta}{r} + \frac{1}{r}\frac{\partial u_r}{\partial \theta} \\ 2D_{\theta z} &= \frac{1}{r}\frac{\partial u_z}{\partial \theta} + \frac{\partial u_\theta}{\partial z}\ ; \quad 2D_{zr} = \frac{\partial u_r}{\partial z} + \frac{\partial u_z}{\partial r} \end{aligned} \qquad (1.4-14)$$

$$\sigma_{rr} = -P + 2\mu D_{rr}\ ; \quad \sigma_{\theta\theta} = -P + 2\mu D_{\theta\theta}\ ; \quad \sigma_{zz} = -P + 2\mu D_{zz} \qquad (1.4-15a)$$

$$\sigma_{r\theta} = 2\mu D_{r\theta}\ ; \quad \sigma_{\theta z} = 2\mu D_{\theta z}\ ; \quad \sigma_{zr} = 2\mu D_{zr} \qquad (1.4-15b)$$

The governing equations are summarized below:

$$\frac{1}{r}\frac{\partial}{\partial r}(ru_r) + \frac{1}{r}\frac{\partial u_\theta}{\partial \theta} + \frac{\partial u_z}{\partial z} = 0 \qquad (1.4-16)$$

$$\rho_0\left(\frac{Du_r}{Dt} - \frac{u_\theta^2}{r}\right) = \rho f_r + \frac{1}{r}\left[\frac{\partial(r\sigma_{rr})}{\partial r} + \frac{\partial \sigma_{r\theta}}{\partial \theta} + \frac{\partial(r\sigma_{zr})}{\partial z}\right] - \frac{\sigma_{\theta\theta}}{r} \qquad (1.4-17a)$$

$$\rho_0\left(\frac{Du_\theta}{Dt} + \frac{u_r u_\theta}{r}\right) = \rho f_\theta + \frac{1}{r}\left[\frac{\partial(r\sigma_{r\theta})}{\partial r} + \frac{\partial \sigma_{\theta\theta}}{\partial \theta} + \frac{\partial(r\sigma_{\theta z})}{\partial z}\right] + \frac{\sigma_{r\theta}}{r} \qquad (1.4-17b)$$

$$\rho_0\left(\frac{Du_z}{Dt}\right) = \rho f_z + \frac{1}{r}\left[\frac{\partial(r\sigma_{zr})}{\partial r} + \frac{\partial \sigma_{z\theta}}{\partial \theta} + \frac{\partial(r\sigma_{zz})}{\partial z}\right] \qquad (1.4-17c)$$

$$\begin{aligned} \rho C_s\left(\frac{\partial T}{\partial t} + u_r\frac{\partial T}{\partial r} + \frac{u_\theta}{r}\frac{\partial T}{\partial \theta} + u_z\frac{\partial T}{\partial z}\right) &= \frac{1}{r}\frac{\partial}{\partial r}\left(rk_{rr}\frac{\partial T}{\partial r}\right) + \frac{1}{r^2}\frac{\partial}{\partial \theta}\left(k_{\theta\theta}\frac{\partial T}{\partial \theta}\right) \\ &+ \frac{\partial}{\partial z}\left(k_{zz}\frac{\partial T}{\partial z}\right) + \Phi + Q \end{aligned} \qquad (1.4-18a)$$

where the stress components are known in terms of the velocity components *via* equations (14) and (15), and the viscous dissipation Φ is given by

$$\Phi = 2\mu\left[\left(\frac{\partial u_r}{\partial r}\right)^2 + \left(\frac{1}{r}\frac{\partial u_\theta}{\partial \theta} + \frac{u_r}{r}\right)^2 + \left(\frac{\partial u_z}{\partial z}\right)^2\right] + \mu\left(\frac{\partial u_\theta}{\partial r} - \frac{u_\theta}{r} + \mu\frac{\partial u_r}{\partial \theta}\right)^2$$

$$+\mu\left(\frac{1}{r}\frac{\partial u_z}{\partial \theta} + \frac{\partial u_\theta}{\partial z}\right)^2 + \mu\left(\frac{\partial u_r}{\partial z} + \frac{\partial u_z}{\partial r}\right)^2 \qquad (1.4-18b)$$

Note that the time derivative of a vector (or tensor) in rotating reference frames is given by (see Reddy and Rasmussen [6], pp. 69–74)

$$\left[\frac{D(\cdot)}{Dt}\right]_{nonrot} = \left[\frac{D(\cdot)}{Dt}\right]_{rot} + \omega \times (\cdot) \qquad (1.4-19)$$

where $\omega = \frac{d\theta}{dt}\hat{\mathbf{e}}_z = \frac{u_\theta}{r}\hat{\mathbf{e}}_z$ is the angular velocity vector of the rotating frame of reference. Therefore, we have

$$\left(\frac{D\mathbf{u}}{Dt}\right)_{nonrot} = \left(\frac{D\mathbf{u}}{Dt}\right)_{rot} + \left(\frac{u_\theta}{r}\hat{\mathbf{e}}_z\right) \times \mathbf{u}$$

$$= \hat{\mathbf{e}}_r\left(\frac{Du_r}{Dt} - u_\theta\frac{u_\theta}{r}\right) + \hat{\mathbf{e}}_\theta\left(\frac{Du_\theta}{Dt} + u_r\frac{u_\theta}{r}\right) - \hat{\mathbf{e}}_z\frac{Du_z}{Dt} \qquad (1.4-20)$$

Equation (20) could be used in conjunction with (17) to provide a description of fluid motion in a rotating cylindrical coordinate system.

1.4–4 Closure

Equations (1)–(5) describe the convective and conductive heat transfer problems of primary interest. Although the velocities, pressure, and temperature were selected as the dependent variables, other choices for the description of the fluid motion are possible. Indeed, equations employing the stream function and vorticity, or the stream function alone, can be derived and used effectively with the energy equation to analyze convection problems. The choice of velocity, pressure, and temperature variables, the so-called "primitive variables", is made here for a number of reasons. The velocity and pressure variables are a natural description for fluid mechanics and permit boundary conditions to be described in a straightforward and intuitive manner. The description of free surface flows, problems in multiply connected domains, and phase change problems is direct and uncomplicated with the primitive variables. Also, the extension to three dimensions is straightforward, as is the inclusion of non-Newtonian and/or porous (Darcy) flow effects. All of the subsequent developments in this book will be carried out using the primitive variable description.

1.5 POROUS FLOW EQUATIONS

Suppose that the fluid domain, Ω_f, is composed of two subregions denoted by Ω_{vf} and Ω_{pf} such that $\Omega_f = \Omega_{vf} \cup \Omega_{pf}$. In the subregion Ω_{vf}, the viscous flow equations (1.4–7)-(1.4–9) are assumed to hold. The subregion Ω_{pf} is assumed to contain a rigid porous medium that is saturated with a viscous, incompressible fluid. The saturating fluid in Ω_{pf} is the same fluid as in Ω_{vf} if the two regions share a common permeable interface; otherwise the fluids in the two regions may be different. If the porous medium is assumed to be homogeneous and isotropic and the fluid and solid are in thermal equilibrium, then the equations describing the fluid motion and energy transport in the region Ω_{pf} can be written for a Cartesian rectangular coordinate system as

$$\frac{\partial u_i}{\partial x_i} = 0 \tag{1.5 – 1}$$

$$\rho_0 \alpha_{ij} \frac{\partial u_j}{\partial t} + \left(\frac{\rho_0 \hat{c}}{\sqrt{\kappa}} ||\vec{u}|| + \frac{\mu}{\kappa} \right) u_i = \frac{\partial}{\partial x_j} \left[-P\delta_{ij} + \mu_e \left(\frac{\partial u_i}{\partial x_j} + \frac{\partial u_j}{\partial x_i} \right) \right] + \rho_0 f_i - \rho_0 g_i \beta (T - T_0) \tag{1.5 – 2}$$

$$(\rho_0 C)_e \left(\frac{\partial T}{\partial t} + u_j \frac{\partial T}{\partial x_j} \right) = \frac{\partial}{\partial x_i} \left(k_e \frac{\partial T}{\partial x_i} \right) + Q \tag{1.5 – 3}$$

Analogous equations are available for other coordinate systems. In equations (1)-(3), κ is the permeablility, $\hat{c}$ is the inertia coefficient, $||\mathbf{u}||$ is the magnitude of the velocity vector, and α_{ij} are the components of the acceleration tensor. The porosity, ϕ, of the medium is defined as the fraction of the total volume of material occupied by interconnected pores (voids). The velocity components in (1)-(3) represent the volume averaged Darcy velocity (also termed the seepage or filtration velocity) which is related to the interstitial pore level velocity divided by the porosity of the medium. The subscript e indicates an effective property; all other symbols are as defined earlier. The effective capacitance and conductivity are generally functions of the fluid and solid matrix properties and the porosity. The effective conductivity can also be a function of the fluid velocity when thermal dispersion effects are important. The acceleration tensor is dependent on the geometry of the porous matrix, though in several cases it is proportional to $\frac{1}{\phi}$. Flow transients in a porous material usually decay very rapidly and the time–dependent term in (2) is often neglected. The effective viscosity is often taken to be $\mu_e = \mu$ although other approximations are possible. Suitable empirical models are available for all of these quantities and can be found in references [11–13].

The equations in (1)-(3) represent a generalization of the standard Darcy equations for nonisothermal flow in a saturated porous medium. This system is sometimes referred to as the *Forchheimer-Brinkman* or *Darcy-Forchheimer-Brinkman model* for porous flow. No attempt will be made here to delineate all the conditions under which such a model is appropriate. However, it should be noted that by selectively including or omitting certain terms, a number of other standard porous flow models can be derived. Thus, if $\hat{c} = 0$, a Brinkman model is obtained, while a prescription of $\hat{c} = 0$ and $\mu_e = 0$ produces a standard Darcy formulation. For further discussion of individual models and their regions of applicability, the reader is referred to [13–16] and the literature on porous media models.

The general form of the porous flow equations is very similar to the equations for the viscous flow of a bulk fluid. By simply redefining certain terms and coefficients in equations (1)–(3), they can, in fact, be obtained from (1.4–7)–(1.4–10). As a result of this simple transformation process it is quite straightforward to include a porous flow model in the general computational framework for nonisothermal flow problems. The region Ω_{pf} may exist in conjunction with Ω_{vf} or in place of Ω_{vf}. The solid region, Ω_s, may or may not be included as the problem dictates. In the following sections specific reference will be made to the porous flow equations only when their treatment differs significantly from the equations for a viscous fluid.

1.6 AUXILIARY TRANSPORT EQUATIONS

The equations governing nonisothermal, viscous flow problems outlined in the previous sections are very general. However, there are still a significant number of flow problems that cannot be described by the stated equations. For flows in which transport processes other than those of a thermal nature are important, additional equations are required. The added equations are of the advection–diffusion type and are generic in the sense that they are not specifically associated with a particular physical process. Possible uses for these additional equations include the description of mass transport in a multicomponent system, the simulation of certain types of chemical reactions, and the prediction of volume fraction or particle orientation for flows containing suspended particles or fibers. Here we give the additional equations describing such flows.

Consider the material region Ω. The transport of a scalar quantity ϕ_i within this region is governed, for fixed i (*i.e.*, no sum on i) by

$$C_i \left(\frac{\partial \phi_i}{\partial t} + u_j \frac{\partial \phi_i}{\partial x_j} \right) = \frac{\partial}{\partial x_j} \left(D_i \frac{\partial \phi_i}{\partial x_j} \right) + Q_i \qquad (1.6-1)$$

In equation (1), C_i is a "capacitance" coefficient [analogous to ρC in equation (1.4–10)], D_i is a diffusion coefficient, and Q_i is a volumetric source term. The subscript i runs between 1 and n, and it indicates that up to n transport equations could be added to the general flow problem. Note that equation (1) is valid in the solid regions of Ω if u_j is set to zero; the transport equation can also be used in a porous region if C_i and D_i are interpreted as "effective" properties of the fluid/matrix system.

In order to fully couple the above transport equation to the general nonisothermal flow problem, certain extensions must be made to the previously stated functional form for the fluid density variation. The possibility of buoyancy forces due to variations in the auxiliary variables ϕ_i requires that the equation of state (1.3–7) be modified to

$$\rho = \rho_0 \left[1 - \beta(T - T_0) - \gamma_1(\phi_1 - \phi_{1_0}) \ldots - \gamma_n(\phi_n - \phi_{n_0}) \right] \qquad (1.6-2)$$

where γ_i is an expansion coefficient and the subscript zero refers to a reference condition. As before this variation in density is only permitted in the body force term. The body force term in (1.4–8) must therefore be altered to include the last n terms in (2) whenever auxiliary transport equations of the appropriate type are included in the problem formulation.

1.7 BOUNDARY CONDITIONS

To complete the description of the general initial– and boundary–value problem posed in the previous sections, suitable boundary and initial conditions are required. Boundary conditions are most easily understood and described by considering the fluid mechanics separate from other transport processes. In the following sections, boundary conditions associated with the viscous flow, porous flow, and thermal transport are discussed, followed by a discussion of the initial conditions.

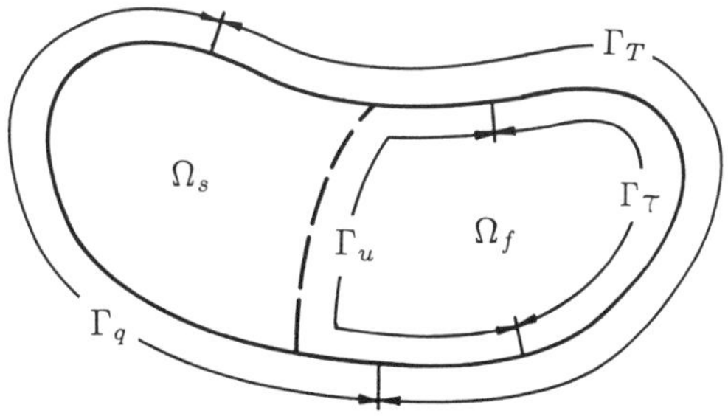

Figure 1.7–1: Schematic for boundary condition definitions.

1.7–1 Viscous Flow Boundary Conditions

For the fluid dynamic part of the problem, either the velocity components or the total surface stress or traction must be specified on the boundary of the fluid region. In general the boundary conditions can be classified into two types (see [17], pp. 28–32), as given below.

Dirichlet or essential boundary conditions:

$$u_i = f_i^u(s,t) \qquad \text{on } \Gamma_u \tag{1.7 – 1a}$$

Neumann or natural boundary conditions:

$$\mathcal{T}_i = \sigma_{ij}(s,t)n_j(s) = f_i^{\mathcal{T}}(s,t) \qquad \text{on } \Gamma_{\mathcal{T}} \tag{1.7 – 1b}$$

where s is the coordinate along the boundary, t is the time, n_i is the outward unit normal to the boundary, and Γ_f is the total boundary enclosing the fluid domain, Ω_f, with $\Gamma_f = \Gamma_u \cup \Gamma_T$ (see Figure 1.7–1). Note that the conditions written in (1) are in component form, *i.e.*, there must be a condition on each component of the velocity. The specified functions f_i^u and f_i^T are generally simple expressions for standard situations where the fluid is contained by fixed boundaries ($f_i^u = 0$) or planes/lines of symmetry ($f_i^T = 0$), or enters/leaves the domain ($f_i^T = constant$) Ω_f. The boundary conditions listed in (1) are adequate for most situations involving a single fluid in an uncomplicated domain.

The general boundary or interface conditions between two immiscible fluids, however, are more complex and are described here for several important types of problems. It is assumed that there is no mass transfer across the interface which allows all components of the velocity field to be continuous across the boundary. Note that this condition is often retained even for change of phase problems in order to simplify the formulation. In concert with this velocity specification, a kinematic condition is also imposed that requires the interface to remain an interface, *i.e.*, the interface is a stream surface. This requirement can be represented by the equation

$$\frac{DF}{Dt} = \frac{\partial F}{\partial t} + u_j \frac{\partial F}{\partial x_j} = 0 \tag{1.7-2}$$

where $F(x_i, t)$ is the equation for the fluid interface. Since the location and shape of the interface between two (or more) fluids is usually not known *a priori*, this type of problem is termed a free boundary problem. For time–dependent free surface flows, equation (2) provides an evolution equation for the free surface that must be solved in conjunction with the viscous flow equations; the fluid domain Ω_f is now a function of time, $\Omega_f(t)$. In the case of a time–independent flow the kinematic condition simplifies to

$$u_i n_i = 0 \tag{1.7-3}$$

which simply states that the free surface is a stream surface. For the steady case there is no explicit equation for the free surface; equation (3) only provides a constraint condition that must be applied along the unknown interface. The numerical treatment of free surface flows will be considered in Chapter 6.

The final condition at the free surface specifies that the change in total stress (traction) between the two fluids is related to surface tension forces. The interfacial stress condition is expressed as

$$\sigma_{ij}^2 n_j - \sigma_{ij}^1 n_j = \gamma \left(\frac{1}{R_1} + \frac{1}{R_2} \right) n_i - (\delta_{ij} - n_i n_j) \frac{\partial \gamma}{\partial x_j} \tag{1.7-4}$$

where γ is the surface tension, R_1 and R_2 are the principal radii of curvature of the interface, and the superscripts '2' and '1' on σ_{ij} refer to the two fluids. The last term in (4) allows variations in the surface tension (*e.g.*, due to temperature) to produce a shear stress in the fluid (Marangoni flows).

The complex conditions expressed in (4) are necessary when the motion in both fluids is of equal importance, *e.g.*, two immiscible liquids. However, in many cases, one of the fluids is a gas and its motion relative to a liquid may be neglected. For this simplified case the continuity condition on the velocity is removed since the gas motion is ignored, and shear stresses (drag forces) on the interface are neglected. If the gas is assumed to be at a uniform pressure, P_{amb}, then (4) becomes

$$\sigma_{ij} n_j = \gamma \left(\frac{1}{R_1} + \frac{1}{R_2} \right) n_i - P_{amb} n_i \qquad (1.7-5)$$

which relates the stress normal to the free surface to the surface tension, surface curvature and ambient external pressure.

Other types of boundary conditions for the fluid domain are possible (*e.g.*, Navier slip condition) and may be used in certain limited situations. The boundary conditions presented here are those most commonly encountered in practice.

1.7–2 Porous Flow Boundary Conditions

The flow conditions that can be applied to the boundary of a fluid-saturated porous medium depend on the specific model used to describe the problem. For the general Forchheimer–Brinkman equation listed in (1.5–2), the admissible boundary conditions are the same as those given for the Navier–Stokes equations, namely, equation (1). However, in cases where the Brinkman terms are excluded from the momentum equation, ($\mu_e = 0$) the order of the equation is reduced and fewer boundary conditions are needed. A Forchheimer or simple Darcy model allows the following types of boundary conditions

$$u_i n_i = f^u(s,t) \qquad \text{on } \Gamma_u \qquad (1.7-6a)$$

$$\mathcal{T}_i n_i = -\hat{P}(s,t)\delta_{ij} n_j(s) n_i = -\hat{P}(s,t) = f^{\mathcal{T}}(s,t) \qquad \text{on } \Gamma_{\mathcal{T}} \qquad (1.7-6b)$$

In essence these conditions state that the fluid velocity normal to the boundary may be specified (outflow/inflow) or the normal force (pressure) on the boundary may be imposed. These simplified models do not allow a tangential velocity to be imposed (*e.g.*, no–slip walls are not admissible) nor the specification of viscous forces. Also, note that only one type of boundary condition can be specified at any point on the boundary.

For many applications a saturated porous layer will adjoin a clear fluid region and certain continuity conditions will be required at the open interface. The applicable conditions vary with the type of porous media model and in many cases have not been rigorously verified. The major difficulty stems from the heterogeneous nature of the porous medium and the fact that a continuum description such as (1.4–1)–(1.4–3) is derived by an averaging process. Clearly, averaging within a distance $\approx \sqrt{\kappa}$ of a boundary or interface is not valid and some loss of information will occur; detailed microscopic analyses or observations must be used to generate valid macroscopic boundary conditions for the continuum models. In both the Darcy and Brinkman formulations the (seepage) velocity normal to the interface is assumed continuous; the Brinkman model assumes all tangential velocity components are also continuous between the porous layer and the clear fluid. Many investigators have assumed that all components of the viscous stress and the pressure are continuous at an interface with a Brinkman formulation, a situation that is very convenient from a computational point of view. However, Nield and Bejan [13] and Nield [14] believe that the stresses and pressure are not continuous, and they advocate the use of a Navier (slip) condition for the tangential stress at the interface. No proposal has been offered for estimating the changes in normal stress or pressure at the interface. The situation for the Darcy model is just as confused since no stress terms are present in this formulation. The standard assumptions in this case are that the pressure is continuous and the shear stress in the bulk fluid is related to the tangential velocity in the porous layer by the Navier type condition proposed by Beavers and Joseph [15],

$$\frac{\partial u_f}{\partial n} = \frac{\alpha}{\sqrt{\kappa}}(u_f - u_p) \qquad (1.7-7)$$

Here u_f is the clear fluid velocity, u_p is the velocity in the porous media, and α is a dimensionless (porous) material parameter. Based on the above discussion it is clear that some caution must be used when considering interface and boundary conditions for a saturated porous material.

1.7–3 Thermal and Transport Boundary Conditions

The thermal part of the boundary value problem for the fluid or solid requires the temperature (Dirichlet or essential condition) or the heat flux (Neumann or natural condition) to be specified on all parts of the boundary enclosing the heat transfer region:

$$T = \hat{T}(s,t) \quad \text{on } \Gamma_T \qquad (1.7-8a)$$

$$\left(k_{ij}\frac{\partial T}{\partial x_j}\right) n_i + q_c + q_r = q(s,t) \quad \text{on } \Gamma_q \qquad (1.7-8b)$$

where Γ_{ht} is the total boundary enclosing the heat transfer region, and $\Gamma_{ht} = \Gamma_T \cup \Gamma_q$. Note that the condition on Γ_q for a fluid or an isotropic solid can be simplified, since k_{ij} reduces to a scalar k. Also, q_c and q_r refer to the convective and radiative components given by

$$q_c = h_c(s,T,t)(T - T_c) \qquad (1.7-9a)$$

$$q_r = h_r(s,T,t)(T - T_r) \qquad (1.7-9b)$$

In equation (9a), h_c is the convective heat transfer coefficient which, in general, depends on the location on the boundary, temperature, and time, and T_c is a reference (or sink) temperature for convective transfer. Also, the effective radiation heat transfer coefficient is $h_r = \mathcal{F}\sigma(T + T_r)(T^2 + T_r^2)$ where σ is the Stefan–Boltzmann constant, $\mathcal{F}$ is a form factor, and T_r the reference temperature for radiative transfer. The form factor, $\mathcal{F}$, is related to the boundary emissivity, ϵ, and the position of the boundary relative to surrounding surfaces (see [18]). Note that this boundary condition is appropriate when a body or surface radiates to a black body environment that can be characterized by a single temperature. A more complex condition of surface to surface or enclosure radiation is described in Section 1.8. As in the case of the fluid boundary conditions, the specified temperature $\hat{T}$ and total flux q are simple functions for the cases normally encountered in practical problems. One of the more complex conditions occurs in change of phase problems and this will be treated separately in the next section.

Another condition that is of concern when considering a material interface between two or more solid regions is the problem of gap or contact resistance. In some situations, the thermal resistance at the interface is attributed to a fictitious material that has a (generally) negligible specific heat and a nonlinear conductivity. The boundary or interface conditions in this situation are the usual continuity conditions on temperature and heat flux since the gap resistance is dictated by property variations. A more mathematical representation of contact resistance provides that the heat flux across the interface be described by an internal boundary condition of the form given in (9):

$$q_{gap} = h_{gap}(s, T_{gap}, t)(T_M - T_S) \qquad (1.7-10)$$

where h_{gap} is an effective heat transfer coefficient for the contact surface, and T_{gap} is an average temperature between T_M and T_S. The subscripts M and S designate the "master" and "slave" sides of the contact surface, a distinction that is important in the numerical implementation of this condition. Note that as h_{gap} becomes large, the temperatures of the two surfaces will essentially become equal, a situation that is useful in some applications.

As a final point on the boundary condition specification note that the boundaries Γ_f and Γ_{ht} need not coincide (see Figure 1.7–1). Indeed, for problems involving regions of convection and solid body conduction the boundary regions are not the same. When both fluid and solid regions exist in a problem, continuity conditions on the temperature and heat flux must prevail along the interface of the two materials. Examples of such applications will be given in a later section.

Although not specifically written here, boundary and initial conditions for the auxiliary transport variables are required if these equations are included in the formulation. The form of these boundary conditions generally follow the mathematical form of the boundary conditions associated with the energy equation for a fluid region. This is due to the close similarity between the transport equation (1.6–1) and the energy equation (1.4–10). Thus the boundary conditions on ϕ_i are the same as those given in equation (8) with T replaced by ϕ_i.

1.7–4 Initial Conditions

For time–dependent problems, a set of initial conditions are necessary for the dependent variables. Very often these conditions consist of a solid body at a uniform temperature and a quiescent fluid at a uniform temperature and hydrostatic pressure. A second common possibility is for the transient motion to be initiated from an established steady–state flow and temperature field. In all cases, the dependent variables must be known for all $\mathbf{x}$ at $t = 0$ and must satisfy the basic conservation equations (*e.g.*, the initial fluid velocity field must be divergence free and have a compatible pressure field).

1.8 CHANGE OF PHASE

Many problems of importance involving conductive and convective heat transfer have the added complication of a material change of phase. Such problems are part of a general class of boundary value problems termed *free* or *moving boundary problems* (see [19]). The major difficulty in analyzing such problems is the occurrence of a boundary (or interface) surface, across which certain jump conditions must be satisfied and whose location is unknown *a priori*. Thus, the location of the moving boundary becomes part of the solution. In this section the equations for a limited class of such problems will be described.

The present discussion will be limited to melting/solidification phase transitions where the solid phase is stationary. Also, it is assumed that density changes upon change of phase may be neglected and that the effects of latent heat release may be accounted for through an appropriate modification of the specific heat. The present development also ignores the explicit modeling of the porous layer or mushy zone that may occur between the solidus and liquidus temperatures in some materials. Methods for treating the mushy zone are easily incorporated into a numerical simulation following the work of Prescott *et al.* [20] and others [21], and primarily involve the use of the porous media equations described previously. With these assumptions, the phase change problem is described by the equations of the previous sections plus a special set of conditions on the melt/solid interface. At the phase change boundary the following conditions are required:

$$u_i\big|_X = 0 \qquad (1.8-1)$$

$$T_f\big|_X = T_s\big|_X \qquad (1.8-2)$$

$$k_f \frac{\partial T}{\partial n}\bigg|_X - k_s \frac{\partial T}{\partial n}\bigg|_X = \rho L \frac{\partial X}{\partial t} \qquad (1.8-3)$$

where L is the latent heat and $X(t)$ is the spatial position of the phase boundary. The function $X(t)$ is unknown and must be determined as part of the solution. Basically, the melt/solid interface is taken to be a no–slip surface [equation (1)], with continuous temperature [equation (2)] and a jump in the heat flux [equation (3)]. Following the work of Bonacini *et al.* [22] and others (see [23–25]), the jump condition in (3) may be rewritten using the so-called "enthalpy method".

By observing that the latent heat, L, corresponds to the isothermal change in the enthalpy, H, for a material at the transition temperature, T_t, the following relation can be introduced

$$H(T) = \int_{T_{ref}}^{T} C(T)dT + L\,\eta(T - T_t) \qquad (1.8-4)$$

with

$$\eta(T) = \begin{cases} 1 & \text{if } T \geq 0 \\ 0 & \text{if } T < 0 \end{cases} \qquad (1.8-5)$$

where η is the Heaviside step function with argument T. The equivalent specific heat, C^*, is then introduced by

$$C^*(T) = \frac{dH}{dT} = C(T) + L\ \delta(T - T_t) \tag{1.8-6}$$

where δ is the Dirac delta function. Through the use of equation (6) latent heat effects may be included *via* the specific heat function, and the jump in the heat flux [see equation (3)] is eliminated from the problem formulation. This particular approach to the problem has a theoretically sound basis, as outlined in [22]. Moreover, it is a computationally effective modification since a two-region problem with a jump condition has been converted to a single-region problem with rapidly varying properties. For use in a finite element model, equation (6) requires additional modification. However, the details of this and other implementation questions will be explained in a later section.

1.9 ENCLOSURE RADIATION

Radiant energy exchange between neighboring surfaces of a region or between a region and its surroundings can produce large effects in the overall heat transfer problem. Though the radiation effects generally enter the heat transfer problem only through the boundary conditions, the coupling is especially strong due to the nonlinear dependence of the radiation on the surface temperature. The present description considers a restricted class of radiation problems that ignore participation of the fluid phase; the radiative exchange only influences the flow and/or solid body conduction problem through modification of the surface temperature distribution.

Enclosure or surface-to-surface radiation is limited to diffuse gray, opaque, surfaces. This assumption implies that all energy emitted or reflected from a surface is diffuse. Further, surface emissivity ϵ, absorbtivity α, and refectivity ρ are independent of wavelength and direction so that $\epsilon(T) = \alpha(T) = 1 - \rho(T)$. Each individual area or surface that is considered in the radiation process must be at a uniform temperature; emitted and reflected energy are uniform over each such surface. Note that the definition of a surface is arbitrary and can be based on geometry alone or be defined to specifically satisfy the uniform temperature criterion.

With the above assumptions, the radiation problem can be approached using the net-radiation method [26]. For purposes of discussion, consider the two-dimensional enclosure made up of N distinct surfaces as shown in Figure 1.9-1. Associated with each surface is a uniform temperature T_j, an area A_j, and a surface emissivity ϵ_j. An energy balance for each surface in the enclosure leads to the following system of equations

$$\sum_{j=1}^{N} \left[\frac{\delta_{kj}}{\epsilon_j} - F_{k-j}\left(\frac{1-\epsilon_j}{\epsilon_j}\right)\right]\frac{Q_j}{A_j} = \sum_{j=1}^{N} (\delta_{kj} - F_{k-j})\,\sigma T_j{}^4 \tag{1.9-1}$$

Equation (1) relates the net energy loss, Q_j, from each surface to the temperature of each surface, where δ_{kj} is the unit tensor, σ is the Stefan-Boltzmann constant, and F_{k-j} are radiation view (configuration) factors. The view factor is defined as the fraction of energy leaving a surface that arrives at a second surface. For surfaces with finite areas, the view factors are defined by

$$F_{k-j} = \frac{1}{A_k} \int_{A_k} \int_{A_j} \frac{\cos\theta_k \cos\theta_j}{\pi S^2} \, dA_j \, dA_k \qquad (1.9-2)$$

where S is the distance from a point on surface A_j to a point on surface A_k. The angles θ_j and θ_k are measured between the line S and the normals to the surface as shown in Figure 1.9–1. It is clear from equation (2) that the view factors are purely geometric quantities that can in principle be evaluated for any given distribution of surfaces. Methods for evaluating F_{k-j} will be outlined later in the book.

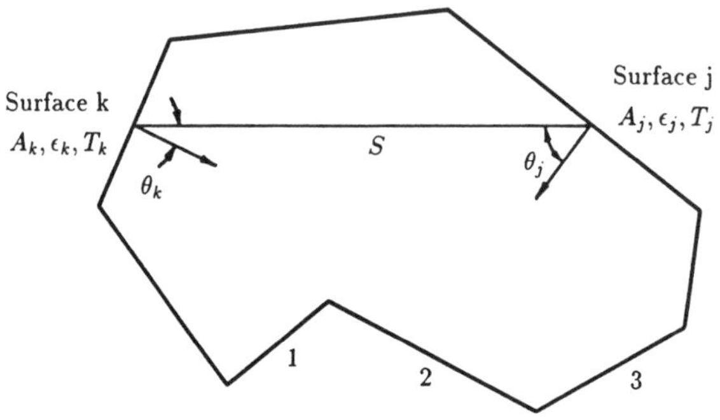

Figure 1.9–1: Nomenclature for enclosure radiation.

For purposes of computation it is convenient to rearrange equation (1) into the following series of equations:

$$\sum_{j=1}^{N} [\delta_{kj} - (1-\epsilon_k)F_{k-j}]\, q_j^o = \epsilon_k \sigma T_k^4 \qquad (1.9-3)$$

and

$$q_k = q_k^o - \sum_{j=1}^{N} F_{k-j} q_j^o \qquad (1.9-4)$$

Equations (3) and (4) are expressed in terms of the outgoing radiative flux for each surface, q_j^o, and the net flux from each surface $q_k = Q_k/A_k$. For known surface temperatures T_k in the enclosure, equation (3) can be solved for the outgoing radiative flux at each surface. Equation (4) then allows the net flux at each surface to be evaluated. The actual method of solution using this formulation in a finite element context will be discussed in the later chapters.

1.10 SUMMARY OF EQUATIONS

All of the governing differential equations, constitutive relations, and boundary conditions of heat transfer and fluid mechanics relevant to the present discussion were presented in the previous sections. It is useful to summarize the complete boundary value problem in one section. Therefore, the complete set of equations are presented here in Cartesian component form for each physical process.

Flows of Viscous Incompressible Fluids:

$$\frac{\partial u_i}{\partial x_i} = 0 \qquad (1.10-1)$$

$$\rho_0 \left(\frac{\partial u_i}{\partial t} + u_j \frac{\partial u_i}{\partial x_j} \right) = \frac{\partial}{\partial x_j} \left[-P\delta_{ij} + \mu \left(\frac{\partial u_i}{\partial x_j} + \frac{\partial u_j}{\partial x_i} \right) \right] + \rho_0 f_i - \rho_0 \hat{g}_i \qquad (1.10-2a)$$

where

$$\hat{g}_i = g_i \left[\beta(T - T_0) + \gamma_1(c_1 - c_{1_0}) + \gamma_2(c_2 - c_{2_0}) + \cdots \right] \qquad (1.10-2b)$$

Porous Flow:

$$\frac{\partial u_i}{\partial x_i} = 0 \qquad (1.10-3)$$

$$\rho_0 \alpha_{ij} \frac{\partial u_j}{\partial t} + \left(\frac{\rho_0 \hat{c}}{\sqrt{\kappa}} ||\vec{\mathrm{u}}|| + \frac{\mu}{\kappa} \right) u_i = \frac{\partial}{\partial x_j} \left[-P\delta_{ij} + \mu_e \left(\frac{\partial u_i}{\partial x_j} + \frac{\partial u_j}{\partial x_i} \right) \right] + \rho_0 f_i - \rho_0 \hat{g}_i \qquad (1.10-4a)$$

where

$$\hat{g}_i = g_i \left[\beta(T - T_0) + \gamma_1(c_1 - c_{1_0}) + \gamma_2(c_2 - c_{2_0}) + \cdots \right] \qquad (1.10-4b)$$

Heat Transfer (Convection):

$$\rho_0 C \left(\frac{\partial T}{\partial t} + u_j \frac{\partial T}{\partial x_j} \right) = \frac{\partial}{\partial x_i} \left(k \frac{\partial T}{\partial x_i} \right) + Q + \Phi \qquad (1.10-5)$$

Heat Transfer (Conduction):

$$\rho_s C_s \left(\frac{\partial T}{\partial t}\right) = \frac{\partial}{\partial x_i}\left(k_{s_{ij}} \frac{\partial T}{\partial x_j}\right) + Q_s \qquad (1.10-6)$$

Auxiliary Transport (i–th transport equation; no sum on i):

$$C_i \left(\frac{\partial \phi_i}{\partial t} + u_j \frac{\partial \phi_i}{\partial x_j}\right) = \frac{\partial}{\partial x_j}\left(D_i \frac{\partial \phi_i}{\partial x_j}\right) + Q_i \qquad (1.10-7)$$

Equations (1)–(7) with the noted boundary and initial conditions form a complete set for the determination of the velocity, pressure, temperature, and auxiliary variables in a fluid, porous, or solid region. Embedded in this description are numerous classes of flow problems that can be recovered through the omission or inclusion of specific equations, the neglect of certain terms within an equation, and the variation of material properties.

EXERCISE PROBLEMS

1.1 Let $\rho(\mathbf{x}, t)$ denote the mass density of a continuous region, Ω. Then conservation of mass for a material region requires that

$$\frac{D}{Dt}\int_\Omega \rho \, d\Omega = 0 \qquad (1)$$

Show that for a fixed region, conservation of mass can also be expressed as

$$\int_\Omega \frac{D\rho}{Dt} \, d\Omega + \int_\Gamma \rho \hat{\mathbf{n}} \cdot \mathbf{u} \, ds = 0 \qquad (2)$$

Interpret the equation in physical terms.

1.2 Newton's second law of motion (or conservation of linear momentum) applied to a continuum states that the rate of change of momentum following a material region Ω of fixed mass is equal to the sum of all the forces acting on the region. This can be expressed as

$$\frac{D}{Dt}\int_\Omega \rho \mathbf{u} \, d\Omega = \int_\Gamma \hat{\mathbf{n}} \cdot \sigma \, ds + \int_\Omega \rho \mathbf{f} \, d\Omega \qquad (1)$$

where σ is the total stress tensor, $\mathbf{f}$ is the body force per unit mass, ρ is the mass density, and $\mathbf{u}$ is the material velocity. Since the material particle mass $\rho d\Omega$ is constant with respect to the material time derivative D/Dt, make use of the divergence theorem and obtain the differential form of Newton's second law of motion:

$$\rho \frac{D\mathbf{u}}{Dt} = \nabla \cdot \sigma + \rho \mathbf{f} \tag{2}$$

1.3 Let ϵ denote the thermodynamic internal energy per unit mass of a material. Then the first law of thermodynamics applied to a material region Ω can be written as

$$\frac{D}{Dt}\int_\Omega \rho(e + \frac{1}{2}\mathbf{u}\cdot\mathbf{u})\, d\Omega = \int_\Gamma \hat{\mathbf{n}}\cdot\sigma\cdot\mathbf{u}\, ds + \int_\Omega \rho\mathbf{f}\cdot\mathbf{u}\, d\Omega - \int_\Gamma \mathbf{q}\cdot\hat{\mathbf{n}}\, ds + \int_\Omega Q\, d\Omega \tag{1}$$

The first two terms on the right–hand side describe the rate of work done on the material region by the surface stresses and the body forces. The third integral describes the net outflow of heat from the region, causing a decrease of energy inside the region, where $\mathbf{q}$ denotes the heat flux vector. The fourth integral describes the internal heat generation in the region, where Q denotes the internal heat generation per unit volume. Show that the differential form of the first law of thermodynamics is given by

$$\rho \frac{De}{Dt} = -\nabla\cdot\mathbf{q} + Q + \sigma : \mathbf{D} \tag{2}$$

Make use of the momentum equation and the identity

$$\text{div}(\sigma\cdot\mathbf{u}) - \mathbf{u}\cdot\text{div}\sigma = \sigma : \mathbf{D} \tag{3}$$

For a viscous incompressible fluid, we take $\rho(De/Dt) = \rho C_v(DT/Dt)$.

1.4 Use equations (1.4–11)-(1.4–13) in equation (1.4–4) and derive the energy equation in the cylindrical coordinate system (r, θ, z).

1.5 For axisymmetric flows of viscous, incompressible fluids (*i.e.*, flow field is independent of $\theta-$ coordinate), show that the Stokes equations (*i.e.*, when the nonlinear terms in the convective parts are neglected) are given by

Conservation of Momentum:

$$\rho\frac{\partial u_r}{\partial t} = \frac{1}{r}\frac{\partial}{\partial r}(r\sigma_{rr}) - \frac{\sigma_{\theta\theta}}{r} + \frac{\partial \sigma_{rz}}{\partial z} + f_r \tag{1a}$$

$$\rho\frac{\partial u_z}{\partial t} = \frac{1}{r}\frac{\partial}{\partial r}(r\sigma_{rz}) + \frac{\partial \sigma_{zz}}{\partial z} + f_z \tag{1b}$$

Conservation of Mass:

$$\frac{1}{r}\frac{\partial}{\partial r}(ru_r) + \frac{\partial u_z}{\partial z} = 0 \tag{2}$$

where

$$\sigma_{rr} = -P + 2\mu\frac{\partial u_r}{\partial r} \ , \quad \sigma_{\theta\theta} = -P + 2\mu\frac{u_r}{r}$$

$$\sigma_{zz} = -P + 2\mu\frac{\partial u_z}{\partial z} \ , \quad \sigma_{rz} = \mu(\frac{\partial u_r}{\partial z} + \frac{\partial u_z}{\partial r}) \tag{3}$$

1.6 Using equations (1.4–11)–(1.4–13) in equations (1.4–1)–(1.4–3), derive equations (1.4–16)–(1.4–18).

1.7 Use the definition of the rate of strain tensor **D** in equation (1.3–10) and equations (1.4–11) and (1.4–13) to verify the relations in equation (1.4–14).

REFERENCES

1. A. Bejan, *Convection Heat Transfer,* John Wiley & Sons, New York (1984).

2. R. B. Bird, W. E. Stewart, and E. N. Lightfoot, *Transport Phenomena,* John Wiley & Sons, New York (1960).

3. J. P. Holman, *Heat Transfer,* Sixth Edition, McGraw–Hill, New York (1986).

4. L. E. Malvern, *Introduction to the Mechanics of a Continuous Medium,* Prentice–Hall, Englewood Cliffs, New Jersey (1969).

5. M. N. Özisik, *Heat Transfer: A Basic Approach,* McGraw–Hill, New York (1985).

6. J. N. Reddy and M. L. Rasmussen, *Advanced Engineering Analysis,* John Wiley & Sons, New York (1982); reprinted by Krieger Publishing, Melbourne, Florida (1991).

7. H. Schlichting, *Boundary-Layer Theory,* (translated by J. Kestin), Seventh Edition, McGraw–Hill, New York (1979).

8. G. K. Batchelor, *An Introduction to Fluid Dynamics,* Cambridge University Press, Cambridge (1967).

9. D. D. Gray and A. Giorgini, "The Validity of the Boussinesq Approximation for Liquids and Gases," *International Journal of Heat and Mass Transfer*, **19**, 545–551 (1976).

10. D. K. Gartling and C. E. Hickox, "A Numerical Study of the Applicability of the Boussinesq Approximation for a Fluid-Saturated Porous Medium," *International Journal of Numerical Methods in Fluids*, **5**, 995–1013 (1985).

11. M. Kaviany, *Principles of Heat Transfer in Porous Media,* Springer–Verlag, New York (1991).

12. D. A. Nield and D. D. Joseph, "Effects of Quadratic Drag on Convection in a Saturated Porous Medium," *Physics of Fluids*, **28**, 995–997 (1985).

13. D. A. Nield and A. Bejan, *Convection in Porous Media*, Springer–Verlag, Berlin (1992).

14. D. A. Nield, "The Limitations of the Brinkman-Forchheimer Equation in Modeling Flow in a Saturated Porous Medium and at an Interface," *International Journal of Heat and Fluid Flow*, **12**, 269–272 (1991).

15. G. S. Beavers and D. D. Joseph, "Boundary Conditions at a Naturally Permeable Wall," *Journal of Fluid Mechanics*, **30**, 197–207 (1967).

16. J. Bear, *Dynamics of Fluids in Porous Media*, American Elsevier, New York (1972).

17. J. N. Reddy, *An Introduction to the Finite Element Method,* Second Edition, McGraw–Hill, New York (1993).

18. V. S. Arpaci, *Conduction Heat Transfer*, Addison-Wesley Publishing, Reading, Massachusetts (1966).

19. J. R. Ockendon and W. R. Hodgkins (Eds.), *Moving Boundary Problems in Heat Flow and Diffusion*, Clarendon Press, Oxford (1975).

20. P. J. Prescott, F. P. Incropera, and W. D. Bennon, "Modeling of Dendrite Solidification Systems: Reassessment of the Continuum Momentum Equation,"*International Journal of Heat and Mass Transfer*, **34**, 2351–2359 (1991).

21. V. R. Voller and C. Prakash, "A Fixed Grid Numerical Modeling Methodology for Convection–Diffusion Mushy Region Phase–Change Problems," *International Journal for Numerical Methods in Engineering,* **30**, 1709–1719 (1987).

22. C. Bonacini, G. Comini, A. Fasano and M. Primicerio, "Numerical Solution of Phase-Change Problems," *International Journal of Heat and Mass Transfer*, **16**, 1825–1832 (1973).

23. G. Comini, S. DelGuidice, R. Lewis, and O. C. Zienkiewicz, "Finite Element Solution of Non–Linear Heat Conduction Problems with Special Reference to Phase Change," *International Journal for Numerical Methods in Engineering,* **8**, 613–624 (1974).

24. D. K. Gartling, "Finite Element Analysis of Convective Heat Transfer Problems with Change of Phase," in *Computer Methods in Fluids*, K. Morgan, C. Taylor and C. Brebbia (Eds), Pentech Press, London (1980).

25. K. Morgan, "A Numerical Analysis of Freezing and Melting with Convection," *Computer Methods in Applied Mechanics and Engineering*, **28**, 275–284 (1981).

26. R. Siegel and J. R. Howell, *Thermal Radiation Heat Transfer,* McGraw–Hill, New York (1972).

Chapter Two

The Finite Element Method: An Overview

2.1 INTRODUCTION

The finite element method is a powerful computational technique for the solution of differential and integral equations that arise in various fields of engineering and applied science. The method is a generalization of the classical variational (*i.e.*, the Rayleigh–Ritz) and weighted–residual (*e.g.*, Galerkin, least-squares, collocation, etc.) methods, which are based on the idea that the solution u of a differential equation can be represented as a linear combination of unknown parameters c_j and appropriately selected functions ϕ_j in the *entire domain* of the problem. The parameters c_j are then determined such that the differential equation is satisfied, often, in a weighted–integral sense. The functions ϕ_j, called the *approximation functions*, are selected such that they satisfy the boundary conditions of the problem. For additional details on the variational and weighted–residual methods, the reader may consult References 1–8 at the end of the chapter.

The traditional variational and weighted–residual methods suffer from one major shortcoming: construction of the approximation functions that satisfy the boundary conditions of the problem to be solved. Most real–world problems are defined on regions that are geometrically complex, and therefore it is difficult to generate approximation functions that satisfy different types of boundary conditions on different portions of the boundary of the complex domain. However, if the domain can be represented as a collection of "simple subdomains" that permit generation of the approximation functions for any arbitrary but physically meaningful boundary conditions, then the traditional variational methods can be used to solve practical engineering problems. The basic idea of the finite element method is to view a given domain as an assemblage of simple geometric shapes, called *finite elements*, for which it is possible to systematically generate the approximation functions needed in the solution of differential equations by any of the variational and weighted–residual methods. The ability to represent domains with irregular geometries by a collection of finite elements makes the method a valuable practical tool for the solution of boundary, initial, and eigenvalue problems arising in various fields of engineering. The approximation functions are often constructed using ideas from interpolation theory, and hence they are also called *interpolation functions*. Thus the finite element method is a piecewise (or elementwise) application of the variational and weighted–residual methods. For a given differential equation, it is possible to develop different finite element approximations (or finite element models), depending on the choice of a particular variational or weighted–residual method.

The primary objective of this chapter is to present an introduction to the finite element method. For elementary introductions to the finite element method, the reader is advised to consult References 7–10. The major steps in the finite element analysis of a typical problem are (see Reddy [7,8]):

1. Discretization of the domain into a set of finite elements (*mesh generation*).
2. Weighted–integral or weak formulation of the differential equation to be analyzed.
3. Development of the finite element model of the problem using its weighted–integral or weak form.
4. Assembly of finite elements to obtain the global system of algebraic equations.
5. Imposition of boundary conditions.
6. Solution of equations.
7. Post–computation of solution and quantities of interest.

In the following sections, we will discuss the first four steps with the help of a model differential equation, namely the equation governing conductive heat transfer in a two–dimensional solid medium. Since this is an overview of the method only the major steps are included. For additional details, the reader is referred to the textbooks by Reddy [7], Becker *et al.* [9], and Burnett [10].

2.2 MODEL DIFFERENTIAL EQUATION

Consider the problem of finding the steady–state temperature $T(x, y)$ distribution in a two–dimensional orthotropic medium Ω, with boundary Γ. The equation governing the temperature distribution is given by setting the time derivative term to zero and $x_1 = x, x_2 = y$ in equation (1.10–6) (also see [11–14]):

$$-\left[\frac{\partial}{\partial x}\left(k_{11}\frac{\partial T}{\partial x}\right) + \frac{\partial}{\partial y}\left(k_{22}\frac{\partial T}{\partial y}\right)\right] = Q \quad \text{in } \Omega \qquad (2.2-1)$$

where k_{11} and k_{22} are conductivities in the x and y directions, respectively, and $Q(x, y)$ is the known internal heat generation per unit volume. For a nonhomogeneous conducting medium, the conductivities k_{ij} are functions of position (x, y). For an isotropic medium, we set $k_{11} = k_{22} = k$ in equation (1) and obtain the Poisson equation

$$-\frac{\partial}{\partial x}(k\frac{\partial T}{\partial x}) - \frac{\partial}{\partial y}(k\frac{\partial T}{\partial y}) = Q \quad \text{in} \quad \Omega \qquad (2.2-2)$$

or in vector form

$$-\nabla \cdot (k\nabla T) = Q \quad \text{in} \quad \Omega \qquad (2.2-3)$$

where ∇ is the gradient operator

$$\nabla = \hat{\mathrm{e}}_x\frac{\partial}{\partial x} + \hat{\mathrm{e}}_y\frac{\partial}{\partial y} \qquad (2.2-4)$$

and $\hat{e}_x$ and $\hat{e}_y$ denote the unit vectors directed along the x and y axes, respectively.

Equation (1) must be solved in conjunction with specified boundary conditions of the problem. The following two types of boundary conditions are assumed in the following development [(see equation (1.7–8)]:

$$T = \hat{T}(s) \quad \text{on } \Gamma_T \qquad (2.2-5a)$$

$$\left(k_{11}\frac{\partial T}{\partial x}n_x + k_{22}\frac{\partial T}{\partial y}n_y\right) + q_c = q(s) \quad \text{on } \Gamma_q \qquad (2.2-5b)$$

where Γ_T and Γ_q are disjoint portions of the boundary Γ such that $\Gamma = \Gamma_T \cup \Gamma_q$, q_c refers to the convective component of heat flux, defined by

$$q_c = h_c(s,T)(T - T_c) \qquad (2.2-6)$$

and (n_x, n_y) denote the direction cosines of the unit normal vector on the boundary. In equation (6) h_c denotes the convective heat transfer coefficient (see Section 1.7).

2.3 FINITE ELEMENT APPROXIMATION

As stated in the introduction to this chapter, the main idea of the finite element method is to generate the approximation functions required in the solution of differential equations by a variational or weighted–integral method. In the finite element method, this is accomplished by subdividing the given domain $\bar{\Omega} = \Omega \cup \Gamma$ into a set of subdomains, called finite elements (see Figure 2.3–1). The phrase *finite element* often refers to both the geometry of the element and degree (or order) of approximation used for the dependent unknown over the element, both of which would be known in the context of the discussion. Any geometric shape for which the approximation functions can be derived uniquely qualifies as an element. We shall discuss various geometric shapes and orders of approximation shortly. To keep the formulative steps very general (*i.e.*, not confine the formulation to a specific geometric shape or order of element), we denote the domain of a typical element by the symbol Ω^e and its boundary by Γ^e. The element Ω^e can be a triangle or quadrilateral in shape, and the degree of interpolation over it can be linear, quadratic, and so on. The non–overlapping sum (or *assembly*) of all elements used to represent the actual domain will be denoted by Ω^h, and it is called the *finite element mesh* of the domain Ω. In general, Ω^h may not equal the actual domain Ω because of the geometric complexity of the actual domain. Of course, for simple geometries (*e.g.*, polygonal domains) the finite element mesh exactly duplicates the actual domain.

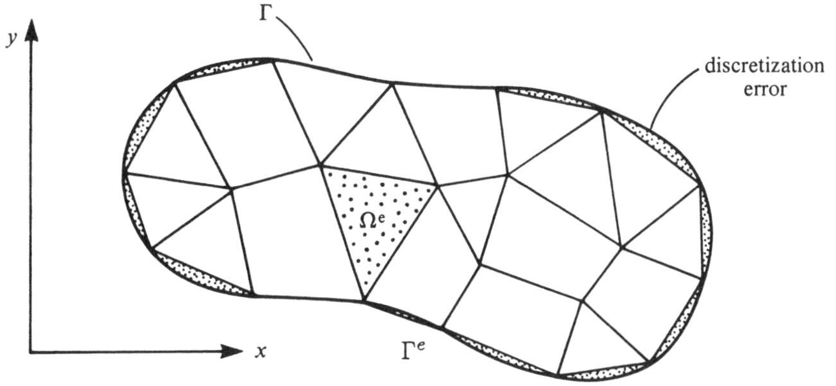

Figure 2.3–1: Finite element discretization of a domain.

For the moment let us assume that it is possible to systematically generate the approximation functions for the element Ω^e. Suppose that the dependent unknown T is approximated over a typical finite element Ω^e by the expression

$$T(x,y) \approx T^e(x,y) = \sum_{j=1}^{n} T_j^e \psi_j^e(x,y) \qquad (2.3-1)$$

where $T^e(x,y)$ represents an approximation of $T(x,y)$ over the element Ω^e, T_j^e denote the values of the function $T^e(x,y)$ at a selected number of points, called *element nodes*, in the element Ω^e, and ψ_j^e are the approximation functions associated with the element. As we shall see shortly, the interpolation functions depend not only on the number of nodes in the element, but also on the shape of the element. The shape of the element must be such that its geometry is uniquely defined by a set of points, which serve as the element nodes in the development of the interpolation functions. A triangle ($n = 3$) is the simplest two–dimensional geometric shape in two dimensions.

Equation (1) is of the same form as that used in the traditional variational and weighted–residual methods, except for the fact that the approximation functions used in the finite element method are often polynomials and the undetermined parameters T_j^e denote the values of the function being approximated. The choice of polynomials for approximation functions is dictated largely by two factors: (1) interpolation theory can be readily used to derive them, and (2) integral expressions of polynomials can be evaluated exactly using numerical integration methods. The use of function values as unknown parameters has the primary objective of imposing continuity of the dependent variable across the boundary between elements. This will be clear when assembly of elements is discussed.

2.4 WEIGHTED–INTEGRAL STATEMENTS AND WEAK FORMS

The n parameters (or nodal values) T_j^e in equation (2.3–1) must be determined such that the approximate solution $T^e(x,y)$ satisfies the governing equation (2.2–1) and specified boundary conditions of the problem. As in the case of a variational and weighted–residual method, we seek to satisfy the governing differential equation in a weighted–integral sense, as described below. Since we are working with a typical element, we satisfy the differential equation in a weighted–integral sense over the element Ω^e. This process leads to n algebraic equations among the nodal values $(T_1^e, T_2^e, \cdots, T_n^e)$. The set of algebraic equations is termed a *finite element model* of the original differential equation. The type of finite element model depends on the weighted–integral form used to generate the algebraic equations. Thus, if one uses a variational form, also called a *weak form*, the resulting model will be different from those obtained with a weighted–residual statement in which the weight function can be any one of the several choices (see [3–8] for additional details). Throughout the present study, we will be primarily concerned with the weak–form finite element models, in which the weight functions are selected to be the same as the approximation functions (the so–called Rayleigh–Ritz–Galerkin models).

The weak form of a differential equation is a weighted–integral statement that is equivalent to both the governing differential equation as well as the associated natural boundary conditions. We shall develop the weak form of equations (2.2–1) and (2.2–5b) over the typical element Ω^e. There are three steps in the development of a weak form. The first step is to take all nonzero expressions in equation (2.2–1) to one side of the equality, multiply the resulting equation with a weight function w, and integrate the equation over the element domain Ω^e:

$$0 = \int_{\Omega^e} w \left[-\frac{\partial}{\partial x}\left(k_{11}\frac{\partial T^e}{\partial x}\right) - \frac{\partial}{\partial y}\left(k_{22}\frac{\partial T^e}{\partial y}\right) - Q(x,y) \right] dxdy \qquad (2.4-1)$$

The expression in the square brackets of the above equation represents a residual of the approximation of the differential equation (2.2–1), because $T^e(x,y)$ is only an approximation of $T(x,y)$. Therefore, equation (1) is called the *weighted–residual statement* of equation (2.2–1). For every choice of the weight function $w(x,y)$, we obtain an algebraic equation from equation (1) among the nodal values T_j^e. For n independent choices of w, we obtain a set of n linearly independent algebraic equations. This set is called a *weighted–residual finite element model.* The drawback of this model for second– and higher–order differential equations is that the approximation functions ψ_j^e should be differentiable as many times as the actual solution T in the differential equation (2.2–1). For the present case, the approximation functions should be twice–differentiable with respect to both x and y. This in turn requires that ψ_j^e be a quadratic or higher–order polynomial in the x and y coordinates. In the weak form, as the name suggests, this continuity requirement is reduced (or weakened) by moving some of the differentiation to the weight function.

In the second step, we distribute the differentiation among T and w equally, so that both T and w are required to be differentiable only once with respect to x and y. To achieve this we use integration–by–parts (or the Green–Gauss theorem) on the first two terms in equation (1). First we note the following identities for any differentiable functions $w(x,y)$, $F_1(x,y)$, and $F_2(x,y)$:

$$\frac{\partial}{\partial x}(wF_1) = \frac{\partial w}{\partial x}F_1 + w\frac{\partial F_1}{\partial x} \quad \text{or} \quad -w\frac{\partial F_1}{\partial x} = \frac{\partial w}{\partial x}F_1 - \frac{\partial}{\partial x}(wF_1) \qquad (2.4-2a)$$

$$\frac{\partial}{\partial y}(wF_2) = \frac{\partial w}{\partial y}F_2 + w\frac{\partial F_2}{\partial y} \quad \text{or} \quad -w\frac{\partial F_2}{\partial y} = \frac{\partial w}{\partial y}F_2 - \frac{\partial}{\partial y}(wF_2) \qquad (2.4-2b)$$

Next, we recall the component form of the gradient (or divergence) theorem,

$$\int_{\Omega^e} \frac{\partial}{\partial x}(wF_1)\, dxdy = \oint_{\Gamma^e} (wF_1)n_x \, ds \qquad (2.4-3a)$$

$$\int_{\Omega^e} \frac{\partial}{\partial y}(wF_2)\, dxdy = \oint_{\Gamma^e} (wF_2)n_y \, ds \qquad (2.4-3b)$$

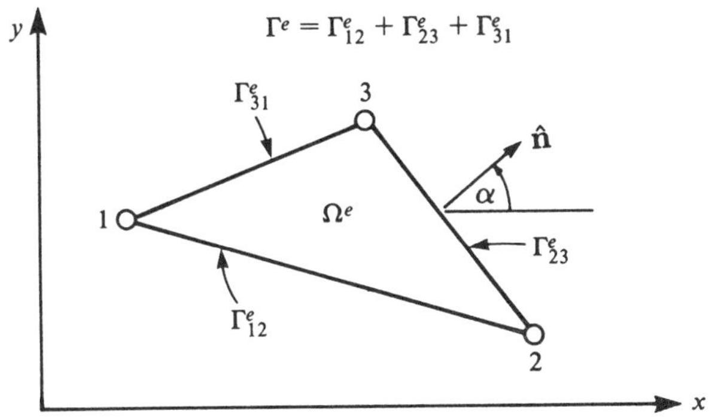

Figure 2.4–1: The unit normal vector on the boundary of an element.

where n_x and n_y are the components (*i.e.*, the direction cosines) of the unit normal vector

$$\hat{\mathbf{n}} = n_x\hat{\mathbf{e}}_x + n_y\hat{\mathbf{e}}_y = \cos\alpha\hat{\mathbf{e}}_x + \sin\alpha\hat{\mathbf{e}}_y \qquad (2.4-4)$$

on the boundary Γ^e, and ds is the arc length of an infinitesimal line element along the boundary (see Figure 2.4–1). Using equations (2) and (3), with

$$F_1 = k_{11}\frac{\partial T}{\partial x}, \quad F_2 = k_{22}\frac{\partial T}{\partial y}$$

we obtain

$$0 = \int_{\Omega^e}\left(\frac{\partial w}{\partial x}k_{11}\frac{\partial T}{\partial x} + \frac{\partial w}{\partial y}k_{22}\frac{\partial T}{\partial y} - wQ\right)dxdy - \oint_{\Gamma^e} w\left(k_{11}\frac{\partial T}{\partial x}n_x + k_{22}\frac{\partial T}{\partial y}n_y\right)ds \tag{2.4-5}$$

From an inspection of the boundary term in equation (5), we note that the specification of T constitutes the essential boundary condition, and T is called the *primary variable.* The specification of the coefficient of the weight function in the boundary expression

$$q_n = k_{11}\frac{\partial T}{\partial x}n_x + k_{22}\frac{\partial T}{\partial y}n_y \tag{2.4-6}$$

constitutes the natural boundary condition, and q_n is the *secondary variable* of the formulation. The function $q_n = q_n(s)$ denotes the projection of the vector $\mathbf{k}\cdot\nabla T$ along the unit normal $\mathbf{n}$. By definition q_n is positive outward from the surface as we move counterclockwise along the boundary Γ^e. The secondary variable q_n denotes the heat flux normal to the boundary of the element (into the element).

The third and last step of the formulation is to use the definition (6) in equation (5) and write it as

$$0 = \int_{\Omega^e}\left(k_{11}\frac{\partial w}{\partial x}\frac{\partial T}{\partial x} + k_{22}\frac{\partial w}{\partial y}\frac{\partial T}{\partial y} - wQ\right)dxdy - \oint_{\Gamma^e} wq_n\, ds \tag{2.4-7}$$

or

$$B(w,T) = \ell(w) \tag{2.4-8}$$

where the *bilinear form* $B(\cdot,\cdot)$ and *linear form* $\ell(\cdot)$ are defined by

$$B(w,T) = \int_{\Omega^e}\left(k_{11}\frac{\partial w}{\partial x}\frac{\partial T}{\partial x} + k_{22}\frac{\partial w}{\partial y}\frac{\partial T}{\partial y}\right)dxdy \tag{2.4-9a}$$

$$\ell(w) = \int_{\Omega^e} wQ\, dxdy + \oint_{\Gamma^e} wq_n\, ds \tag{2.4-9b}$$

The word 'bilinear' indicates that $B(w,T)$ is linear in *both* w and T.

This completes the three–step procedure of constructing the weak form of a differential equation. The weak form in equation (7) or (8) forms the basis of the finite element model of equation (2.2–1). By substituting for T from equation (2.3–1) and setting $w = \psi_1, \psi_2, ..., \psi_n$ (for the Rayleigh–Ritz–Galerkin model), we obtain n algebraic equations from equation (7) or (8), as shown in the next section.

The statement in equation (8) is called the *variational problem* associated with equations (2.2–1) and (2.2–5b). Whenever the bilinear form is symmetric in its arguments (w, T),

$$B(w,T) = B(T,w),$$

it is possible to construct the associated quadratic functional from the formula

$$I(T) = \frac{1}{2}B(T,T) - \ell(T) \tag{2.4 - 10}$$

However, it is *not necessary* to have a quadratic functional in order to develop a finite element model. What is needed is a weighted–integral statement or a weak form of the differential equation to be solved. It is always possible to construct the weighted–integral statement of *any* differential equation. The weak form exists for any second– and higher–order equations, because for such equations it is possible to trade differentiation from the dependent unknown to the weight function and hence include the natural boundary condition (or the secondary variable) into the weighted–integral statement. These observations hold for linear as well as for nonlinear problems. For example, the Navier–Stokes equations governing the flow of a viscous incompressible fluid do not admit a functional; however, a weak form can be constructed for the equations, as will be shown later in this book.

2.5 FINITE ELEMENT MODEL

The weak form in equation (2.4–7) requires that the approximation chosen for T should be at least linear in both x and y so that there are no terms in (2.4–7) that become identically zero. Suppose that T is approximated over a typical finite element Ω^e by the expression [cf. equation (2.3–1)]

$$T(x,y) \approx T^e(x,y) = \sum_{j=1}^{n} T_j^e \psi_j^e(x,y) \tag{2.5 - 1}$$

where T_j^e is the value of $T^e(x,y)$ at the j-th node with coordinates (x_j, y_j) of the element, and ψ_j^e are the interpolation functions, with the property (see Reddy [7])

$$\psi_i^e(x_j, y_j) = \delta_{ij} \tag{2.5 - 2}$$

The specific form of ψ_i^e will be presented for triangular and rectangular elements in Section 2.6. Additional discussion of the elements is given in Chapter 3.

Substituting the finite element approximation (2) for T into the weak form (2.4–7), we obtain

$$0=\int_{\Omega^e}\left[\frac{\partial w}{\partial x}\left(k_{11}\sum_{j=1}^{n}T_j^e\frac{\partial\psi_j^e}{\partial x}\right)+\frac{\partial w}{\partial y}\left(k_{22}\sum_{j=1}^{n}T_j^e\frac{\partial\psi_j^e}{\partial y}\right)-wQ\right]dxdy-\oint_{\Gamma^e}wq_n ds \tag{2.5 – 3}$$

This equation must hold for any weight function w. Since we need n independent algebraic equations to solve for the n unknowns, $T_1^e, T_2^e, ..., T_n^e$, we choose n independent functions for w : $w = \psi_1^e, \psi_2^e, ..., \psi_n^e$. This particular choice of weight functions is a natural one when the weight function is viewed as a virtual variation of the dependent unknown (*i.e.*, $w = \delta T^e = \sum_{i=1}^{n} \delta T_i^e \psi_i^e$). For this choice, the solution of (3) is termed the "best approximation" (see [1,8]) in the variational sense.

For each choice of w we obtain an algebraic relation among $(T_1^e, T_2^e, ..., T_n^e)$. We label the algebraic equation resulting from substitution of ψ_1^e for w into equation (2.4–7) as the first algebraic equation. The $i-th$ algebraic equation is obtained by substituting $w = \psi_i^e$ into equation (2.4–7):

$$\sum_{j=1}^{n} K_{ij}^e T_j^e = Q_i^e + q_i^e \tag{2.5 – 4}$$

where the coefficients $K_{i,j}^e, Q_i^e$, and q_i^e are defined by

$$K_{ij}^e=\int_{\Omega^e}\left(k_{11}\frac{\partial\psi_i^e}{\partial x}\frac{\partial\psi_j^e}{\partial x}+k_{22}\frac{\partial\psi_i^e}{\partial y}\frac{\partial\psi_j^e}{\partial y}\right)dxdy \tag{2.5 – 5a}$$

$$Q_i^e=\int_{\Omega^e}Q\psi_i^e\,dxdy,\quad q_i^e=\oint_{\Gamma^e}q_n\psi_i^e\,ds \tag{2.5 – 5b}$$

In matrix notation, equation (4) takes the form

$$[K^e]\{T^e\}=\{Q^e\}+\{q^e\} \tag{2.5 – 6}$$

The matrix $[K^e]$ is called the *coefficient matrix*, or conductivity matrix in the present context. We note that $K_{ij}^e = K_{ji}^e$ (*i.e.*, $[K]$ is symmetric). The symmetry of the coefficient matrix is due to the symmetry of the bilinear form in equation (2.4–9a), which in turn is due to the weak form development. Equation (6) is called the finite element model of equation (2.2–1) or its weak form (2.4–7).

This completes the finite element model development. Before we discuss assembly of elements, it is informative to consider the interpolation functions ψ_i^e for certain basic elements.

2.6 INTERPOLATION FUNCTIONS

The finite element approximation $T^e(x,y)$ of $T(x,y)$ over an element Ω^e must satisfy the following conditions in order for the approximate solution to converge to the true solution:

1. $T^e(x,y)$ must be continuous as required in the weak form of the problem (*i.e.*, all terms in the weak form are represented as nonzero values).

2. The polynomials used to represent $T^e(x,y)$ must be complete (*i.e.*, all terms, beginning with a constant term up to the highest–order used in the polynomial should be included in the expression of $T^e(x,y)$).

3. All terms in the polynomial should be linearly independent.

The number of linearly independent terms in the representation of T^e dictates the shape and number of degrees of freedom of the element. Here we review the interpolation functions of linear triangular and rectangular elements.

An examination of the variational form (2.4–7) and the finite element matrices in equations (2.5–5a) shows that the ψ_i^e should be at least linear functions of x and y. The polynomial

$$T^e(x,y) = c_1^e + c_2^e x + c_3^e y \qquad (2.6-1)$$

is the lowest–order polynomial that meets the requirements. It contains three linearly independent terms, and it is linear in both x and y. The polynomial is complete because the lower–order term, namely the constant term, is included. To write the three constants (c_1^e, c_2^e, c_3^e) in terms of the nodal values of T^e, we must identify three points or nodes in the element Ω^e. The three nodes must be such that they uniquely define the geometry of the element and allow the imposition of interelement continuity of the variable $T^e(x,y)$. Obviously, the geometric shape defined by three points in a two-dimensional domain is a triangle. Thus the polynomial (1) is associated with a triangular element and the three nodes are identified as the vertices of the triangle.

On the other hand, the polynomial

$$T^e(x,y) = c_1^e + c_2^e x + c_3^e y + c_4^e xy \qquad (2.6-2)$$

contains four linearly independent terms, and is linear in x and y, with a bilinear term in x and y. This polynomial requires an element with four nodes. It is a rectangle with nodes at the four corners of the rectangle, or a triangle with three nodes at vertices and the fourth at the centroid of the triangle. Note that the four–node triangular element does not meet the interelement continuity of the functions approximated.

The interpolation functions for linear triangular and rectangular elements are given below. Higher–order two-dimensional elements (*i.e.*, element with higher–order interpolation polynomials) will be discussed in Section 2.12.

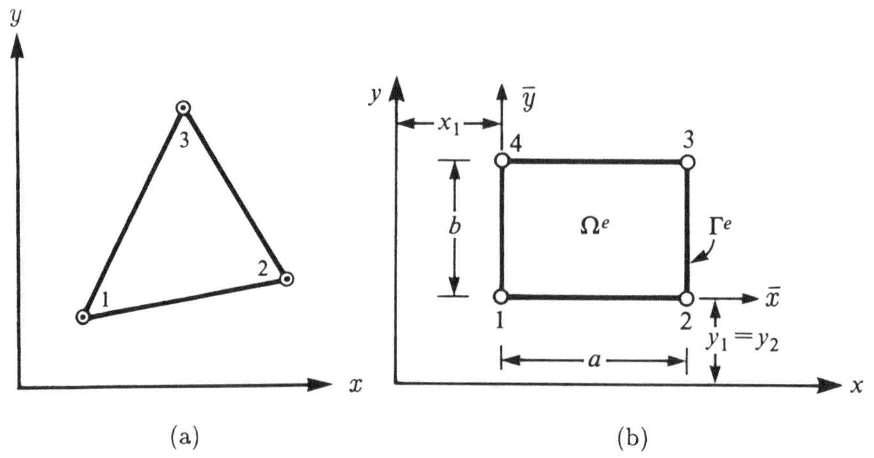

Figure 2.6–1: The linear triangular and rectangular elements.

Linear triangular element

The linear interpolation functions for the three–node triangle (see Figure 2.6–1a) are (see Reddy [7, pp. 304–307])

$$\psi_i^e = \frac{1}{2A_e}(\alpha_i^e + \beta_i^e x + \gamma_i^e y), \quad (i = 1, 2, 3) \tag{2.6 – 3}$$

where A_e is the area of the triangle, and α_i^e, β_i^e, and γ_i^e are geometric constants known in terms of the nodal coordinates (x_i, y_i):

$$\alpha_i^e = x_j y_k - x_k y_j \ ; \quad \beta_i^e = y_j - y_k \ ; \quad \gamma_i^e = -(x_j - x_k) \tag{2.6 – 4}$$

Here the subscripts are such that $i \neq j \neq k$, and i, j, and k permute in a natural order. Note that (x, y) are the coordinates used in the governing equation (2.2–1) over the domain Ω, and are called *global coordinates*. Since the element interpolation functions are restricted to an element, it is possible to express them in a *local coordinate system*. Such local coordinate systems are useful in the numerical evaluation of the integral expressions in equations (2.5–5). We will return to these issues in Sections 2.12 and 2.13.

The interpolation functions $\psi_i^e (i = 1, 2, ..., n)$ have the following properties:

$$\psi_i^e(x_j, y_j) = \delta_{ij}, \quad (i, j = 1, 2, 3) \tag{2.6 – 5a}$$

$$\sum_{i=1}^{3} \psi_i^e = 1 \tag{2.6-5b}$$

Note that use of the linear interpolation functions ψ_i^e of a triangle will result in the approximation of the curved surface $T(x, y)$ by a planar function $T^e = \sum_{i=1}^3 T_i^e \psi_i^e$.

The integrals in the definition of K_{ij}^e and Q_i^e can be evaluated for given data: k_{11}, k_{22}, and Q. For example, for element-wise constant values of the data, *i.e.*, $k_{11} = k_{11}^e$, $k_{22} = k_{22}^e$, and $Q = Q^e$, we have (see Reddy [7, pp. 311–313]) the following results:

$$K_{ij}^e = \frac{1}{4A_e}(k_{11}^e \beta_i^e \beta_j^e + k_{22}^e \gamma_i^e \gamma_j^e) \tag{2.6-6a}$$

$$Q_i^e = \frac{Q_e A_e}{3} \tag{2.6-6b}$$

where A_e is the area of the triangular element, and β^e and γ^e are known in terms of the global nodal coordinates of the element nodes, as given in equation (4). For a right-angled triangular element with base a and height b, $[K^e]$ takes the form (see Reddy [7, p. 387])

$$[K^e] = \frac{k_{11}^e}{2}\begin{bmatrix} \alpha & -\alpha & 0 \\ -\alpha & \alpha & 0 \\ 0 & 0 & 0 \end{bmatrix} + \frac{k_{22}^e}{2}\begin{bmatrix} \beta & 0 & -\beta \\ 0 & 0 & 0 \\ -\beta & 0 & \beta \end{bmatrix} \tag{2.6-6c}$$

where $\alpha = \frac{b}{a}$ and $\beta = \frac{a}{b}$. Of course, for cases in which the conductivities are functions of (x, y), numerical integration can be used to evaluate the coefficients (see Section 2.13).

The evaluation of boundary integrals of the type [see equation (2.5–5b)]

$$q_i^e = \oint_{\Gamma^e} q_n^e \psi_i^e(s)\, ds \tag{2.6-7}$$

where q_n^e is a known function of the distance s along the boundary Γ^e, involves evaluation of line integrals. It is necessary to compute such integrals only when Γ^e, or a portion of it, coincides with the boundary Γ_q of the total domain Ω on which the flux is specified. On portions of Γ^e that are in the interior of the domain Ω, q_n^e on side (i, j) of element Ω^e cancels with q_n^f on side (p, q) of element Ω^f when sides (i, j) of element Ω^e and (p, q) of element Ω^f are the same (*i.e.*, at the interface of elements Ω^e and Ω^f). This can be viewed as the balance of the internal flux. When Γ^e falls on the boundary Γ_T of the domain Ω, q_n^e is not known there and can be determined in the post-computation. Note that the primary variable T is specified on Γ_T. For additional details, see Reddy [7, pp. 313–318].

Linear rectangular element

For a linear rectangular element, we have

$$T(\bar{x},\bar{y}) = \sum_{i=1}^{4} T_i^e \psi_i^e \qquad (2.6-8)$$

where (see Reddy [7, pp. 308–311])

$$\psi_1^e = (1-\frac{\bar{x}}{a})(1-\frac{\bar{y}}{b}), \quad \psi_2^e = \frac{\bar{x}}{a}(1-\frac{\bar{y}}{b})$$

$$\psi_3^e = \frac{\bar{x}}{a}\frac{\bar{y}}{b}, \quad \psi_4^e = (1-\frac{\bar{x}}{a})\frac{\bar{y}}{b} \quad , \qquad (2.6-9)$$

and $(\bar{x},\bar{y})$ denote the local coordinates with origin located at node 1 of the element, and (a,b) denote the horizontal and vertical dimensions of the rectangle (see Figure 2.6–1b).

The integrals in the definition of K_{ij}^e and Q_i^e can be easily evaluated over a rectangular element of sides a and b. For example, for element–wise constant values of the data, *i.e.*, $k_{11} = k_{11}^e$, $k_{22} = k_{22}^e$, and $Q = Q^e$, we have (see Reddy [7, p. 313; p. 387]) the following results:

$$[K^e] = k_{11}^e[S^{11}] + k_{22}^e[S^{22}] \qquad (2.6-10a)$$

$$Q_i^e = \frac{Q_e ab}{4} \qquad (2.6-10b)$$

where

$$[S^{11}] = \frac{1}{6}\begin{bmatrix} 2\alpha & -2\alpha & -\alpha & \alpha \\ -2\alpha & 2\alpha & \alpha & -\alpha \\ -\alpha & \alpha & 2\alpha & -2\alpha \\ \alpha & -\alpha & -2\alpha & 2\alpha \end{bmatrix} \qquad (2.6-10c)$$

$$[S^{22}] = \frac{1}{6}\begin{bmatrix} 2\beta & \beta & -\beta & -2\beta \\ \beta & 2\beta & -2\beta & -\beta \\ -\beta & -2\beta & 2\beta & \beta \\ -2\beta & -\beta & \beta & 2\beta \end{bmatrix} \qquad (2.6-10d)$$

and α and β are defined in equation (6c). Again, for cases in which the conductivities are functions of (x,y), numerical integration is used to evaluate the coefficients, as discussed in Section 2.13. When the element is nonrectangular, *i.e.*, a quadrilateral, we use coordinate transformations to represent the integrals over a square geometry (see Section 3.12) and then use numerical integration to evaluate them.

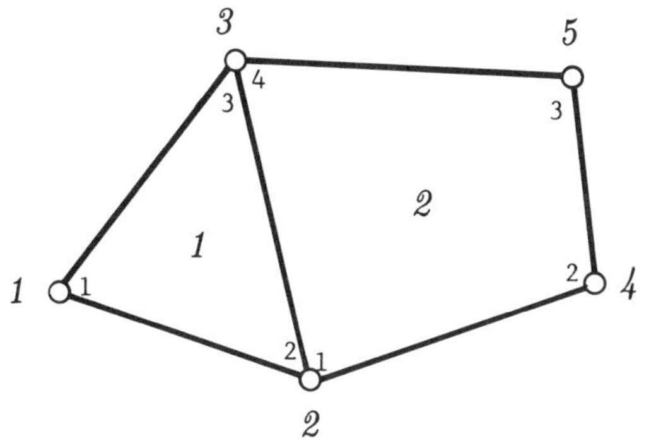

Figure 2.7–1: Global–local correspondence of nodes for assembly of elements.

2.7 ASSEMBLY OF ELEMENTS

The assembly of finite elements to obtain the equations of the entire domain is based on the following two rules:

1. Continuity of the primary variable (*i.e.*, temperature)

2. Balance of secondary variables(*i.e.*, heat flux)

We illustrate the assembly procedure by considering a finite element mesh consisting of a triangular element and a quadrilateral element (see Figure 2.7–1). Let $K^1_{ij}(i, j = 1, 2, 3)$ denote the coefficient matrix corresponding to the triangular element, and let $K^2_{ij}(i, j = 1, 2, 3, 4)$ denote the coefficient matrix corresponding to the quadrilateral element. The nodes of the finite element mesh are called *global nodes*. From the mesh shown in Figure 2.7–1, it is clear that the following correspondence between the global and element nodes exists: nodes 1, 2, and 3 of element 1 correspond to global nodes 1, 2, and 3, respectively. Nodes 1, 2, 3, and 4 of element 2 correspond to global nodes 2, 4, 5, and 3, respectively. Hence, the correspondence between the local and global nodal values of temperature is

$$T^1_1 = T_1 \ , \ T^1_2 = T^2_1 = T_2 \ , \ T^1_3 = T^2_4 = T_3 \ , \ T^2_2 = T_4 \ , \ T^2_3 = T_5 \qquad (2.7-1)$$

which amounts to imposing the continuity of the primary variables at the nodes common to elements 1 and 2. Note that the continuity of the primary variables at the interelement nodes guarantees the continuity of the primary variable along the entire interelement boundary.

Next, we consider the balance of secondary variables. At the interface between the two elements, the flux from the two elements should be equal in magnitude and opposite in sign. For the two elements in Figure 2.7–1, the interface is along the side connecting global nodes 2 and 3. Hence, the internal flux q_n^1 on side 2–3 of element 1 should balance the flux q_n^2 on side 4–1 of Element 2 (recall the sign convention on q_n^e):

$$(q_n^1)_{2-3} = (q_n^2)_{4-1} \text{ or } (q_n^1)_{2-3} = (-q_n^2)_{1-4} \tag{2.7 – 2}$$

In the finite element method, the above relation is imposed in a weighted–integral sense:

$$\int_{h_{23}^1} q_n^1 \psi_2^1 \, ds = -\int_{h_{14}^2} q_n^2 \psi_1^2 \, ds \tag{2.7 – 3a}$$

$$\int_{h_{23}^1} q_n^1 \psi_3^1 \, ds = -\int_{h_{14}^2} q_n^2 \psi_4^2 \, ds \tag{2.7 – 3b}$$

where h_{pq}^e denotes length of the side connecting node p to node q of element Ω^e.

Now we are ready to assemble the element equations to obtain the equations for the two–element mesh. The element equations of the two elements are written separately first. For the triangular element, the element equations are of the form

$$\begin{aligned}
K_{11}^1 T_1^1 + K_{12}^1 T_2^1 + K_{13}^1 T_3^1 &= Q_1^1 + q_1^1 \\
K_{21}^1 T_1^1 + K_{22}^1 T_2^1 + K_{23}^1 T_3^1 &= Q_2^1 + q_2^1 \\
K_{31}^1 T_1^1 + K_{32}^1 T_2^1 + K_{33}^1 T_3^1 &= Q_3^1 + q_3^1
\end{aligned} \tag{2.7 – 4a}$$

For the rectangular element the element equations are given by

$$\begin{aligned}
K_{11}^2 T_1^2 + K_{12}^2 T_2^2 + K_{13}^2 T_3^2 + K_{14}^2 T_4^2 &= Q_1^2 + q_1^2 \\
K_{21}^2 T_1^2 + K_{22}^2 T_2^2 + K_{23}^2 T_3^2 + K_{24}^2 T_4^2 &= Q_2^2 + q_2^2 \\
K_{31}^2 T_1^2 + K_{32}^2 T_2^2 + K_{33}^2 T_3^2 + K_{34}^2 T_4^2 &= Q_3^2 + q_3^2 \\
K_{41}^2 T_1^2 + K_{42}^2 T_2^2 + K_{43}^2 T_3^2 + K_{44}^2 T_4^2 &= Q_4^2 + q_4^2
\end{aligned} \tag{2.7 – 4b}$$

In order to impose the balance of secondary variables in equations (3), it is required that we add the second equation of element 1 to the first equation of element 2, and also add the third equation of element 1 to the fourth equation of element 2:

$$\begin{aligned}
(K_{21}^1 T_1^1 + K_{22}^1 T_2^1 + K_{23}^1 T_3^1) + (K_{11}^2 T_1^2 + K_{12}^2 T_2^2 + K_{13}^2 T_3^2 + K_{14}^2 T_4^2) \\
= (Q_2^1 + q_2^1) + (Q_1^2 + q_1^2)
\end{aligned} \tag{2.7 – 5a}$$

and

$$(K_{31}^1 T_1^1 + K_{32}^1 T_2^1 + K_{33}^1 T_3^1) + (K_{41}^2 T_1^2 + K_{42}^2 T_2^2 + K_{43}^2 T_3^2 + K_{44}^2 T_4^2)$$
$$= (Q_3^1 + q_3^1) + (Q_4^2 + q_4^2) \qquad (2.7-5b)$$

Using the local–global nodal variable correspondence in equation (1), we can rewrite the above equations as

$$K_{21}^1 T_1 + (K_{22}^1 + K_{11}^2) T_2 + (K_{23}^1 + K_{14}^2) T_3 + K_{12}^2 T_4 + K_{13}^2 T_5$$
$$= Q_2^1 + Q_1^2 + (q_2^1 + q_1^2) \qquad (2.7-6a)$$

$$K_{31}^1 T_1 + (K_{32}^1 + K_{41}^2) T_2 + (K_{33}^1 + K_{44}^2) T_3 + K_{42}^2 T_4 + K_{43}^2 T_5$$
$$= Q_3^1 + Q_4^2 + (q_3^1 + q_4^2) \qquad (2.7-6b)$$

Now we can impose the conditions in equations (3) by setting appropriate portions of the expressions in parenthesis on the right–hand side of the above equations to zero (or a specified nonzero value). In general, when several elements are connected, the assembly of the elements is carried out by putting element coefficients K_{ij}^e, Q_i^e and q_i^e into proper locations of the global coefficient matrix and right hand column vectors. This is done by means of the connectivity relations, *i.e.*, correspondence of the local node number to the global node number.

The assembly procedure described above can be used to assemble elements of any shape and type. The procedure can be implemented in a computer with the help of the local–global nodal correspondence.

This completes the first four steps in the finite element modeling of the model equation (2.2–1). The remaining three steps of the analysis, namely, the imposition of boundary conditions, solution of equations, and postprocessing of the solution will not be discussed here, but will be reserved for subsequent chapters. It may be recalled that the derivatives of $T^e(x, y)$ will not be continuous at interelement boundaries because continuity of the derivatives is not imposed in the model. The weak form of the equation suggests that the primary variable is T, and therefore it should be made continuous across the interelement boundaries. If additional variables, such as the derivatives of temperature, are carried as nodal variables in the interest of making them continuous across interelement boundaries, the degree of interpolation (or order of the element) increases. In addition, the continuity of the derivatives, which are not identified as the primary variables in the weak formulation of the problem, may violate the physical principles of the problem.

2.8 TIME–DEPENDENT PROBLEMS

2.8–1 Introduction

In this section we present the finite element model of time–dependent heat transfer problems in two dimensions.

$$\rho C \frac{\partial T}{\partial t} - \left[\frac{\partial}{\partial x}\left(k_{11}\frac{\partial T}{\partial x}\right) + \frac{\partial}{\partial y}\left(k_{22}\frac{\partial T}{\partial y}\right)\right] = Q \quad \text{in } \Omega \qquad (2.8-1)$$

We wish to analyze equation (1) under appropriate boundary and initial conditions. The boundary conditions for this equation are given by equations (2.2–5), and the initial condition on the temperature is given by

$$T(\mathbf{x}, 0) = T_0(\mathbf{x}) \qquad (2.8-2)$$

The numerical solution of the initial–boundary value problem described by equation (1) involves two stages of approximation. The first stage, called *spatial discretization*, involves the development of the weak form of the equation over an element and the spatial approximation of the dependent variable (*i.e.*, temperature) T of the problem. The three–step procedure presented in Section 2.4 for the development of the weak form is still applicable to time–dependent problems. This step results in a set of ordinary differential equations in time among the nodal values T_j^e of the dependent variable. The second stage consists of a time approximation, called *temporal approximation*, of the ordinary differential equations (*i.e.*, numerical integration of the equations) by finite difference schemes. This step leads to a set of algebraic equations involving the nodal values T_j^e at time t_{n+1} [$= (n+1)\Delta t$, where n is an integer and Δt is the time increment] in terms of known values from the previous time step. The finite element model of equation (1) using the *time–space finite elements* can also be developed (see Reddy [7, p. 401] and Argyris and Scharpf [15]). In this procedure, time is treated as another spatial coordinate. Consequently, one should know the solution at the initial as well as final times (*i.e.*, at the time boundaries). Here we discuss only the two–stage approximation. The semidiscretization is discussed first.

2.8–2 Semidiscretization

We assume, as usual, that the conduction heat transfer region Ω is discretized into an appropriate collection of finite elements. The weak form of equation (1) over an element Ω^e is obtained by the standard procedure: multiply equation (1) with the weight function $w(x, y)$ and integrate over the element, integrate–by–parts (spatially) those terms which involve higher–order derivatives using the gradient or divergence theorem, and replace the coefficient of the weight function in the boundary integral with the secondary variable [*i.e.*, use equation (2.2–5b)]. We obtain

$$0 = \int_{\Omega^e} \left[w\left(\rho C \frac{\partial T}{\partial t} - Q_s\right) + k_{11}\frac{\partial w}{\partial x}\frac{\partial T}{\partial x} + k_{22}\frac{\partial w}{\partial y}\frac{\partial T}{\partial y}\right] dxdy - \oint_{\Gamma^e} (q_n - q_c) w \, ds \qquad (2.8-3)$$

Note that the procedure to obtain the weak form for time-dependent problems is not much different from that used for steady-state problems in Section 2.4. The difference is that all terms of the equations may be functions of time. Also, no integration by parts with respect to time is used, and the weight function w is not a function of time.

The semidiscrete finite element model is obtained from equation (3) by substituting a finite element approximation for the dependent variable, T. In selecting the approximation for T, we assume that the time dependence can be separated fr the spatial variation,

$$T(\mathbf{x},t) \simeq \sum_{j=1}^{n_e} T_j^e(t)\psi_j^e(\mathbf{x}) \qquad (2.8-4)$$

The i–th differential equation (in time) of the finite element model is obtained by substituting $w = \psi_i^e(\mathbf{x})$ and replacing T by the expression in equation (4):

$$0 = \sum_{j=1}^{n_e}\left(M_{ij}^e \frac{dT_j^e}{dt} + K_{ij}^e T_j^e\right) - Q_i^e - q_i^e \qquad (2.8-5)$$

Equation (5) can be expressed in matrix form as

$$[M^e]\{\dot{T}^e\} + [K^e]\{T^e\} = \{Q^e\} + \{q^e\} \qquad (2.8-6)$$

where a superposed dot on T denotes a derivative with time ($\dot{T} = \partial T/\partial t$), and

$$M_{ij}^e = \int_{\Omega^e} \rho C\psi_i\psi_j \; dxdy, \; K_{ij}^e = \int_{\Omega^e} (k_{11}\frac{\partial\psi_i}{\partial x}\frac{\partial\psi_j}{\partial x} + k_{22}\frac{\partial\psi_i}{\partial y}\frac{\partial\psi_j}{\partial y}) \; dxdy \quad (2.8-7a)$$

$$Q_i^e = \int_{\Omega^e} \psi_i Q(x,y,t) \; dxdy \; , \quad q_i^e = \oint_{\Gamma^e} \psi_i(q_n - q_c) \; ds \qquad (2.8-7b)$$

The element label 'e' on ψ is omitted for brevity. It should be noted that the capacitance (or mass) matrix, $[M]$, in equation (7a) is not a diagonal matrix. It is called a *consistent (mass) matrix* because it is defined consistent with the weak formulation (3).

2.8–3 Temporal Approximation

The ordinary differential equations (6) should be integrated with respect to time to obtain the transient response. Since it is not possible, in general, to integrate these equations analytically, they are further approximated in time to obtain a set of algebraic equations in terms of the nodal temperatures (*i.e.*, a fully discretized model is obtained). In principle, it is possible to use any of the standard methods for the solution of ordinary differential equations. However, practical considerations, such as the computational cost, dictate that only the simpler time integration (or approximation) methods be considered.

The most commonly used time integration methods for equation (6) are all part of a one–parameter family, called the θ–family of approximation,

$$\{T\}_{n+1} = \{T\}_n + \Delta t[(1-\theta)\{\dot{T}\}_n + \theta\{\dot{T}\}_{n+1}] \; , \; 0 \le \theta \le 1 \qquad (2.8-8)$$

Using equation (8) in (6), we can transform the ordinary differential equations into a set of algebraic equations at time t_{n+1} :

$$[\hat{K}]_{n+1}\{T\}_{n+1} = \{\hat{F}\}_{n,n+1} \qquad (2.8-9)$$

$$[\hat{K}]_{n+1} = [M] + a_1[K]_{n+1} \qquad (2.8-10a)$$

$$\{\hat{F}\}_{n,n+1} = \Delta t(\theta\{F\}_{n+1} + (1-\theta)\{F\}_n) + ([M] - a_2[K]_n)\{T\}_n \qquad (2.8-10b)$$

$$a_1 = \theta\Delta t \ , \ a_2 = (1-\theta)\Delta t \qquad (2.8-10c)$$

Equations (9), after assembly and imposition of boundary conditions, are solved at each time step for the nodal values T_j at time $t_n = (n+1)\Delta t$. At time $t = 0$ (*i.e.*, $n = 0$), the right-hand side is computed using the initial values $\{T\}_0$; the vector $\{F\}$, which is the sum of the source vector $\{Q^e\}$ and internal flux vector $\{q^e\}$, is always known, for both times t_n and t_{n+1}, at all nodes at which the solution is unknown (because $Q(x, y, t)$ is a known function of time and the sum of q_j^e at these nodes is zero).

For different values of the parameter θ, we obtain several well-known time approximation schemes:

$$\begin{aligned}
&\theta = 0, \text{ the forward difference scheme (conditionally stable)}\\
&\theta = 0.5, \text{ the Crank-Nicolson scheme (unconditionally stable)}\\
&\theta = \frac{2}{3}, \text{ the Galerkin scheme (unconditionally stable)}\\
&\theta = 1, \text{ the backward difference scheme (unconditionally stable)}
\end{aligned} \qquad (2.8-11)$$

For $\theta \geq 0.5$, the scheme is stable, and for $\theta < 0.5$ the scheme is stable only if the time step meets certain restriction (*i.e.*, ***conditionally stable schemes***) (see [7]). For the forward difference scheme the stability requirement is

$$\Delta t < \Delta t_{cr} = \frac{2}{(1-2\theta)\lambda_{\max}} \ , \ \theta < \frac{1}{2} \qquad (2.8-12)$$

where $\lambda_{\max}$ is the largest eigenvalue of the eigenvalue problem associated with the matrix equation (6):

$$|\hat{\mathbf{K}} - \lambda\mathbf{I}| = 0 \qquad (2.8-13)$$

This completes the development of the finite element model of a transient heat transfer problem in two dimensions. Further details on time-dependence will be covered in Chapter 3.

2.9 AXISYMMETRIC PROBLEMS

In studying heat transfer problems involving cylindrical geometries, it is convenient to use the cylindrical coordinate system (r, θ, z) (see Figure 2.9–1) to formulate the problem. If the geometry, data, and boundary conditions of the problem are independent of the angular coordinate θ, the problem solution will also be independent of θ. Consequently, a three-dimensional problem is reduced to a two-dimensional one in (r, z) coordinates. Here we present the weak form for an axisymmetric problem.

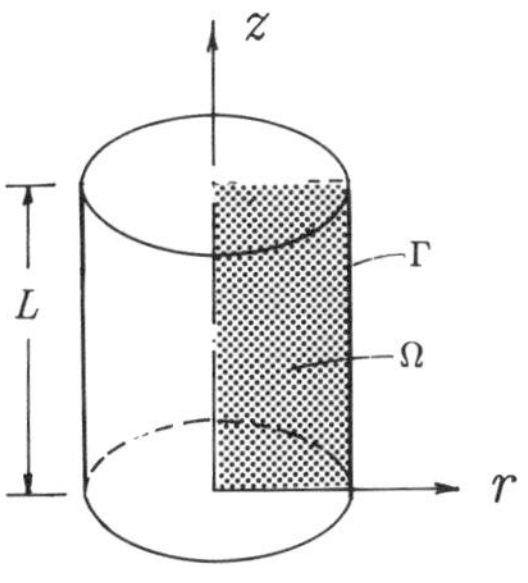

Figure 2.9–1: The cylindrical coordinate system for an axisymmetric problem.

Consider the partial differential equation governing heat transfer in an axisymmetric geometry [*cf.*, equation (1.4–18a)],

$$-\frac{1}{r}\frac{\partial}{\partial r}(rk_{rr}\frac{\partial T}{\partial r}) - \frac{\partial}{\partial z}(k_{zz}\frac{\partial T}{\partial z}) = Q(r, z) \qquad (2.9-1)$$

where $(k_{rr},\ k_{zz})$ and Q are the conductivities and internal heat generation per unit volume, respectively. In developing the weak form, we integrate over the elemental volume of the axisymmetric geometry: $rdrd\theta dz$. Since the solution is independent of the $\theta-$ coordinate, the integration with respect to θ yields a multiplicative constant, 2π.

Following the three–step procedure, we write the weak form of equation (1):

$$0 = 2\pi \int_{\Omega^e} w[-\frac{1}{r}\frac{\partial}{\partial r}(rk_{rr}\frac{\partial T}{\partial r}) - \frac{\partial}{\partial z}(k_{zz}\frac{\partial T}{\partial z}) - Q]\ rdrdz$$

Integrating the highest–order terms by parts, we obtain

$$0 = 2\pi \int_{\Omega^e} [\frac{\partial w}{\partial r} \cdot rk_{rr}\frac{\partial T}{\partial r} + \frac{\partial w}{\partial z} \cdot rk_{zz}\frac{\partial T}{\partial z} - w \cdot rQ]\ drdz$$

$$- 2\pi \oint_{\Gamma^e} w(rk_{rr}\frac{\partial T}{\partial r}n_r + rk_{zz}\frac{\partial T}{\partial z}n_z)\ ds$$

or

$$0 = 2\pi \int_{\Omega^e} [k_{rr} \frac{\partial w}{\partial r}\frac{\partial T}{\partial r} + k_{zz}\frac{\partial w}{\partial z}\frac{\partial T}{\partial z} - wQ]\, rdrdz - \oint_{\Gamma^e} wq_n\, ds \qquad (2.9-2)$$

where w is the weight function and q_n is the normal flux,

$$q_n = 2\pi r(k_{rr}\frac{\partial T}{\partial r}n_r + k_{zz}\frac{\partial T}{\partial z}n_z) \qquad (2.9-3)$$

Note that the weak form (2) does not differ significantly from that developed for equation (2.2–1). The only difference is the presence of r in the integrand. Consequently, equation (2) can be obtained as a special case of equation (2.4–7) for $k_{11} = 2\pi k_{rr} \cdot x$, $k_{22} = 2\pi k_{zz} \cdot x$, and Q is replaced by $2\pi Q \cdot x$; the coordinates r and z are treated like x and y, respectively.

Let us assume that $T(r,z)$ is approximated by the finite element interpolation T^e over the element Ω^e :

$$T \approx T^e(r,z) = \sum_{j=1}^{n} T_j^e \psi_j^e(r,z) \qquad (2.9-4)$$

The interpolation functions $\psi^e(r,z)$ are the same as those given in equations (2.6–3) and (2.6–9) for linear triangular and rectangular elements, with $x = r$ and $y = z$. Substitution of equation (4) for T and ψ_i^e for w into the weak form gives the i-th equation of the finite element model

$$0 = 2\pi \sum_{j=1}^{n} \int_{\Omega^e} \left[\frac{\partial \psi_i^e}{\partial r}\left(k_{rr}T_j^e\frac{\partial \psi_j^e}{\partial r}\right) + \frac{\partial \psi_i^e}{\partial z}\left(k_{zz}T_j^e\frac{\partial \psi_j^e}{\partial z}\right) - wQ\right] rdrdz - \oint_{\Gamma^e} \psi_i^e q_n ds \qquad (2.9-5)$$

or

$$0 = \sum_{j=1}^{n} K_{ij}^e T_j^e - Q_i^e - q_i^e \qquad (2.9-6)$$

where

$$K_{ij}^e = 2\pi \int_{\Omega^e} \left(k_{rr}\frac{\partial \psi_i^e}{\partial r}\frac{\partial \psi_j^e}{\partial r} + k_{zz}\frac{\partial \psi_i^e}{\partial z}\frac{\partial \psi_j^e}{\partial z}\right) rdrdz \qquad (2.9-7)$$

$$Q_i^e = 2\pi \int_{\Omega^e} \psi_i^e Q\, rdrdz, \quad q_i^e = \oint_{\Gamma^e} q_n \psi_i^e\, ds \qquad (2.9-8)$$

This completes the development of the time–independent finite element model of an axisymmetric problem; the formulation for a time–dependent model follows exactly the steps as outlined for the planar case.

2.10 CONVECTIVE BOUNDARY CONDITIONS

For heat conduction problems that involve convection heat transfer at the boundary, *i.e.*, when heat is transfered from one medium to the surrounding medium (often, a fluid) by convection, the finite element model developed earlier requires some modification. The model to be presented allows the computation of the additional contributions to the coefficient matrix and source vector whenever the element has the convection boundary condition. For a convection boundary, the natural boundary condition is a balance of energy transfer across the boundary due to conduction and/or convection (*i.e.*, Newton's law of cooling):

$$(k_{11}\frac{\partial T}{\partial x}n_x + k_{22}\frac{\partial T}{\partial y}n_y) + h_c(T - T_c) = q_n \qquad (2.10-1)$$

where h_c is the convective conductance [or the convective heat transfer coefficient], T_c is the (ambient) temperature of the surrounding fluid medium, and q_n is the specified heat flow. The first term accounts for heat transfer by conduction, the second by convection, and the third accounts for the specified heat flux, if any. It is the presence of the term $h_c(T - T_c)$ that requires some modification of the weak form in equation (2.4–7). To include the convective boundary condition (1), the boundary integral in equation (2.4–5) should be modified. Instead of replacing the coefficient of w in the boundary integral with q_n, we use equation (1):

$$\begin{aligned} 0 &= \int_{\Omega^e}\left(k_{11}\frac{\partial w}{\partial x}\frac{\partial T}{\partial x} + k_{22}\frac{\partial w}{\partial y}\frac{\partial T}{\partial y} - wQ\right)dxdy - \oint_{\Gamma^e} w\left(k_{11}\frac{\partial T}{\partial x}n_x + k_{22}\frac{\partial T}{\partial y}n_y\right)ds \\ &= \int_{\Omega^e}\left(k_{11}\frac{\partial w}{\partial x}\frac{\partial T}{\partial x} + k_{22}\frac{\partial w}{\partial y}\frac{\partial T}{\partial y} - wQ\right)dxdy - \oint_{\Gamma^e} w[q_n - h_c(T - T_c)]\,ds \end{aligned}$$

or

$$B(w,T) = \ell(w) \qquad (2.10-2)$$

where w is the weight function, and $B(\cdot,\cdot)$ and $\ell(\cdot)$ are the bilinear and linear forms

$$B(w,T) = \int_{\Omega^e}\left(k_{11}\frac{\partial w}{\partial x}\frac{\partial T}{\partial x} + k_{22}\frac{\partial w}{\partial y}\frac{\partial T}{\partial y}\right)dxdy + \int_{\Gamma^e} h_c wT\,ds \qquad (2.10-3a)$$

$$\ell(w) = \int_{\Omega^e} wQ\,dxdy + \int_{\Gamma^e} h_c wT_c\,ds + \int_{\Gamma^e} wq_n\,ds \qquad (2.10-3b)$$

Note that the unknown surface temperature in the convective boundary condition has been made part of the bilinear form $B(\cdot,\cdot)$ while all the known quantities remain part of the linear form $\ell(\cdot)$. The finite element model of equation (2) is straightforward (see Problem 2.1; also, see Reddy [7, pp. 341–346]).

The finite element model of equation (2) is valid for conductive and convective heat transfer boundary conditions. Radiative heat transfer boundary conditions are nonlinear and will be considered in the later chapters. For problems with no convective boundary conditions, the convective contributions to the element coefficients are omitted. Indeed, these contributions have to be included only for those elements whose sides fall on the boundary with specified convection heat transfer. The contribution due to convective boundaries to the element coefficient matrix and source vector can be computed by evaluating line integrals, as discussed in [7, pp. 342–345].

2.11 LIBRARY OF FINITE ELEMENTS

2.11–1 Introduction

The objective of this section is to present a library of two–dimensional triangular and rectangular elements of the Lagrange family, *i.e.*, elements over which only the function—not its derivatives—are interpolated. Once we have elements of different shapes and order at our disposal, we can choose appropriate elements and associated interpolation functions for a given problem. The interpolation functions are developed here for regularly shaped elements, called *master elements.* These elements can be used for numerical evaluation of integrals defined on irregularly shaped elements. This requires a transformation of the geometry from the actual element shape to its associated master element. We will discuss the numerical evaluation of integrals in Section 2.12.

2.11–2 Triangular Elements

The three–noded triangular element was developed in Section 2.6. Higher–order triangular elements (*i.e.*, triangular elements with interpolation functions of higher degree) can be systematically developed with the help of the so-called *area coordinates.* For triangular elements, it is possible to construct three nondimensional coordinates $L_i (i = 1, 2, 3)$, which vary in a direction normal to the sides directly opposite each node (see Figure 2.11–1). The coordinates are defined such that

$$L_i = \frac{A_i}{A}, \quad A = \sum_{i=l}^{3} A_i \tag{2.11-1}$$

where A_i is the area of the triangle formed by nodes j and k and an arbitrary point P in the element, and A is the total area of the element. For example A_1 is the area of the shaded triangle, which is formed by nodes 2 and 3 and point P. The point P is at a perpendicular distance of s from the side connecting nodes 2 and 3. We have $A_1 = bs/2$ and $A = bh/2$. Hence

$$L_1 = \frac{A_1}{A} = \frac{s}{h} \tag{2.11-2}$$

Clearly, L_1 is zero on side 2–3 (hence, zero at nodes 2 and 3) and has a value of unity at node 1. Thus, L_1 is the interpolation function associated with node 1. Similarly, L_2 and L_3 are the interpolation functions associated with nodes 2 and 3, respectively. In summary, we have

$$\psi_i = L_i \tag{2.11-3}$$

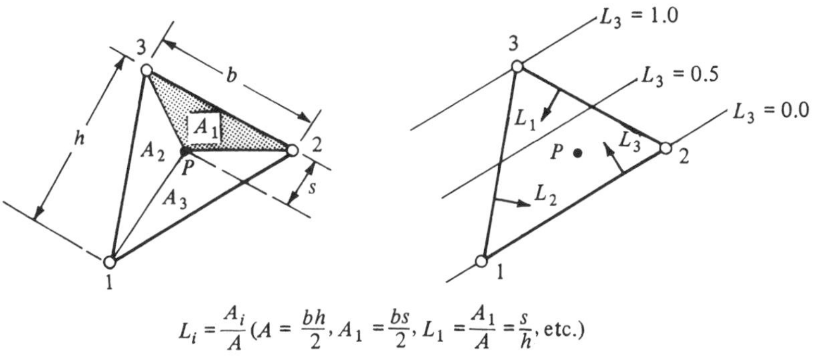

Figure 2.11–1: Definition of area coordinates L_i used for triangular elements.

The area coordinates L_i can be used to construct interpolation functions for higher–order triangular elements. For example, a higher–order element with k nodes per side (equally spaced on each side) has a total of n nodes,

$$n = \sum_{i=0}^{k-1}(k-i) = k + (k-1) + \cdots + 1 = \frac{k}{2}(k+1) \tag{2.11-4}$$

and its degree is equal to $k-1$. The explicit form of the interpolation functions for the linear and quadratic elements are recorded below:

$$\{\Psi^e\} = \begin{Bmatrix} L_1 \\ L_2 \\ L_3 \end{Bmatrix} \; ; \; \{\Psi^e\} = \begin{Bmatrix} L_1(2L_1-1) \\ L_2(2L_2-1) \\ L_3(2L_3-1) \\ 4L_1L_2 \\ 4L_2L_3 \\ 4L_3L_1 \end{Bmatrix} \tag{2.11-5}$$

Note that the order of the interpolation functions in the above arrays corresponds to the node numbers shown in Figure 2.11–2. Thus, the first three rows of equation (5) correspond to the first three nodes of the quadratic element, which correspond to the three vertices of the triangular element.

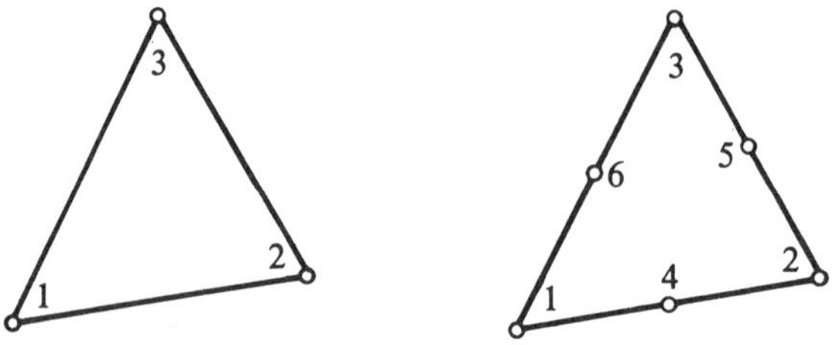

Figure 2.11–2: Linear and quadratic triangular elements.

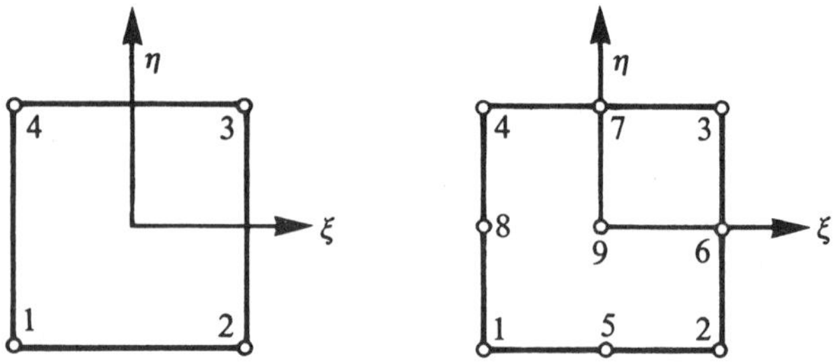

Figure 2.11–3: Linear and quadratic rectangular Lagrange elements.

2.11–3 Rectangular Elements

The Lagrange interpolation functions associated with rectangular elements can be obtained from the tensor product of corresponding one–dimensional Lagrange interpolation functions. We take a local coordinate system (ξ, η) such that $-1 \leq (\xi, \eta) \leq 1$. This choice of local coordinate system is dictated by the Gauss quadrature rule used in the numerical evaluation of integrals over the element (see Section 2.12).

The linear and quadratic interpolation functions are given below (see Figure 2.11–3).

$$\{\Psi^e\} = \frac{1}{4}\begin{Bmatrix} (1-\xi)(1-\eta) \\ (1+\xi)(1-\eta) \\ (1+\xi)(1+\eta) \\ (1-\xi)(1+\eta) \end{Bmatrix} \qquad (2.11-6)$$

$$\{\Psi^e\} = \frac{1}{4}\begin{Bmatrix} (1-\xi)(1-\eta)(-\xi-\eta-1)+(1-\xi^2)(1-\eta^2) \\ (1+\xi)(1-\eta)(\xi-\eta-1)+(1-\xi^2)(1-\eta^2) \\ (1+\xi)(1+\eta)(\xi+\eta-1)+(1-\xi^2)(1-\eta^2) \\ (1-\xi)(1+\eta)(-\xi+\eta-1)+(1-\xi^2)(1-\eta^2) \\ 2(1-\xi^2)(1-\eta)-(1-\xi^2)(1-\eta^2) \\ 2(1+\xi)(1-\eta^2)-(1-\xi^2)(1-\eta^2) \\ 2(1-\xi^2)(1+\eta)-(1-\xi^2)(1-\eta^2) \\ 2(1-\xi)(1-\eta^2)-(1-\xi^2)(1-\eta^2) \\ 4(1-\xi^2)(1-\eta^2) \end{Bmatrix} \qquad (2.11-7)$$

The *serendipity elements* are tho[...] [...]ents which have no interior nodes. These elements have fewe[...] [...]he higher–order Lagrange elements. The interpolation fur[...] [...]ents are not complete, and they cannot be obtaine[...] [...]imensional Lagrange interpolation functions. [...] [...] be employed, as discussed in [7]. The in[...] [...]pity element are given in equation (8[...] [...]not complete, the serendipity e[...] [...]ions.

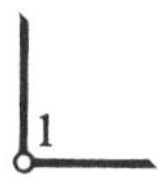

Figure 2.11–4: Quadratic rectangular serendipity element.

$$\{\Psi^e\} = \frac{1}{4}\begin{Bmatrix} (1-\xi)(1-\eta)(-\xi-\eta-1) \\ (1+\xi)(1-\eta)(\xi-\eta-1) \\ (1+\xi)(1+\eta)(\xi+\eta-1) \\ (1-\xi)(1+\eta)(-\xi+\eta-1) \\ 2(1-\xi^2)(1-\eta) \\ 2(1+\xi)(1-\eta^2) \\ 2(1-\xi^2)(1+\eta) \\ 2(1-\xi)(1-\eta^2) \end{Bmatrix} \tag{2.11-8}$$

2.12 NUMERICAL INTEGRATION

2.12–1 Preliminary Comments

An accurate representation of irregular domains (*i.e.*, domains with curved boundaries) can be accomplished by the use of refined meshes and/or irregularly shaped curvilinear elements. For example, a non–rectangular region cannot be represented using rectangular elements; however, it can be represented by quadrilateral elements. Since the interpolation functions are easily derivable for a rectangular element and it is easier to evaluate integrals over rectangular geometries, we transform the finite element integral statements defined over quadrilaterals to a rectangle. The transformation results in complicated expressions for the integrands in terms of the coordinates used for the rectangular element. Therefore, numerical integration is used to evaluate such complicated integrals. The numerical integration schemes, such as the Gauss–Legendre numerical integration scheme, require the integral to be evaluated on a specific domain or with respect to a specific coordinate system. Gauss quadrature, for example, requires the integral to be expressed over a square region $\hat{\Omega}$ of dimension 2 by 2 with respect to the coordinate system, (ξ,η) to be such that $-1 \le (\xi,\eta) \le 1$. The transformation of the geometry and the variable coefficients of the differential equation from the problem coordinates (x,y) to the local coordinates (ξ,η) results in algebraically complex expressions, and they preclude analytical (*i.e.*, exact) evaluation of the integrals. Thus, the transformation of a given integral expression, defined over element Ω^e, to one on the domain $\hat{\Omega}$ facilitates the numerical integration. Each element of the finite element mesh is transformed to $\hat{\Omega}$, only for the purpose of numerically evaluating the integrals. The element $\hat{\Omega}$ is called a *master element*. For example, every quadrilateral element can be transformed to a square element with side of length 2 and $-1 \le (\xi,\eta) \le 1$ that facilitates the use of Gauss-Legendre quadrature to evaluate integrals defined over the quadrilateral element.

The transformation between the actual element Ω^e and the master element $\hat{\Omega}$ [or equivalently, between (x,y) and (ξ,η)] is accomplished by a coordinate transformation of the form,

$$x = \sum_{j=1}^{m} x_j^e \hat{\psi}_j^e(\xi, \eta) \ , \quad y = \sum_{j=1}^{m} y_j^e \hat{\psi}_j^e(\xi, \eta) \tag{2.12-1}$$

where $\hat{\psi}_j$ denote the finite element interpolation functions of the master element $\hat{\Omega}$. Consider, as an example, the master element shown in Figure 2.12–1. The coordinates in the master element are chosen to be the natural coordinates (ξ, η) such that $-1 \leq (\xi, \eta) \leq 1$. This choice is dictated by the limits of integration in the Gauss quadrature rule, which is used to evaluate the integrals. For this case, the $\hat{\psi}_j^e$ denote the interpolation functions of the four–node rectangular element shown in Figure 2.11–3 (*i.e.*, $m = 4$). The transformation (1) maps a point (ξ, η) in the master element $\hat{\Omega}$ onto a point (x, y) in element Ω^e, and vice versa if the Jacobian of the transformation is positive-definite. The transformation maps the line $\xi = 1$ in $\hat{\Omega}$ to the line defined parametrically by $x = x(1, \eta)$ and $y = y(1, \eta)$ in the xy-plane. In other words, the master element $\hat{\Omega}$ is transformed, under the linear transformation, into a quadrilateral element (*i.e.*, a four–sided element whose sides are not parallel) in the xy-plane. Conversely, every quadrilateral element of a mesh can be transformed to the same four–noded square (master) element $\hat{\Omega}$ in the (ξ, η)-plane.

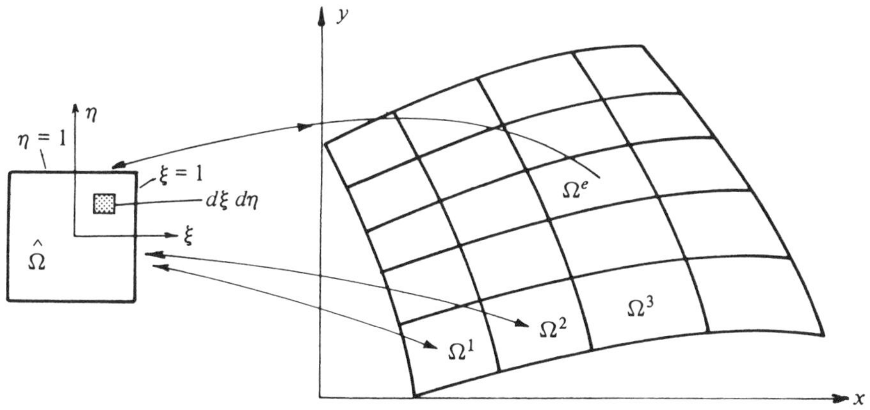

Figure 2.12–1: Transformation of arbitrarily shaped quadrilateral elements to the master rectangular element for numerical evaluation of integral expressions.

In general, the dependent variable(s) of the problem are approximated by expressions of the form

$$u(x,y) = \sum_{j=1}^{n} u_j^e \psi_j^e(x,y) \qquad (2.12-2)$$

The interpolation functions ψ_j^e used for the approximation of the dependent variable, in general, are different from $\hat{\psi}_j^e$ used in the approximation of the geometry. Depending on the relative degree of approximations used for the geometry [see equation (1)] and the dependent variable(s) [see equation (2)], the finite element formulations are classified into three categories:

1. *Superparametric* ($m > n$). The approximation used for the geometry is higher order than that used for the dependent variable.

2. *Isoparametric* ($m = n$). Equal degree of approximation is used for both geometry and dependent variables.

3. *Subparametric* ($m < n$). Higher–order approximation of the dependent variable is used.

In the present context, we use only the isoparametric formulations.

2.12–2 Coordinate Transformations

It should be noted that the transformation of a quadrilateral element of a mesh to the master element $\hat{\Omega}$ is solely for the purpose of numerically evaluating the integrals (see Figure 2.12–1). *No transformation of the physical domain or elements is involved in the finite element analysis.* The resulting algebraic equations of the finite element formulation are always in terms of the nodal values of the physical domain. Different elements of the finite element mesh can be generated from the same master element by assigning appropriate global coordinates to each of the elements. Master elements of different order define different transformations and hence different collections of finite elements within the mesh. For example, a quadratic rectangular master element can be used to generate a mesh of quadratic curvilinear quadrilateral elements. The transformations of a master element should be such that no spurious gaps exist between elements, and no element overlaps occur. For example, consider the element coefficients

$$K_{ij}^e = \int_{\Omega^e} \left[k_{11}(x,y) \frac{\partial \psi_i^e}{\partial x} \frac{\partial \psi_j^e}{\partial x} + k_{22}(x,y) \frac{\partial \psi_i^e}{\partial y} \frac{\partial \psi_j^e}{\partial y} + c(x,y) \psi_i^e \psi_j^e \right] dx dy \qquad (2.12-3)$$

The integrand (*i.e.*, the expression in the square brackets under the integral) is a function of the global coordinates x and y. We must rewrite it in terms of ξ and η using the transformation (1). Note that the integrand contains not only functions but also derivatives with respect to the global coordinates (x, y). Therefore, we must relate $(\frac{\partial \psi_i^e}{\partial x}, \frac{\partial \psi_i^e}{\partial y})$ to $(\frac{\partial \psi_i^e}{\partial \xi}, \frac{\partial \psi_i^e}{\partial \eta})$ using the transformation (1).

The functions $\psi_i^e(x, y)$ can be expressed in terms of the local coordinates ξ and η by means of equation (1). Hence, by the chain rule of partial differentiation, we have

$$\frac{\partial \psi_i^e}{\partial \xi} = \frac{\partial \psi_i^e}{\partial x}\frac{\partial x}{\partial \xi} + \frac{\partial \psi_i^e}{\partial y}\frac{\partial y}{\partial \xi}$$

$$\frac{\partial \psi_i^e}{\partial \eta} = \frac{\partial \psi_i^e}{\partial x}\frac{\partial x}{\partial \eta} + \frac{\partial \psi_i^e}{\partial y}\frac{\partial y}{\partial \eta}$$

$$\begin{Bmatrix} \frac{\partial \psi_i^e}{\partial \xi} \\ \frac{\partial \psi_i^e}{\partial \eta} \end{Bmatrix} = \begin{bmatrix} \frac{\partial x}{\partial \xi} & \frac{\partial y}{\partial \xi} \\ \frac{\partial x}{\partial \eta} & \frac{\partial y}{\partial \eta} \end{bmatrix} \begin{Bmatrix} \frac{\partial \psi_i^e}{\partial x} \\ \frac{\partial \psi_i^e}{\partial y} \end{Bmatrix} \tag{2.12 – 4}$$

which gives the relation between the derivatives of ψ_i^e with respect to the global and local coordinates. The matrix in equation (4) is called the *Jacobian matrix* of the transformation (1):

$$[J] = \begin{bmatrix} \frac{\partial x}{\partial \xi} & \frac{\partial y}{\partial \xi} \\ \frac{\partial x}{\partial \eta} & \frac{\partial y}{\partial \eta} \end{bmatrix} \tag{2.12 – 5}$$

Note from the expression given for K_{ij}^e in equation (3) that we must relate $(\frac{\partial \psi_i^e}{\partial x}, \frac{\partial \psi_i^e}{\partial y})$ to $(\frac{\partial \psi_i^e}{\partial \xi}, \frac{\partial \psi_i^e}{\partial \eta})$, whereas equation (4) provides the inverse relations. Therefore, equation (4) must be inverted by inverting the Jacobian matrix:

$$\begin{Bmatrix} \frac{\partial \psi_i^e}{\partial x} \\ \frac{\partial \psi_i^e}{\partial y} \end{Bmatrix} = [J]^{-1} \begin{Bmatrix} \frac{\partial \psi_i^e}{\partial \xi} \\ \frac{\partial \psi_i^e}{\partial \eta} \end{Bmatrix} \tag{2.12 – 6}$$

This requires that the Jacobian matrix $[J]$ be nonsingular.

Using the transformation (1), we can write

$$\frac{\partial x}{\partial \xi} = \sum_{j=1}^{m} x_j \frac{\partial \hat{\psi}_j^e}{\partial \xi}, \quad \frac{\partial y}{\partial \xi} = \sum_{j=1}^{m} y_j \frac{\partial \hat{\psi}_j^e}{\partial \xi} \tag{2.12 – 7a}$$

$$\frac{\partial x}{\partial \eta} = \sum_{j=1}^{m} x_j \frac{\partial \hat{\psi}_j^e}{\partial \eta}, \quad \frac{\partial y}{\partial \eta} = \sum_{j=1}^{m} y_j \frac{\partial \hat{\psi}_j^e}{\partial \eta} \tag{2.12 – 7b}$$

and by means of equation (5) one can compute the Jacobian matrix and then its inverse. Thus, given the global coordinates (x_j, y_j) of element nodes and the interpolation functions $\hat{\psi}_j^e$ used for geometry, the Jacobian matrix can be evaluated using equation (5). A necessary and sufficient condition for $[J]^{-1}$ to exist is that the determinant J, called the Jacobian, be nonzero at every point (ξ, η) in $\hat{\Omega}$:

$$J \equiv \det[J] = \frac{\partial x}{\partial \xi}\frac{\partial y}{\partial \eta} - \frac{\partial x}{\partial \eta}\frac{\partial y}{\partial \xi} \neq 0. \tag{2.12 – 8}$$

From equation (8) it is clear that the functions $\xi = \xi(x,y)$ and $\eta = \eta(x,y)$ must be continuous, differentiable, and invertible. Moreover, the transformation should be algebraically simple so that the Jacobian matrix can be easily evaluated. Transformations of the form in equation (1) satisfy these requirements and the requirement that no spurious gaps between elements or overlapping of elements occur.

Returning to numerical evaluation of integrals, we have from equation (6),

$$\left\{ \begin{array}{c} \frac{\partial \psi_i^e}{\partial x} \\ \frac{\partial \psi_i^e}{\partial y} \end{array} \right\} = [J]^{-1} \left\{ \begin{array}{c} \frac{\partial \psi_i^e}{\partial \xi} \\ \frac{\partial \psi_i^e}{\partial \eta} \end{array} \right\} \equiv [J^*] \left\{ \begin{array}{c} \frac{\partial \psi_i^e}{\partial \xi} \\ \frac{\partial \psi_i^e}{\partial \eta} \end{array} \right\} \tag{2.12-9}$$

where J_{ij}^* is the element in position (i,j) of the inverse of the Jacobian matrix $[J]$. The element area $dA = dxdy$ in element Ω^e is transformed to

$$dA = |J| \, d\xi d\eta \tag{2.12-10}$$

in the master element $\hat{\Omega}$.

Equations (7)–(10) provide the necessary relations to transform integral expressions on any element Ω^e to an associated master element $\hat{\Omega}$. For instance, consider the integral expression in equation (3), where k_{11}, k_{22}, and c are functions of x and y. Suppose that the finite element Ω^e can be generated by the master element $\hat{\Omega}^e$. Under the transformation (1) we can write

$$\begin{aligned} K_{ij}^e &= \int_{\Omega^e} \left[k_{11}(x,y) \frac{\partial \psi_i^e}{\partial x} \frac{\partial \psi_j^e}{\partial x} + k_{22}(x,y) \frac{\partial \psi_i^e}{\partial y} \frac{\partial \psi_j^e}{\partial y} + c(x,y) \psi_i^e \psi_j^e \right] dxdy \\ &= \int_{\hat{\Omega}^e} F_{ij}(\xi,\eta) \, d\xi d\eta \end{aligned} \tag{2.12-11}$$

The discussion presented above is valid for master elements of both rectangular and triangular geometry.

2.12–3 Integration Over a Master Rectangular Element

Integrals defined over a rectangular master element $\hat{\Omega}_R$ can be numerically evaluated using the Gauss–Legendre quadrature formulas

$$\int_{\hat{\Omega}_R} F(\xi,\eta) \, d\xi d\eta = \int_{-1}^{1} \int_{-1}^{1} F(\xi,\eta) \, d\xi d\eta \approx \sum_{I=1}^{M} \sum_{J=1}^{N} F(\xi_I, \eta_J) \, W_I W_J \tag{2.12-12}$$

where M and N denote the number of Gauss quadrature points, (ξ_I, η_J) denote the Gauss point coordinates, and W_I and W_J denote the corresponding Gauss weights as shown in Table 2.12–1 (from Table 7.2 in [7]). The selection of the number of Gauss points is based on the following formula: a polynomial of degree p is integrated exactly employing $N = int[(p+1)/2]+1$ integration points. In most cases, the interpolation functions are of the same degree in both ξ and η, and therefore one has $M = N$. When the integrand is of different degree in ξ and η, the number of Gauss points is selected on the basis of the largest–degree polynomial. The minimum allowable quadrature rule is one that computes the volume of the element exactly. The maximum degree of the polynomial refers to the degree of the highest polynomial in ξ or η that is present in the integrands of the element matrices of the type in equation (3). Note that the polynomial degree of coefficients as well as J_{ij} should be accounted for in determining the total polynomial degree of the integrand. Of course, the coefficients k_{11}, k_{22}, and c and J_{ij} in general may not be polynomials. In those cases, their functional variations must be approximated by a suitable polynomial (for example, by a binomial series) in order to determine the polynomial degree of the integrand.

Table 2.12–1: Quadrature weights and points for rectangular elements.

$$\int_{-1}^{1} F(\xi)d\xi = \sum_{i=1}^{r} F(\xi_i)w_i$$

Points ξ_i	r	Weights w_i
0.0000000000	One-point formula	2.0000000000
±0.5773502692	Two-point formula	1.0000000000
0.0000000000	Three-point formula	0.8888888889
±0.7745966692		0.5555555555
±0.3399810435	Four-point formula	0.6521451548
±0.8611363116		0.3478548451
0.0000000000	Five-point formula	0.5688888889
±0.5384693101		0.4786286705
±0.9061798459		0.2369268850
±0.2386191861	Six-point formula	0.4679139346
±0.6612093865		0.3607615730
±0.9324695142		0.1713244924

2.12–4 Integration Over a Master Triangular Element

In the preceding section we discussed numerical integration on quadrilateral elements which can be used to represent very general geometries as well as field variables in a variety of problems. Here we discuss numerical integration on triangular elements. Since quadrilateral elements can be geometrically distorted, it is possible to distort a quadrilateral element to obtain a required triangular element by moving the position of the corner nodes, and the fourth corner in the quadrilateral is merged with one of the neighboring nodes. In actual computation, this is achieved by assigning the same global node number to two corner nodes of the quadrilateral element. Thus, master triangular elements can be obtained in a natural way from associated master rectangular elements. Here we discuss the transformations from a master triangular element to an arbitrary triangular element.

Table 2.12–2: Quadrature weights and points for triangular elements.

Number of integration points	Degree of polynomial order of the residual	Location of integration points L_1	L_2	L_3	W	Geometric locations	
1	1 $O(h^2)$	$\frac{1}{3}$	$\frac{1}{3}$	$\frac{1}{3}$	1	a	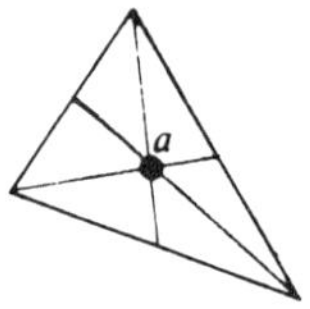
3	2 $O(h^3)$	1/2 1/2 0	0 1/2 1/2	1/2 0 1/2	$\frac{1}{3}$ (all)	a b c	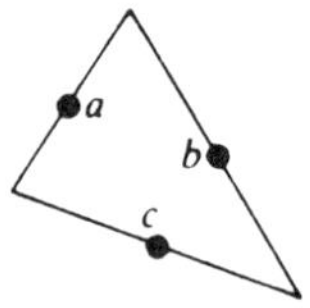
4	3 $O(h^4)$	1/3 0.6 0.2 0.2	1/3 0.2 0.6 0.2	1/3 0.2 0.2 0.6	-27/48 25/48 25/48 25/48	a b c d	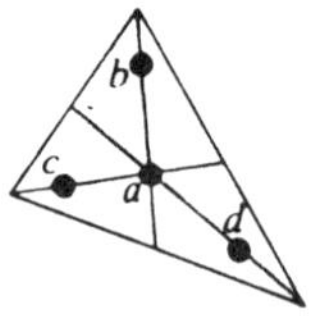

We choose the unit right isosceles triangle (see Table 2.12–2) as the master element. An arbitrary triangular element Ω^e can be generated from the master triangular element $\hat{\Omega}_T$ by transformation of the form (1). The derivatives of ψ_i with respect to the global coordinates can be computed from equation (6), which take the form

$$\left\{ \begin{matrix} \frac{\partial \psi_i^e}{\partial x} \\ \frac{\partial \psi_i^e}{\partial y} \end{matrix} \right\} = [J]^{-1} \left\{ \begin{matrix} \frac{\partial \psi_i^e}{\partial L_1} \\ \frac{\partial \psi_i^e}{\partial L_2} \end{matrix} \right\} \tag{2.12 – 13a}$$

$$[J] = \begin{bmatrix} \frac{\partial x}{\partial L_1} & \frac{\partial y}{\partial L_1} \\ \frac{\partial x}{\partial L_2} & \frac{\partial y}{\partial L_2} \end{bmatrix} \tag{2.12 – 13b}$$

Note that only L_1 and L_2 are treated as linearly independent coordinates because $L_3 = 1 - L_1 - L_2$.

After transformation, integrals on $\hat{\Omega}_T$ have the form

$$\int_{\Omega^e} G(\xi, \eta)\, d\xi d\eta = \int_{\Omega^e} G(L_1, L_2, L_3)\, dL_1 dL_2 \tag{2.12-14}$$

which can be approximated by the quadrature formula

$$\int_{\Omega^e} G(L_1, L_2, L_3)\, dL_1 dL_2 \approx \sum_{I=1}^{N} G(\mathbf{S}_I) W_I \tag{2.12-15}$$

where W_I and $\mathbf{S}_I$ denote the weights and integration points of the quadrature rule. Table 2.12–2 (from Table 9.3 of [7]) contains the location of integration points and weights for one-point, three-point, and four-point quadrature rules over triangular elements. For evaluation of integrals whose integrands are polynomials of degree equal to or higher than 5 (in any of the area coordinates) the reader must consult books on numerical integration.

2.13 MODELING CONSIDERATIONS

Here we discuss some aspects of finite element model development. Guidelines concerning mesh generation, boundary flux representation, and imposition of boundary conditions are discussed here in very general terms (see Reddy [7] for further details).

2.13–1 Mesh Generation

Generation of a finite element mesh for a given domain should follow the guidelines listed below:

1. The mesh should represent the geometry of the computational domain and boundary flux representation accurately.
2. The mesh should be such that large gradients in the solution (temperature and velocities) are adequately represented.
3. The mesh should not contain elements with very large aspect ratios (*i.e.*, ratio of the largest side to the smallest side of an element) and/or angular distortions, especially in regions of large gradients.

Within the above guidelines, the mesh used can be *coarse* (*i.e.*, few elements) or *refined* (*i.e.*, many elements), and may consist of one or more orders and types of elements (*e.g.*, linear and quadratic, triangular and quadrilateral). It should be noted that the choice of element type and mesh is problem dependent. An analyst with physical insight into the process being simulated can make a better choice of elements and mesh for the problem at hand than one who does not have the physical insight.

One should evaluate the results obtained in the light of physical understanding and approximate analytical and/or experimental information.

Generation of meshes of single element type (*i.e.*, linear elements or quadratic elements) is easy, because elements of the same degree are compatible with each other. Mesh refinements involve several options. Refine the mesh by subdividing existing elements into two or more elements of the same type (see Figure 2.13–1a). This is called the *h–version mesh refinement.* Refine the mesh by replacing existing elements by elements of higher order (see Figure 2.13–1b). This type of refinement is called the *p–version mesh refinement.* The *(h, p)–version mesh refinement* is one in which elements are subdivided into two or more elements in some places and replaced by higher–order elements in other places. Generally, local mesh refinements should be such that very small elements are not placed adjacent to very large elements.

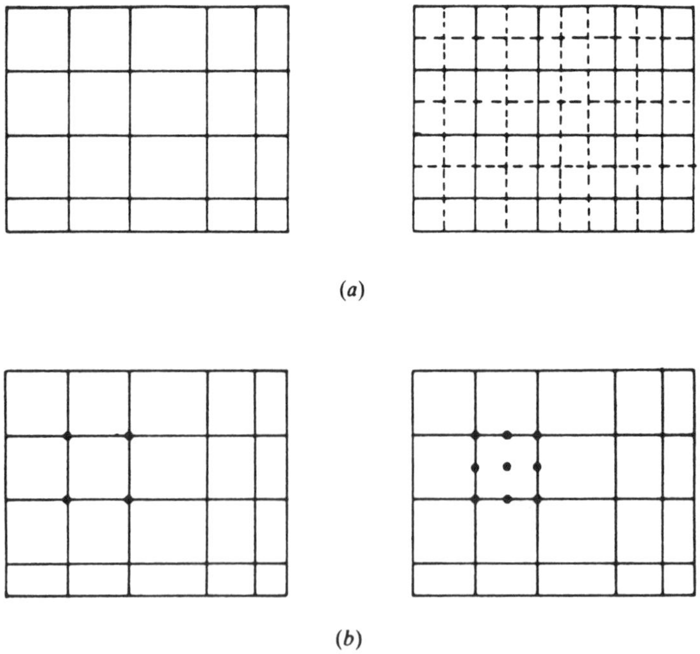

Figure 2.13–1: The two types of mesh refinements in finite element analyses: (a) h-refinement (refinement with the same order of elements); (b) p-refinement (refinement with higher–order elements).

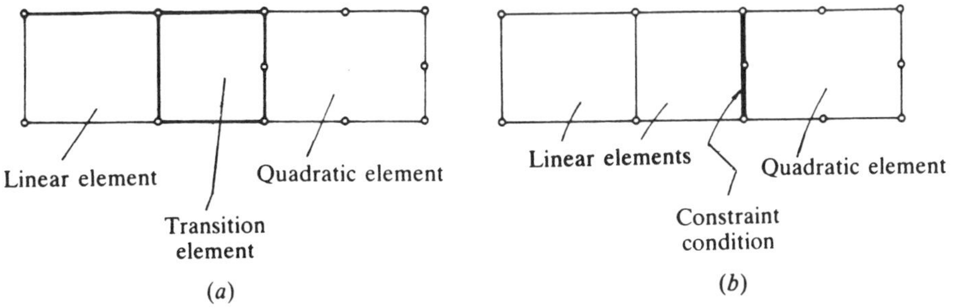

Figure 2.13–2: Examples of local mesh refinements. Mesh refinement with (a) transition elements, and (b) linear constraint equations.

To accomplish local mesh refinements it may be necessary to combine elements of different orders, say linear and quadratic elements. There are two ways to combine elements of different orders. One way is to use transition elements, which have different numbers of nodes on different sides of the element (see Figure 2.13–2a). Another way is to impose a condition that constrains the midside node to have the same value as that at the node of the lower-order element (see Figure 2.13–2b). Neither combination enforces interelement continuity of the solution.

2.13–2 Representation of Boundary Flux

When the temperature distribution in the domain is approximated by a set of finite elements, the boundary flux distribution, if known, must be replaced by a set of equivalent nodal fluxes. The accuracy of the solution depends on the element type and mesh used to represent the domain and the representation of actual flux distribution. Use of linear elements, for example, to represent a curved boundary will change the actual distribution (see Figure 2.13–3). Of course, mesh refinements by h-version or p-version will improve the representation of the specified boundary flux. Also, the vector of nodal fluxes due to a distributed heat flux should be computed using the interpolation functions of the element used to represent the solution. For example, if the temperature is approximated using the nine–node quadratic elements, the nodal contributions of a distributed heat flux should not be evaluated using four–node linear elements. This would be an inconsistent computation of nodal fluxes. Mesh refinements make the flux representation more accurate, and hence convergence of the numerical solution to the actual solution can be expected.

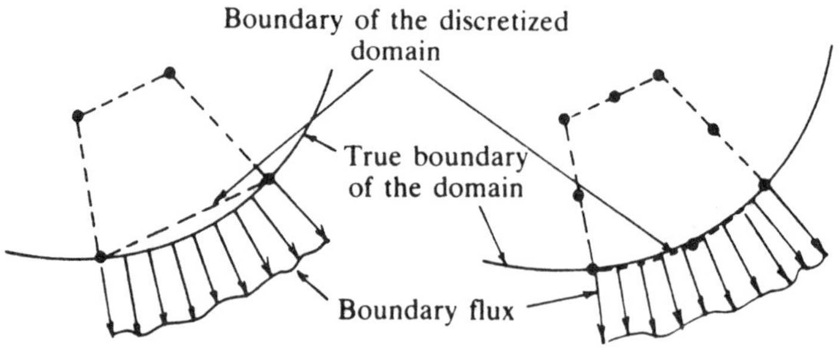

Figure 2.13–3: Representation of boundary flux in finite element analyses.

2.13–3 Imposition of Boundary Conditions

In most problems one encounters situations where the portion of the boundary on which flux is specified has points in common with the portion of the boundary on which the temperature is specified. In other words, at a few nodal points of the mesh, both the heat input and temperature may be specified, *e.g.*, at an exterior corner. Such points are called ***singular points***. Obviously, one cannot impose both boundary conditions at the same point. As a general rule, one should impose the essential (*i.e.*, temperature) boundary condition at the singular points and disregard the flux boundary condition. Of course, if the true situation in a problem is that the flux boundary condition is imposed and the temperature boundary condition is a result of it, then consideration must be given to the former one.

Another type of singularity one encounters in the analysis of thermal boundary-value problems is the specification of two different values of a temperature or heat input at the same boundary point. In the finite element analysis, one must make a choice between the two values or take a weighted average of the two. In any case, one must note that the true boundary condition is replaced by an approximate condition. The closeness of the approximate boundary condition to the true one depends on the size of the element containing the point. It is often necessary to make a mesh refinement in the vicinity of the singular point to obtain an acceptable solution.

In most practical problems a variety of approximations are introduced before reaching the finite element model; additional approximations, some of which were described here, are introduced by the computational model. It is important to keep these various sources of error in perspective. A feel for the relative proportions and directions of various errors introduced into the analysis helps the finite element practitioner to make a decision on selecting a mesh and representing the boundary conditions. In summary, scientific (or engineering) knowledge of and experience with the problem being analyzed is an essential part of any approximate analysis.

2.14 ILLUSTRATIVE EXAMPLES

A sample finite element program that reflects the ideas presented in this chapter is included in Appendix A. This is a simplified version of the program FEM2DV2 from [7]. The program is limited to linear two-dimensional problems. Plane as well as axisymmetric (see Section 2.9) geometries can be modeled, both conductive and convective boundary conditions can be handled, and steady as well as transient heat transfer problems can be solved using the program.

In this section simple examples of applications of the finite element method are discussed. In the interest of simplicity, the examples are limited to heat transfer problems on rectangular domains and the use of triangular and rectangular elements. Additional examples can be found in [7]. More complicated problems, which reflect practical situations, will be addressed in the subsequent chapters.

2.14–1 Example 1

Problem description

Consider the conduction heat transfer in a square plate of dimension 2 cm by 2 cm, conductivity k $(W/m/^{\circ}C)$, and uniform internal heat generation of Q_0 (W/m^3). The edges of the plate are maintained at a temperature of $T_0 = 0$ $(^{\circ}C)$ (see Figure 2.14–1). We wish to determine the steady temperature distribution inside the plate using the finite element method.

The problem can be expressed mathematically as one of solving the equation,

$$-k\nabla^2 T = Q_0 \quad \text{or} \quad -k(\frac{\partial^2 T}{\partial x^2} + \frac{\partial^2 T}{\partial y^2}) = Q_0 \quad \text{in} \quad \Omega \tag{2.14-1a}$$

$$T = T_0 \quad \text{on} \quad \Gamma \tag{2.14-1b}$$

Equation (1) has an analytical (series) solution, which can be obtained using the method of separation of variables. We wish to compare the finite element solution with the analytical solution obtained with various meshes and elements.

The problem has symmetries that can be exploited in the finite element analysis. A problem possesses symmetry of the solution about a line only when there exists a symmetry of the (1) geometry, (2) material properties, (3) source (*i.e.*, heat generation), and (d) boundary conditions of the problem. Such a line is called the *line of symmetry.* When a line of symmetry exists in a problem, then it is sufficient to model the domain on either side of the line. Then the line of symmetry becomes a part of the boundary of the computational domain. On such boundaries, the normal derivative of the solution (*i.e.*, derivative of the solution with respect to the coordinate normal to the line of symmetry) is zero. In the context of heat transfer problems, this implies that the temperature gradient (or heat flux) across the line of symmetry is zero:

$$q_n \equiv \frac{\partial T}{\partial n} = \hat{\mathbf{n}} \cdot \nabla T = \frac{\partial T}{\partial x} n_x + \frac{\partial T}{\partial y} n_y = 0 \tag{2.14-2}$$

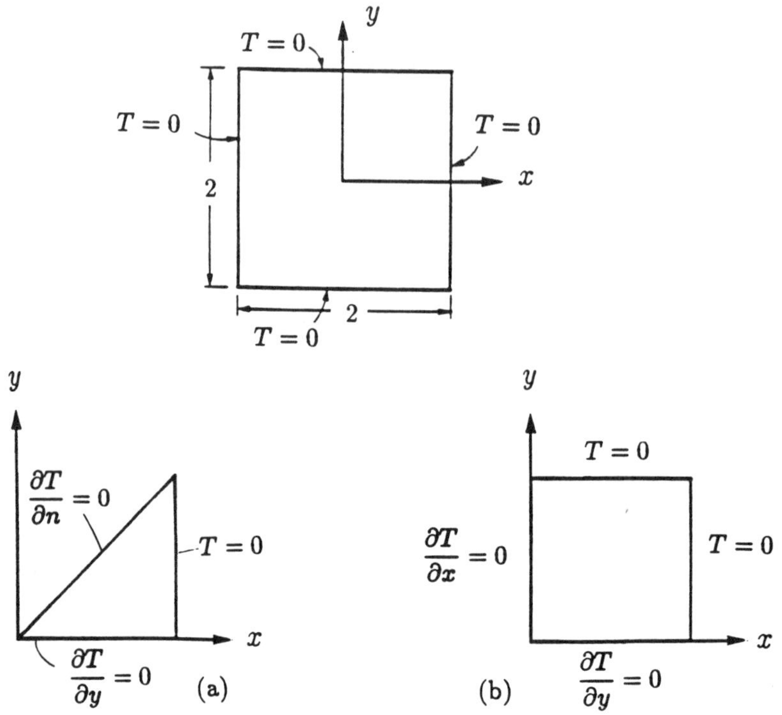

Figure 2.14–1: Heat transfer in a square domain. Geometry and boundary conditions (note use of lines of symmetry): (a) computational domain for use of triangular elements; (b) computational domain for use of rectangular elements.

The problem at hand has the symmetry about the $x = 0$ and $y = 0$ axes; it is also symmetric about the diagonal line $x = y$ (see Figure 2.14–1a). Thus, we can use a quadrant of the domain for meshes of rectangular elements and an octant of the domain for meshes of triangular elements to analyze the problem. Of course, it is possible to mix triangular and rectangular elements to represent the domain as well as the solution.

Solution by linear triangular elements

Due to the symmetry along the diagonal $x = y$, we model the triangular domain shown in Figure 2.14–1a. As a first choice we use a uniform mesh of four linear triangular elements to represent the domain (see Figure 2.14–2a), and then a refined mesh (see Figure 2.14–2b) to compare the solutions. In the present case, there is no discretization error involved in the problem because the geometry is exactly represented.

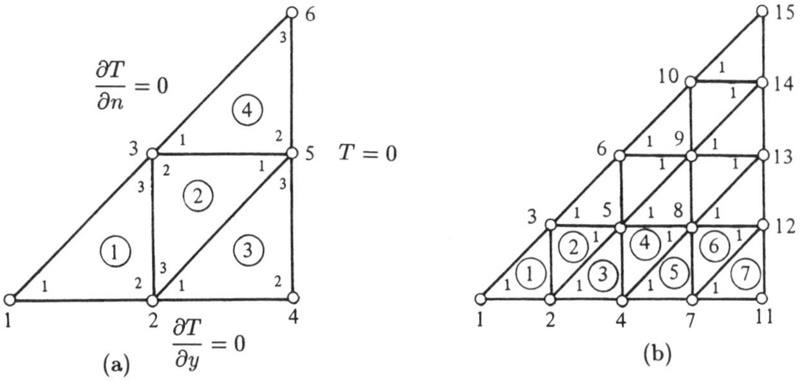

Figure 2.14–2: Discretization of the computational domain with triangular elements: (a) mesh of four elements; (b) mesh of sixteen elements.

We consider element 1 as the typical element with its local coordinate system $(\bar{x}, \bar{y})$. Suppose that the element dimensions, *i.e.*, length and height are a and b, respectively. The coordinates of the element nodes are

$$(\bar{x}_1, \bar{y}_1) = (0,0);\ (\bar{x}_2, \bar{y}_2) = (a,0),\ (\bar{x}_3, \bar{y}_3) = (a,b) \qquad (2.14-3)$$

Hence the parameters α_i, β_i, and γ_i are given by

$$\alpha_1 = \bar{x}_2\bar{y}_3 - \bar{x}_3\bar{y}_2 = ab,\ \alpha_2 = \bar{x}_3\bar{y}_1 - \bar{x}_1\bar{y}_3 = 0,\ \alpha_3 = \bar{x}_1\bar{y}_2 - \bar{x}_2\bar{y}_1 = 0,$$

$$\beta_1 = \bar{y}_2 - \bar{y}_3 = -b,\ \beta_2 = \bar{y}_3 - \bar{y}_1 = b,\ \beta_3 = \bar{y}_1 - \bar{y}_2 = 0,$$

$$\gamma_1 = -(\bar{x}_2 - \bar{x}_3) = 0,\ \gamma_2 = -(\bar{x}_3 - \bar{x}_1) = -a,\ \gamma_3 = -(\bar{x}_1 - \bar{x}_2) = a \qquad (2.14-4)$$

For the mesh shown in Figure 2.14–2a, we have

$$[K^1] = [K^2] = [K^3] = [K^4];\ \{Q^1\} = \{Q^2\} = \{Q^3\} = \{Q^4\} \qquad (2.14-5)$$

Therefore, the element coefficients K^e_{ij} and Q^e_i $(e = 1,2,3,4)$ are given by (note that a denotes the length of side connecting element nodes 1 and 2)

$$[K^e] = \frac{k}{2}\begin{bmatrix} \alpha & -\alpha & 0 \\ -\alpha & \gamma & -\beta \\ 0 & -\beta & \beta \end{bmatrix}, \tag{2.14-6a}$$

$$\{Q^e\} = \frac{Q_0 ab}{6}\begin{Bmatrix} 1 \\ 1 \\ 1 \end{Bmatrix} \tag{2.14-6b}$$

where $\alpha = \frac{b}{a}$, $\beta = \frac{a}{b}$, and $\gamma = \alpha + \beta$. The element matrix in (6a) is valid for the Laplace operator $-\nabla^2$ on any right-angle triangle with sides a and b in which the right angle is at node 2, and the diagonal line of the triangle connects node 3 to node 1 [also see equation (2.6–6)]. Note that the off–diagonal coefficient associated with the nodes on the diagonal line is zero for a right–angled triangle. These observations can be used to write the element matrix associated with the Laplace operator on any right-angled triangle, *i.e.*, for any element–node numbering system.

The assembled coefficient matrix for the finite element mesh is 6×6, because there are six global nodes with one unknown per node. The assembled matrix can be obtained directly by using the correspondence between the global nodes and the local nodes, as explained earlier.

The specified boundary conditions on the primary degrees of freedom of the problem are

$$T_4 = T_5 = T_6 = T_0 = 0 \tag{2.14-7}$$

The specified secondary degrees of freedom at nodes 1, 2, and 3 are all zero because the flux is zero there:

$$q_1^1 = 0, \quad q_2^1 + q_3^2 + q_1^3 = 0, \quad q_3^1 + q_2^2 + q_1^4 = 0 \tag{2.14-8}$$

Since T_4, T_5, and T_6 are known, the secondary variables at these nodes are unknown, and they can be obtained in the postcomputation.

Solution by linear rectangular elements

Note that we cannot exploit the symmetry along the diagonal $x = y$ to our advantage when we use a mesh of rectangular elements. Therefore, we use a 2×2 uniform mesh of four linear rectangular elements and a refined 4×4 mesh (see Figure 2.14–3a) to discretize a quadrant of the domain. Once again, no discretization error is introduced in the present case.

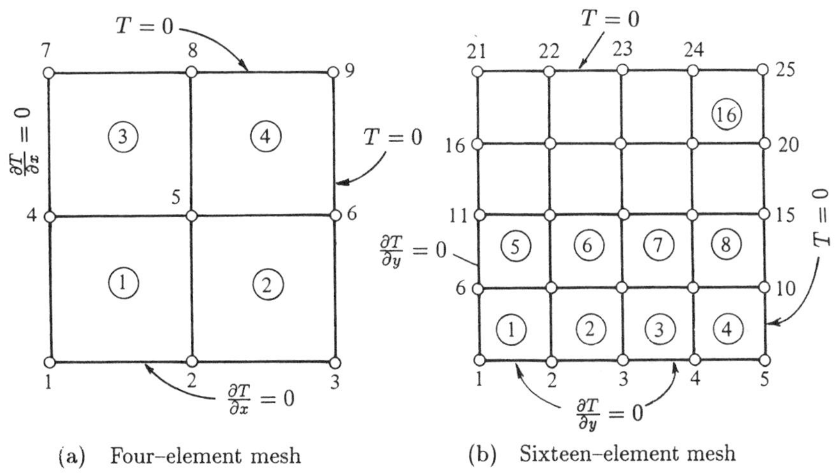

Figure 2.14–3: Discretization of the computational domain with rectangular elements: (a) mesh of four elements; (b) mesh of sixteen elements.

All elements in the mesh are identical. Therefore, we have the following element matrices [see equation (2.6–10)] for a typical element of the mesh ($\alpha = \frac{b}{a}$, $\beta = \frac{a}{b}$):

$$[K^e] = \frac{k}{6}\begin{bmatrix} 2(\alpha+\beta) & -2\alpha+\beta & -(\alpha+\beta) & \alpha-2\beta \\ -2\alpha+\beta & 2(\alpha+\beta) & \alpha-2\beta & -(\alpha+\beta) \\ -(\alpha+\beta) & \alpha-2\beta & 2(\alpha+\beta) & -2\alpha+\beta \\ \alpha-2\beta & -(\alpha+\beta) & -2\alpha+\beta & 2(\alpha+\beta) \end{bmatrix}, \qquad (2.14-9a)$$

$$\{Q^e\} = \frac{Q_0 ab}{4}\begin{Bmatrix} 1 \\ 1 \\ 1 \\ 1 \end{Bmatrix} \qquad (2.14-9b)$$

The boundary conditions on the secondary variables are

$$q_1^1 = 0, \; q_2^1 + q_1^2 = 0, \; q_4^1 + q_1^3 = 0 \qquad (2.14-10a)$$

and the balance of secondary variables at global node 5 requires

$$q_3^1 + q_4^2 + q_2^3 + q_1^4 = 0 \tag{2.14-10b}$$

Table 2.14–1: Convergence of the finite element solutions to the analytical solution of equation (2.14–1).

	Triangular elements		**Rectangular elements**		**Series solution**
y	**4 elements**	**16 elements**	**4 elements**	**16 elements**	
0.0	0.3125	0.3013	0.3107	0.2984	0.2947
0.25	0.2709†	0.2805	0.2759†	0.2824	0.2789
0.50	0.2292	0.2292	0.2411	0.2322	0.2293
0.75	0.1146†	0.1393	0.1205†	0.1414	0.1397
1.0	0.0000	0.0000	0.0000	0.0000	0.0000

† Interpolated values.

Discussion of the results

The finite element solutions obtained by two different meshes of triangular elements and two different meshes of rectangular elements are compared in Table 2.14–1 at $x = 0$ for varying y for $k = 1, Q_0 = 1$, and $T_0 = 0$. The finite element solution obtained by 16 triangular elements (in an octant) is the most accurate when compared to the series solution. The accuracy of the triangular element mesh is due to the larger number of elements it has compared to the number of elements in the rectangular element mesh for the same size of the domain.

The finite element nodal temperatures obtained with the 4×4 mesh of linear triangular elements in a quadrant is compared in Table 2.14–2 with the finite difference solution obtained with a 4×4 rectangular grid and with the exact (or series) solution. The node numbering scheme is the same as that used in Figure 2.14–3(b) for a rectangular element mesh. The central difference approximation is used to approximate the second derivatives in the equation. The finite element nodal temperatures are slightly more accurate, except at node 1, than those obtained with the finite difference method. One should remember that the finite element method minimizes the residual of the approximation of the differential equation over each element—not the error in the nodal values. In other words, the nodal values are adjusted such that the weighted–integral of the error in the approximation of the differential equation is minimized over each element.

Table 2.14–2: Comparison of the finite element solution with the finite difference and analytical solutions of equation (2.14–1). A 4×4 mesh of linear triangular elements is used in the finite element method and a comparable mesh is used in the finite difference method.

Node	Series solution	Finite difference solution	Error	Finite element solution (steady)	Error
1	0.2947	0.2911	0.0036	0.3013	−0.0066
2	0.2789	0.2755	0.0034	0.2805	−0.0016
3	0.2293	0.2266	0.0027	0.2292	0.0001
4	0.1397	0.1381	0.0016	0.1392	0.0005
5	0.0000	0.0000	0.0000	0.0000	0.0000
7	0.2642	0.2609	0.0033	0.2645	−0.0003
8	0.2178	0.2151	0.0027	0.2172	0.0006
9	0.1333	0.1317	0.0016	0.1327	0.0006
10	0.0000	0.0000	0.0000	0.0000	0.0000
13	0.1811	0.1787	0.0024	0.1801	0.0010
14	0.1127	0.1110	0.0017	0.1117	0.0010
15	0.0000	0.0000	0.0000	0.0000	0.0000
19	0.0728	0.0711	0.0017	0.0715	0.0013
20	0.0000	0.0000	0.0000	0.0000	0.0000
25	0.0000	0.0000	0.0000	0.0000	0.0000

The solution T and its gradient can be computed at any interior point of the domain using the interpolation,

$$T^e(x,y) = \sum_{j=1}^{n} u_j^e \psi_j^e(x,y) \qquad (2.14-11)$$

Note that for a linear triangular element the components $(\frac{\partial T}{\partial x}, \frac{\partial T}{\partial y})$ of the temperature gradient vector are constants over an entire element, whereas $\frac{\partial T}{\partial x}$ is linear in y and $\frac{\partial T}{\partial y}$ is linear in x for a linear rectangular element.

2.14–2 Example 2

Consider a cylinder of height $L = 1$ cm, radius $R = 1$ cm, thermal conductivity $k = 25\ W/(m\ ^\circ C)$, and constant internal heat generation of $Q_0 = 5 \times 10^8\ W/m^3$. The top and bottom faces of the cylinder are assumed to be insulated and the boundary surface is maintained at $T_0 = 100^\circ C$. The assumption of an insulated boundary condition at the top and bottom of the cylinder makes the heat flow one dimensional along the radial direction. For this case, the steady temperature along the radial direction is given by

$$T(r) = T_0 + \frac{Q_0}{4k}(R^2 - r^2) \qquad (2.14-12)$$

Table 2.14–3 contains the finite element solution obtained with various finite element meshes along the $r-$coordinate. Only one element is used in the $z-$direction. The 5 × 1 mesh of nine–node quadratic elements gives exact values at the nodes.

Table 2.14–3: The finite element solutions of an axisymmetric heat transfer problem (1–D).

$r \times 10^{-2}$	5LR	10LR	5QR*
0.0	611.92	603.56	600.00
0.1	-	596.89	595.00
0.2	585.25	581.33	580.00
0.3	-	556.00	555.00
0.4	523.03	520.76	520.00
0.5	-	475.58	475.00
0.6	421.69	420.43	420.00
0.7	-	355.30	355.00
0.8	280.74	280.19	280.00
0.9	-	195.09	195.00
1.0	100.00	100.00	100.00

*5LR = Five linear rectangular elements, 10RL = Ten linear rectangular elements, 5QR = Five quadratic rectangular elements.

Next we consider a cylinder of height $L = 2$ cm, radius $R = 1$ cm, thermal conductivity $k = 25\ W/(m\ ^\circ C)$, and constant internal heat generation of $Q_0 = 5 \times 10^8\ W/m^3$. The top and bottom faces and the surface of the cylinder are maintained at $T_0 = 100$°C. For this case, the steady solution is two dimensional. Using the symmetry about $z = 1$ cm, a 10 × 10 mesh of linear rectangular elements is used. The results are presented in Table 2.14–4.

2.14–3 Example 3

Consider a plane wall of height $b = 1$ cm, thickness $a = 1$ cm, thermal conductivity $k = 18\ W/(m\ ^\circ C)$, and constant internal heat generation of $Q_0 = 7.2 \times 10^7\ W/m^3$. The boundaries at $y = 0, b$ of the domain are assumed to be insulated, the boundary at $x = 0$ is maintained at $T_0 = 50°C$, and the boundary at $x = a$ is exposed to ambient temperature $T_c = 100°C$. The film coefficient is $h_c = 200\ W/(m^2\ °C)$. The assumption of an insulated boundary condition at $y = 0, b$ of the domain makes the heat flow one–dimensional along the x–direction. For this case, the temperature is given by

$$T(x) = 50 + 5\frac{x}{a} + 200\left(1.9 - \frac{x}{a}\right)\frac{x}{a} \qquad (2.14-13)$$

Table 2.14–4: The finite element solutions of an axisymmetric heat transfer problem (2–D).

$r\ (z = 0.001)$	Temp.	$r\ (z = 0.01)$	Temp.	$z\ (r = 0.0)$	Temp.
0.000	195.86	0.000	504.64	0.000	100.00
0.001	195.03	0.001	499.82	0.001	195.86
0.002	193.06	0.002	488.52	0.002	274.19
0.003	189.78	0.003	469.93	0.003	337.35
0.004	185.07	0.004	443.70	0.004	387.55
0.005	178.74	0.005	409.43	0.005	426.72
0.006	170.50	0.006	366.68	0.006	456.54
0.007	159.91	0.007	314.92	0.007	478.35
0.008	146.20	0.008	253.61	0.008	493.20
0.009	127.81	0.009	182.16	0.009	501.81
0.010	100.00	0.010	100.00	0.010	504.64

For this problem, any finite element mesh would give the exact solution at the nodes. For example, the temperature values at $x/a = 0.2, 0.4, 0.6, 0.8$, and 1.0 are $119, 172, 209, 230$, and 235, respectively.

Next we consider a square domain, $a = b = 1$ cm, thermal conductivity $k = 18\ W/(m\ °C)$, and constant internal heat generation of $Q_0 = 7.2 \times 10^7\ W/m^3$. The boundaries at $y = 0, b$ and $x = 0$ are maintained at $T_0 = 50°C$, and the boundary at $x = a$ is exposed to ambient temperature $T_c = 100°C$. The film coefficient is $h_c = 200\ W/(m^2\ °C)$. Using the symmetry about $y = 0.005$ m, a 10×5 mesh of linear rectangular elements is used to solve the problem. The results are presented in Table 2.14–5.

Table 2.14–5: The finite element solutions of a heat transfer (2–D) problem with convective boundary condition.

x ($y = 0.001$)	Temp.	x ($y = 0.003$)	Temp.	x ($y = 0.005$)	Temp.
0.000	50.000	0.000	50.000	0.000	50.000
0.001	55.768	0.001	61.396	0.001	63.002
0.002	59.303	0.002	69.702	0.002	72.765
0.003	61.717	0.003	75.714	0.003	79.991
0.004	63.416	0.004	80.055	0.004	85.279
0.005	64.623	0.005	83.178	0.005	89.109
0.006	65.481	0.006	85.405	0.006	91.848
0.007	66.083	0.007	86.964	0.007	93.763
0.008	66.500	0.008	88.015	0.008	95.038
0.009	66.796	0.009	88.664	0.009	95.789
0.010	67.135	0.010	88.960	0.010	96.067

EXERCISE PROBLEMS

The following problems are designed to test the understanding of the ideas presented in this chapter. Some of the exercise problems are reproduced from the textbook by Reddy [7]. For additional examples and exercise problems, the reader may consult Reference 7.

2.1 Develop the weak form of the vector equation (2.2–3) over an element

$$-\nabla \cdot (k\nabla T) = Q \quad \text{in} \quad \Omega \tag{1}$$

subjected to the boundary conditions of the form

$$T = \hat{T} \text{ on } \Gamma_T \quad \text{and} \quad \hat{\mathbf{n}} \cdot (k\nabla T) + q_c = q \quad \text{on } \Gamma_q \tag{2}$$

Solution:

$$\begin{aligned} 0 &= \int_{\Omega^e} w\,[-\nabla \cdot (k\nabla T) - Q]\,d\mathbf{x} \\ &= \int_{\Omega^e} (k\nabla w \cdot \nabla T - wQ)\,d\mathbf{x} - \oint_{\Gamma^e} w\hat{\mathbf{n}} \cdot (k\nabla T) ds \\ &= \int_{\Omega^e} (k\nabla w \cdot \nabla T - wQ)\,d\mathbf{x} - \oint_{\Gamma^e} w(q - q_c) ds \end{aligned}$$

2.2 Use the weak form given in equation (2.10–2) to develop the finite element model of heat transfer problems with convective boundary conditions. Show that the finite element model is of the form

$$[K^e + H^e]\{T^e\} = \{Q^e\} + \{P^e\} + \{q^e\} \tag{1a}$$

$$K_{ij}^e = \int_{\Omega^e} \left(k_{11} \frac{\partial \psi_i^e}{\partial x} \frac{\partial \psi_j^e}{\partial x} + k_{22} \frac{\partial \psi_i^e}{\partial y} \frac{\partial \psi_j^e}{\partial y} \right) dxdy \tag{1b}$$

$$H_{ij}^e = \int_{\Gamma^e} h_c \psi_i^e \psi_j^e ds, \quad Q_i^e = \int_{\Omega^e} \psi_i^e Q \, dxdy \tag{1c}$$

$$q_i^e = \oint_{\Gamma^e} q_n \psi_i^e ds, \quad P_i^e = \oint_{\Gamma^e} h_c T_c^e \psi_i^e ds \tag{1d}$$

2.3 Modify the weak form (2.9–2) of an axisymmetric problem to include convective heat transfer boundary condition, and develop the finite element model. Note that the finite element model will be the same as in Problem 2.2, except for the definitions of $[H^e]$ and P^e.

2.4 Evaluate the coefficients H_{ij}^e and P_i^e (due to the convective boundary condition) in Problem 2.2 for the case of a linear triangular element.

2.5 Repeat Problem 2.4 for a linear rectangular element.

2.6 Solve Example 1 of Section 2.14–1 with the mesh shown in Figure P2.6. Use the element matrices given in equations (2.14–6) and (2.14–9).

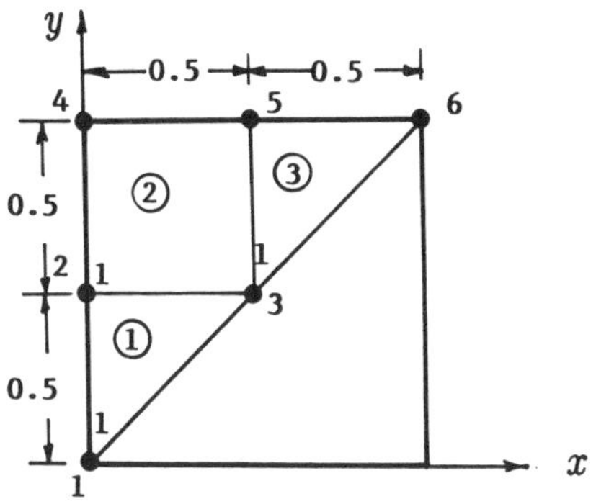

Figure P6

2.7 Consider steady–state heat conduction in a square region of side $2a$. Assume that the medium has a conductivity of $k = 30\ W/(m\ ^\circ C)$ and a uniform heat generation of $Q_0 = 10^7\ W/m^3$. For the boundary conditions and mesh shown in Figure P2.7, write the finite element algebraic equations for nodes 1, 3, and 7. Take $h_c = 60\ W/(m^2\ ^\circ C)$, $T_c = 0.0^\circ C$, $T_0 = 100^\circ C$, $q_0 = 2 \times 10^5 W/m^2$, and $a = 1cm$.

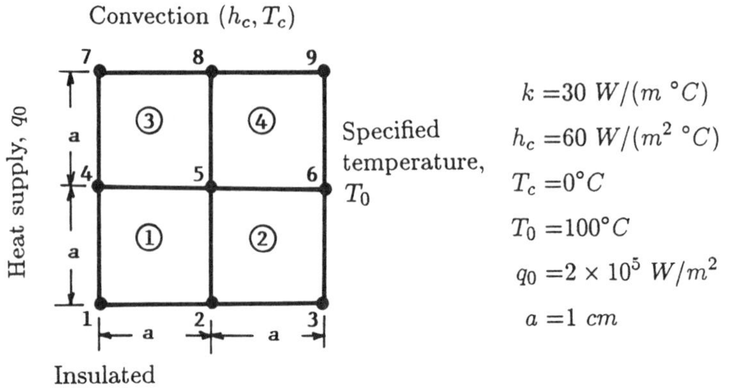

Figure P7

Solution:

The algebraic equations associated with nodes 1, 3, and 7 are

$$K_{11}T_1 + K_{12}T_2 + K_{14}T_4 + K_{15}T_5 = F_1$$

$$K_{32}T_2 + K_{33}T_3 + K_{35}T_5 + K_{36}T_6 = F_3$$

$$K_{74}T_4 + K_{75}T_5 + K_{77}T_7 + K_{78}T_8 = F_8$$

where

$$K_{11} = K_{11}^1,\ K_{12} = K_{12}^1,\ K_{14} = K_{14}^1,\ K_{15} = K_{13}^1,\ F_1 = \frac{Q_0 a^2}{4} + \frac{q_0 a}{2}$$

$$K_{32} = K_{21}^2,\ K_{33} = K_{22}^2,\ K_{35} = K_{24}^2,\ K_{36} = K_{23}^2,\ F_3 = \frac{Q_0 a^2}{4} + Q_2^2,\ T_3 = T_6 = T_0$$

$$K_{74} = K_{41}^3,\ K_{75} = K_{42}^3,\ K_{77} = K_{44}^3 + H_{44}^3,\ K_{78} = K_{43}^3 + H_{43}^3,\ F_7 = \frac{Q_0 a^2}{4} + \frac{q_0 a}{2} + P_4^3$$

Note that the equation associated with node 3 will be used to compute Q_2^2. The coefficients H_{ij}^e and P_i^e are defined in equations (1c,d) of Problem 2.1. We have

$$H_{44}^e = \frac{ah_c}{3}, \; H_{43}^e = \frac{ah_c}{6}, \; P_4^e = \frac{aT_c h_c}{2}$$

$$K_{11} = \frac{4}{6}k, \; K_{12} = -\frac{k}{6}, \; K_{13} = -\frac{2}{6}k, \; K_{14} = -\frac{k}{6}, \; K_{22} = \frac{4}{6}k,$$

$$K_{23} = -\frac{k}{6}, \; K_{24} = -\frac{2}{6}k, \; K_{33} = \frac{4}{6}k, \; K_{34} = -\frac{k}{6}, \; K_{44} = \frac{4}{6}k$$

2.8 Repeat Problem 2.7 for nodes 4, 5, and 9.

2.9 For the convection heat transfer problem shown in Figure P2.9, write the four finite element equations for the unknown temperatures. Assume that the thermal conductivity of the material is $k = 5\ W/(m\ ^\circ C)$, the convection heat transfer coefficient on the left surface is $h_c = 28\ W/(m^2\ ^\circ C)$, and the internal heat generation is zero.

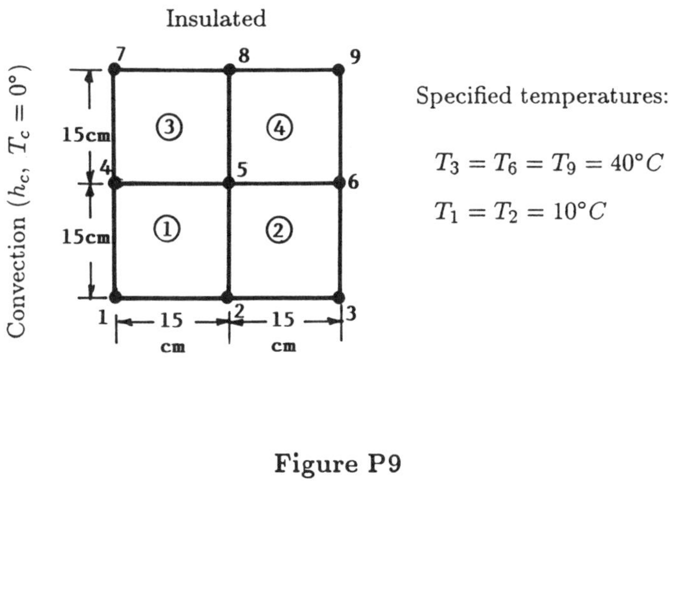

Figure P9

Solution:

There are four algebraic equations associated with nodal unknowns T_4, T_5, T_7 and T_8:

$$(K_{44}^1 + K_{11}^4 + H_{44}^1 + H_{11}^4)T_4 + (K_{43}^1 + K_{12}^4)T_5 + (K_{14}^4 + H_{14}^4)T_7 = -(K_{41}^1 + H_{41}^1)T_1 - K_{42}^1 T_2$$

$$(K_{34}^1 + K_{21}^4)T_4 + (K_{33}^1 + K_{44}^2 + K_{11}^3 + K_{22}^4)T_5 + K_{24}^4 T_7 + (K_{23}^4 + K_{14}^3)T_8 =$$
$$-K_{13}^3 T_9 - K_{31}^1 T_1 - (K_{32}^1 + K_{41}^2)T_2 - K_{42}^2 T_3 - (K_{43}^2 + K_{12}^3)T_0$$

$$(K_{41}^4 + H_{41}^4)T_4 + K_{42}^4 T_5 + (K_{44}^4 + H_{44}^4)T_7 + K_{43}^4 T_8 = 0$$

$$K_{31}^4 T_4 + (K_{32}^4 + K_{41}^3)T_5 + K_{34}^4 T_7 + (K_{33}^4 + K_{44}^3)T_8 = -K_{42}^3 T_6 - K_{43}^3 T_9$$

The heat at node 2 is given from the assembled equation associated with the node

$$Q_2 = K_{21}^1 T_1 + K_{12}^2 T_3 + K_{24}^1 T_4 + (K_{23}^1 + K_{14}^2) T_5 + K_{13}^2 T_6$$

2.10 The fin shown in Figure P2.10 has its base maintained at 300°C and is exposed to convection on its remaining boundary. Write the finite element equations at nodes 7 and 10. Take $h_c = 40\ W/(m^2\ ^\circ C), T_c = 20^\circ C$, and $k = 5\ W/(m\ ^\circ C)$.

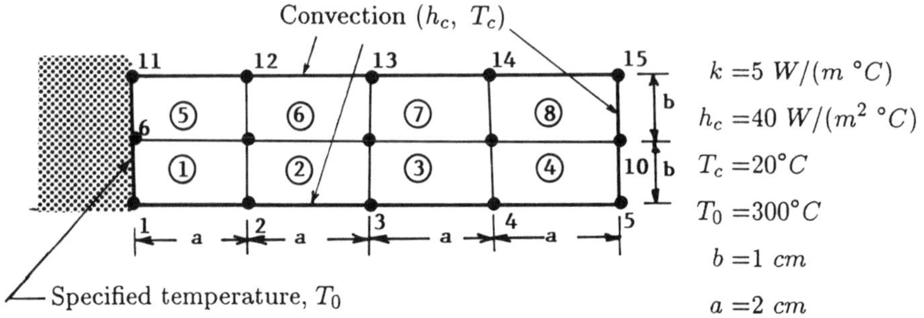

Figure P10

Solution:

The equations for nodes 7 and 10 are given by

$$(K_{32}^1 + K_{41}^2)U_2 + K_{42}^2 U_3 + (K_{43}^2 + K_{12}^6)U_8 + (K_{23}^5 + K_{14}^6)U_{12} + K_{13}^6 U_{13} =$$

$$-K_{31}^1 U_1 - (K_{34}^1 + K_{21}^5)U_6 - K_{24}^5 U_{11}$$

or

$$\frac{16}{6}U_7 - \frac{1}{6}U_2 - \frac{2}{6}U_3 - \frac{1}{6}U_8 - \frac{1}{6}U_{12} - \frac{2}{16}U_{13} = (100 + 50 + 100)$$

and

$$K_{31}^4 U_4 + (K_{32}^4 + H_{32}^4)U_5 + (K_{34}^4 + K_{21}^8)U_9 + (K_{33}^4 + K_{22}^8 + H_{33}^4 + H_{22}^8)U_{10} + K_{24}^8 U_{14} +$$

$$+(K_{23}^8 + H_{23}^8)U_{15} = P_3^4 + P_2^8$$

2.11 Solve the 2–D problem in Example 2 of Section 2.14–2 for the case in which the boundary $z = L$ is insulated. Use 5 × 10 mesh of linear rectangular elements in the full domain.

2.12 Repeat Problem 2.11 for the case where the surface $r = R$ is exposed to an ambient temperature of $T_c = 100°C$. Take the film coefficient to be $h_c = 200\ W/(m^2\ °C)$.

2.13 (*Galerkin time appproximation*) Consider the differential equation in time,

$$a\frac{du}{dt} + bu = f, \quad t_n < t < t_{n+1}$$

where a and b are constants and f is a function of time. Using linear approximation of the form, $u(t) = u_n\psi_n(t) + u_{n+1}\psi_{n+1}(t)$, where $\psi_n(t) = (t_{n+1} - t)/\Delta t$ and $\psi_{n+1}(t) = (t - t_n)/\Delta t$, derive the Galerkin finite element equations for a typical time element, (t_n, t_{n+1}). Assuming that u_n is known, compute the solution u_{n+1} from the element equations, and then compare the result with that obtained with the θ–family of approximation in equations (2.8–9) and (2.8–10).

Solution:

The weighted–integral statement is given by

$$0 = \int_{t_n}^{t_{n+1}} w\left(a\frac{du}{dt} + bu - f\right) dt \tag{1}$$

Substituting the interpolation for $u = u_n\psi_n + u_{n+1}\psi_{n+1}$ and taking $w = \psi_n$ and $w = \psi_{n+1}$, we obtain the equations for the time interval $[t_n, t_{n+1}]$:

$$\left(\frac{a}{2}\begin{bmatrix} -1 & 1 \\ -1 & 1 \end{bmatrix} + \frac{b\Delta t}{6}\begin{bmatrix} 2 & 1 \\ 1 & 2 \end{bmatrix}\right)\left\{\begin{matrix} u_n \\ u_{n+1} \end{matrix}\right\} = \frac{\Delta t}{6}\begin{bmatrix} 2 & 1 \\ 1 & 2 \end{bmatrix}\left\{\begin{matrix} f_n \\ f_{n+1} \end{matrix}\right\} \tag{2}$$

where f is also interpolated as $f = f_n\psi_n + f_{n+1}\psi_{n+1}$. Now we assume that the solution at time t_n is known and we wish to determine that at t_{n+1}. Thus we solve the second of the two equations in (2) for u_{n+1} in terms of u_n, f_n, and f_{n+1}:

$$\left(a + \frac{2}{3}b\Delta t\right) u_{n+1} = \left(a - \frac{1}{3}b\Delta t\right) u_n + \frac{\Delta t}{3}(f_n + 2f_{n+1}) \tag{3}$$

Next let us apply the θ–family of approximation to the equation. We obtain [see equations (2.8–9) and (2.8–10) on page 48],

$$(a + \theta\Delta t b)\, u_{n+1} = [a - (1-\theta)\Delta t b]\, u_n + \Delta t\,[(1-\theta)f_n + \theta f_{n+1}] \tag{4}$$

Comparing equation (3) with equation (4), we note that they are the same for $\theta = 2/3$. Thus, the Galerkin time approximation is a special case of the θ–family of approximation. Note that if we solve the first equation in (2) for u_{n+1}, we obtain the scheme corresponding to $\theta = 1/3$.

2.14 Repeat Problem 13 for the case of quadratic interpolation between the stations, (t_{n-1}, t_n, t_{n+1}). Solve the third equation of the resulting finite element equations for u_{n+1} in terms of $u_{n-1}, u_n, t_{n-1}, t_n$, and t_{n+1}.

2.15 (*Collocation time appproximation*) Consider the differential equation in time,

$$a\frac{du}{dt} + bu = f, \quad t_n < t < t_{n+1}$$

where a and b are constants and f is a function of time. Using linear approximation of the form $u(t) = u_n\psi_n(t) + u_{n+1}\psi_{n+1}(t)$, where $\psi_n(t) = (t_{n+1} - t)/\Delta t$ and $\psi_{n+1}(t) = (t - t_n)/\Delta t$, derive the one–point collocation finite element equations (*i.e.*, set $w = \delta(t - t_0)$, where $\delta(\cdot)$ denotes the Dirac delta function, in equation (1) of Problem 13) for a typical time element, (t_n, t_{n+1}), and compare the result with that obtained with the θ–family of approximation. Use the collocation point to be $t_0 = (1-\theta)t_n + \theta t_{n+1}$.

REFERENCES

1. S. G. Mikhlin, *Variational Methods in Mathematical Physics*, (translated from the Russian by T. Boddington), Pergamon Press, Oxford (1964).

2. S. G. Mikhlin, *The Numerical Performance of Variational Methods*, (translated from the Russian by R. S. Anderssen), Wolters–Noordhoff, The Netherlands (1971).

3. B.A. Finlayson, *The Method of Weighted Residuals and Variational Principles*, Academic Press, New York (1972).

4. J. T. Oden and J. N. Reddy, *Variational Methods in Theoretical Mechanics*, Springer–Verlag, Berlin (1976); second edition (1983).

5. J. N. Reddy and M. L. Rasmussen, *Advanced Engineering Analysis*, John Wiley & Sons, New York (1982); reprinted by Krieger Publishing, Melbourne, Florida (1990).

6. J. N. Reddy, *Energy and Variational Methods in Applied Mechanics*, John Wiley & Sons, New York (1984).

7. J. N. Reddy, *An Introduction to the Finite Element Method*, Second Edition, McGraw–Hill, New York (1993).

8. J. N. Reddy, *Applied Functional Analysis and Variational Methods in Engineering*, McGraw–Hill, New York (1986); reprinted by Krieger Publishing, Melbourne, Florida (1991).

9. E. B. Becker, G. F. Carey, and J. T. Oden, *Finite Elements, an Introduction*, Vol. I, Prentice–Hall, Englewood Cliffs, New Jersey (1981).

10. D. S. Burnett, *Finite Element Analysis*, Addison–Wesley, Reading, Massachusetts (1987).

11. H. S. Carslaw and J. C. Jaeger, *Conduction of Heat in Solids*, Second Edition, Oxford University Press, Fairlawn, New Jersey (1959).

12. J. P. Holman, *Heat Transfer*, Sixth Edition, McGraw–Hill, New York (1986).

13. G. E. Myers, *Analytical Methods in Conduction Heat Transfer*, McGraw–Hill, New York (1971).

14. M. N. Özisik, *Heat Transfer: A Basic Approach*, McGraw–Hill, New York (1985).

15. J. H. Argyris and O. W. Scharpf, "Finite Elements in Time and Space," *Aeronautical Journal of the Royal Society,* **73**, 1041–1044 (1969).

Chapter Three

CONDUCTION HEAT TRANSFER

3.1 INTRODUCTION

In this chapter we present the finite element analysis of problems in conduction heat transfer and related problems in diffusion. Though conduction is the least complex mode of energy transport, it plays a major role in industrial design, manufacturing, and engineering. The prediction of temperature distributions, thermal performance, and thermal stressing in complex geometries has become a routine part of engineering analysis due in large part to the finite element method and the availability of general purpose computer programs. The conduction problem also provides a convenient framework to discuss various advanced aspects of the finite element method and numerical algorithms. The introduction to the finite element method presented in Chapter 2 aids the development presented here for the general heat conduction problem.

As described in the previous chapters, the appropriate mathematical description of thermal conduction in a solid region, Ω, is given by equation (1.4–4). In the interest of brevity, the subscript 's', which refers to a solid region, is omitted throughout the chapter. The Cartesian component form of the equation is [see equation (1.4–10)]

$$\rho C \frac{\partial T}{\partial t} = \frac{\partial}{\partial x_i}\left(k_{ij}\frac{\partial T}{\partial x_j}\right) + Q \qquad (3.1-1)$$

where T is the temperature, ρ is the density, C is the specific heat, k_{ij} are the Cartesian components of the conductivity tensor (symmetric), and Q is the internal heat generation per unit volume. Summation on repeated subscripts is assumed (*i.e.*, i and j are summed over the range of $i, j = 1, 2, 3$). All of the variables may be functions of position $\mathbf{x} = (x_1, x_2, x_3)$ and time t. We wish to solve equation (1) under appropriate boundary and initial conditions. The boundary conditions for this equation are given by the equations [see equation (1.7–8)]:

$$T = \hat{T}(s,t) \qquad \text{on } \Gamma_T, \quad \text{for } t > 0 \qquad (3.1-2a)$$

$$\left(k_{ij}\frac{\partial T}{\partial x_j}\right) n_i + q_c + q_r = q_n(s,t) \qquad \text{on } \Gamma_q, \quad \text{for } t > 0 \qquad (3.1-2b)$$

and

$$q_c = h_c(s,T,t)(T - T_c)$$

$$q_r = h_r(s,T,t)(T - T_r) \qquad (3.1-2c)$$

Here s denotes the coordinate of a point on the boundary surface. The initial condition on the temperature is given by

$$T(\mathbf{x}, 0) = T_0(\mathbf{x}) \tag{3.1-3}$$

The heat conduction problem described by equations (1)–(3) differs from that discussed in Chapter 2 in several respects. First, the problem considered here is for general three–dimensional geometries. Second, the boundary conditions are general enough to include conduction, convection, and radiation heat transfer between the solid regions and the surrounding environment. Third, the material coefficients, ρ, C, k_{ij}, h_c, and h_r can be functions of temperature, making the problem a nonlinear one.

As described in Section 2.8, the finite element model of the initial–boundary value problem described by equations (1)–(3) is developed in two steps: (1) *spatial discretization*, in which the weak form of equation (1) over a typical element is developed and the spatial approximation of the dependent variable (*i.e.*, temperature) T of the problem is assumed to obtain a set of ordinary differential equations in time among the nodal values T_j^e of the dependent variable; (2) *temporal approximation* of the ordinary differential equations obtained in the first step is carried out using some appropriate method such as a finite difference approximation. This step leads to a set of algebraic equations involving the nodal values T_j^e at time $t_{n+1}[= (n+1)\Delta t$, where n is an integer and Δt is the time increment] in terms of known values from the previous time step. See Section 2.8 for an account of these two steps for two–dimensional problems. Here we develop the finite element models of three–dimensional, transient heat conduction problems using the two–step procedure described above. We begin with the development of the semidiscrete finite element model of equations (1)–(2).

3.2 SEMIDISCRETE FINITE ELEMENT MODEL

Suppose that the conduction heat transfer region Ω is discretized into an appropriate collection of finite elements (see Figure 3.2–1). The weak form of equations (3.1–1) and (3.1–2) over an element Ω^e is obtained by multiplying equation (3.1–1) with the weight function $w(\mathbf{x})$ and integrating over the element, integrating–by–parts (spatially) those terms which involve higher–order derivatives, and replacing the coefficient of the weight function in the boundary integral with the secondary variable [*i.e.*, use equation (3.1–2b)]. We obtain

$$0 = \int_{\Omega^e} \left[w \left(\rho C \frac{\partial T}{\partial t} - Q \right) + k_{ij} \frac{\partial w}{\partial x_i} \frac{\partial T}{\partial x_j} \right] d\mathbf{x} - \oint_{\Gamma^e} (q_n - q_c - q_r) w \, ds \tag{3.2-1}$$

where the subscript 's' on the variables is omitted for brevity and summation on repeated subscripts, i and j, is used. Note that no integration by parts with respect to time is used, and w is not a function of time.

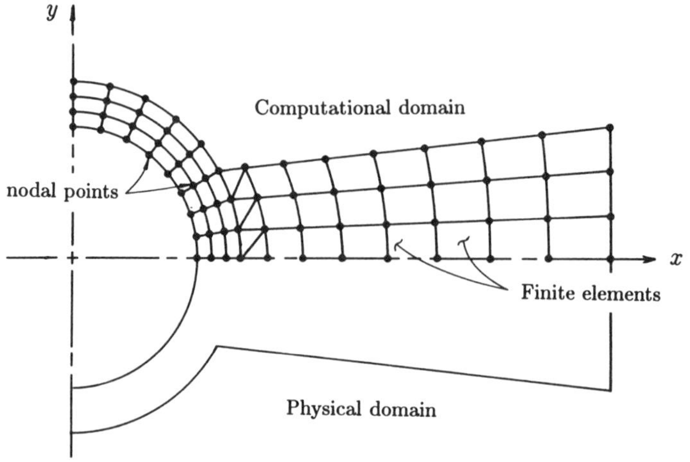

Figure 3.2–1: The finite element discretization of a region.

The semidiscrete finite element model is obtained from equation (1) by substituting a finite element approximation for the dependent variable, T. In selecting the approximation for T, we assume that the time dependence can be separated from the spatial variation,

$$T(\mathbf{x}, t) \simeq \sum_{j=1}^{n_e} T_j^e(t)\psi_j^e(\mathbf{x}) \tag{3.2-2a}$$

which can be written in matrix notation as

$$T(\mathbf{x}, t) = (\mathbf{\Psi}^e)^T \mathbf{T}^e \tag{3.2-2b}$$

where $(\mathbf{\Psi}^e)^T$ is a $(1 \times n_e)$ vector of interpolation functions of the element Ω^e,

$$(\mathbf{\Psi}^e)^T = \{\psi_1^e \ \psi_2^e \ \cdots \ \psi_{n_e}^e\} \tag{3.2-2c}$$

$\mathbf{T}^e$ is a $(n_e \times 1)$ vector of nodal temperatures, superscript $(\cdot)^T$ denotes the transpose of a vector or matrix, and n_e is the number of nodal points in an element. The i–th differential equation (in time) of the semidiscrete, weak–form finite element model is obtained by substituting $w(\mathbf{x}) = \psi_i^e(\mathbf{x})$ and replacing T by the interpolation in equation (2):

$$0 = \sum_{j=1}^{n_e} \left(M_{ij}^e \frac{dT_j^e}{dt} + K_{ij}^e T_j^e \right) - Q_i^e - q_i^e \tag{3.2 – 3}$$

Equation (3) can be expressed in matrix form as

$$[M^e]\{\dot{T}^e\} + [K^e]\{T^e\} = \{Q^e\} + \{q^e\} \tag{3.2 – 4a}$$

or

$$\mathbf{M}^e(\mathbf{T}^e)\dot{\mathbf{T}}^e + \mathbf{K}^e(\mathbf{T}^e)\mathbf{T}^e = \mathbf{Q}^e(\mathbf{T}^e) + \mathbf{q}^e(\mathbf{T}^e) \tag{3.2 – 4b}$$

where a superposed dot on T denotes a derivative with time ($\dot{T} = \partial T/\partial t$), and

$$M_{ij}^e = \int_{\Omega^e} \rho C \psi_i \psi_j \, d\mathbf{x} \ , \ K_{ij}^e = \int_{\Omega^e} k_{rs} \frac{\partial \psi_i}{\partial x_r} \frac{\partial \psi_j}{\partial x_s} \, d\mathbf{x} \tag{3.2 – 5a}$$

$$Q_i^e = \int_{\Omega^e} \psi_i Q(\mathbf{x}, t) \, d\mathbf{x} \ , \ q_i^e = \oint_{\Gamma^e} \psi_i (q_n - q_c - q_r) \, ds \tag{3.2 – 5b}$$

The element label 'e' on ψ is omitted for brevity. The element coefficient matrices and vectors are written in the vector form as

$$\mathbf{M}^e = \int_{\Omega^e} \rho C \mathbf{\Psi} \mathbf{\Psi}^T \, d\mathbf{x} \ , \ \mathbf{K}^e = \int_{\Omega^e} \frac{\partial \mathbf{\Psi}}{\partial x_r} k_{rs} \frac{\partial \mathbf{\Psi}^T}{\partial x_s} \, d\mathbf{x} \tag{3.2 – 6a}$$

$$\mathbf{Q}^e = \int_{\Omega^e} \mathbf{\Psi} Q \, d\mathbf{x} \ , \ \mathbf{q}^e = \oint_{\Gamma^e} \mathbf{\Psi} \, (q_n - q_c - q_r) \, ds \tag{3.2 – 6b}$$

Equation (4) is written in a form to show the most general nonlinear case in which material properties, boundary conditions, and volumetric sources depend on temperature. The implementation of variable material properties and coefficients will be discussed in a subsequent section.

The process of assembly for two–dimensional problems was discussed in Chapter 2; the assembly process for three–dimensional geometries is completely analogous. The assembled matrices are symbolically represented as

$$\mathbf{M} = \sum_e \mathbf{M}^e; \quad \mathbf{K} = \sum_e \mathbf{K}^e; \quad \mathbf{F} = \sum_e \mathbf{F}^e, \ \mathbf{F}^e = \mathbf{Q}^e + \mathbf{q}^e \tag{3.2 – 7}$$

In equation (7) the sum is taken over all the elements in the mesh, and the element matrices are defined by equation (6). Once the form of the element interpolation functions $\mathbf{\Psi}^e$ is known and the element geometry is specified, the integrals in equation

(6) can be evaluated, and the global (assembled) equations are then constructed through use of equation (7):

$$\mathbf{M}(\mathbf{T})\dot{\mathbf{T}} + \mathbf{K}(\mathbf{T})\mathbf{T} = \mathbf{F}(\mathbf{T}) \qquad (3.2-8)$$

Before we consider the time or temporal approximation of equations (4) or (8), it is informative to discuss the geometries of elements and interpolation functions used in equation (2) and the numerical evaluation of the coefficients defined in equations (5) or (6). We will return to equation (8) in Section 3.6 to discuss its solution for steady–state and transient problems.

3.3 INTERPOLATION FUNCTIONS

3.3–1 Preliminary Comments

The equations for an individual element, as indicated by equation (3.2–6), require the specification of the element interpolation functions $\psi_i{}^e(\mathbf{x})$ for the approximation of the temperature and a description of the element geometry. In practical applications, elements based on linear and quadratic polynomials are generally used. The use of an isoparametric formulation is also standard in practice; this allows complex geometric regions to be easily and accurately described. Two–dimensional linear and quadratic elements were presented in Chapter 2. In this section, a number of commonly used three–dimensional elements are presented (see also [1–5]).

3.3–2 Hexahedral (Brick) Elements

Brick elements represent the most commonly used finite elements for three–dimensional analyses, and the straight–sided, linear eight–node brick element is the most cost–effective choice. The interpolation functions of the linear element (see Figure 3.3–1) are given in terms of the normalized coordinates (ξ, η, ζ) as follows:

$$\{\Psi^e\} = \frac{1}{8}\begin{Bmatrix} (1-\xi)(1-\eta)(1-\zeta) \\ (1+\xi)(1-\eta)(1-\zeta) \\ (1+\xi)(1+\eta)(1-\zeta) \\ (1-\xi)(1+\eta)(1-\zeta) \\ (1-\xi)(1-\eta)(1+\zeta) \\ (1+\xi)(1-\eta)(1+\zeta) \\ (1+\xi)(1+\eta)(1+\zeta) \\ (1-\xi)(1+\eta)(1+\zeta) \end{Bmatrix} \qquad (3.3-1)$$

The quadratic shape functions for the twenty–node serendipity element (see Figure 3.3–2) are given by

$$
\{\Psi^e\} = \frac{1}{8}\left\{\begin{array}{l}
(1-\xi)(1-\eta)(1-\zeta)(-\xi-\eta-\zeta-2) \\
(1+\xi)(1-\eta)(1-\zeta)(\xi-\eta-\zeta-2) \\
(1+\xi)(1+\eta)(1-\zeta)(\xi+\eta-\zeta-2) \\
(1-\xi)(1+\eta)(1-\zeta)(-\xi+\eta-\zeta-2) \\
(1-\xi)(1-\eta)(1+\zeta)(-\xi-\eta+\zeta-2) \\
(1+\xi)(1-\eta)(1+\zeta)(\xi-\eta+\zeta-2) \\
(1+\xi)(1+\eta)(1+\zeta)(\xi+\eta+\zeta-2) \\
(1-\xi)(1+\eta)(1+\zeta)(-\xi+\eta+\zeta-2) \\
2(1-\xi^2)(1-\eta)(1-\zeta) \\
2(1+\xi)(1-\eta^2)(1-\zeta) \\
2(1-\xi^2)(1+\eta)(1-\zeta) \\
2(1-\xi)(1-\eta^2)(1-\zeta) \\
2(1-\xi)(1-\eta)(1-\zeta^2) \\
2(1+\xi)(1-\eta)(1-\zeta^2) \\
2(1+\xi)(1+\eta)(1-\zeta^2) \\
2(1-\xi)(1+\eta)(1-\zeta^2) \\
2(1-\xi^2)(1-\eta)(1+\zeta) \\
2(1+\xi)(1-\eta^2)(1+\zeta) \\
2(1-\xi^2)(1+\eta)(1+\zeta) \\
2(1-\xi)(1-\eta^2)(1+\zeta)
\end{array}\right\} \tag{3.3-2}
$$

As in the case of the two–dimensional quadratic (nine–node) Lagrange element, a brick element with 27 nodes may also be constructed; the shape functions are not shown for this element.

The transformation between the actual element Ω^e and the master element $\hat{\Omega}$ [or equivalently, between $(x_1, x_2, x_3) = (x, y, z)$ and (ξ, η, ζ)] is accomplished by a coordinate transformation of the form

$$
x = \sum_{j=1}^{m} x_j^e \hat{\psi}_j^e(\xi,\eta,\zeta), \quad y = \sum_{j=1}^{m} y_j^e \hat{\psi}_j^e(\xi,\eta,\zeta), \quad z = \sum_{j=1}^{m} z_j^e \hat{\psi}^e w_j(\xi,\eta,\zeta) \tag{3.3-3}
$$

For an isoparametric formulation, we take $\hat{\psi}_j^e = \psi_j^e$.

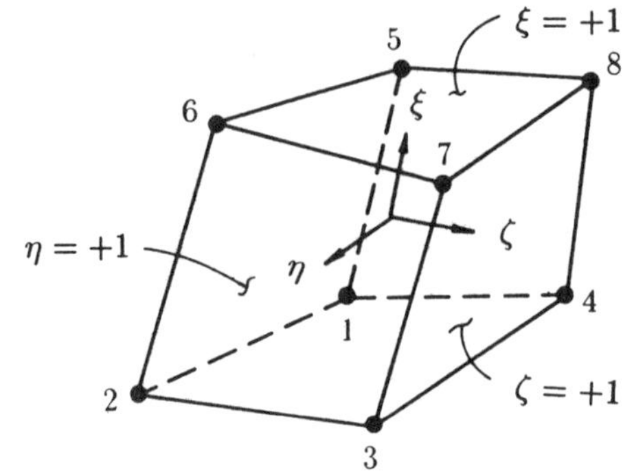

Figure 3.3–1: The linear (eight–node) brick element.

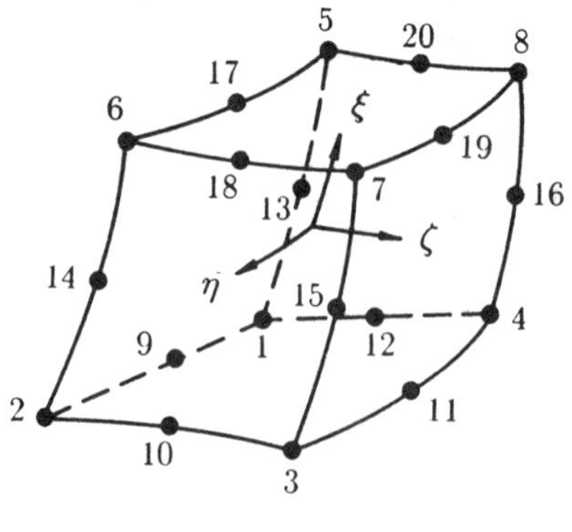

Figure 3.3–2: The (twenty–node) quadratic brick element.

3.3–3 Prism Elements (3–D)

A prism or wedge element is often useful in three–dimensional geometries. The shape functions for the six–node linear element (see Figure 3.3–3) are given by

$$\{\Psi^e\} = \frac{1}{2}\begin{Bmatrix} L_1(1-\zeta) \\ L_2(1-\zeta) \\ L_3(1-\zeta) \\ L_1(1+\zeta) \\ L_2(1+\zeta) \\ L_3(1+\zeta) \end{Bmatrix} \qquad (3.3-4)$$

The interpolation functions for the fifteen–node quadratic prism element (see Figure 3.3–4) are

$$\{\Psi^e\} = \frac{1}{2}\begin{Bmatrix} L_1[(2L_1-1)(1-\zeta)-(1-\zeta^2)] \\ L_2[(2L_2-1)(1-\zeta)-(1-\zeta^2)] \\ L_3[(2L_3-1)(1-\zeta)-(1-\zeta^2)] \\ L_1[(2L_1-1)(1+\zeta)-(1-\zeta^2)] \\ L_2[(2L_2-1)(1+\zeta)-(1-\zeta^2)] \\ L_3[(2L_3-1)(1+\zeta)-(1-\zeta^2)] \\ 4L_1L_2(1-\zeta) \\ 4L_2L_3(1-\zeta) \\ 4L_3L_1(1-\zeta) \\ 2L_1(1-\zeta^2) \\ 2L_2(1-\zeta^2) \\ 2L_3(1-\zeta^2) \\ 4L_1L_2(1+\zeta) \\ 4L_2L_3(1+\zeta) \\ 4L_3L_1(1+\zeta) \end{Bmatrix} \qquad (3.3-5)$$

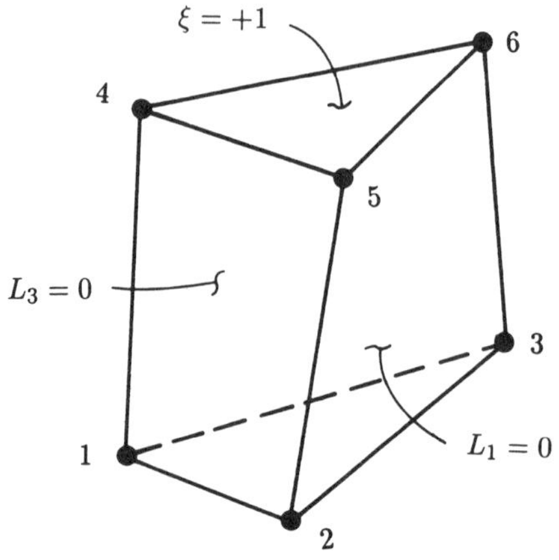

Figure 3.3–3: The (six–node) linear prism element.

The area coordinates, L_i (see Section 2.11), are used to describe the functional variation in the triangular cross section of the prism elements, while a standard normalized coordinate, ζ, describes the variation in the axial direction. Note that $L_1 + L_2 + L_3 = 1$.

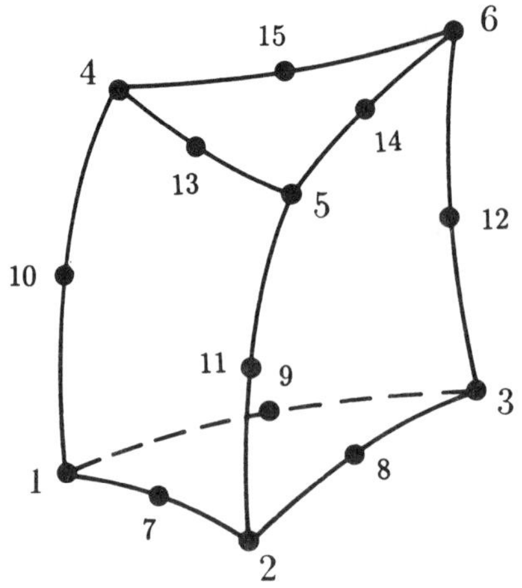

Figure 3.3–4: The (fifteen–node) quadratic prism element.

3.3–4 Tetrahedral Elements (3–D)

The tetrahedral elements are a three–dimensional version of the triangular elements. The three–node and ten–node tetrahedral elements are shown in Figure 3.3–5, and their interpolation functions are given below:

$$\{\Psi^e\} = \frac{1}{2}\begin{Bmatrix} L_1 \\ L_2 \\ L_3 \\ L_4 \end{Bmatrix}; \quad \{\Psi^e\} = \begin{Bmatrix} L_1(2L_1 - 1) \\ L_2(2L_2 - 1) \\ L_3(2L_3 - 1) \\ L_4(2L_4 - 1) \\ 4L_1L_2 \\ 4L_2L_3 \\ 4L_3L_1 \\ 4L_1L_4 \\ 4L_2L_4 \\ 4L_3L_4 \end{Bmatrix} \qquad (3.3-6)$$

Figure 3.3–5: Linear and quadratic tetrahedral elements.

The volume coordinates, L_i, are used to describe the interpolation functions for linear and quadratic elements, where $L_1 + L_2 + L_3 + L_4 = 1$.

3.4 NUMERICAL INTEGRATION

The evaluation of the various finite element coefficient matrices in (3.2–5) requires the integration of combinations of the interpolation functions and their spatial derivatives over the area or volume of the element. The numerical integration ideas described in Section 2.13 can be easily extended to three dimensions. Here we give the pertinent equations for the three–dimensional case.

For an isoparametric formulation [6], the following relations, based on the chain rule of differentiation, can be derived for brick elements:

$$\begin{Bmatrix} \frac{\partial\psi_i}{\partial\xi} \\ \frac{\partial\psi_i}{\partial\eta} \\ \frac{\partial\psi_i}{\partial\zeta} \end{Bmatrix} = \begin{bmatrix} \frac{\partial x}{\partial\xi} & \frac{\partial y}{\partial\xi} & \frac{\partial z}{\partial\xi} \\ \frac{\partial x}{\partial\eta} & \frac{\partial y}{\partial\eta} & \frac{\partial z}{\partial\eta} \\ \frac{\partial x}{\partial\zeta} & \frac{\partial y}{\partial\zeta} & \frac{\partial z}{\partial\zeta} \end{bmatrix} \begin{Bmatrix} \frac{\partial\psi_i}{\partial x} \\ \frac{\partial\psi_i}{\partial y} \\ \frac{\partial\psi_i}{\partial z} \end{Bmatrix} = \begin{bmatrix} J_{11} & J_{12} & J_{13} \\ J_{21} & J_{22} & J_{23} \\ J_{31} & J_{32} & J_{33} \end{bmatrix} \begin{Bmatrix} \frac{\partial\psi_i}{\partial x} \\ \frac{\partial\psi_i}{\partial y} \\ \frac{\partial\psi_i}{\partial z} \end{Bmatrix} \tag{3.4-1}$$

where $[J]$ is the Jacobian of the transformation from global coordinates (x, y, z) to the local element coordinates (ξ, η, ζ). The parametric transformation defined in equation (3.3–3) can be used to define the components of $[J]$. Inverting equation (1), we obtain the global spatial derivatives of the interpolation functions in terms of the local derivatives:

$$\begin{Bmatrix} \frac{\partial\psi_i}{\partial x} \\ \frac{\partial\psi_i}{\partial y} \\ \frac{\partial\psi_i}{\partial z} \end{Bmatrix} = \begin{bmatrix} J^*_{11} & J^*_{12} & J^*_{13} \\ J^*_{21} & J^*_{22} & J^*_{23} \\ J^*_{31} & J^*_{32} & J^*_{33} \end{bmatrix} \begin{Bmatrix} \frac{\partial\psi_i}{\partial\xi} \\ \frac{\partial\psi_i}{\partial\eta} \\ \frac{\partial\psi_i}{\partial\zeta} \end{Bmatrix} \tag{3.4-2}$$

where $[J^*]$ is the inverse of the Jacobian matrix $[J]$. The components J^*_{ij} are complicated functions of the components of $[J]$ that can in principle be obtained by analytically inverting the 3×3 Jacobian matrix. In practice, the Jacobian is usually inverted numerically at each integration point.

In performing numerical integration over the element volume, it is necessary to transform the integrand and limits of integration from the global coordinates to the local element coordinates. The differential elemental volume transforms according to

$$dxdydz = \|[J]\|\, d\xi d\eta d\zeta \tag{3.4-3}$$

The integration limits for the integrals transform to the limits on the local coordinates (ξ, η, ζ), *i.e.*, -1 to $+1$. In the previous equations the (ξ, η, ζ) coordinates for a brick element were used for purposes of explanation. Similar relations for a tetrahedral element can be derived by replacing (ξ, η, ζ) with (L_1, L_2, L_3). The variable L_4 does not enter the formulas due to the relation $L_1 + L_2 + L_3 + L_4 = 1$. Hybrid coordinates, such as those utilized in the prism element, are treated in an analogous manner [5].

To illustrate further the development of a computational form for typical entries in the element equations (3.2–5) or (3.2–6), we consider a cross component of the diffusion matrix $\mathbf{K}^e$ given by

$$K^e_{12} = \int_{\Omega^e} k_{rs} \frac{\partial\psi_1}{\partial x_r}\frac{\partial\psi_2}{\partial x_s} dxdydz \tag{3.4-4}$$

which will be evaluated for a three–dimensional, brick element. From the definitions in (2) and (3) then

$$K_{12}^e = \int_{-1}^{+1}\int_{-1}^{+1}\int_{-1}^{+1} k_{rs}\left[J_{r1}^*\frac{\partial\psi_1}{\partial\xi} + J_{r2}^*\frac{\partial\psi_1}{\partial\eta} + J_{r3}^*\frac{\partial\psi_1}{\partial\zeta}\right]\cdot$$

$$\left[J_{s1}^*\frac{\partial\psi_2}{\partial\xi} + J_{s2}^*\frac{\partial\psi_2}{\partial\eta} + J_{s3}^*\frac{\partial\psi_2}{\partial\zeta}\right]|J|\; d\xi d\eta d\zeta \qquad (3.4-5)$$

Note that since the thermal conductivity could vary over the element, it must be treated as a function of (ξ, η, ζ) and be included in the evaluation of the integral. Variable coefficients are discussed further in Section 3.8. The above coefficient is typical of those encountered in building finite element equations. Each component of the integral in (5) is of the form

$$I = \int_{-1}^{+1}\int_{-1}^{+1}\int_{-1}^{+1} F(\xi,\eta,\zeta)\; d\xi d\eta d\zeta \qquad (3.4-6)$$

The Gauss quadrature procedure discussed in Section 2.13 can be readily extended to three dimensions to evaluate the integral expression, I:

$$\int_{-1}^{+1}\int_{-1}^{+1}\int_{-1}^{+1} F(\xi,\eta,\zeta)\; d\xi d\eta d\zeta \approx \sum_{I=1}^{M}\sum_{J=1}^{N}\sum_{K=1}^{P} F(\xi_I,\eta_J,\zeta_K)\; W_I W_J W_K \qquad (3.4-7)$$

For brick elements it is typical to employ a product rule as shown in (7) with $M = N = P = 2$ or 3. Other element types (*e.g.*, tetrahedrals) are also evaluated using quadrature formulas similar to (7) but slightly different in form [see equation (2.12–15)]. Hybrid elements, such as the prism, generally utilize combination quadrature formulas such as (2.12–15) in the triangular plane and a one–dimensional Gauss formula along the axis. For additional discussion of quadrature rules and their utilization, the reader may consult References 2, 3, 5, and 7–13.

3.5 COMPUTATION OF SURFACE FLUXES

The nodal fluxes q_i^e due to applied boundary conditions need only be computed for those element sides that coincide with the boundary of the problem domain; contributions from interior element boundaries are canceled by adjoining elements. The surface flux vector is given by

$$\mathbf{q}^e = \oint_{\Gamma^e} \mathbf{\Psi} q_i n_i\; ds = \oint_{\Gamma^e} \mathbf{\Psi}(q_n - q_c - q_r)\; ds \qquad (3.5-1)$$

where Γ^e denotes the surface of the element and $q_i n_i$ is the heat flux normal to the surface.

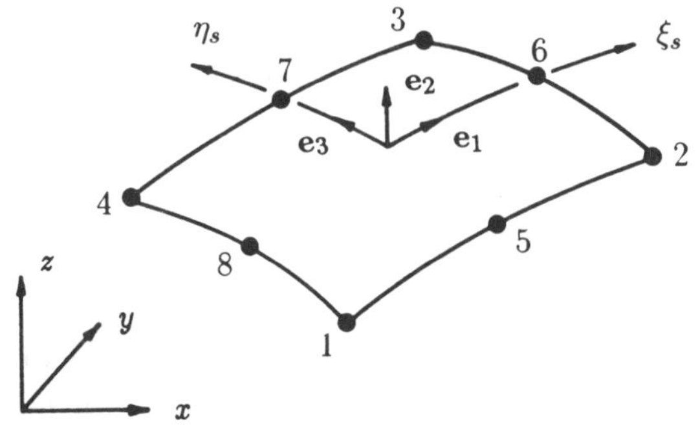

Figure 3.5–1: Nomenclature for surface flux computation.

The computation of the indicated surface integrals is most easily carried out in the normalized or natural coordinate system for the face (edge) of an element. This requires that the elemental surface area (edge length) ds, be related to the local surface coordinates. Consider the typical quadrilateral element face shown in Figure 3.5–1, where the vectors $\mathbf{e}_1$ and $\mathbf{e}_2$ are defined as being tangent to the curvilinear coordinates, ξ_s and η_s. The $\mathbf{e}$ vectors are not necessarily unit vectors; the ξ_s and η_s coordinates are assumed to be the natural coordinates for the element face. The elemental area ds in terms of the global coordinates (x, y, z) is related to an elemental area in surface coordinates by

$$ds = |\mathbf{J}_s| \, d\xi_s d\eta_s \qquad (3.5-2)$$

where $\mathbf{J}_s$ is the Jacobian of the coordinate transformation and $|\cdot|$ indicates the determinant. The determinant of the Jacobian can be written in terms of the vectors $(\mathbf{e}_1, \mathbf{e}_2)$ as

$$|\mathbf{J}_s| = |\mathbf{e}_1 \times \mathbf{e}_2| = \left[(\mathbf{e}_1 \cdot \mathbf{e}_1)(\mathbf{e}_2 \cdot \mathbf{e}_2) - (\mathbf{e}_1 \cdot \mathbf{e}_2)^2\right]^{1/2} \qquad (3.5-3)$$

The vectors $\mathbf{e}_i$ can be expressed in terms of the global coordinates by

$$\mathbf{e}_1 = \begin{Bmatrix} \frac{\partial x}{\partial \xi_s} \\ \frac{\partial y}{\partial \xi_s} \\ \frac{\partial z}{\partial \xi_s} \end{Bmatrix}, \quad \mathbf{e}_2 = \begin{Bmatrix} \frac{\partial x}{\partial \eta_s} \\ \frac{\partial y}{\partial \eta_s} \\ \frac{\partial z}{\partial \eta_s} \end{Bmatrix} \qquad (3.5-4)$$

Using the transformation (3.3–3) in equation (4), we can write

$$\mathbf{e}_1 = \begin{Bmatrix} \sum_{j=1}^{m} x_j^e \frac{\partial \hat{\psi}_j^e}{\partial \xi_s} \\ \sum_{j=1}^{m} y_j^e \frac{\partial \hat{\psi}_j^e}{\partial \xi_s} \\ \sum_{j=1}^{m} z_j^e \frac{\partial \hat{\psi}_j^e}{\partial \xi_s} \end{Bmatrix}; \quad \mathbf{e}_2 = \begin{Bmatrix} \sum_{j=1}^{m} x_j^e \frac{\partial \hat{\psi}_j^e}{\partial \eta_s} \\ \sum_{j=1}^{m} y_j^e \frac{\partial \hat{\psi}_j^e}{\partial \eta_s} \\ \sum_{j=1}^{m} z_j^e \frac{\partial \hat{\psi}_j^e}{\partial \eta_s} \end{Bmatrix} \tag{3.5-5}$$

where the ˆ indicates the restriction of the interpolation functions to an element face (edge). The functions $\hat{\psi}_j$ may be either linear or quadratic, depending on the type of mapping used to describe the element geometry. Equation (5) provides a means for computing $|\mathbf{J}_s|$, which can be used in equation (2).

With the above definitions, the terms in equation (1) can be considered individually. For the case of an applied heat flux $\hat{q}_n$, the contribution to $\mathbf{q}^e$ is given by

$$\mathbf{q}_n = \oint_{\Gamma_e} \hat{\mathbf{\Psi}} \hat{q}_n \, ds \tag{3.5-6}$$

where $\hat{\mathbf{\Psi}}$ is the vector of interpolation functions that are restricted to the element boundary (face or edge). Employing the relations defined above, equation (6) can be written in a more explicit form

$$\mathbf{q}_n = \int_{-1}^{1} \int_{-1}^{1} \hat{\mathbf{\Psi}}(\xi, \eta) \hat{q}_n(\xi, \eta) |\mathbf{J}_s| \, d\xi_s d\eta_s \tag{3.5-7}$$

Once the distribution of the normal flux $\hat{q}_n(\xi, \eta)$ is known for the element, equation (7) can be evaluated using numerical quadrature as discussed in Section 2.12. The order of the quadrature used to evaluate $\mathbf{q}_n$ is typically the same as that used on the volume (area) integrals defining the conductivity $[K^e]$ and capacitance $[M^e]$ matrices. For a three–dimensional element, the surface shape functions $\hat{\mathbf{\Psi}}$ correspond directly to the interpolation functions of the two–dimensional triangular and quadrilateral elements. In the two–dimensional case, $\hat{\mathbf{\Psi}}$ corresponds to the one–dimensional interpolation functions

$$\{\hat{\mathbf{\Psi}}^e\} = \frac{1}{2} \begin{Bmatrix} (1-\xi) \\ (1+\xi) \end{Bmatrix} \tag{3.5-8}$$

for the linear case and

$$\{\hat{\mathbf{\Psi}}^e\} = \frac{1}{2} \begin{Bmatrix} (1-\xi)\xi \\ 2(1-\xi^2) \\ (1+\xi)\xi \end{Bmatrix} \tag{3.5-9}$$

for the quadratic case, where ξ is the normalized coordinate along the edge. For the two–dimensional case, the integral in equation (7) simplifies to an integral over the edge.

The contributions to $\mathbf{q}^e$ due to convection and radiation are treated in an analogous manner. For the convection boundary condition, we have

$$\mathbf{q}_c = -\oint_{\Gamma_e} \hat{\mathbf{\Psi}} q_c \, ds = -\oint_{\Gamma_e} \hat{\mathbf{\Psi}} h_c (T - T_c) \, ds \tag{3.5-10}$$

where the definition in equation (3.1–2c) has been used. The temperature on the element surface is given by

$$T(\xi, \eta) = \left(\hat{\mathbf{\Psi}}(\xi, \eta) \right)^T \mathbf{T} \tag{3.5-11}$$

where $\hat{\mathbf{\Psi}}$ are surface or edge interpolation functions. Substituting this relation into equation (10) and using the definition for ds yields

$$\mathbf{q}_c = -\left(\int_{-1}^{1} \int_{-1}^{1} h_c(\xi, \eta) \hat{\mathbf{\Psi}}(\xi, \eta) \hat{\mathbf{\Psi}}^T(\xi, \eta) |\mathbf{J}_s| \, d\xi_s d\eta_s \right) \mathbf{T}$$

$$+ \int_{-1}^{1} \int_{-1}^{1} h_c(\xi, \eta) \hat{\mathbf{\Psi}}(\xi, \eta) T_c(\xi, \eta) |\mathbf{J}_s| \, d\xi_s d\eta_s \tag{3.5-12a}$$

$$= -\mathbf{CT} + \mathbf{F}_{hc} \tag{3.5-12b}$$

where

$$\mathbf{C} = \int_{-1}^{1} \int_{-1}^{1} h_c \hat{\mathbf{\Psi}} \hat{\mathbf{\Psi}}^T |\mathbf{J}_s| \, d\xi_s d\eta_s \tag{3.5-13a}$$

$$\mathbf{F}_{hc} = \int_{-1}^{1} h_c T_c \hat{\mathbf{\Psi}} |\mathbf{J}_s| \, d\xi_s d\eta_s \tag{3.5-13b}$$

For a given variation of h_c and T_c, the matrix $\mathbf{C}$ and vector $\mathbf{F}_{hc}$ can be computed. Note that in equation (12b) the first term $\mathbf{CT}$ contains the vector of unknown nodal temperatures and will thus be moved to the left hand side of the matrix equation (3.2–4). When h_c is a function of temperature, the problem is nonlinear. More discussion on the solution of nonlinear problems will be presented in the sequel.

A computation similar to the convective boundary condition may be carried out for the radiation boundary condition. We obtain

$$\mathbf{q}_r = -\mathbf{RT} + \mathbf{F}_{hr} \tag{3.5-14}$$

where

$$\mathbf{R} = \int_{-1}^{1} \int_{-1}^{1} h_r \hat{\mathbf{\Psi}} \hat{\mathbf{\Psi}}^T |\mathbf{J}_s| \, d\xi_s d\eta_s \tag{3.5-15a}$$

$$\mathbf{F}_{hr} = \int_{-1}^{1} \int_{-1}^{1} h_r T_r \hat{\Psi} |\mathbf{J}_s| \, d\xi_s d\eta_s \qquad (3.5-15b)$$

Again, the term $\mathbf{RT}$ will be moved to the left–hand side of equation (3.3–4) while $\mathbf{F}_{hr}$ is retained in the source vector. EquationS (15) are highly nonlinear due to the fact that both $\mathbf{R}$ and $\mathbf{F}_{hr}$ are functions of the unknown temperature field $\mathbf{T}$, since the heat transfer coefficient, h_r, is an explicit function of temperature.

3.6 SEMIDISCRETE FINITE ELEMENT MODEL

Returning to the finite element model (3.2–4), we have

$$\mathbf{M}^e(\mathbf{T}^e)\dot{\mathbf{T}}^e + \mathbf{K}^e(\mathbf{T}^e)\mathbf{T}^e = \mathbf{Q}^e(\mathbf{T}^e) + \mathbf{q}^e(\mathbf{T}^e) \qquad (3.6-1)$$

or using the definition of the boundary fluxes in equations (3.5–1), (3.5–7), (3.5–12) and (3.5–14)

$$\mathbf{M}\dot{\mathbf{T}} + \mathbf{KT} = \mathbf{Q} + \mathbf{q}_n - \mathbf{CT} + \mathbf{F}_{hc} - \mathbf{RT} + \mathbf{F}_{hr} \qquad (3.6-2)$$

Rearranging the terms yields the following matrix equation of ordinary differential equations in time:

$$\mathbf{M}\dot{\mathbf{T}} + \hat{\mathbf{K}}\mathbf{T} = \hat{\mathbf{F}} \qquad (3.6-3)$$

with

$$\hat{\mathbf{K}} = \mathbf{K} + \mathbf{C} + \mathbf{R} \qquad (3.6-4a)$$

$$\hat{\mathbf{F}} = \mathbf{Q} + \mathbf{q}_n + \mathbf{F}_{hc} + \mathbf{F}_{hr} \qquad (3.6-4b)$$

In general, $\mathbf{M}, \hat{\mathbf{K}}$, and $\hat{\mathbf{F}}$ are functions of temperature and/or time.

3.7 SOLUTION OF NONLINEAR EQUATIONS

3.7–1 Preliminary Comments

The discussion of solution procedures for the discretized form of the heat conduction problem naturally divides itself into two sections according to the time–independent (steady–state) or transient nature of the problem. Solution algorithms for the two types of problems are described here. For most practical conduction problems, the finite element model ultimately involves the solution of a large, sparse, symmetric system of nonlinear algebraic equations. In most cases, the nonlinear equations are linearized first and then solved, often, via a direct method (*e.g.*, Gauss elimination), which is tailored to the finite element formulation (*e.g.*, frontal [14,15] or skyline [16,17] methods). The methods are termed direct methods because they produce a solution to the matrix system in a fixed number of steps; the solution would be exact if infinite precision arithmetic were possible. However, despite the historical popularity of direct methods, there has been a renewed interest in iterative methods. Iterative methods provide an approximate solution to the matrix problem that usually improves with continued iteration; the number of iterations required to achieve a specified accuracy is not known *a priori*. The need to analyze extremely large three-dimensional problems and the move to parallel computer architectures has intensified the exploration of iterative methods, such as the preconditioned conjugate gradient (PCG) method. These techniques will certainly become the standard matrix solution algorithms in the near future. The matrix solution problem will not be discussed here but may be found in books on linear algebra and matrices (*e.g.*, [18,19]).

In the following sections, a number of linearization algorithms for nonlinear algebraic equations will be described for both steady and transient problems. The iterative methods discussed here for nonlinear problems are basically linearization procedures, and these methods should not be confused with the iterative methods used for the solution of linear algebraic equations.

3.7–2 Steady–State Problems

For problems that are independent of time, the matrix equation (3.6–3) is simplified to

$$\hat{\mathbf{K}}(\mathbf{T})\mathbf{T} = \hat{\mathbf{F}}(\mathbf{T}) \qquad (3.7-1)$$

which is a system of nonlinear algebraic equations. Consider first the case where $\hat{\mathbf{K}}$ and $\hat{\mathbf{F}}$ are not functions of the temperature. Then equation (1) reduces to a linear matrix problem, which can be solved directly, without iteration, by the standard matrix methods discussed above.

When equation (1) is nonlinear (for example, due to temperature–dependent properties, heat source, and/or boundary conditions), it should be linearized. For many mildly nonlinear conduction problems, a simple successive substitution iteration scheme, also known as the *Picard method*, is suitable. The Picard scheme is given by

$$\hat{\mathbf{K}}(\mathbf{T}^n)\mathbf{T}^{n+1} = \hat{\mathbf{F}}(\mathbf{T}^n) \qquad (3.7-2)$$

where superscript n denotes the iteration number. By evaluating the coefficient matrix $\hat{\mathbf{K}}$ and right–hand side vector $\hat{\mathbf{F}}$ using the solution $\mathbf{T}^n$ from the previous iteration, we have linearized the problem. The linearized problem is solved for $\mathbf{T}^{n+1}$ (using any matrix solution method), which is then used to evaluate the coefficient matrix and right–hand vector for the next iteration. This procedure is continued until the root-mean–square value of the difference between the solution vectors at two consecutive iterations (normalized with respect to the current solution) is reduced to a value less than a preassigned tolerance, ϵ:

$$\frac{(\mathbf{T}^{n+1} - \mathbf{T}^n)^T(\mathbf{T}^{n+1} - \mathbf{T}^n)}{(\mathbf{T}^{n+1})^T(\mathbf{T}^{n+1})} < \epsilon^2 \qquad (3.7-3)$$

The value of ϵ may be taken to be in the range of $10^{-2} - 10^{-4}$.

Typically, problems with slow variations of the material properties, heat transfer coefficients, or heat sources are effectively solved using the Picard method; convergence is normally obtained in a few iterations for problems of this type. An improvement in convergence rate can sometimes be realized by use of a relaxation formula where

$$\hat{\mathbf{K}}(\mathbf{T}^n)\mathbf{T}^* = \hat{\mathbf{F}}(\mathbf{T}^n) \tag{3.7-4a}$$

and

$$\mathbf{T}^{n+1} = \gamma\mathbf{T}^n + (1-\gamma)\mathbf{T}^* \qquad 0 \leq \gamma \leq 1 \tag{3.7-4b}$$

Conduction problems with strong nonlinearities, such as those with radiation boundary conditions, boiling heat transfer coefficients, and Arrhenius type heat sources require iterative techniques that are more powerful than the Picard method. The application of second–order iteration schemes, such as Newton's method (see Section 4.6), is one such approach. However, for some types of nonlinearities, Newton's method produces an unsymmetric (tangent or Jacobian) matrix. The additional computational cost and complexity of solving an unsymmetric system of equations has discouraged the use of such procedures. The most popular alternative for strongly nonlinear problems is to solve the steady–state heat transfer problem via a "false transient" procedure. Generally, implicit time integration formulas, such as those described below, are used to advance the solution through time and asymptotically approach the required steady–state solution. The cost–effectiveness of such a procedure relies heavily on the judicious choice of the integration time step; the step must be large enough to reach steady state quickly but small enough to maintain stability of the solution procedure. The combination of a simple, first–order accurate backward Euler method (see Section 4.6) with an adaptive time step control [20] would be an appropriate scheme for most nonlinear conduction problems of this type.

3.7–3 Transient Problems

The semidiscretization process employed in the finite element method produces a system of coupled, nonlinear ordinary differential equations of the form shown in (3.6–3) for the time–dependent conduction problem. There are a number of possible methods for solving or integrating the system in (3.6–3), including many of the classical methods used for ordinary differential equations (ODE). However, the (usually) large size of the equation system and the need to take advantage of the special (banded) form of the matrices in (3) eliminate many integration methods from consideration.

Finite element time integration methods are usually direct integration methods that employ some relatively simple discrete approximation for the time derivative. Mode superposition methods [21] are also applicable to the system in (3.6–3), but unlike direct integration, this technique works with a transformed equation set. Mode superposition as an integration method will not be discussed here, though some aspects will be utilized when discussing the stability of integration methods. The discrete time approximation needed for a direct integration method may be obtained via a finite element or finite difference technique or by introducing more formal ODE ideas such as the linear multistep methods (LMS) [22]. Finite elements in time have been described by Argyris and Scharpf [23], Donea [24], Bettencourt *et al.* [25], and others. These techniques result in many of the same integration methods produced by finite difference approximations and will not be considered here. A few comments on LMS methods will be reserved until the end of this section.

The procedure for developing a finite difference based direct integration method begins by dividing the integration interval of interest ($0 \leq t \leq t_{final}$) into a number of discrete segments denoted by $\Delta t = t^{n+1} - t^n$, where the superscript denotes the time step number. Within each interval Δt, the temperature and temperature rate are allowed to vary according to some prescription that can be expressed as a simple difference formula. For example, assume that the temperature varies linearly over the interval Δt; the time–rate–of change of the temperature over the interval Δt can be approximated by the backward difference $\dot{\mathbf{T}}^{n+1} = \left(\mathbf{T}^{n+1} - \mathbf{T}^n\right)/\Delta t$. Evaluating (3.6–3) at the $n+1$ time step and using the above definition produces

$$\frac{1}{\Delta t}\mathbf{M}\left(\mathbf{T}^{n+1} - \mathbf{T}^n\right) + \hat{\mathbf{K}}\mathbf{T}^{n+1} = \hat{\mathbf{F}}^{n+1} \qquad (3.7-5)$$

Equation (5) represents a system of algebraic equations for the unknown temperatures $\mathbf{T}^{n+1}$ that can be successively solved at each new time plane.

Other difference approximations for $\dot{\mathbf{T}}$ are possible and lead to other types of recursive formulas for the solution of the temperature field as a function of time. The most popular method used for diffusion problems was introduced in Chapter 2 and consists of the following approximation:

$$\dot{\mathbf{T}}^a = \left(\mathbf{T}^{n+1} - \mathbf{T}^n\right)/\Delta t$$
$$\mathbf{T}^a = (1-\theta)\mathbf{T}^n + \theta\mathbf{T}^{n+1} \qquad (3.7-6)$$

for $0 \leq \theta \leq 1$. When equation (3.6–3) is evaluated at $\mathbf{T}^a$ and the above definitions are used, the resulting algorithm, which goes by a variety of names including the generalized trapezoid method, the generalized Crank–Nicolson method, or the generalized midpoint rule, is given by

$$\frac{1}{\Delta t}\mathbf{M}(\mathbf{T}^a)\left(\mathbf{T}^{n+1} - \mathbf{T}^n\right) + \theta\hat{\mathbf{K}}(\mathbf{T}^a)\mathbf{T}^{n+1} = \hat{\mathbf{F}}(\mathbf{T}^a) - (1-\theta)\hat{\mathbf{K}}(\mathbf{T}^a)\mathbf{T}^n \qquad (3.7-7)$$

where superscript n indicates the time level, Δt is the time step, and θ is a parameter that determines where in the time interval, t_n to $t_n + \Delta t = t_{n+1}$, the equation will be evaluated.

The choice of $\theta = 0$ produces the Euler (forward difference) method, which is an explicit method that can be rewritten for the general nonlinear case as

$$\frac{1}{\Delta t}\mathbf{M}(\mathbf{T}^n)\mathbf{T}^{n+1} = \frac{1}{\Delta t}\mathbf{M}(\mathbf{T}^n)\mathbf{T}^n - \hat{\mathbf{K}}(\mathbf{T}^n)\mathbf{T}^n + \hat{\mathbf{F}}(\mathbf{T}^n) \qquad (3.7-8)$$

and may be rearranged to clearly show its explicit nature

$$\mathbf{T}^{n+1} = \mathbf{T}^n + \Delta t \left[\mathbf{M}^{-1}(\mathbf{T}^n)\hat{\mathbf{F}}(\mathbf{T}^n) - \mathbf{M}^{-1}(\mathbf{T}^n)\hat{\mathbf{K}}(\mathbf{T}^n)\mathbf{T}^n\right] \tag{3.7-9}$$

For effective implementation, the algorithm in equation (9) requires that the capacitance matrix, **M**, be easily invertible, *i.e.,* be diagonal. The reduction of the consistent capacitance matrix to a diagonal or "lumped" form can be accomplished by various procedures (*e.g.*, row–sum technique [2,5,21,22]) for some types of finite elements (generally linear elements) but is not a universally available option. Also, the explicit nature of equation (9) implies a stability condition in terms of a maximum allowable time step, the thermal diffusivity of the material, and the finite element mesh spacing (see [22,26]); the stability condition can produce a prohibitively small time step for some problems. The Euler method is first–order accurate in time (*i.e.*, $O(\Delta t)$). Despite these limitations, explicit procedures are in common use in conduction problems, especially for analyses involving short–time transient phenomena, such as thermal shocks. The main attraction of an explicit scheme is the fact that a matrix solution is not required and computer storage requirements are therefore minimal.

For the cases for which $\theta > 0$, equation (7) produces a family of implicit methods which require the solution of a matrix problem at each time step. For example, when $\theta = 1$ (backward difference method), equation (7) reduces to

$$\left[\frac{1}{\Delta t}\mathbf{M}(\mathbf{T}^{n+1}) + \hat{\mathbf{K}}(\mathbf{T}^{n+1})\right]\mathbf{T}^{n+1} = \frac{1}{\Delta t}\mathbf{M}(\mathbf{T}^{n+1})\mathbf{T}^n + \hat{\mathbf{F}}(\mathbf{T}^{n+1}) \tag{3.7-10}$$

which is a fully implicit scheme. This is an unconditionally stable method (*i.e.*, there is no time step restriction; see page 114 for a discussion of convergence and stability of numerical integration methods), and is first–order accurate in time. In the general nonlinear case, the matrix problem presented in (10) requires an iterative solution for $\mathbf{T}^{n+1}$ within each time step; the linear case can be solved without nonlinear iteration, though a matrix solution is still required. The Picard scheme would be an acceptable choice for iteratively solving equation (10) at time t_{n+1}. In treating the nonlinear form of (10), the use of predictor–corrector schemes, extrapolation methods, and quasilinearization can often reduce the computational effort at each step without significantly reducing the accuracy of the method. For mild nonlinearities and modest time steps, a quasilinearization of equation (10) yields

$$\left[\frac{1}{\Delta t}\mathbf{M}(\mathbf{T}^n) + \hat{\mathbf{K}}(\mathbf{T}^n)\right]\mathbf{T}^{n+1} = \frac{1}{\Delta t}\mathbf{M}(\mathbf{T}^n)\mathbf{T}^n + \hat{\mathbf{F}}(\mathbf{T}^n) \tag{3.7-11}$$

which is now a single-step, non–iterative method for the solution of $\mathbf{T}^{n+1}$. An extrapolation procedure of the form

$$\mathbf{T}^* = \frac{3}{2}\mathbf{T}^n - \frac{1}{2}\mathbf{T}^{n-1} \tag{3.7-12}$$

h

$$\Big[\ldots (\mathbf{T}^*) + \hat{\mathbf{K}}(\mathbf{T}^*)\Big] \mathbf{T}^{n+1} = \frac{1}{\Delta t}\mathbf{M}(\mathbf{T}^*)\mathbf{T}^n + \hat{\mathbf{F}}(\mathbf{T}^*) \qquad (3.7-13)$$

ıce a better solution than (11) and allow a larger time step. This form is attractive as a false transient procedure for mildly nonlinear problems. For strong nonlinearities the coupling of a forward Euler predictor (9) with (10) as a corrector is a viable option (see [20]). Such methods are often used in practice with good success though the experience of the analyst is a crucial element in selecting the proper method.

We close this discussion with a brief outline of the only second–order accurate method in this family of procedures — the Crank–Nicolson method with $\theta = \frac{1}{2}$. Like the backward difference method, this is an unconditionally stable method, and it requires an iterative solution at each time step for the general nonlinear case. From equation (7) we have

$$\left[\frac{2}{\Delta t}\mathbf{M}(\mathbf{T}^a) + \hat{\mathbf{K}}(\mathbf{T}^a)\right] \mathbf{T}^a = \frac{2}{\Delta t}\mathbf{M}(\mathbf{T}^a)\mathbf{T}^n + \hat{\mathbf{F}}(\mathbf{T}^a) \qquad (3.7-14a)$$

with the new solution recovered from the definition

$$\mathbf{T}^{n+1} = 2\mathbf{T}^a - \mathbf{T}^n \qquad (3.7-14b)$$

Though (14) has no time step restriction for stability, it does suffer from "noise" or temporal oscillations in the solution when the time step exceeds a critical value (see the discussion on pages 114–116). Several modifications have been proposed to correct or filter this problem (see [4,27,28]). Predictor–corrector, extrapolation, and quasilinearization schemes can also be used with (14) to reduce the computational effort at each time step. The Crank–Nicolson scheme and its variants are probably the most frequently used procedures for heat conduction problems.

Predictor–Corrector Methods

The time integration methods outlined above are single–step methods that work effectively when an appropriate time step Δt is selected. The time step selection, however, is crucial to accuracy though in many physical problems there is little guidance available to precisely determine a suitable value for Δt. A cost–effective solution to time step selection and a general increase in integrator performance is available through use of predictor–corrector methods. The method described here was originally developed by Gresho *et al.* [20].

An implicit integration method that is second–order accurate in time can be developed by combining an explicit prediction method with the Crank–Nicolson (trapezoid) method described previously. A second–order, variable step, explicit method is the Adams–Bashforth predictor given by

$$\mathbf{T}_p^{n+1} = \mathbf{T}^n + \frac{\Delta t_n}{2}\left[\left(2 + \frac{\Delta t_n}{\Delta t_{n-1}}\dot{\mathbf{T}}^n\right) - \left(\frac{\Delta t_n}{\Delta t_{n-1}}\dot{\mathbf{T}}^{n-1}\right)\right] \qquad (3.7-15)$$

where $\Delta t_n = t_{n+1} - t_n$ and $\Delta t_{n-1} = t_n - t_{n-1}$. This formula can be used to predict the temperature solution vector given two "acceleration" vectors from previous timeplanes; no matrix solution is required.

A compatible corrector formula for use with (15) is available in the form of the trapezoid rule. When applied to the conduction equation, the trapezoid rule produces

$$\left[\frac{2}{\Delta t_n}\mathbf{M}(\mathbf{T}^a) + \hat{\mathbf{K}}(\mathbf{T}^a)\right]\mathbf{T}^{n+1} = \frac{2}{\Delta t_n}\mathbf{M}(\mathbf{T}^a)\mathbf{T}^n + \mathbf{M}(\mathbf{T}^a)\dot{\mathbf{T}}^n + \hat{\mathbf{F}}(\mathbf{T}^a) \qquad (3.7-16)$$

which is a rearranged form of the Crank–Nicolson method given in (14). Equation (16) is observed to be a nonlinear, algebraic system for the vector $\mathbf{T}^{n+1}$ and can again be solved using an iterative procedure such as Picard's method.

The major steps in the time integration procedure utilizing (15) and (16) are outlined here. At the beginning of each time step it is assumed that all of the required solution and "acceleration" vectors are known and the time increment for the next step has been selected. To advance the solution from time t_n to time t_{n+1} then requires the following steps:

1. A tentative solution vector, $\mathbf{T}_p^{n+1}$, is computed using the predictor equation (15).

2. The corrector equation (16) is solved for the "true" solution, $\mathbf{T}^{n+1}$. This involves the iterative solution of (16) via Picard's method. The predicted values $\mathbf{T}_p^{n+1}$ are used to estimate $\mathbf{T}^a$ and initialize the equation for the iteration procedure.

3. The "acceleration" vectors are updated using the new solution $\mathbf{T}^{n+1}$ and the "inverted" forms of the corrector formulas, that is

$$\dot{\mathbf{T}}^{n+1} = \frac{2}{\Delta t_n}\left(\mathbf{T}^{n+1} - \mathbf{T}^n\right) - \dot{\mathbf{T}}^n \qquad (3.7-17)$$

4. A new integration time step is computed. The time step selection process is based on an analysis of the time truncation errors in the predictor and corrector formulas as described in the next section. If a constant time step is being used, this step is omitted.

5. Return to step 1 for next time increment.

In actual implementation the Picard iteration process in step 2 is not carried to absolute convergence. Rather, a one–step correction is employed as advocated in [20]. This procedure is quite efficient and can be very accurate provided the time step is suitably controlled.

Time Step Control

The implicit time integration procedures outlined above can be used with a fixed, user specified time step or a time step that changes only at certain points during the integration interval. However, the *a priori* selection and modification of a reasonable integration time step can be a difficult task, especially for a complex problem. One of the benefits of using the predictor–corrector algorithms described here is that it provides a rational basis for dynamically selecting the time step.

The detailed derivation of the time step selection formula is omitted here. The reader interested in further details is referred to [20]. The general ideas for the time step selection process come from the well–established procedures for solving ordinary differential equations. By comparing the time truncation errors for two time integration methods of comparable order, a formula can be developed to predict the next time step based on a user specified error tolerance. In the present case, the time truncation errors for the explicit predictor and implicit corrector steps are analyzed.

The time step estimation formula is given by Gresho *et al.* [20] as

$$\Delta t_{n+1} = \Delta t_n \left(b \cdot \frac{\epsilon^t}{d_{n+1}} \right)^m \tag{3.7 - 18}$$

where $m = 1/3$, $b = 3(1 + \Delta t_{n-1}/\Delta t_n)$ for this second–order scheme. The user–specified error tolerance for the integration process is ϵ^t, which typically has a default value of 0.001. The quantity d_{n+1} is an appropriate norm on the integration error, which is defined as the difference between the predicted solution and the corrected value. Generally, the following norm is used

$$d^T_{n+1} = \frac{1}{T_{max}} \left[\sum_{i=1}^{N} \left(T_i^{(n+1)} - T_{ip}^{(n+1)} \right)^2 \right]^{\frac{1}{2}} \tag{3.7 - 19}$$

Initialization

The predictor equation (15) requires that one or more acceleration vectors be available at each timeplane in order to estimate a new solution vector. At the beginning of a transient solution these vectors are not generally available and thus a special starting procedure must be used. The approach normally taken is to use the dissipative backward Euler method for the first few steps and then switch to the standard second–order predictor–corrector methods. This procedure has the advantage that any nonphysical features of the numerical model are quickly damped by the backward Euler scheme.

For the first time step, the implicit, backward Euler scheme is used alone; the second step uses a forward Euler predictor and backward Euler corrector. Both of these steps use a fixed, user–supplied time step. At the third step, the usual predictor/corrector integration procedure begins and automatic time step selection is started. The initial time step supplied by the user to start the problem should be very conservative to prevent large time step reductions when the automatic selection procedure takes control.

Linear Multi–Step Methods

Most of the time integration methods described above were obtained by introducing relatively simple finite difference expressions for the time derivative into the semidiscrete finite element equation. This approach resulted in single step integration formulas in which the solution at the new time plane relies only on the solution data from the previous time plane. Multi–step methods that rely on data from more than one previous time step are also possible. These methods may be derived from "higher-order" finite difference approximations for $\dot{\mathbf{T}}$. However, a more formal approach with rigorous mathematical support involves the linear multi–step (LMS) methods developed for the solution of ordinary differential equations. Here we will outline a few applications of LMS to the finite element conduction problems and illustrate how some of the previous methods can be obtained as special cases of LMS. Further details on LMS are available in [29].

Consider the first–order system of equations expressed by

$$\mathbf{y} = \mathbf{f}(\mathbf{y}, t) \tag{3.7 – 20}$$

A general LMS of order k for the solution of (20) is written as

$$\sum_{i=0}^{k} \{\alpha_i \, \mathbf{y}^{n-i} + \Delta t \; \beta_i \; \mathbf{f}(\mathbf{y}^{n-i}, t_{n-i})\} \tag{3.7 – 21}$$

Through the selection of appropriate values of the coefficients α_i and β_i and the order k, a number of well–known methods may be described. Note that if $\beta_0 = 0$, the resulting method is *explicit*, and if $\beta_0 \neq 0$ then the integration method is *implicit*. For single–step cases where $k = 1$, the selection of $\alpha_0 = 1$, $\alpha_1 = -1$, $\beta_0 = 0$, $\beta_1 = -1$ produces the explicit method of equation (9), while the choice $\alpha_0 = 1$, $\alpha_1 = -1$, $\beta_0 = -1$, $\beta_1 = 0$ corresponds to the implicit backward Euler scheme in (10).

Predictor–corrector methods are closely associated with LMS methods due to the commonality with the form in (21) for explicit and implicit integration methods. As noted previously, the Adams–Bashforth–Moulton family of methods has been used in a finite element context by Gresho [20]. An explicit two step ($k = 2$), second–order form of (21) selects $\alpha_0 = 1$, $\alpha_1 = -1$, $\alpha_2 = 0$, $\beta_0 = 0$, $\beta_1 = -3/2$, $\beta_2 = 1/2$, which produces

$$\mathbf{y}^{n+1} = \mathbf{y}^n + \frac{3}{2}\Delta t \; \mathbf{f}(\mathbf{y}^n, t_n) - \frac{1}{2}\Delta t \; \mathbf{f}(\mathbf{y}^{n-1}, t_{n-1}) \tag{3.7 – 22}$$

Equation (22) is the constant time step form of (15) and would function as the predictor equation. The implicit, second–order companion to (22) is a single step method ($k = 1$) with coefficients $\alpha_0 = 1$, $\alpha_1 = -1$, $\beta_0 = -1/2$, $\beta_1 = -1/2$. This leads to

$$\mathbf{y}^{n+1} = \mathbf{y}^n + \frac{\Delta t}{2} \{\mathbf{f}(\mathbf{y}^n, t_n) + \mathbf{f}(\mathbf{y}^{n+1}, t_{n+1})\} \tag{3.7 – 23}$$

which is recognized as the trapezoidal rule. Rewriting (3.6–3) in the form of (20), we obtain

$$\dot{\mathbf{T}} = -\mathbf{M}^{-1}\mathbf{K}\mathbf{T} + \mathbf{M}^{-1}\mathbf{F} \tag{3.7 – 24}$$

and substituting into the formulas in (22) and (23) produce the predictor–corrector equations shown in (15) and (16).

Other first– and second–order methods can be derived from (21) as well as higher–order schemes. However, practical limits on function evaluations and stability considerations limit most LMS methods to those outlined here. The most significant aspect of the LMS formula is its use in the area of predictor–corrector procedures.

Convergence and Stability

To conclude this section on time approximation methods, a few comments on convergence of the numerical solution to the actual solution are required. The time integration methods of the previous sections can all be written in symbolic form as

$$\mathbf{T}^{n+1} = -\mathbf{G}\mathbf{T}^{n} + \mathbf{H} \tag{3.7 – 25}$$

where $\mathbf{G}$ represents the discretized form of the relevant differential operator. A time integration method is said to be *convergent* if, for fixed time t_n and time step Δt, the numerical solution $\mathbf{T}^n$ converges to its true value $\mathbf{T}(t_n)$ as $\Delta t \rightarrow 0$. To prove convergence of (25), the well-known Lax Equivalence Theorem [30] may be invoked, which allows the consistency and stability of $\mathbf{G}$ to be necessary and sufficient conditions for convergence. The numerical scheme (25) is said to be *consistent* with the original continuous problem if the errors (*e.g.*, round–off and truncation errors) go to zero in the limit as $\Delta t \rightarrow 0$. For the diffusion problems considered here consistency can be proved as shown in [22,26].

Since a time integration scheme is approximate, error is introduced into the solution $\mathbf{T}^{n+1}$ at each time step. The error in the solution at one time step feeds to the next time step due to the recursive use of equation (25). If the error is ultimately bounded, the time integration scheme is said to be *stable.* Thus, the stability of the time integration method in (25) implies that as $\mathbf{G}$ is recursively applied to each new vector $\mathbf{T}^n$, any errors that occur ultimately decay. That is, $\mathbf{G}$ is a bounded operator with a bounded spectral radius. For this discussion, the source of error is unimportant though we recognize that it may occur due to approximation of the differential equations and/or boundary conditions (truncation error) or due to finite precision computation (round–off errors).

There are a variety of techniques for assessing the stability of methods such as (25). For finite element applications, the spectral or modal methods are generally preferred. Here we will only sketch the procedure, without proof, and illustrate the stability conditions for the θ–family of integration methods.

The basic semidiscrete equation

$$\mathbf{M}\dot{\mathbf{T}} + \mathbf{K}\mathbf{T} = \mathbf{F} \qquad (3.7-26)$$

is difficult to analyze since it represents a large number of equations (degrees of freedom) that are coupled through the **M** and **K** matrices. A standard approach to this difficulty is to rewrite (26) in terms of its normal modes and thereby produce a system of uncoupled equations that are more easily studied. The generalized eigenvalue problem associated with (26) is (no sum on i)

$$\mathbf{K}\phi_i - \lambda_{(i)}\mathbf{M}\phi_i = \mathbf{0} \qquad (3.7-27)$$

where λ_i is the eigenvalue corresponding to the eigenvector ϕ_i. Assume that the eigenvectors are orthonormalized with respect to the capacitance matrix **M** such that

$$\phi_i^T\mathbf{M}\phi_j = \delta_{ij} \qquad (3.7-28)$$

where δ_{ij} are the components of the unit tensor (Kronecker delta). Then multiplying equation (27) by ϕ_j^T and using equation (28) yields

$$\phi_j^T\mathbf{K}\phi_i - \lambda_{(i)}\phi_j^T\mathbf{M}\phi_i = \mathbf{0} \qquad (3.7-29)$$

or

$$\phi_j^T\mathbf{K}\phi_i - \lambda_{(i)}\delta_{ij} = \mathbf{0} \qquad (3.7-30)$$

which shows that the eigenvectors are also orthogonal with respect to the diffusion matrix **K**.

Since the eigenvectors form a basis for the semidiscrete system in (26), the solution for **T** may be represented as a linear combination of the eigenvectors,

$$\mathbf{T}(\mathbf{x},t) = \sum_i \alpha_i(t)\phi_i(\mathbf{x}) \qquad (3.7-31)$$

where α_i are the generalized coordinates. Substituting equation (31) into equation (26), premultiplying by ϕ_j^T, and using the orthogonal properties in equations (28) and (30) leads to the result

$$\dot{\alpha}_i\left(\phi_j^T\mathbf{M}\phi_i\right) + \alpha_i\left(\phi_j^T\mathbf{K}\phi_i\right) = \phi_j^T\mathbf{F} \qquad (3.7-32)$$

and

$$\dot{\alpha}_i\delta_{ij} + \alpha_i\lambda_{(i)}\delta_{ij} = f_j \qquad (3.7-33)$$

or

$$\dot{\alpha}_j + \alpha_j\lambda_{(j)} = f_j \qquad (3.7-34)$$

Note that (34) represents the decoupled modal equation for each degree of freedom (and eigenvalue) in the original system (26). If equation (34) can be integrated in time for each α_i, then the solution $\mathbf{T}$ can be constructed from the eigenvector expression in (31). This is the basis for mode superposition solution techniques [21,22].

The single degree of freedom equations in (34) serve another purpose and allow study of the stability properties of various time integration methods. To demonstrate this idea, the time derivative and difference approximations for the θ method, which are defined in equation (6), can be applied to equation (34) to give

$$\frac{\alpha_i^{n+1} - \alpha_i^n}{\Delta t} + \lambda_{(i)}(1-\theta)\alpha_i^n + \lambda_{(i)}\theta\alpha_i^{n+1} = f_i^a \qquad (3.7-35)$$

This can be rearranged as

$$\alpha_i^{n+1} = \frac{1 - \lambda_{(i)}\Delta t(1-\theta)}{1 + \lambda_{(i)}\Delta t\theta}\alpha_i^n + \frac{\Delta t f_i^a}{1 + \lambda_{(i)}\Delta t\theta} \qquad (3.7-36)$$

which is of the same form as given in (25) with

$$G_i = \frac{1 - \lambda_{(i)}\Delta t(1-\theta)}{1 + \lambda_{(i)}\Delta t\theta} \qquad (3.7-37)$$

As stated previously the stability of the integration method is proved by showing that the amplification matrix $\mathbf{G}$ is a bounded operator. For the single degree of freedom equations, the analogous requirement is that the amplification factor, G_i, be less than unity, *i.e.*, $|G_i| < 1$. The restriction on G_i then is

$$-1 < \frac{1 - \lambda_i\Delta t(1-\theta)}{1 + \lambda_i\Delta t\theta} < 1 \qquad (3.7-38)$$

where θ must be between 0 and 1, and (38) must hold for all the modes, λ_i of the system. The right–hand inequality imposes no restrictions on the values $\lambda_i, \Delta t$, or θ; the left–hand inequality requires that $\lambda_i\Delta t < 2/(1-2\theta)$ when $\theta < 1/2$ but has no restrictions for $\theta \geq 1/2$. When there are no restrictions on the integration time step the method is termed *unconditionally stable* or simply *stable*. This analysis shows that all θ methods with $\theta \geq 1/2$ are unconditionally stable and these include the backward difference method ($\theta = 1$) in equation (10) and the Crank–Nicolson method ($\theta = 1/2$) in equation (14). The conditionally stable θ method is the Euler method ($\theta = 0$) in equation (8), where the time step is limited by $\lambda_i\Delta t_{crit} < 2$.

To ensure stability for all modes in the system, the time step must be limited by the largest eigenvalue in the system or $\Delta t_{crit} < 2/\lambda_{max}$. Recall that the λ_i are related to the system matrices $\mathbf{M}$ and $\mathbf{K}$; it can be shown that $\lambda_{max} \sim \alpha/h_{min}^2$, where $\alpha = k/\rho c$ is the thermal diffusivity (k is the thermal conductivity, ρ is the density, and c is the specific heat) and h_{min} is the characteristic dimension of the smallest finite element in the mesh. Thus for an explicit method the time step is limited by $\Delta t_{crit} < Ch_{min}^2$ (where C is a constant), which shows the strong influence of mesh size (refinement) on performance of the time integration method. Though the unconditionally stable methods ($\theta \geq 1/2$) have no stability restrictions on the time step, there are time step ranges which will produce a nonphysical oscillatory response in the solution. The oscillation limit corresponds to the point $\lambda_i \Delta t_{osc}$ where $G_i = 0$. As an example, for the Crank–Nicolson method ($\theta = 1/2$) the oscillation limit is $\lambda_i \Delta t_{osc} = 2$, and for $\Delta t > 2/\lambda_i$ the solution will exhibit an oscillatory decay. Oscillatory behavior can be filtered to produce acceptable solutions as shown by Wood and Lewis [27] and Wood [28].

3.8 RADIATION SOLUTION ALGORITHM

The enclosure radiation problem was outlined in Section 1.9. An energy balance for each surface in the enclosure leads to the following system of equations [see equations (1.9–3) and (1.9–4)]

$$\sum_{j=1}^{N} [\delta_{kj} - (1 - \epsilon_k) F_{k-j}]\, q_j^o = \epsilon_k \sigma T_k^4 \qquad (3.8-1)$$

$$q_k = q_k^o - \sum_{j=1}^{N} F_{k-j} q_j^o \qquad (3.8-2)$$

where q_j^o is the outgoing radiative flux for each surface, q_k is the net flux from each surface $q_k = Q_k/A_k$, Q_j is the net energy loss from each surface, δ_{kj} is the unit tensor, σ is the Stefan–Boltzmann constant, and F_{k-j} are radiation view (configuration) factors. When the surface temperatures of all surfaces are known, equation (1) forms a set of linear algebraic equations for the unknown, outgoing surface fluxes, q_j^o. That is, equation (1) can be written as

$$\mathbf{A}(\mathbf{T})\mathbf{q} = \mathbf{F}(\mathbf{T}) \qquad (3.8-3)$$

where $\mathbf{A}$ is a function of $\mathbf{T}$ due to the possible dependence of surface emissivities on temperature. The matrix $\mathbf{A}$ is a full matrix due to the surface to surface coupling represented by the view factors F_{k-j}. This characteristic, along with the possible temperature dependencies, suggests the use of an iterative solution method for (3) rather than a direct matrix factorization.

A standard approach employs a Gauss-Seidel method to solve (3) for the components of $\mathbf{q}$, though other iterative schemes are possible. When these values are available, equation (2) is then used to compute the effective flux to the surface. The surface fluxes provide boundary conditions to the finite element model for the conduction process. When new surface temperatures are computed, due to either a new time step or iteration cycle, the above process is repeated to obtain a new surface flux condition. The surface temperatures used in the above computation must be uniform over each surface in order to satisfy the conditions of the radiation model. In most applications, it is assumed that each surface in the radiation problem corresponds to a face or edge of a finite element. The uniform surface temperature needed for use in (1) is obtained by combining (averaging) the nodal point temperatures on the appropriate element face or edge.

View factors for use in enclosure radiation computations were defined in Section 1.9 as purely geometric factors. The computation of such factors is generally performed by specialized programs (*e.g.*,[31, 32]) that are external to the basic finite element program. For plane, two-dimensional geometries with no obstructing (shadowing) surfaces, the view factor computation can be done with simplified analytic methods such as Hottel's cross–string method. More complex geometries in two and three dimensions require the use of numerical procedures, such as double area integration, hemicube, or Monte Carlo techniques. A comparative study of a number of view factor algorithms is presented in [33].

The above solution procedure, which decouples the enclosure radiation solution from the conduction solution, is a standard approach for this type of problem. When the simulation is time dependent this type of algorithm is very successful, since the time stepping procedure usually produces only "small" changes in the temperature field within each time step. Incremental changes in surface temperature lead to incremental changes in surface flux., and rapid convergence of both the radiation and conduction solutions, *i.e.*, the two equation sets are always "close" to equilibrium. Predictor–corrector techniques can also be used to advantage with this decoupled approach.

For steady solutions that are approached iteratively, the decoupled algorithm tends to be less sucessful. Large changes in the temperature field between iterations are typical and can easily be amplified by large variations in surface flux due to the strong (fourth power) dependence on surface temperature. Very slow convergence or more often, divergence, of the algorithm is observed. Severe underrelaxation of the solution vectors [equation (3.7–4)] may improve the solution process as will incrementation of the boundary conditions and/or material properties. Other solution approaches, in which the conduction and radiation problems are more fully coupled, are also possible. Such a coupling would permit use of a Newton–Raphson type iteration and a significant increase in the rate of convergence. The penalty for a coupled algorithm is a dramatic (and in many cases, prohibitive) increase in size of the matrix problem and the cost of each iteration. In many cases, the most cost–effective method for steady, enclosure radiation problems is the decoupled method with a time–dependent approach (false transient) to steady state.

3.9 SOME VARIABLE PROPERTY COMPLICATIONS

In all of the previous discussion we have mentioned the possible occurrence of variable material properties, boundary conditions, and heat sources. The previous sections on solution methods showed how such temperature dependence produced nonlinearities in the matrix problem and how various solution algorithms were set up to account for this behavior. In the present section three other aspects of variable properties will be discussed—-methods for the implementation of variable coefficients into the finite element matrix formulation, the treatment of latent heat effects during phase change, and the use of anisotropic conductivities.

Consider first the problem of evaluating finite element matrices that contain a variable coefficient. To focus the discussion, a typical diffusion term from the conduction equation will be examined. From equation (3.2–6) the xx–component of the diffusion matrix is

$$\mathbf{K}_{xx} = \int_{\Omega^e} \frac{\partial \Psi}{\partial x} k \frac{\partial \Psi^T}{\partial x} \, d\mathbf{x} \tag{3.9-1}$$

where for simplicity an isotropic conductivity has been assumed. As seen in Section 3.4, it is typical to evaluate element integrals, such as the one in equation (1), by use of numerical quadrature. When the conductivity is constant the matrix in (1) is independent of temperature and may be constructed once and stored as a two-dimensional array for subsequent use in a solution procedure. However, when the conductivity is variable, the matrix in (1) must be reevaluated each time new values of the temperature are produced during the solution procedure. One standard method for treating this variable coefficient simply involves recomputing the matrix for each new set of conductivities. For each element the temperature is evaluated at the quadrature points via the basis functions, thus permitting the conductivity to be evaluated at the quadrature locations. The matrix coefficients are then computed by the standard quadrature procedure. The work involved in such a procedure can become large when considering anisotropic conductivities (three conductivity matrices in two dimensions, six conductivity matrices in three dimensions) and/or higher-order elements (higher-order quadrature schemes).

A second approach to this problem makes use of an interpolation function for the variable coefficient. Let the conductivity be approximated by

$$k = \Psi^T \mathbf{k} \tag{3.9-2}$$

where Ψ is an appropriate set of element basis functions and $\mathbf{k}$ are nodal point values of the conductivity. Inserting (2) into (1) gives

$$\mathbf{K}_{xx} = \int_{\Omega_e} \frac{\partial \Psi}{\partial x} (\Psi^T \mathbf{k}) \frac{\partial \Psi^T}{\partial x} \, d\mathbf{x} \tag{3.9-3}$$

The above integral may be evaluated (via quadrature) once and stored as a triply subscripted array (hypermatrix). During the solution process, the conductivity is evaluated at each node as a function of the nodal point temperature, and a simple inner product in (3) produces the needed two-dimensional diffusion matrix. Such a procedure eliminates the need for recomputing element-level matrices, albeit at the expense of some additional storage and I/O operations.

Though we are concerned in the present example with the treatment of variable material properties it is important to realize that both of the above procedures can be used with any type of variable coefficient. In the heat conduction problem variable heat capacities, convective and radiative heat transfer coefficients, and volumetric heat sources are all candidates for either of the described methods.

One particularly difficult type of property variation is the latent heat effect that occurs in the simulation of phase change problems. As described in Section 1.8, latent heat effects can be adequately treated by use of a temperature-dependent specific heat. That is, from equation (1.8-6) we have

$$C^*(T) = C(T) + L\,\delta(T - T_t) \qquad (3.9-4)$$

where C^* is the effective specific heat, C is the normal specific heat, L is the latent heat, δ is the Dirac delta function, and T_t is the transition temperature. For application in a finite element procedure equation (4) is replaced by

$$C^*(T) = C(T) + L\,\delta^*(T - T_t, \Delta T) \qquad (3.9-5)$$

where δ^* is the delta form function; δ^* has a large but finite value in the interval centered about T_t and is zero outside the interval (see Figure 3.9–1).

The interval ΔT is often referred to as the "mushy" zone and corresponds to the difference between the liquidus, T_l, and solidus, T_s, temperatures for the material. Note that (5) is thus an approximation for the behavior of pure materials that change phase at a specified temperature, T_t, but is accurate for nonpure substances that have truly distinct liquidus and solidus temperatures.

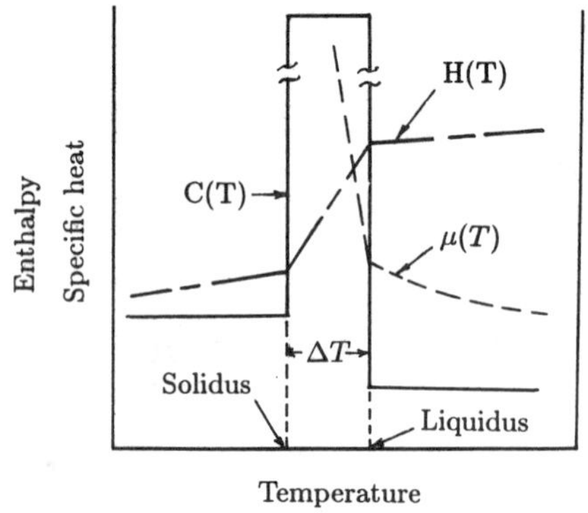

Figure 3.9–1: Definition of material properties for phase change computation.

Though the effective capacitance model described above is useful in accounting for latent heat effects, a few words of caution should be issued with regard to the time integration of this type of phase change model. Since the transition temperature interval, ΔT, is often small compared to the overall temperature variation in a conduction problem it is important to note that there are some severe practical limitations on the time integration procedure. In general, the time–stepping algorithm must be controlled such that every node that "changes phase" is forced to attain a temperature value in the interval bracketed by ΔT. If a nodal point does not "land" in the ΔT range but simply steps over this temperature interval, the latent heat effect is lost for that node and an incorrect temperature response and energy balance will result. Various methods for dealing with this difficulty have been proposed (see [34,35]) including broadening the ΔT range, placing a limit on the maximum temperature change that can occur during a time step, and using special forms of a predictor–corrector algorithm. The problem is a numerically delicate one and is still an area of active research.

A number of useful alternatives to the effective capacitance method make direct use of the enthalpy, H, versus temperature curve and differ only in how they evaluate the slope of H versus T. In one method, H is first computed at the nodes of an element (knowing T); the heat capacity at an integration point within the element is then found from

$$C_p = \left[\frac{\nabla H \cdot \nabla H}{\nabla T \cdot \nabla T}\right]^{1/2} \qquad (3.9-6)$$

where ∇H and ∇T are evaluated via the element shape functions. This technique will maintain its accuracy as long as the phase boundary passes through each element and does not skip over an element. This is certainly a more robust technique than the effective capacitance method.

In closing this section, we mention briefly the treatment of anisotropic thermal conductivities. In two dimensions the most general, physically realistic, material model has an orthotropic conductivity such that $k_{11} \neq k_{22}$ and $k_{12} = k_{21} = 0$ when written in terms of the principal material axes. Since the global finite element coordinate frame will generally not be aligned with the material axes, a coordinate transformation is required. Referring to the sketch in Figure 3.9–2 the global components of the conductivity matrix can be obtained from

$$[k] = \begin{bmatrix} \cos\alpha & \sin\alpha \\ -\sin\alpha & \cos\alpha \end{bmatrix} \begin{bmatrix} k_{11} & 0 \\ 0 & k_{22} \end{bmatrix} \begin{bmatrix} \cos\alpha & -\sin\alpha \\ \sin\alpha & \cos\alpha \end{bmatrix} \qquad (3.9-7a)$$

or when multiplied out

$$[k] = \begin{bmatrix} k_{11}\cos^2\alpha + k_{22}\sin^2\alpha & -k_{11}\sin\alpha\cos\alpha + k_{22}\sin\alpha\cos\alpha \\ -k_{11}\sin\alpha\cos\alpha + k_{22}\sin\alpha\cos\alpha & k_{11}\sin^2\alpha + k_{22}\cos^2\alpha \end{bmatrix} \qquad (3.9-7b)$$

These last expressions give the needed conductivity components for use in the finite element diffusion matrix **K** defined in equation (3.2–5a). In the most general two-dimensional case three matrices are defined by **K** for the xx, yy, and xy components of the conductivity tensor as given in equation (7). It is of course possible for the principal material conductivity components, k_{11} and k_{22}, to be independent functions of temperature and for the local orientation angle, α, to vary from material to material or even from from element to element in the finite element mesh.

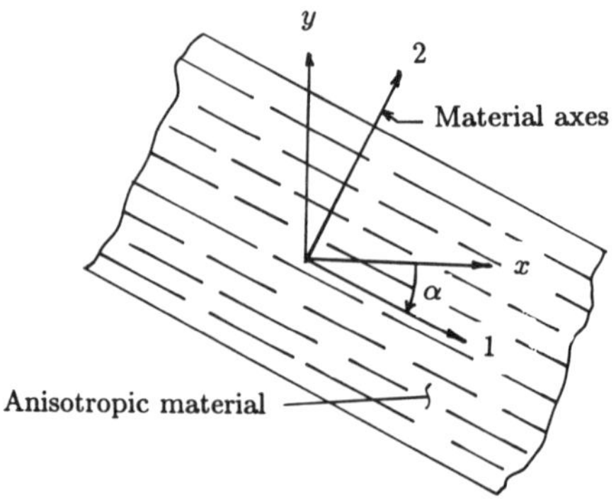

Figure 3.9–2: Notation for anisotropic conductivity.

The three–dimensional transformation for an orthotropic material has the same form but is slightly more complex. In tensor notation, the conductivity components in the global coordinate system are obtained from

$$k_{ij} = C_{in} k^p_{nm} C_{jm} \qquad (3.9-8)$$

where k^p_{nm} is the conductivity tensor in the material coordinate frame and C_{in} are the direction cosines between the axes of the two coordinate systems. For an orthotropic material, k^p_{nm} is diagonal and (8) can be written as

$$[k] = \begin{bmatrix} C_{11} & C_{21} & C_{31} \\ C_{12} & C_{22} & C_{32} \\ C_{13} & C_{23} & C_{33} \end{bmatrix} \begin{bmatrix} k_{11} & 0 & 0 \\ 0 & k_{22} & 0 \\ 0 & 0 & k_{33} \end{bmatrix} \begin{bmatrix} C_{11} & C_{12} & C_{13} \\ C_{21} & C_{22} & C_{23} \\ C_{31} & C_{32} & C_{33} \end{bmatrix} = \mathbf{C}^T\mathbf{k}\mathbf{C} \quad (3.9-9)$$

which has the same form as the two-dimensional case in (7). Note that the three–dimensional form of k_{ij} now requires six matrices to define the diffusion operator $\mathbf{K}$; the principal conductivities may again be temperature dependent and their orientation may vary over the domain.

3.10 POST–COMPUTATION OF HEAT FLUX

An important post–processing operation in many heat conduction problems involves the computation of the heat flux from various surfaces or bodies. The diffusion fluxes associated with the conduction equation can be computed on an element–by–element basis. In the following only the heat fluxes associated with two–dimensional problems will be specifically considered since the fluxes in the three–dimensional case are completely analogous. Fourier's law provides the definition of the conductive heat flux as ($k_{xx} = k_{11}, k_{xy} = k_{12}, k_{yy} = k_{22}$, and $k_{yx} = k_{21}$)

$$q_x = -k_{xx}\frac{\partial T}{\partial x} - k_{xy}\frac{\partial T}{\partial y} \qquad (3.10-1a)$$

$$q_y = -k_{yx}\frac{\partial T}{\partial x} - k_{yy}\frac{\partial T}{\partial y} \qquad (3.10-1b)$$

The component fluxes in (1) are computed by using the standard finite element approximations for T and the components of the conductivity tensor, k_{ij}

$$T(x_i, t) = \mathbf{\Psi}^T(x_i)\mathbf{T}(t) \qquad (3.10-2)$$

Also needed are the relations for the local temperature derivatives in terms of the local element coordinates and the isoparametric description of the element geometry. That is,

$$\left\{ \begin{array}{c} \frac{\partial \Psi}{\partial x} \\ \frac{\partial \Psi}{\partial y} \end{array} \right\} = \begin{bmatrix} J^*_{11} & J^*_{12} \\ J^*_{21} & J^*_{22} \end{bmatrix} \left\{ \begin{array}{c} \frac{\partial \Psi}{\partial \xi} \\ \frac{\partial \Psi}{\partial \eta} \end{array} \right\} \qquad (3.10-3a)$$

where $J^*_{ij} = \hat{J}^*_{ij}/|J|$, and

$$\hat{J}^*_{11} = \frac{\partial y}{\partial \eta}\ ,\ \hat{J}^*_{12} = -\frac{\partial y}{\partial \xi}\ ,\ \hat{J}^*_{21} = -\frac{\partial x}{\partial \eta}\ ,\ \hat{J}^*_{22} = \frac{\partial x}{\partial \xi} \qquad (3.10-3b)$$

Note that (x, y) are functions of the local element coordinates, (ξ, η) through the parametric mapping (3.3–3)(see Section 3.4).

Using these definitions the flux components can be expressed as

$$\begin{aligned} q_x = &- k_{xx}\left(J^*{}_{11}\frac{\partial \mathbf{\Psi}^T}{\partial \xi}\mathbf{T} + J^*{}_{12}\frac{\partial \mathbf{\Psi}^T}{\partial \eta}\mathbf{T}\right) \\ &- k_{xy}\left(J^*{}_{21}\frac{\partial \mathbf{\Psi}^T}{\partial \xi}\mathbf{T} + J^*{}_{22}\frac{\partial \mathbf{\Psi}^T}{\partial \eta}\mathbf{T}\right) \end{aligned} \qquad (3.10-4a)$$

$$\begin{aligned} q_y = &- k_{yx}\left(J^*{}_{11}\frac{\partial \mathbf{\Psi}^T}{\partial \xi}\mathbf{T} + J^*{}_{12}\frac{\partial \mathbf{\Psi}^T}{\partial \eta}\mathbf{T}\right) \\ &- k_{yy}\left(J^*{}_{21}\frac{\partial \mathbf{\Psi}^T}{\partial \xi}\mathbf{T} + J^*{}_{22}\frac{\partial \mathbf{\Psi}^T}{\partial \eta}\mathbf{T}\right) \end{aligned} \qquad (3.10-4b)$$

In addition to the local components of the flux vector, the heat flux normal to the element edge is often of importance. By definition

$$q_n = \mathbf{q} \cdot \mathbf{n} \qquad (3.10-5a)$$

$$\mathbf{q} = q_x\mathbf{e_x} + q_y\mathbf{e_y} \qquad (3.10-5b)$$

$$\mathbf{n} = n_x\mathbf{e_x} + n_y\mathbf{e_y} \qquad (3.10-5c)$$

and thus

$$q_n = q_x n_x + q_y n_y \tag{3.10-6}$$

In order to employ (6), the components of the normal vector are also required; these may also be computed from the element geometry and isoparametric mapping

$$n_x = \frac{1}{\Delta}\frac{\partial y}{\partial s}\ ; \quad n_y = -\frac{1}{\Delta}\frac{\partial x}{\partial s} \tag{3.10-7}$$

where

$$\Delta = \left[\left(\frac{\partial x}{\partial s}\right)^2 + \left(\frac{\partial y}{\partial s}\right)^2\right]^{1/2} \tag{3.10-8}$$

The definitions in (6) are sufficient to define the flux components at any point (ξ_0, η_0) in the element; for optimal accuracy the 2×2 Gauss points are usually selected for the evaluation of derivatives. Relations analogous to the above may be derived for the three–dimensional case where fluxes normal to an element surface are to be evaluated.

3.11 EXAMPLES OF DIFFUSION PROBLEMS

3.11–1 Introduction

In this section a number of example diffusion problems are presented to illustrate the finite element analysis procedure. The first several examples are relatively simple, two–dimensional solutions that illustrate the use of anisotropic and nonlinear material properties as well as the phase change algorithm. The remaining examples are more complex and represent typical engineering analyses. Most of the problems described here were analyzed using the finite element code, COYOTE (see [36]), or one of its derivatives.

3.11–2 Temperature–Dependent Conductivity

Many materials of engineering interest have thermophysical properties, such as conductivity, that vary with temperature. This example illustrates the difference in results that can be expected when conductivity variations are included in the model. Figure 3.11–1 contains a schematic and mesh for a simple planar geometry. Each half of the region is occupied by a different isotropic material; the first material has constant conductivity, k_0, while the second material has a conductivity that varies with temperature as $k = k_0 + aT$. A constant heat flux is applied along the left edge of the domain. The upper and lower boundaries are insulated ($q_n = \mathbf{q} \cdot \mathbf{n} = 0$) as is the plane of symmetry between the two materials. All other surfaces are convectively cooled with a constant h_c and T_c. Four–node bilinear elements were used to model the problem.

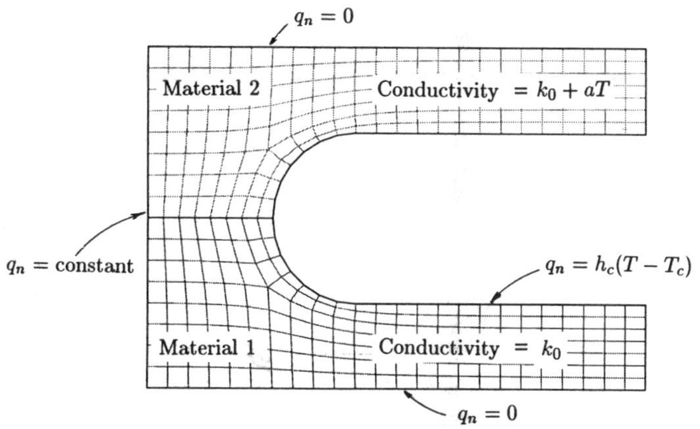

Figure 3.11–1: Schematic and mesh for the temperature–dependent conductivity problem.

Due to the variable conductivity this problem is nonlinear and the steady–state solution was obtained via a Picard iteration. A converged solution was obtained in four iterations. Figure 3.11–2 contains a plot of the isotherms. Since the plotted isotherm bands are the same for the two materials, the difference in the temperature fields is readily observed. At the high temperature end (applied flux surface) of the region, the conductivity for the variable material is high and the temperature gradient (isotherm spacing) is therefore lower than for the constant conductivity case. At the low temperature end of the region where the conductivities are comparable, the temperature gradients are similar.

3.11–3 Anisotropic Conductivity

As explained in Section 3.9, the inclusion of an anisotropic conductivity into a finite element model is relatively straightforward and may be important for the simulation of laminated, fibrous, or wire–wrapped materials. A planar square with a tilted square insert is used to demonstrate the effects of an orthotropic conductivity. Referring to Figure 3.11–3, the outer square is an isotropic material with conductivity, k. Constant temperature boundary conditions are imposed on the vertical edges of the square while the horizontal edges are insulated. The insert material is orthotropic with $k_{11} = k$ and $k_{22} = k/10$; the orientation of the material axes with respect to the global coordinate axes will be specified for each simulation.

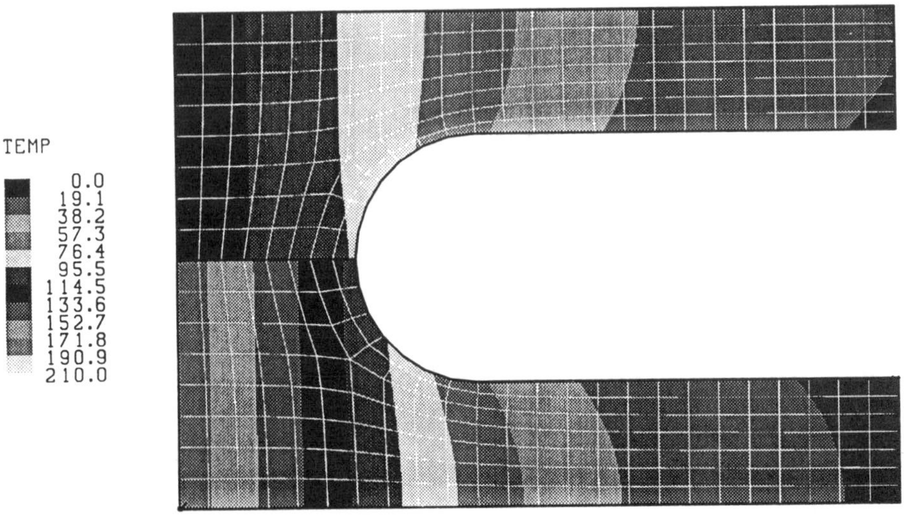

Figure 3.11–2: Isotherms for the temperature–dependent conductivity problem.

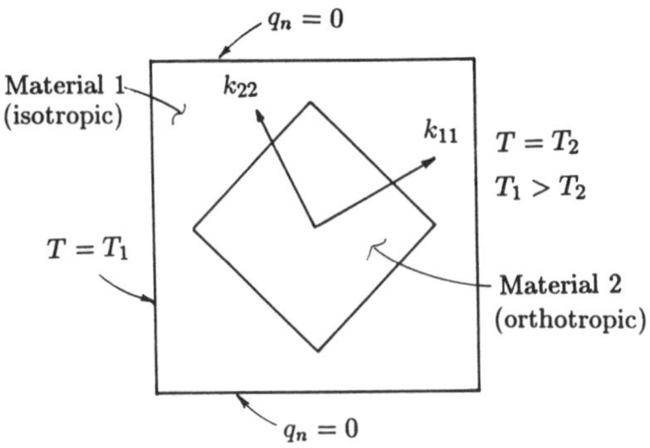

Figure 3.11–3: Geometry of the anisotropic conductivity problem.

Since the problem is linear, it may be solved directly without iteration. The first solution (see Figure 3.11-4a) corresponds to the case where the material axes for the insert material are aligned with the coordinate axes. As expected, the solution is one–dimensional since the k_{11} component for the insert is the same magnitude as the surrounding isotropic material; the k_{22} component plays no role in the solution since there is no temperature gradient imposed in the transverse direction. Figure 3.11–4b shows the temperature field for the case where the material axes of the insert are inclined upward 45° with respect to the global axes (the material axes are aligned with the mesh lines in the insert). The distortion of the temperature field due to the anisotropy is very obvious. Finally, Figure 3.11–4c illustrates the solution for the case as in (b) but with both components of the insert conductivity reduced by a factor of 10. This accentuates the difference between the two materials and leads to higher gradients and more distortion of the isotherms within the insert.

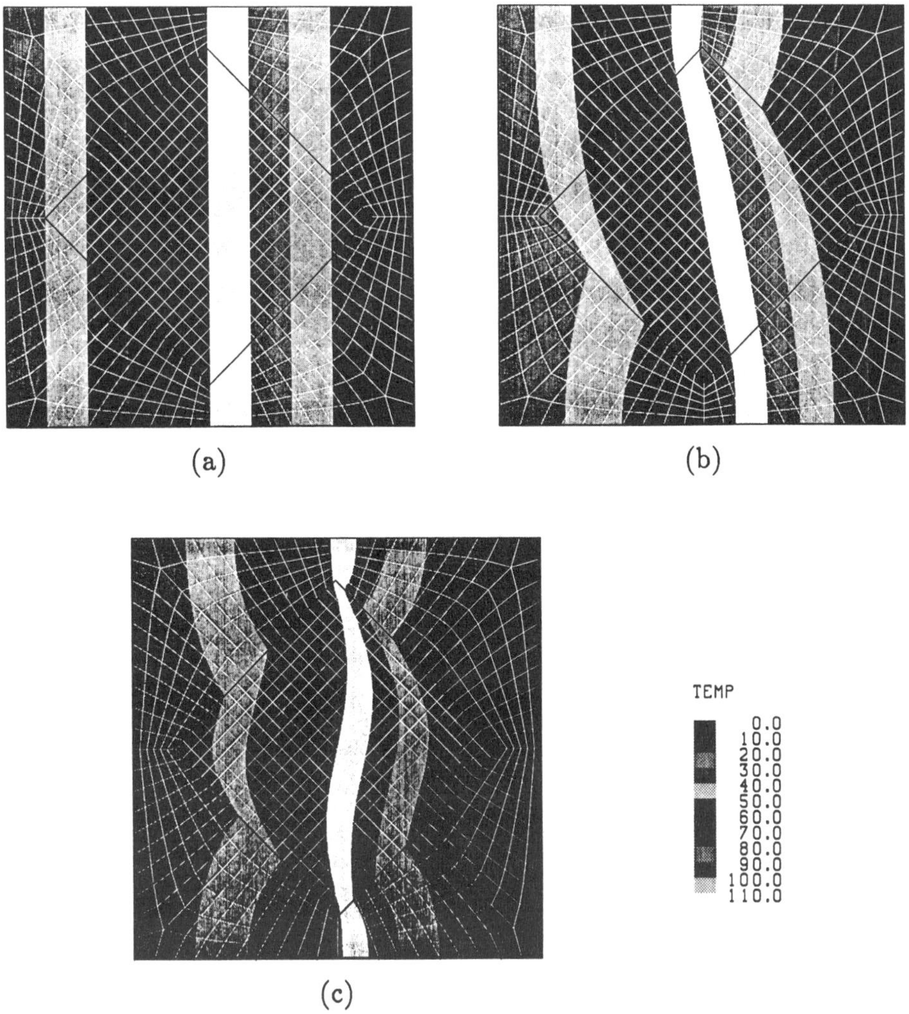

Figure 3.11–4: Temperature fields for the anisotropic conductivity problem.

3.11–4 One–Dimensional Stefan Problem

A standard test problem for phase change algorithms is the one–dimensional Stefan problem (see Carslaw and Jaeger [37]). The problem consists of a material region $0 \leq x \leq 4$, held initially at a uniform temperature greater than the liquidus temperature. At time zero, one face of the region ($x = 0$) is lowered to a temperature below the solidus temperature, causing a solidification front to propagate into the material. The present example is solved using a mesh of eight–node (biquadratic) elements and the Crank–Nicolson time integration procedure. The schematic of the problem is shown in Figure 3.11–5, which also contains the finite element mesh and boundary conditions. The following material properties are used: $\rho = 1.0$, $C = 1.0$, $k = 1.08$, $L = 70.26$, $\Delta T = 2.0, 0.5$, $T_s = -0.15$, and $T_l = -2.15, -0.65$.

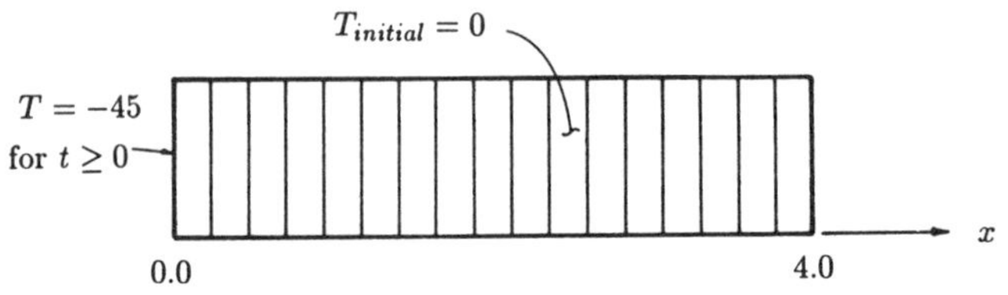

Figure 3.11–5: Schematic of the one–dimensional Stefan problem.

Figure 3.11–6 shows the computed temperature response for a point one unit from the cooled face. Computations were made for two values of the phase change temperature interval, ΔT, and are compared to the exact solution. The numerical solution is of reasonable accuracy with the smaller ΔT case being slightly closer to the exact solution. Figure 3.11–7 contains a plot of the location of the phase boundary as a function of time for the cases studied. Early in time, the numerical solution overpredicts the location of the phase boundary but comes into good agreement with the analytical solution for $t \geq 1.0$. Note that the analytical solution was developed for a single–phase transition temperature while the numerical solution requries a finite transition interval, ΔT. As ΔT gets smaller, the numerical solution becomes more difficult and more costly.

3.11–5 Drag Bit Analysis

The second example deals with the thermal (and stress) analysis of a polycrystalline diamond compact (PDC) drag bit that is commonly used in the oil and gas drilling industry. A planar, two–dimensional model of a single stud in the drag bit body is considered (see Figure 3.11–8). The thermal boundary conditions for the model are also indicated in the sketch. Heat is input to the stud at the wearflat due to frictional heating; the heating function is given as an empirical function of cutter, rock, and operating conditions. The remaining surfaces of the stud are convectively

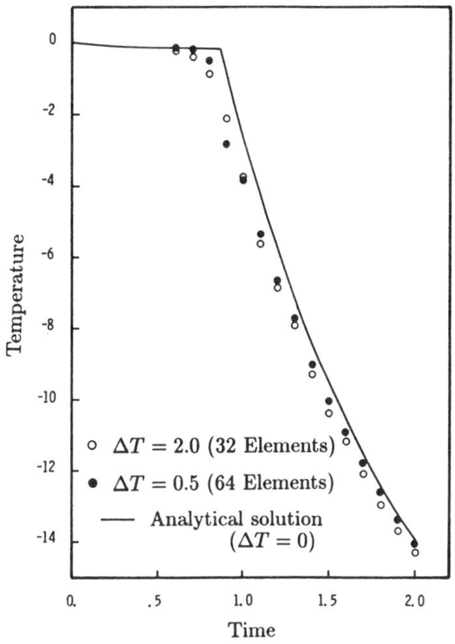

Figure 3.11–6: Temperature versus time at $x = 1.0$ for the Stefan problem.

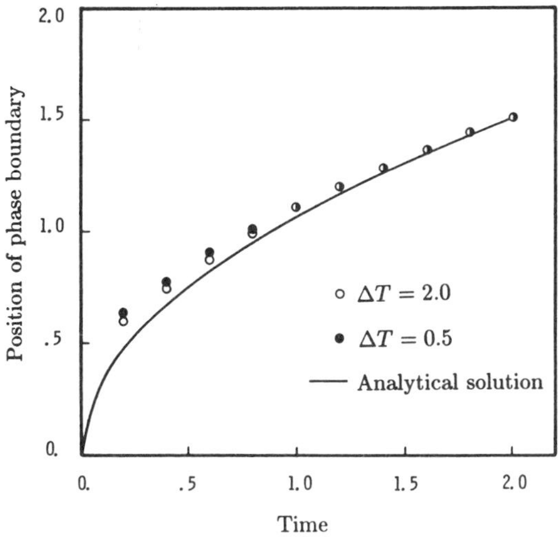

Figure 3.11–7: Location of the phase boundary versus time for the Stefan problem.

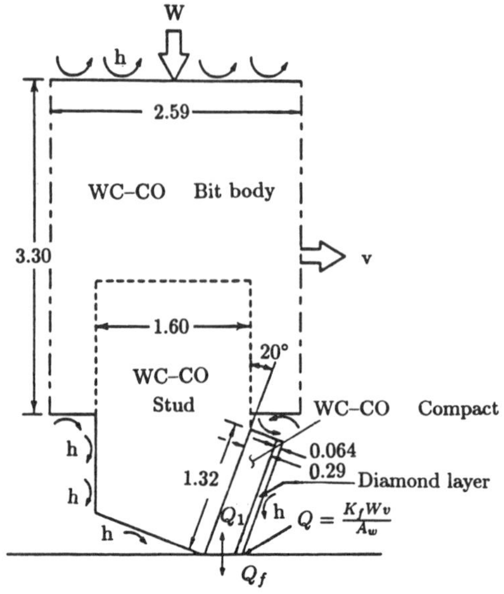

Figure 3.11–8: Schematic of the polycrystalline diamond compact (PDC) drag bit.

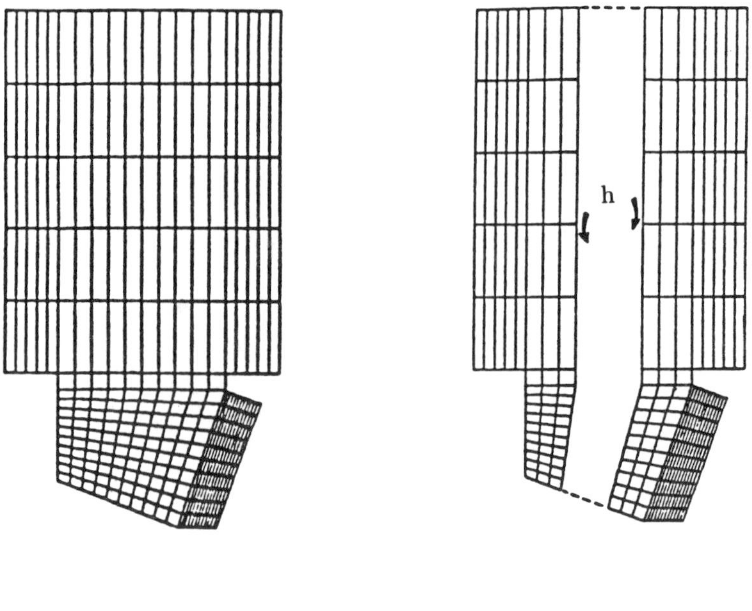

Figure 3.11–9: Finite element meshes used for the analysis of the PDC drag bit.

cooled by the drilling fluid. Boundary conditions on the bit body are taken to be adiabatic since they represent planes of symmetry with adjoining cutters. Figure 3.11–9 contains two finite element meshes of eight–node quadrilateral elements used for this problem; the second mesh shows the location of a cooling nozzle and a convection boundary condition used in some of the parametric studies.

A wide variety of boundary and geometric conditions in both the steady–state and transient regimes [38] have been examined for the problem. Figures 3.11–10 and 3.11–11 contain plots of isotherms for two steady–state cases, which differ in the assumed convective heat transfer coefficient for the drilling fluid. The high temperature gradients caused by the high cooling rate are clearly seen in the figures. As is quite common in many applications, a finite element stress analysis followed the heat transfer study. Figures 3.11–12 and 3.11–13 show contour plots of the predicted plastic strains in the cutter corresponding to the temperature fields of Figures 3.11–10 and 3.11–11 and an appropriate set of mechanical loads. Studies such as this have been used to predict the life of drill bits, improve cutter design, and provide operational limits for improved drilling performance.

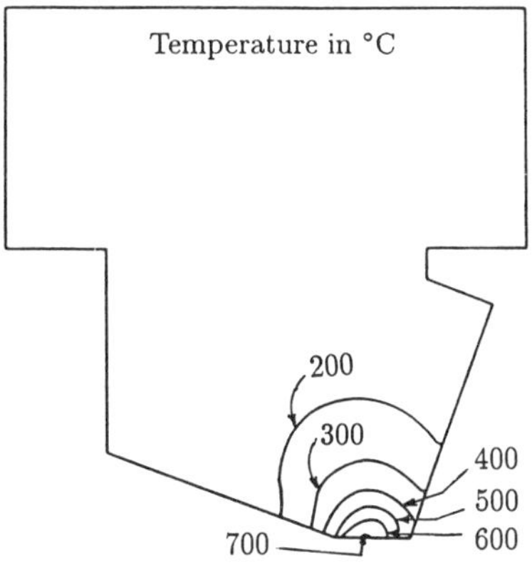

Figure 3.11–10: Steady–state isotherms for the drag bit, $h_c = 5.00\ W/cm^2\ °C$.

3.11–6 Brazing and Welding Analyses

The next two examples come from the analyses of two manufacturing processes for electrical components. Figure 3.11–14 shows a schematic and finite element mesh for one-half (upper half) of the axisymmetric geometry of interest; a three–dimensional mesh used in a subsequent analysis to verify the two–dimensional results is also shown. In manufacturing this component, two halves (upper and lower) are brought together under a modest pressure. An electrical current is passed through the outer flange to resistively heat the braze rings and bond the two halves of the device. In the simplified thermal model considered here the brazing process is modeled using a time–dependent (pulsed) volumetric heating of the flange and braze ring. All exterior boundaries are assumed to be adiabatic since the entire process takes place on a very short time scale. This problem was solved using four–node, bilinear elements and a modified Crank-Nicolson procedure.

Figure 3.11–15 shows a series of isotherm plots for two times during a one–pulse brazing cycle. Heating occurs over the first 0.5 seconds, during which time the temperature on the flat of the braze ring reaches $\sim 2100°F$. Later in time the temperature in the braze ring decreases as energy is conducted toward the centerline of the component. A companion stress analysis was used to identify high stress regions and potential failure (cracking) sites in the component.

Like brazing, welding is a process that is often used in the assembly of electrical components. Figure 3.11–16 shows the finite element representation of a battery case and attached header; the header and case are welded using a gas–tungsten arc (GTA) weld. The metallic connecting pins in the battery header are fixed to the header with glass–to–metal seals and must retain an hermetic seal during the GTA process. The figure shows an enlargement of the connecting pins, seals and header assembly. A study was performed to evaluate the benefits of providing various types of heat sinks to the header and case during the weld process. Details of the parametric study and weld parameters and modeling are available in [39]. Figure 3.11–17 contains one heat sink configuration; the mesh for this problem contained approximately 27,000 eight–node, brick elements.

Figures 3.11–18 and 3.11–19 show temperature fields in the header at the conclusion of one revolution of the GTA welder. It is clear from the large differences in temperature at the pins that heat sinking has a strong positive benefit. The temperature gradients in the glass for the no heat sink case are sufficient to cause a tension failure (cracking) of the glass.

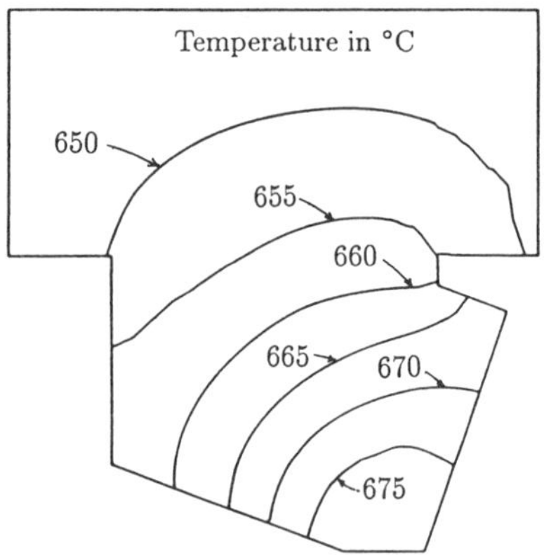

Figure 3.11–11: Steady–state isotherms for the drag bit, $h_c = 0.01\ W/cm^2\ °C$.

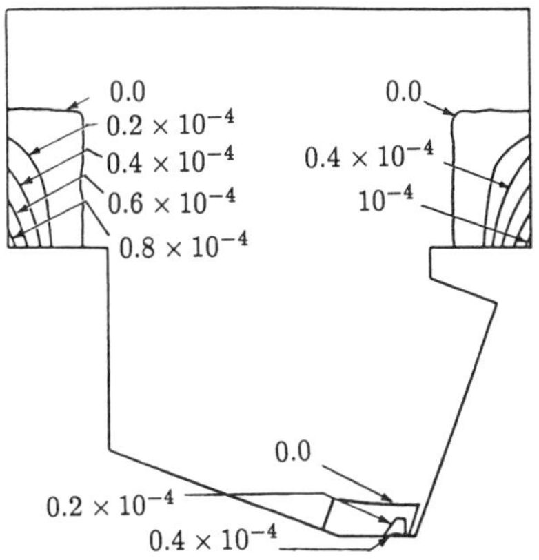

Figure 3.11–12: Plastic strain contours for the drag bit, corresponds to thermal field in Figure 3.11–10.

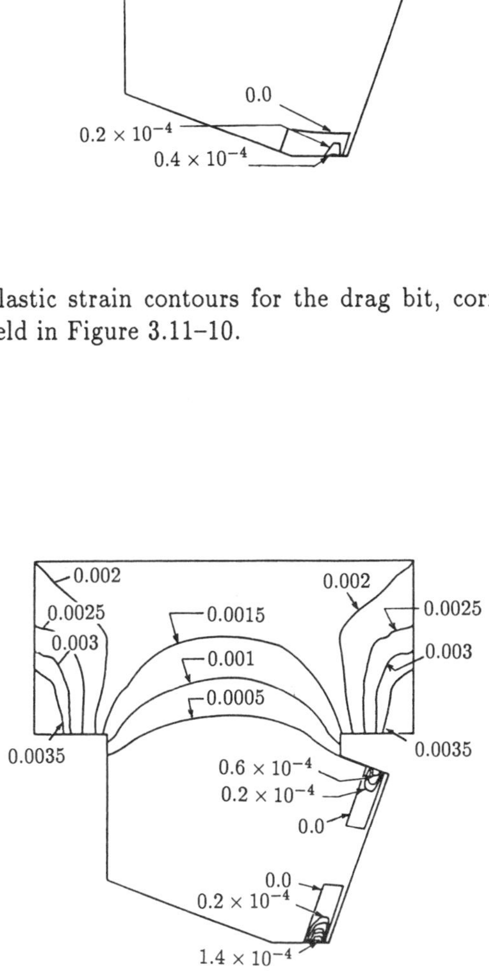

Figure 3.11–13: Plastic strain contours for the drag bit, corresponds to thermal field in Figure 3.11–11.

3.11–7 Investment Casting

A heat transfer analysis of the cooling phase of an investment casting can provide important guidance to the design of the metal distribution system (gates, runners, *etc.*) for the casting and indicate potential casting problems (*e.g.*, shrinkage). Shown in Figure 3.11–20 is a finite element mesh of a structural part to be produced in stainless steel by the investment casting process. The mesh contains ~2400 eight-node brick elements. For purposes of this preliminary analysis, the mold for the casting is not considered and the metal distribution system is also ignored. The casting model is initially assumed to be at the metal pour temperature and loses energy

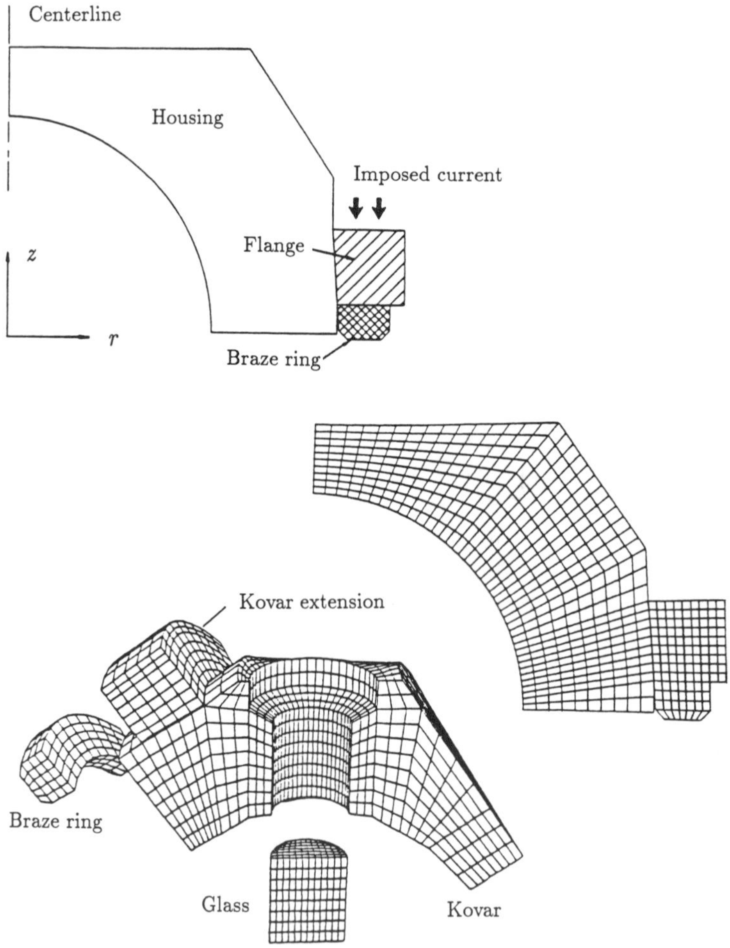

Figure 3.11–14: Schematic and finite element mesh for the brazing problem.

energy to the furnace walls through radiation; investment castings are poured in a vacuum. The number of element surfaces participating in radiative transfer is ~4500, which leads to approximately 20 million view factors. The view factors were computed using both the double area integration and the hemicube method. A further review of some of the numerical and modeling problems encountered in this type of simulation is available in [40].

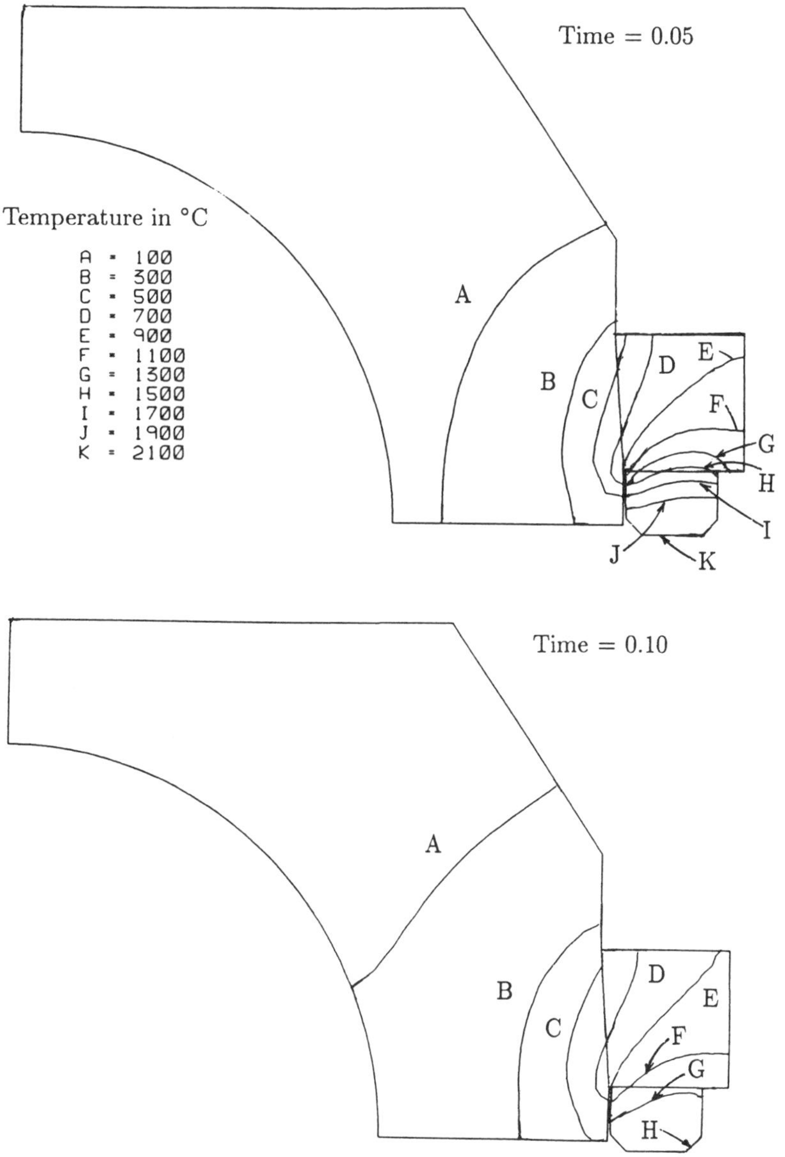

Figure 3.11–15: Isotherms at two times during the brazing analysis.

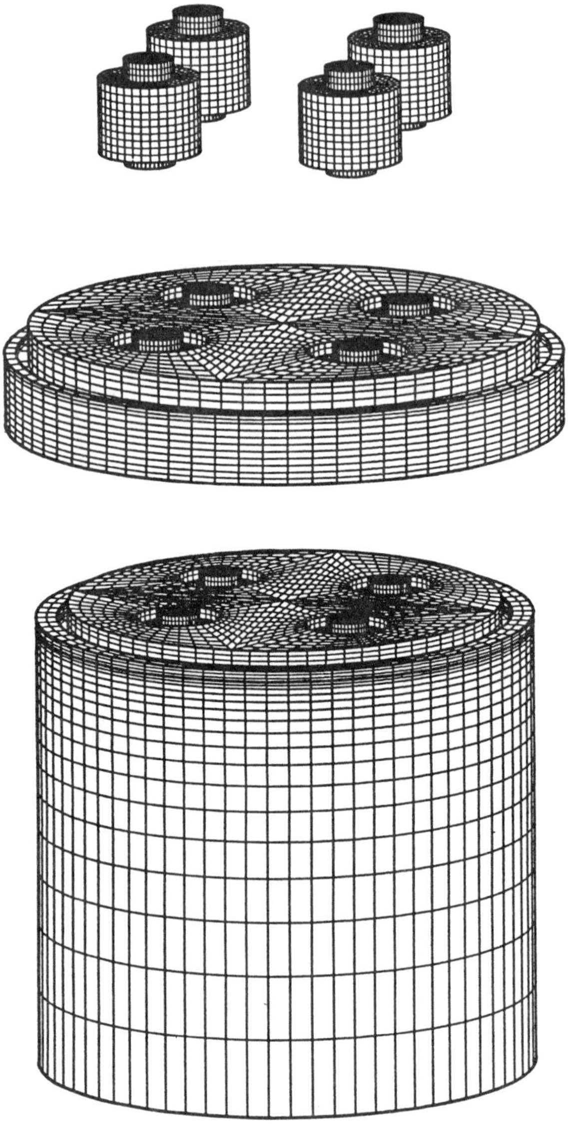

Figure 3.11–16: Finite element mesh for the battery header welding problem.

Shown in Figure 3.11–21 are surface temperature plots at two times during the cooling process. The temperatures in Figure 3.11–21 were computed using the full radiation model; a simplified radiation boundary condition that did not use view factors but employed a uniform σT^4 heat loss on the surface produced a significantly different temperature field.

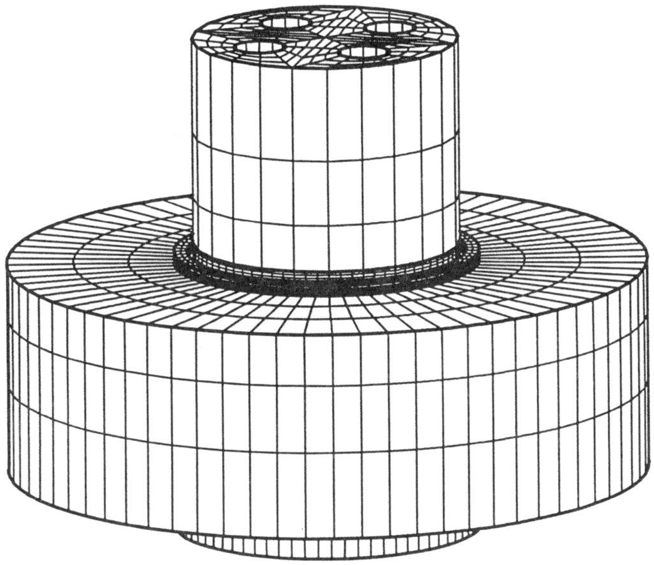

Figure 3.11–17: Finite element mesh for the heat sink used in the battery header weld.

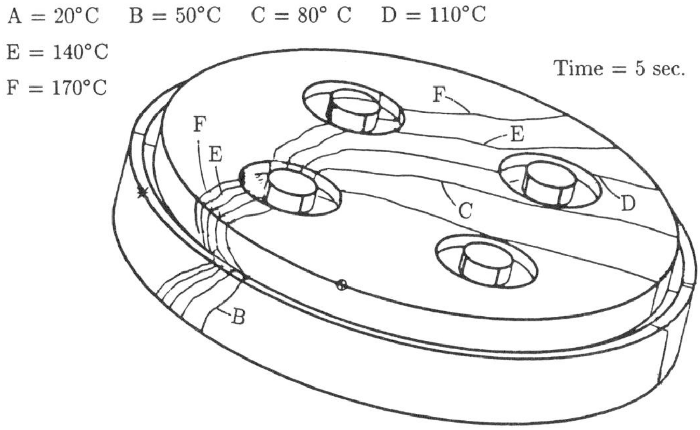

Figure 3.11–18: Temperature field in the header following weld – without heat sink.

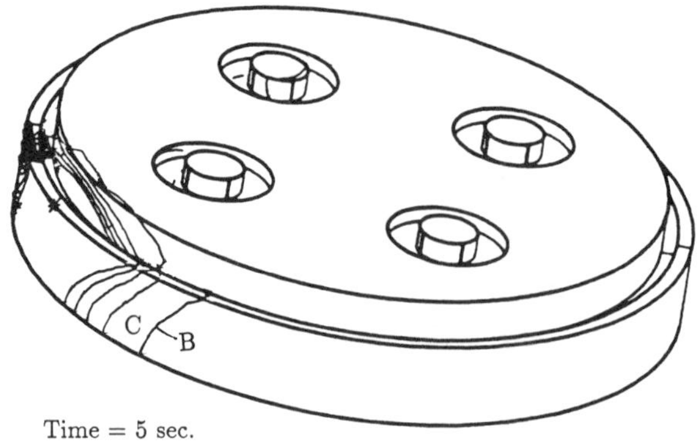

Figure 3.11–19: Temperature field in the header following weld – with heat sink.

Figure 3.11–22 contains plots showing the extent of the mushy zone (liquidus to solidus temperature interval) at two times during cooling. Such information, coupled with cooling rate data, can provide an indication of the metal microstructure and part quality.

3.11–8 Magnetic Diffusion

For two–dimensional geometries there is a direct analogy between the physical problems of thermal diffusion and the diffusion of a magnetic field. The present example considers the problem of predicting the magnetic field and thermal field in a pulsed power device subjected to a specified electric current. The thermal field is coupled to the magnetic field through the eddy currents and resistive (Joule) heating.

Shown in Figure 3.11–23 is a schematic of the upper half of the axisymmetric component with the various materials labeled. The regions labeled Coil 1,2,3 are coils that carry a specified current density **J** that is a function of time. The magnetic potential for each region is described in planar geometries by

$$\sigma\frac{\partial \mathbf{A}}{\partial t} = \frac{\partial}{\partial x_i}\left(\frac{1}{\mu}\frac{\partial \mathbf{A}}{\partial x_i}\right) + \mathbf{J} \qquad (3.11-1)$$

which is recognized as being of the same form as the heat conduction equation. In the above equation, **A** is the vector magnetic potential, σ is the conductivity, and μ is the magnetic permeability. The thermal conduction equation must also be used

$$\rho C\frac{\partial T}{\partial t} = \frac{\partial}{\partial x_i}(k\frac{\partial T}{\partial x_i}) + Q \qquad (3.11-2)$$

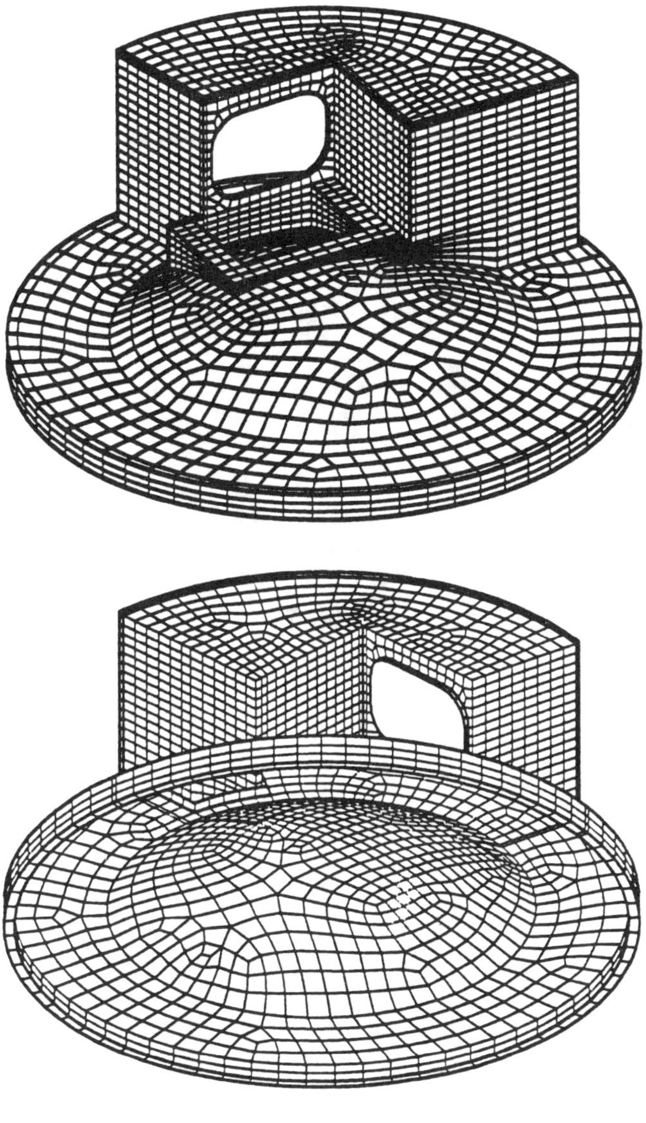

Figure 3.11–20: Finite element mesh for a cast structural part.

where $Q = \mathbf{J} \cdot \mathbf{J}\sigma$. For rapidly varying currents such as are used in the present example, thermal conduction is unimportant and is neglected (*i.e.*, the conductivity is set to zero). The above two equations are solved together along with the additional relations

$$\mathbf{B} = \nabla \times \mathbf{A} \; ; \; \mathbf{J} = \nabla \times (\frac{\mathbf{B}}{\mu}) \; ; \; \mathbf{F} = \mathbf{J} \times \mathbf{B} \qquad (3.11-3)$$

where **B** is the magnetic flux vector and **F** is the Lorentz force vector on the solid. The result of the analysis is a time–dependent temperature and force field that allows a stress analysis to be performed for the structure.

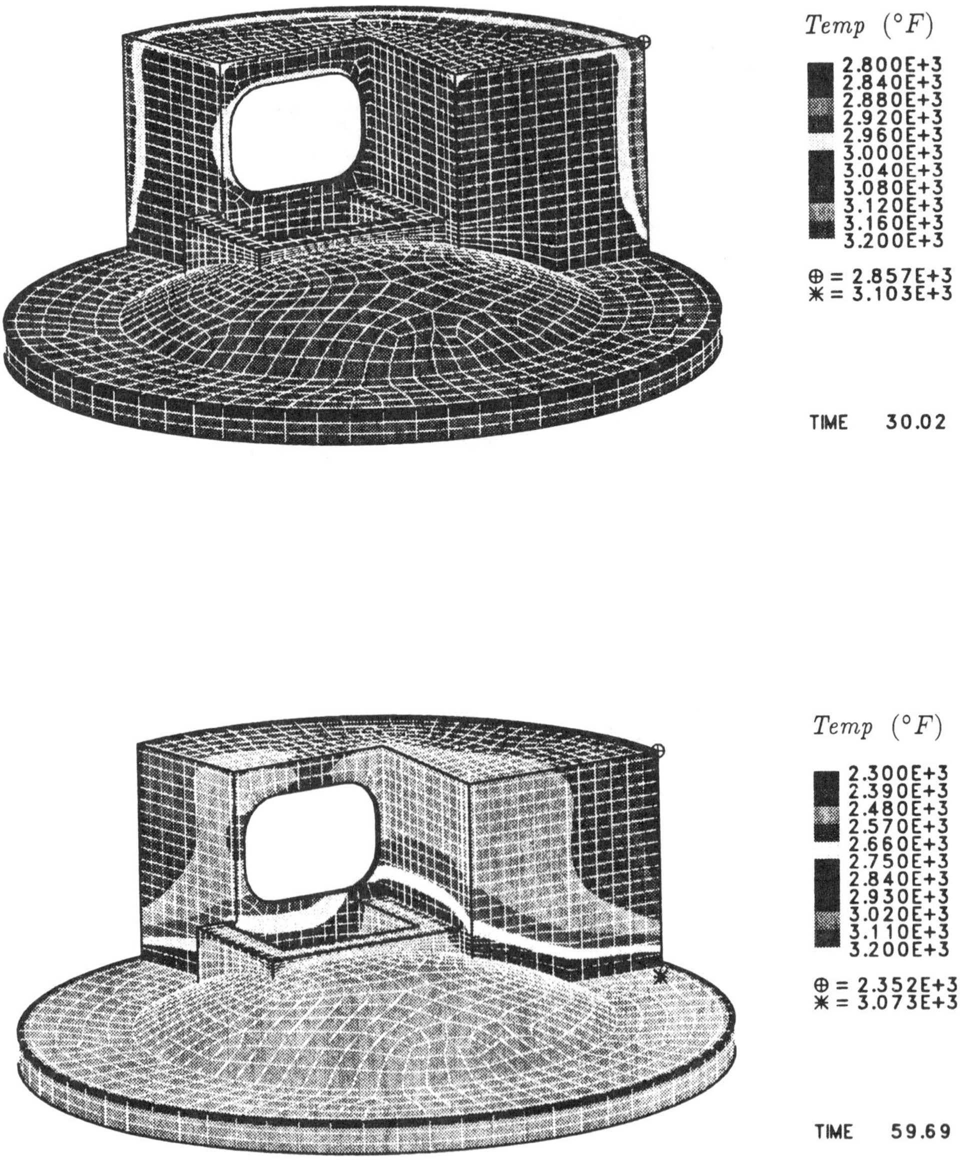

Figure 3.11–21: Temperature field during cooling of a casting using an enclosure radiation model. Temperature is in °F.

Figure 3.11–24 shows various regions of the finite element mesh used in the problem; four-node bilinear elements were used for both temperature and the magnetic potential. Figures 3.11–25 and 3.11–26 illustrate the potential field and

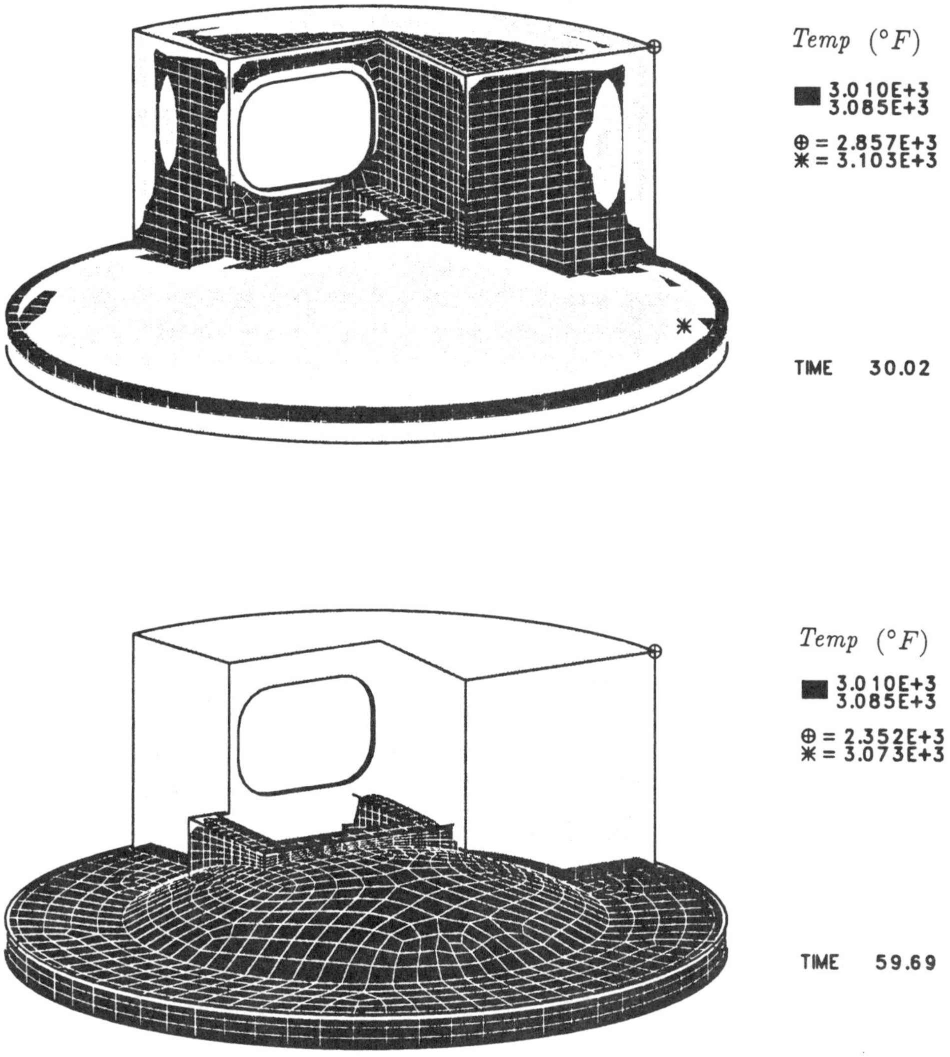

Figure 3.11–22: Liquidus to solidus temperature interval (mushy zone) during cooling of a casting. Temperature is in °F.

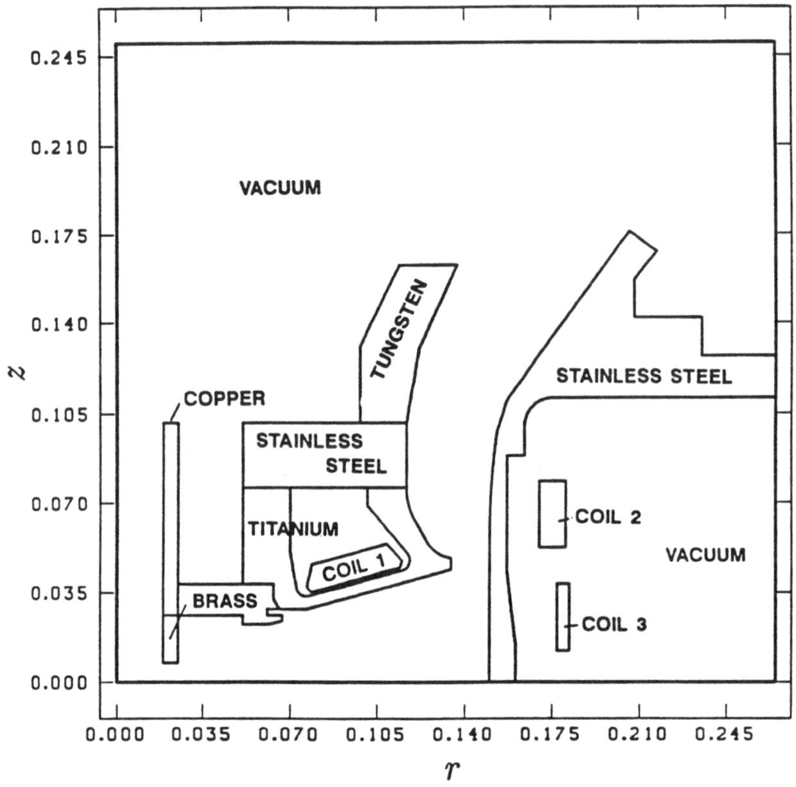

Figure 3.11–23: Schematic of axisymmetric pulsed power diode.

thermal field at two times during the current pulse. Figure 3.11–25 is for a time after two coils have been activated but before the third coil is pulsed. Figure 3.11–26 shows the fields toward the end of the current pulse when the temperature is at a maximum. This analysis was followed by a stress analysis of the diode components and the magnetic pressures and temperature gradients produced by the pulse sequence were shown to lead to unacceptable deformations in the structure.

EXERCISE PROBLEMS

3.1 Show that the weak form of equation (3.1–1) is given by equation (3.2–1). Use the following component form of the Green–Gauss theorem to integrate–by–parts,

$$\int_{\Omega^e} \frac{\partial}{\partial x_i}(wF_i)\ d\mathbf{x} = \oint_{\Gamma^e} (wF_i)n_i\ ds$$

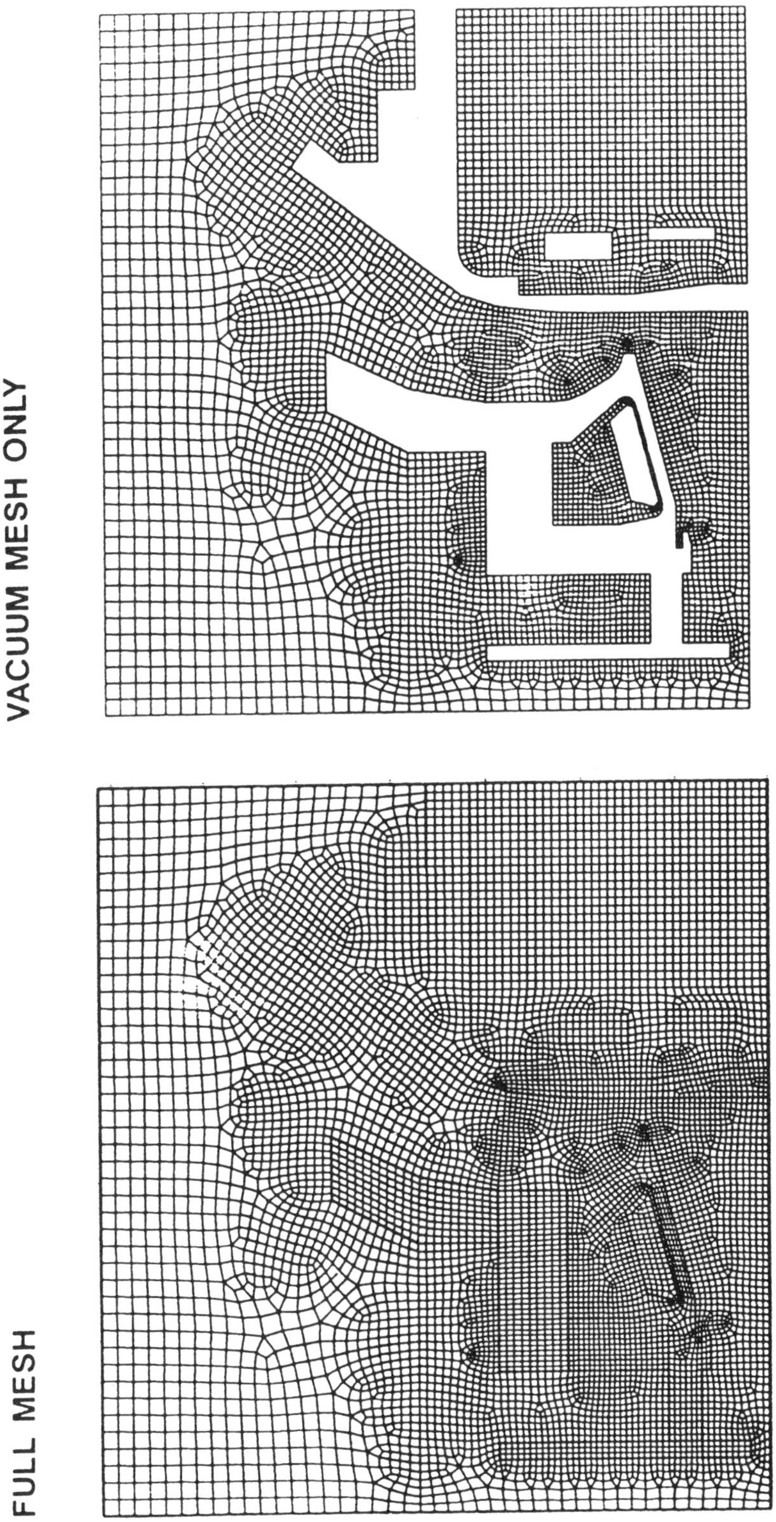

Figure 3.11–24: Finite element mesh for pulsed power diode.

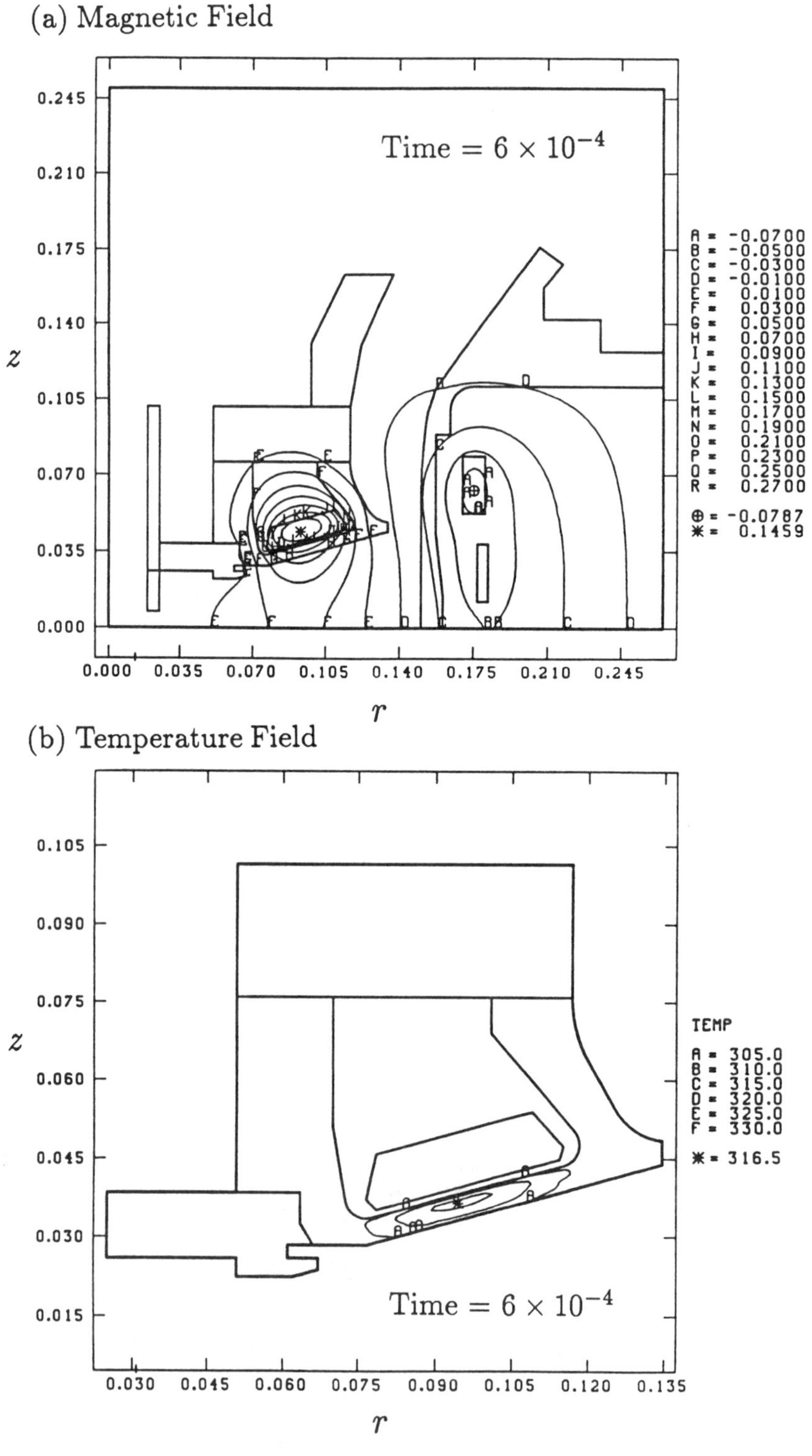

Figure 3.11–25: Magnetic and temperature fields during an early part of the current pulse.

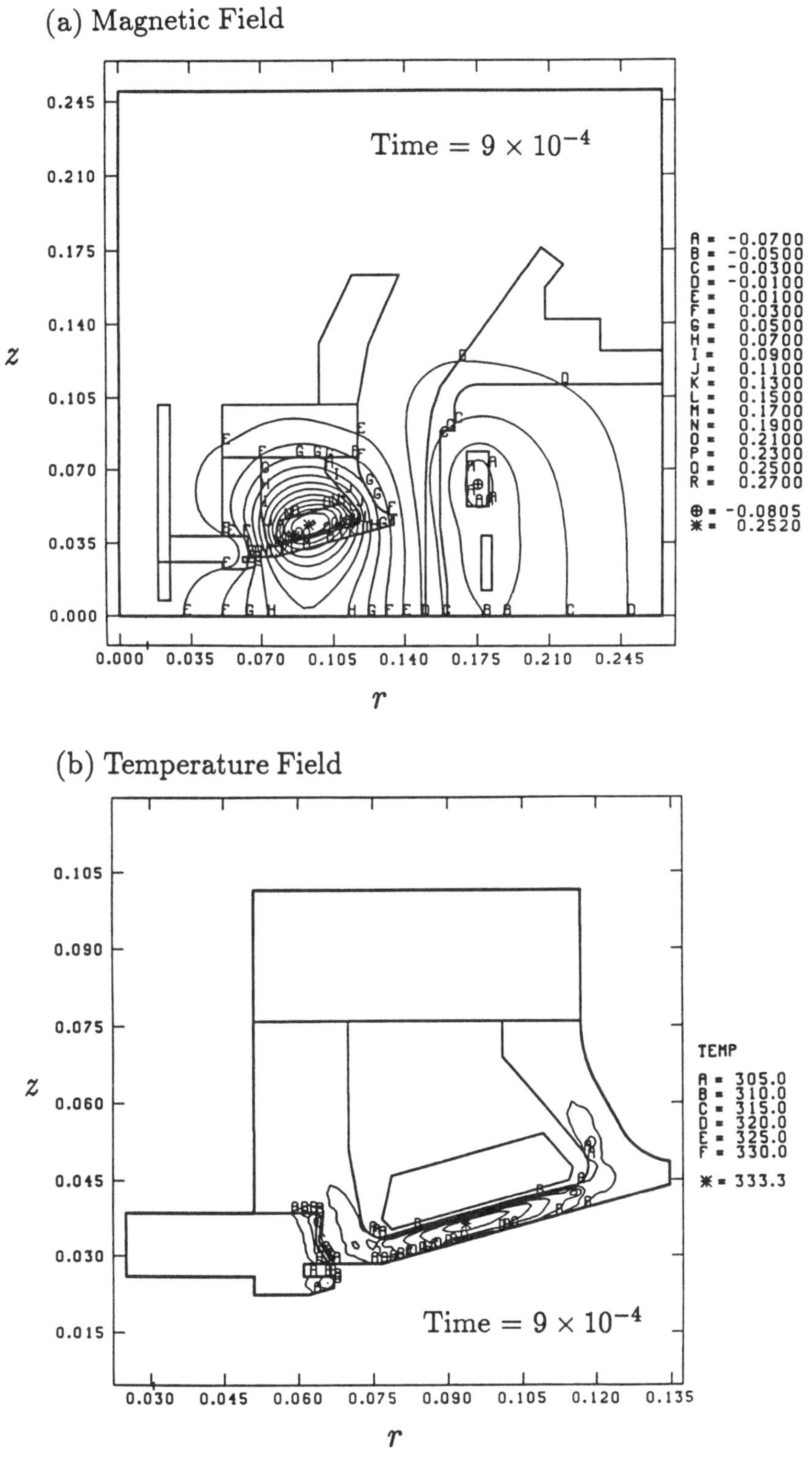

Figure 3.11–26: Magnetic and temperature fields at the end of the current pulse.

3.2 Using the $\theta-$ family of approximation,

$$\{T\}^{n+1} = \{T\}^n + \Delta t[(1-\theta)\{\dot{T}\}^n + \theta\{\dot{T}\}^{n+1}] \, , 0 \leq \theta \leq 1$$

in equation (3.2–4a), derive the fully discretized set of equations in the form

$$[\hat{K}(T^{n+1})]^{n+1}\{T\}^{n+1} = \{\hat{F}(T^n)\}$$

Define $[\hat{K}(T^{n+1})]$ and $\{\hat{F}(T^n)\}$ in terms of the original matrices, time increment Δt, and the parameter θ.

3.3 Specialize the results of Problem 3.2 to the case of $\theta = 0$ and verify equation (3.6–7).

3.4 Specialize the results of Problem 3.2 to the case of $\theta = 1$ and verify equation (3.6–9).

3.5 (*Newton's Method*) Rewrite equation (3.6–1) as

$$\mathbf{R}(\mathbf{T}) \equiv \hat{\mathbf{K}}(\mathbf{T})\mathbf{T} - \hat{\mathbf{F}}(\mathbf{T}) = \mathbf{0} \tag{1}$$

Newton's method is based on a truncated Taylor's series expansion of $\mathbf{R}(\mathbf{T})$ about the known solution $\mathbf{T}^r$, where r denotes the iteration number:

$$\mathbf{0} = \mathbf{R}(\mathbf{T}^r) + \left.\frac{\partial \mathbf{R}}{\partial \mathbf{T}}\right|_{\mathbf{T}}^{r} \delta\mathbf{T} + O(\delta\mathbf{T})^2 \tag{2}$$

Here $\delta\mathbf{T} = (\mathbf{T}^{r+1} - \mathbf{T}^r)$. Omitting the terms of order two and higher, write

$$\mathbf{R}(\mathbf{T}^r) = -\left.\frac{\partial \mathbf{R}}{\partial \mathbf{T}}\right|_{\mathbf{T}}^{r} (\mathbf{T}^{r+1} - \mathbf{T}^r) \equiv -\mathcal{J}(T^r)(\mathbf{T}^{r+1} - \mathbf{T}^r) \tag{3}$$

where $\mathcal{J}$ is the Jacobian matrix (also known as the tangent matrix),

$$\mathcal{J} = \left.\frac{\partial \mathbf{R}}{\partial \mathbf{T}}\right|_{\mathbf{T}}^{r} \tag{4}$$

Show that

$$\mathbf{T}^{r+1} = \mathbf{T}^r - \mathcal{J}^{-1}(\mathbf{T}^r)\mathbf{R}(\mathbf{T}^r) \tag{5}$$

3.6 Derive the tangent matrix for the case where $\hat{\mathbf{K}}$ is defined by equation (3,2–6a) and k_{rs} are given by

$$k_{rs} = \delta_{rs}(k_0 + k_1 T^m)$$

where k_0 and k_1 are constants, and m is an integer.

REFERENCES

1. E. B. Becker, G. F. Carey, and J. T. Oden, *Finite Elements, an Introduction*, Vol. I, Prentice–Hall, Englewood Cliffs, New Jersey (1981).

2. J. N. Reddy, *An Introduction to the Finite Element Method*, Second Edition, McGraw–Hill, New York (1993).

3. D. S. Burnett, *Finite Element Analysis*, Addison–Wesley, Reading, Massachusetts (1987).

4. O. C. Zienkiewicz and R. L. Taylor, *The Finite Element Method, Vol. 1: Linear Problems*, McGraw–Hill, New York (1989).

5. T. J. R. Hughes, *The Finite Element Method, Linear Static and Dynamic Finite Element Analysis*, Prentice–Hall, Englewood Cliffs, New Jersey (1987).

6. I. Ergatoudis, B. M. Irons, and O. C. Zienkiewicz, "Curved, Isoparametric, 'Quadrilateral', Elements for Finite Element Analysis," *International Journal of Solids and Structures*, **4**, 31–42 (1968).

7. B. M. Irons, "Engineering Applications of Numerical Integration in Stiffness Methods," *AIAA Journal*, **14**, 2035–2037 (1966).

8. B. M. Irons, "Quadrature Rules for Brick-Based Finite Elements," *International Journal for Numerical Methods in Engineering*, **3**, 293–294 (1971).

9. G. R. Cowper, "Gaussian Quadrature Formulas for Triangles," *International Journal for Numerical Methods in Engineering*, **7**, 405–408 (1973).

10. P. C. Hammer, O. P. Marlowe, and A. H. Stroud, "Numerical Integration over Simplexes and Cones," *Mathematics Tables Aids Computation*, National Research Council, Washington, D. C., **10**, 130–137 (1956).

11. A. N. Loxan, N. Davids, and A. Levenson, "Table of the Zeros of the Legendre Polynomials of Order 1–16 and the Weight Coefficients for Gauss' Mechanical Quadrature Formula," *Bulletin of the American Mathematical Society*, **48**, 739–743 (1942).

12. C. T. Reddy and D. J. Shippy, "Alternative Integration Formulae for Triangular Finite Elements, *International Journal for Numerical Methods in Engineering*, **17**, 133–139 (1981).

13. B. Carnahan, H. A. Luther, and J. O. Wilkes, *Applied Numerical Methods*, John Wiley & Sons, New York (1969).

14. B. M. Irons, "A Frontal Solution Program for Finite Element Analysis," *International Journal for Numerical Methods in Engineering*, **2**, 5–32 (1970).

15. P. Hood, "Frontal Solution Program for Unsymmetric Matrices," *International Journal for Numerical Methods in Engineering*, **10**, 379–399 (1976).

16. Y. Hasbani and M. Engelman, "Out–of–Core Solution of Linear Equations with Non–Symmetric Coefficient Matrix," *Computers and Fluids*, **7**, 13–31 (1979).

17. R. L. Taylor, E. L. Wilson, and S. J. Sackett, "Direct Solution of Equations by Frontal and Variable Band, Active Column Methods," in *Proceedings of the U. S.-European Workshop on Nonlinear Finite Element Analysis in Structural Mechanics*, Bochum, W. Germany (1980).

18. G. E. Forsythe and C. B. Meler, *Computer Solution of Linear Algebraic Systems*, Prentice–Hall, Englewood Cliffs, New Jersey (1967).

19. G. Strang, *Linear Algebra and Its Applications*, Second Edition, Academic Press, New York (1980).

20. P. Gresho, R. Lee, R. Sani, and T. Stullich, "On the Time–Dependent FEM Solution of the Incompressible Navier-Stokes Equations in Two and Three Dimensions," in *Recent Advances in Numerical Methods in Fluids, Vol 1*, Pineridge Press, Swansea (1980).

21. K. J. Bathe, *Numerical Methods in Finite Element Analysis*, Prentice–Hall, Englewood Cliffs, New Jersey (1976).

22. T. Belytschko and T. J. R. Hughes (Eds.), *Computational Methods for Transient Analysis*, North–Holland, Amsterdam (1983).

23. J. H. Argyris and O. W. Scharpf, "Finite Elements in Time and Space," *The Aeronautical Journal of the Royal Society*, **73**, 1041–1044 (1969).

24. J. Donea, "On the Accuracy of Finite Element Solutions to the Transient Heat Conduction Equation," *International Journal for Numerical Methods in Engineering*, **8**, 103–110 (1974).

25. J. M. Bettencourt, O. C. Zienkiewicz, and G. Cantin, "Consistent Use of Finite Elements in Time and the Performance of Various Recurrence Schemes for the Heat Diffusion Equation," *International Journal for Numerical Methods in Engineering*, **17**, 931–938 (1981).

26. T. J. R. Hughes, "Unconditionally Stable Algorithms for Nonlinear Heat Conduction," *Computer Methods in Applied Mechanics and Engineering*, **10**, 135–139 (1977).

27. W. L. Wood and R. W. Lewis, "A Comparison of Time Marching Schemes for the Transient Heat Conduction Equation," *International Journal for Numerical Methods in Engineering*, **9**, 679–689 (1975).

28. W. L. Wood, "Control of Crank–Nicolson Noise in the Numerical Solution of the Heat Conduction Equation," *International Journal for Numerical Methods*

in Engineering, **11**, 1059–1065 (1977).

29. C. W. Gear, *Numerical Initial Value Problems in Ordinary Differential Equations*, Prentice–Hall, Englewood Cliffs, New Jersey (1971).

30. P. D. Lax and R. D. Richtmyer, "Survey of the Stability of Linear Finite Difference Equations," *Communications in Pure and Applied Mathematics*, **9**, 267–293 (1956).

31. A. B. Shapiro, "FACET—A Radiation View Factor Computer Code for Axisymmetric, 2D Planar and 3D geometries with Shadowing," Lawrence Livermore National Laboratory, UCID-19887, Livermore, CA (1983).

32. A. F. Emery, "VIEWC User's Manual," University of Washington, Seattle, Washington (1988).

33. A. F. Emery, D. Johansson, M. Lobo, and A. Abrous, "A Comparative Study of Methods for Computing the Diffuse Radiation Viewfactors for Complex Structures," *Journal of Heat Transfer*, **113**, 413–422 (1991).

34. K. Morgan, R. W. Lewis, and O. C. Zienkiewicz, "An Improved Algorithm for Heat Conduction Problems with Phase Change," *International Journal for Numerical Methods in Engineering*, **12**, 1191–1195 (1978).

35. W. Randolph and K. J. Bathe, "An Efficient Algorithm for Analysis of Nonlinear Heat Transfer with Phase Changes," *International Journal for Numerical Methods in Engineering*, **18**, 119–134 (1982).

36. D. K. Gartling, "COYOTE — A Finite Element Computer Program for Nonlinear Heat Conduction Problems," Sandia National Laboratories Report, SAND77-1332 (Revised), Albuquerque, New Mexico (1982).

37. H. S. Carslaw and J. C. Jaeger, *Conduction of Heat in Solids*, Oxford University Press, Oxford (1959).

38. D. A. Glowka and C. M. Stone,"Thermal Response of Polycrystalline Diamond Compact Cutters Under Simulated Downhole Conditions," *Society of Petroleum Engineering Journal*, 143–156 (1985).

39. G. A. Knorovsky and S. N. Burchett, "The Effect of Heat Sinks in GTA Microwelding," in *Proceedings of the Second International Conference on Trends in Welding*, American Welding Society, Gatlinburg, Tennessee (1989).

40. R. E. Hogan, M. W. Glass, P. R. Schunk, and D. K. Gartling, "Thermal Analysis of the Cooling and Solidification of Investment Castings," in *Proceedings of the First International Conference on Transport Phenomena in Processing*, Pacific Institute for Thermal Engineering, Honolulu, Hawaii (1992).

Chapter Four

VISCOUS INCOMPRESSIBLE FLOWS

4.1 INTRODUCTION

4.1–1 Background

The field of fluid mechanics is concerned with the motion of fluids and their effect on the surroundings. A fluid state of matter is often characterized by the relative mobility of the molecules that constitute the matter. Fluids exist either in the form of a gas or a liquid; more complex materials, such as mixtures, may also have a fluid behavior. In this chapter, finite element models are developed based on the weak formulation of the equations of viscous, incompressible fluids under isothermal conditions.

The motion of a fluid is governed by the global laws of conservation of mass, momenta, and energy (see Chapter 1). These equations consist of a set of coupled, nonlinear, partial differential equations in terms of the velocity components, temperature, and pressure. When temperature effects are not important, the energy equation is uncoupled from the momentum (*i.e.*, the Navier–Stokes) equations. Thus for isothermal flows, we need to solve only the Navier–Stokes equations and continuity equation. When the Reynolds number for the flow is very low, the nonlinear terms due to inertial effects can be neglected, resulting in a linear boundary value problem. Such a flow is termed *Stokes flow.*

4.1–2 Governing Equations

The laws governing the flow of Newtonian fluids were reviewed in Chapter 1. The equations are specialized here for viscous fluids that are subject to the assumption of incompressibility. Both incompressible fluids and incompressible flows are distinguished by changes in the fluid density being negligibly small compared to the base density. We will use the phrases incompressible fluid and incompressible flow interchangeably, while recognizing the subtle differences in meaning. The equations are written here for a fluid region Ω both in vector as well as Cartesian component form $\mathbf{x} = (x_1, x_2, x_3)$ in an Eulerian reference frame; time is denoted by t.

Conservation of Mass:

$$\nabla \cdot \mathbf{u} = 0 \tag{4.1-1a}$$

$$\frac{\partial u_i}{\partial x_i} = 0 \tag{4.1-1b}$$

Conservation of Momentum:

$$\rho_0 \left(\frac{\partial \mathbf{u}}{\partial t} + \mathbf{u} \cdot \nabla \mathbf{u} \right) - \nabla \cdot \sigma + \rho_0 \mathbf{f} = 0 \tag{4.1 – 2a}$$

$$\rho_0 \left(\frac{\partial u_i}{\partial t} + u_j \frac{\partial u_i}{\partial x_j} \right) - \frac{\partial \sigma_{ij}}{\partial x_j} + \rho_0 f_i = 0 \tag{4.1 – 2b}$$

Constitutive Equations:

$$\sigma = \tau - P\mathbf{I} \; ; \quad \tau = 2\mu \mathbf{D}$$

$$\mathbf{D} = \frac{1}{2} \left[(\nabla \mathbf{u}) + (\nabla \mathbf{u})^T \right] \tag{4.1 – 3a}$$

$$\sigma_{ij} = \tau_{ij} - P\delta_{ij} \; ; \quad \tau_{ij} = \mu D_{ij}$$

$$D_{ij} = \frac{1}{2} \left(\frac{\partial u_i}{\partial x_j} + \frac{\partial u_j}{\partial x_i} \right) \tag{4.1 – 3b}$$

where $\mathbf{u}$ denotes the velocity vector, σ is the total stress tensor, τ is the viscous stress tensor, P is the pressure, $\mathbf{f}$ is the body force vector (per unit mass), ρ_0 is the density, and μ is the shear viscosity of the fluid. The Cartesian components of the dependent variables are obvious from the above equations. In writing the Cartesian component form of the equations, the index notation with summation on repeated indices is used. The operator D/Dt denotes the material (or convective) time derivative,

$$\frac{D}{Dt} \equiv \left[\frac{\partial}{\partial t} + \mathbf{u} \cdot \nabla \right] \tag{4.1 – 4a}$$

$$\frac{D}{Dt} \equiv \left[\frac{\partial}{\partial t} + u_j \frac{\partial}{\partial x_j} \right] \tag{4.1 – 4b}$$

Equation (1) is also known as the *continuity equation*, incompressibility constraint, or divergence–free condition on the velocity field. Equations (2), with the constitutive relation indicated in (3), are known as the *Navier–Stokes equations*.

The boundary conditions are given by

$$\mathbf{u} = \hat{\mathbf{u}} \quad \text{or} \quad u_i = \hat{u}_i(s,t) \qquad \text{on } \Gamma_u \tag{4.1 – 5a}$$

$$\mathcal{T} \equiv \hat{\mathbf{n}} \cdot \sigma = \hat{\mathcal{T}} \quad \text{or} \quad \mathcal{T}_i \equiv \sigma_{ij}(s,t) n_j(s) = \hat{\mathcal{T}}_i(s,t) \qquad \text{on } \Gamma_{\mathcal{T}} \tag{4.1 – 5b}$$

where $\hat{\mathbf{n}}$ is the unit normal to the boundary and Γ_u and Γ_{calT} are defined in Section 1.7.

The vector form of the equations in (1–3) can be written in any coordinate system that is suitable for the description and solution of the problem under consideration (see Section 1.4). Using the constitutive equations and deformation rate–velocity relations (3), equations (1) and (2) can be expressed solely in terms of the velocity components and pressure. These are summarized below.

$$\frac{\partial u_i}{\partial x_i} = 0 \tag{4.1-6}$$

$$\rho_0\left(\frac{\partial u_i}{\partial t} + u_j\frac{\partial u_i}{\partial x_j}\right) - \frac{\partial}{\partial x_j}\left[-P\delta_{ij} + \mu\left(\frac{\partial u_i}{\partial x_j} + \frac{\partial u_j}{\partial x_i}\right)\right] + \rho_0 f_i = 0 \tag{4.1-7}$$

It is possible to express the equations in terms of the stream function and vorticity functions. For example, in the two–dimensional case, we could write equations (6) and (7) in terms of the *stream function* ψ and *vorticity* ω which are defined in a two–dimensional case by the relations:

$$u_2 \equiv -\frac{\partial \psi}{\partial x_1}, \quad u_1 \equiv \frac{\partial \psi}{\partial x_2}, \quad \omega \equiv \frac{\partial u_1}{\partial x_2} - \frac{\partial u_2}{\partial x_1} = \nabla^2 \psi \tag{4.1-8}$$

Here we chose to use the standard notation for the stream function and vorticity. The symbol ψ should not be confused with ψ_i used for interpolation functions. Of course, it would be clear in the context of the discussion.

Note that the stream function is defined in such a way that the continuity equation (6) is identically satisfied, and the vorticity satisfies the momentum equations (7). The introduction of these variables in the two–dimensional case reduces the number of equations from three to two. For the three–dimensional case, the stream function and vorticity are vector functions with three components each, and thus there is no reduction in the number of variables. Since these alternative formulations are not as computationally convenient as the original equations, especially for the imposition of boundary conditions, we will confine our attention here to the development of finite element models for equations (6) and (7).

In the present study we shall consider two different finite element models of equations (6) and (7). The first one is a natural formulation in which the weak forms of equations (6) and (7) are used to construct the finite element model. The resulting finite element model is termed the *velocity–pressure model* or *mixed model.* The phrase 'mixed' is used because velocity variables are mixed with the force–like variable, pressure, and both types of variables are retained in a single formulation. The second model is based on the interpretation that the continuity equation (6) is an additional relation among the velocity components (*i.e.*, a constraint among the u_i), and the constraint is satisfied in a least–squares (*i.e.*, approximate) sense. This particular method of including the constraint in the formulation is known as the *penalty function method*, and the model is termed the *penalty–finite element model* (see [1–4]). In this case, the pressure variable is effectively eliminated from the formulation. It is informative to note that the velocity–pressure (or mixed) formulation is the same as the Lagrange multiplier formulation, wherein the constraint is included by means of the Lagrange multiplier method. The Lagrange multiplier turns out to be the negative of the pressure.

4.2 MIXED FINITE ELEMENT MODEL

4.2–1 Weak Form

The starting point for the development of the finite element models of equations (4.1–6) and (4.1–7) is their weak forms. As discussed in Section 2.4, the weak forms are developed using a three–step procedure. These steps are briefly reviewed here. For convenience, we denote expressions on the left side of equations (4.1–6) and (4.1–7) by f_1 and $\mathbf{f}_2$, respectively. The weighted–integral statements of the two equations over a typical element Ω^e are given by

$$\int_{\Omega^e} Q f_1 \, d\mathbf{x} = 0 \tag{4.2 – 1}$$

$$\int_{\Omega^e} \mathbf{w} \cdot \mathbf{f}_2 \, d\mathbf{x} = 0 \tag{4.2 – 2}$$

where $(Q, \mathbf{w})$ are weight functions, which will be equated, in the Rayleigh–Ritz–Galerkin finite element models, to the interpolation functions used for $(P, \mathbf{u})$, respectively (see Reddy [1] for details).

To obtain the weak form of equation (4.1–7), we use integration–by–parts to equally distribute integration between the dependent variables and the weight functions. However, in any problem, such trading of differentiability is subjected to the restriction that the resulting boundary expressions are physically meaningful. Otherwise, the secondary variables of the formulation may not be the quantities the physical problem admits as the boundary conditions. An examination of the boundary stress components $\mathcal{T}_i$ in equation (4.1–5) shows that the pressure term occurs as a part of the expression [see equation (4.1–3)]. For example, $\mathcal{T}_1$ is the x_1-component of the total boundary stress, which is the sum of the viscous boundary stress and the hydrostatic boundary stress,

$$\mathcal{T}_1 = \mu\left[2\frac{\partial u_1}{\partial x_1}n_1 + (\frac{\partial u_1}{\partial x_2} + \frac{\partial u_2}{\partial x_1})n_2 + (\frac{\partial u_1}{\partial x_3} + \frac{\partial u_3}{\partial x_1})n_3\right] + (-Pn_1) \tag{4.2 – 3}$$

Here (n_1, n_2, n_3) denote the components of the unit normal vector on the boundary. It is clear from the above expression that the entire expression in the square brackets of equation (4.1–7) must be integrated by parts in the momentum equations to obtain $\mathcal{T}_i$ in the boundary expression. No integration–by–parts is used in the continuity equation (4.1–6) because no relaxation of differentiability on u_i can be accomplished; further, the resulting boundary conditions will not be physically justifiable.

Keeping the above comments in mind, we carry out the remaining two steps of the weak formulation, and obtain the following integral statements:

$$0 = \int_{\Omega^e} Q(\frac{\partial u_i}{\partial x_i})d\mathbf{x} \tag{4.2 – 4}$$

$$0 = \int_{\Omega^e} \left[\rho_0 \left(w_i \frac{\partial u_i}{\partial t} + w_i u_j \frac{\partial u_i}{\partial x_j} \right) + \frac{\partial w_i}{\partial x_j} \left(-P\delta_{ij} + \mu (\frac{\partial u_i}{\partial x_j} + \frac{\partial u_j}{\partial x_i}) \right) - \rho_0 w_i f_i \right] d\mathbf{x}$$
$$- \oint_{\Gamma^e} w_i T_i \, ds \qquad (4.2-5)$$

This completes the weak form development.

4.2–2 Finite Element Model

Since we are developing the Rayleigh–Ritz–Galerkin finite element models, the choice of the weight functions is restricted to the spaces of approximation functions used for the pressure and velocity fields. Suppose that the dependent variables (u_i, P) are approximated by expansions of the form

$$u_i(\mathbf{x}, t) = \sum_{m=1}^{M} \psi_m(\mathbf{x}) u_i^m(t) = \mathbf{\Psi}^T \mathbf{u}_i \qquad (4.2-6a)$$

$$P(\mathbf{x}, t) = \sum_{l=1}^{L} \phi_l(\mathbf{x}) P_l(t) = \mathbf{\Phi}^T \mathbf{P} \qquad (4.2-6b)$$

where $\mathbf{\Psi}$ and $\mathbf{\Phi}$ are (column) vectors of interpolation (or shape) functions, and $\mathbf{u}_i$ and $\mathbf{P}$ are vectors of nodal values of velocity components and pressure, respectively. The weight functions $(Q, \mathbf{w})$ have the following correspondence (see [1] for further details),

$$Q \approx \phi_l, \;\; \mathbf{w} \approx \psi_m \qquad (4.2-7)$$

Substitution of equations (6) and (7) into equations (4) and (5) results in the following finite element equations:

Continuity:

$$-\left[\int_{\Omega^e} \mathbf{\Phi} \frac{\partial \mathbf{\Psi}^T}{\partial x_i} d\mathbf{x} \right] \mathbf{u}_i = 0 \qquad (4.2-8)$$

i-th Momentum:

$$\left[\int_{\Omega^e} \rho_0 \mathbf{\Psi} \mathbf{\Psi}^T d\mathbf{x} \right] \dot{\mathbf{u}}_i + \left[\int_{\Omega^e} \rho_0 \mathbf{\Psi} (\mathbf{\Psi}^T \mathbf{u}_j) \frac{\partial \mathbf{\Psi}^T}{\partial x_j} d\mathbf{x} \right] \mathbf{u}_i + \left[\int_{\Omega^e} \mu \frac{\partial \mathbf{\Psi}}{\partial x_j} \frac{\partial \mathbf{\Psi}^T}{\partial x_j} d\mathbf{x} \right] \mathbf{u}_i$$
$$+ \left[\int_{\Omega^e} \mu \frac{\partial \mathbf{\Psi}}{\partial x_j} \frac{\partial \mathbf{\Psi}^T}{\partial x_i} d\mathbf{x} \right] \mathbf{u}_j - \left[\int_{\Omega^e} \frac{\partial \mathbf{\Psi}}{\partial x_i} \mathbf{\Phi}^T d\mathbf{x} \right] \mathbf{P}$$

$$= \left[\int_{\Omega^e} \rho_0 \Psi f_i dx\right] + \left\{\oint_{\Gamma^e} T_i \Psi \, ds\right\} \qquad (4.2-9)$$

where superscript $(\cdot)^T$ denotes a transpose of the enclosed vector or matrix. The above equations can be written symbolically in matrix form as

Continuity:

$$-\mathbf{Q}^{\mathbf{T}}\mathbf{u} = 0 \qquad (4.2-10)$$

Momentum:

$$\mathbf{M}\dot{\mathbf{u}} + \mathbf{C}(\mathbf{u})\mathbf{u} + \mathbf{K}\mathbf{u} - \mathbf{Q}\mathbf{P} = \mathbf{F} \qquad (4.2-11)$$

where the superposed dot represents a time derivative and $\mathbf{u} = \{\mathbf{u}_1, \mathbf{u}_2, \mathbf{u}_3\}^T$. For the two–dimensional case, equations (10) and (11) have the following explicit matrix form [the continuity equation (10) and momentum equations (11) are combined into one]:

$$\begin{bmatrix} \mathbf{M} & 0 & 0 \\ 0 & \mathbf{M} & 0 \\ 0 & 0 & 0 \end{bmatrix} \begin{Bmatrix} \dot{\mathbf{u}}_1 \\ \dot{\mathbf{u}}_2 \\ \dot{\mathbf{P}} \end{Bmatrix} + \begin{bmatrix} \mathbf{C}(\mathbf{u}) & 0 & 0 \\ 0 & \mathbf{C}(\mathbf{u}) & 0 \\ 0 & 0 & 0 \end{bmatrix} \begin{Bmatrix} \mathbf{u}_1 \\ \mathbf{u}_2 \\ \mathbf{P} \end{Bmatrix} +$$

$$\begin{bmatrix} 2\mathbf{K}_{11} + \mathbf{K}_{22} & \mathbf{K}_{21} & -\mathbf{Q}_1 \\ \mathbf{K}_{12} & \mathbf{K}_{11} + 2\mathbf{K}_{22} & -\mathbf{Q}_2 \\ -\mathbf{Q}_1^T & -\mathbf{Q}_2^T & 0 \end{bmatrix} \begin{Bmatrix} \mathbf{u}_1 \\ \mathbf{u}_2 \\ \mathbf{P} \end{Bmatrix} = \begin{Bmatrix} \mathbf{F}_1 \\ \mathbf{F}_2 \\ 0 \end{Bmatrix} \qquad (4.2-12)$$

The coefficient matrices shown in equations (12) are defined by

$$\mathbf{M} = \int_{\Omega^e} \rho_0 \Psi \Psi^T \, d\mathbf{x} \; ; \quad \mathbf{C}(\mathbf{u}) = \int_{\Omega^e} \rho_0 \Psi (\Psi^T \mathbf{u}_j) \frac{\partial \Psi^T}{\partial x_j} \, d\mathbf{x}$$

$$\mathbf{K}_{ij} = \int_{\Omega^e} \mu \frac{\partial \Psi}{\partial x_i} \frac{\partial \Psi^T}{\partial x_j} \, d\mathbf{x} \; ; \quad \mathbf{Q}_i = \int_{\Omega^e} \frac{\partial \Psi}{\partial x_i} \Phi^T \, d\mathbf{x}$$

$$\mathbf{F}_i = \int_{\Omega^e} \rho_0 \Psi f_i \, d\mathbf{x} + \oint_{\Gamma^e} \Psi T_i \, ds \qquad (4.2-13)$$

where a sum on repeated indices is implied in the definition of $\mathbf{C}$.

The two sets of interpolation functions used in equation (6) should be of the Lagrange type, *i.e.*, derived by interpolating only the values of the functions—and not their derivatives. There are two different finite elements associated with the two field variables (u_i, P), and hence there are two different finite element meshes corresponding to the two variables over the same domain, Ω. If one of the meshes contains the other mesh as a subset, then we choose to display the first mesh and indicate the nodal degrees of freedom associated with the nodes of a typical element of the mesh.

The interpolation used for the pressure variable should be different from that used for the velocities, because the weak form in equations (4) and (5) contain the first derivatives of the velocities u_i and no derivatives of the pressure P. In addition, the essential boundary conditions of the formulation do not include specification of the pressure; it enters the boundary conditions as a part of the natural boundary conditions. This implies that the pressure variable need not be carried as a variable that is continuous across interelement boundaries. These observations lead to the conclusion that the pressure variable should be interpolated with functions that are one order less than those used for the velocity field and that the approximation may be discontinuous (*i.e.*, not continuous from one element to other). Thus, quadratic interpolation of u_i and discontinuous linear interpolation of P are admissible. Models that use equal interpolation of the velocities and pressure with this formulation are known to give inaccurate results (see [5–15]). Further comments on appropriate interpolation functions can be found in Section 4.5–2.

Returning to equations (10) and (11), they can be combined into one,

$$\begin{bmatrix} \mathbf{M} & 0 \\ 0 & 0 \end{bmatrix} \begin{Bmatrix} \dot{\mathbf{u}} \\ \dot{\mathbf{P}} \end{Bmatrix} + \begin{bmatrix} \mathbf{C}(\mathbf{u}) + \mathbf{K}(\mathbf{u}) & -\mathbf{Q} \\ -\mathbf{Q}^{\mathrm{T}} & 0 \end{bmatrix} \begin{Bmatrix} \mathbf{u} \\ \mathbf{P} \end{Bmatrix} = \begin{Bmatrix} \mathbf{F} \\ 0 \end{Bmatrix} \tag{4.2-14}$$

or in a more symbolic format as

$$\overline{\mathbf{M}}\dot{\mathbf{U}} + \overline{\mathbf{K}}\mathbf{U} = \overline{\mathbf{F}} \tag{4.2-15a}$$

where

$$\mathbf{U} = \{\mathbf{u}_1, \mathbf{u}_2, \mathbf{u}_3, \mathbf{P}\}^T \tag{4.2-15b}$$

Note that the general form of equation (15) is the same as the nonlinear diffusion equation (3.5–18). Therefore, the time–approximations discussed in Section 3.6–3 are readily applicable to the ordinary differential equations in (14).

4.3 PENALTY FINITE ELEMENT MODELS

4.3–1 Introduction

The penalty function method, like the Lagrange multiplier method, allows us to reformulate a problem with constraints as one without constraints (see [1–4, 12–21]). The problem at hand, namely, the flow of a viscous incompressible fluid, must first be restated as a variational problem subjected to a constraint. For the purpose of describing the penalty function method, we consider the steady Stokes flow problem (*i.e.*, without time–dependent and nonlinear terms) in two dimensions:

Conservation of mass:

$$\frac{\partial u_1}{\partial x_1} + \frac{\partial u_2}{\partial x_2} = 0 \tag{4.3-1}$$

Conservation of momentum:

$$-\frac{\partial}{\partial x_1}(2\mu\frac{\partial u_1}{\partial x_1}) - \frac{\partial}{\partial x_2}[\mu(\frac{\partial u_1}{\partial x_2} + \frac{\partial u_2}{\partial x_1})] + \frac{\partial P}{\partial x_1} - \rho_0 f_1 = 0 \qquad (4.3-2a)$$

$$-\frac{\partial}{\partial x_1}[\mu(\frac{\partial u_1}{\partial x_2} + \frac{\partial u_2}{\partial x_1})] - \frac{\partial}{\partial x_2}(2\mu\frac{\partial u_2}{\partial x_2}) + \frac{\partial P}{\partial x_2} - \rho_0 f_2 = 0 \qquad (4.3-2b)$$

The weak form of these equations, also known as the *variational problem*, is given by (see Reddy [1,2]):

$$B((w_1, w_2, Q), (u_1, u_2, P)) = \ell(w_1, w_2, Q) \qquad (4.3-3)$$

where (w_1, w_2, Q) are the weight functions used for the momentum and continuity equations, respectively, $B(\cdot,\cdot)$ is a *bilinear form* [*i.e.*, an expression that is linear in (w_1, w_2, Q) and (u_1, u_2, P)] and $\ell(\cdot)$ is a *linear form*,

$$B(\cdot,\cdot) = \mu\int_{\Omega^e}\left[2\left(\frac{\partial w_1}{\partial x_1}\frac{\partial u_1}{\partial x_1} + \frac{\partial w_2}{\partial x_2}\frac{\partial u_2}{\partial x_2}\right) + \left(\frac{\partial w_1}{\partial x_2} + \frac{\partial w_2}{\partial x_1}\right)\left(\frac{\partial u_1}{\partial x_2} + \frac{\partial u_2}{\partial x_1}\right)\right] d\mathbf{x}$$

$$-\int_{\Omega^e}\left[(\frac{\partial w_1}{\partial x_1} + \frac{\partial w_2}{\partial x_2})P + (\frac{\partial u_1}{\partial x_1} + \frac{\partial u_2}{\partial x_2})Q\right] d\mathbf{x} \qquad (4.3-4a)$$

$$\ell(\cdot) = \int_{\Omega^e}\rho_0(f_1 w_1 + f_2 w_2)\, d\mathbf{x} + \oint_{\Gamma^e}(\mathcal{T}_1 w_1 + \mathcal{T}_2 w_2)\, ds \qquad (4.3-4b)$$

and $(\mathcal{T}_1, \mathcal{T}_2)$ are the boundary stress components,

$$\mathcal{T}_1 = (2\mu\frac{\partial u_1}{\partial x_1} - P)n_1 + \mu(\frac{\partial u_1}{\partial x_2} + \frac{\partial u_2}{\partial x_1})n_2$$

$$\mathcal{T}_2 = \mu(\frac{\partial u_1}{\partial x_2} + \frac{\partial u_2}{\partial x_1})n_1 + (2\mu\frac{\partial u_2}{\partial x_2} - P)n_2 \qquad (4.3-5)$$

The finite element model based on the weak form (3) is a special case of the mixed finite element model given in equation (4.2–14). Equation (4.2–14) is more general than the problem at hand in that equation (4.2–14) is valid for time–dependent Navier–Stokes equations. In the interest of completeness, the penalty function method is described first using the steady Stokes problem. The results will then be generalized to the time–dependent Navier–Stokes equations. We describe two different penalty finite element models of flows of viscous, incompressible fluids, namely, the reduced integration penalty model and the consistent penalty model.

4.3–2 Penalty Function Method

Suppose that the velocity field (u_1, u_2) is constrained to satisfy the continuity equation (1) identically. Then the weight functions (w_1, w_2) (being virtual variations of the velocity components) also satisfy the constraint condition (continuity equation)

$$\frac{\partial w_1}{\partial x_1} + \frac{\partial w_2}{\partial x_2} = 0 \tag{4.3-6}$$

As a result, the second integral expression in the bilinear form (4a) drops out, and the pressure (and hence the weight function Q) does not appear explicitly in the variational problem (3). The resulting variational problem now can be stated as follows: among all vectors $\mathbf{u} = (u_1, u_2)$ which satisfy the continuity equation (1), find the one that satisfies the variational problem

$$B_0((w_1, w_2), (u_1, u_2)) = \ell_0(w_1, w_2) \tag{4.3-7}$$

for all admissible weight functions $\mathbf{w} = (w_1, w_2)$[*i.e.*, those which satisfy the condition in equation (6)]. The bilinear and linear forms are defined by

$$B_0(\mathbf{w}, \mathbf{u}) = \mu \int_{\Omega^e} \left[2 \left(\frac{\partial w_1}{\partial x_1}\frac{\partial u_1}{\partial x_1} + \frac{\partial w_2}{\partial x_2}\frac{\partial u_2}{\partial x_2} \right) + \left(\frac{\partial w_1}{\partial x_2} + \frac{\partial w_2}{\partial x_1} \right) \left(\frac{\partial u_1}{\partial x_2} + \frac{\partial u_2}{\partial x_1} \right) \right] d\mathbf{x} \tag{4.3-8a}$$

$$\ell_0(\mathbf{w}) = \int_{\Omega^e} \rho_0 (f_1 w_1 + f_2 w_2)\, d\mathbf{x} + \oint_{\Gamma^e} (T_1 w_1 + T_2 w_2)\, ds \tag{4.3-8b}$$

Equation (7) is called a constrained variational problem, because the solution vector $\mathbf{u}$ is constrained to satisfy the continuity equation. We note that the bilinear form in equation (8a) is symmetric: $B_0(\mathbf{w}, \mathbf{u}) = B_0(\mathbf{u}, \mathbf{w})$,

$$B_0((w_1, w_2), (u_1, u_2)) = B_0((u_1, u_2), (w_1, w_2)) \tag{4.3-9}$$

Whenever the bilinear form of a variational problem is symmetric in its arguments, it is possible to construct a quadratic functional such that the minimum of the quadratic functional is equivalent to the variational problem (see Reddy [1,2]). The quadratic functional for the problem at hand is given by the expression

$$I_0(\mathbf{u}) = \frac{1}{2} B_0(\mathbf{u}, \mathbf{u}) - \ell_0(\mathbf{u}) \tag{4.3-10}$$

where the velocity vector $\mathbf{u}$ satisfies the continuity equation (1). Now we can state that equations (1) and (2) governing the steady flow of viscous incompressible fluids are equivalent to minimizing the quadratic functional $I_0(\mathbf{u})$ subjected to the constraint

$$G(\mathbf{u}) \equiv \frac{\partial u_1}{\partial x_1} + \frac{\partial u_2}{\partial x_2} = 0 \qquad (4.3-11)$$

At this point it should be remembered that the discussion presented in this section thus far is to reformulate the problem as one of a constrained problem so that the penalty function method can be used. As can be seen from the discussion, the advantage of the formulation is that the pressure variable does not appear in the model. However, there are certain computational complications associated with the penalty finite element model. These will be discussed shortly.

In the penalty function method, the constrained problem is reformulated as an unconstrained problem as follows: minimize the modified functional

$$I_P(\mathbf{u}) \equiv I_0(\mathbf{u}) + \frac{\gamma_e}{2}\int_{\Omega^e} [G(\mathbf{u})]^2 d\mathbf{x} \qquad (4.3-12)$$

where γ_e is called the *penalty parameter.* Note that the constraint is included in a least–squares sense into the functional. Seeking the minimum of the modified functional $I_P(\mathbf{u})$ is equivalent to seeking the minimum of both $I_0(\mathbf{u})$ and $G(\mathbf{u})$, the latter with respect to the weight γ_e. The larger the value of γ_e, the more exactly the constraint is satisfied. The necessary condition for the minimum of I_P is

$$\delta I_P = 0 \qquad (4.3-13)$$

We have

$$0 = \int_{\Omega^e} \left[2\mu \frac{\partial \delta u_1}{\partial x_1}\frac{\partial u_1}{\partial x_1} + \mu \frac{\partial \delta u_1}{\partial x_2}\left(\frac{\partial u_1}{\partial x_2} + \frac{\partial u_2}{\partial x_1}\right)\right] d\mathbf{x} - \int_{\Omega^e} \rho_0 f_1 \delta u_1 \, d\mathbf{x} - \oint_{\Gamma^e} \delta u_1 \, T_1 \, ds$$
$$+ \int_{\Omega^e} \gamma_e \frac{\partial \delta u_1}{\partial x_1}\left(\frac{\partial u_1}{\partial x_1} + \frac{\partial u_2}{\partial x_2}\right) d\mathbf{x} \qquad (4.3-14a)$$

and

$$0 = \int_{\Omega^e} \left[2\mu \frac{\partial \delta u_2}{\partial x_2}\frac{\partial u_2}{\partial x_2} + \mu \frac{\partial \delta u_2}{\partial x_1}\left(\frac{\partial u_1}{\partial x_2} + \frac{\partial u_2}{\partial x_1}\right)\right] d\mathbf{x} - \int_{\Omega^e} \rho_0 f_2 \delta u_2 \, d\mathbf{x} - \oint_{\Gamma^e} \delta u_2 \, T_2 \, ds$$
$$+ \int_{\Omega^e} \gamma_e \frac{\partial \delta u_2}{\partial x_2}\left(\frac{\partial u_1}{\partial x_1} + \frac{\partial u_2}{\partial x_2}\right) d\mathbf{x} \qquad (4.3-14b)$$

These two statements provide the weak forms for the penalty finite element model with $\delta u_1 = w_1$ and $\delta u_2 = w_2$. We note that the pressure does not appear explicitly in the weak form (14), although it is a part of the boundary stresses (5). An approximation for the pressure can be post–computed from the relation (see [1,2,4,5,14,15,22,23])

$$P = -\gamma_e \left(\frac{\partial u_1}{\partial x_1} + \frac{\partial u_2}{\partial x_2}\right) \qquad (4.3-15)$$

where $\mathbf{u} = \mathbf{u}(\gamma_e)$ is the solution of equations (14).

The time derivative terms and nonlinear terms can be added to equations (14) without affecting the above discussion. For the general three–dimensional case, the weak forms are given by

$$\begin{aligned}0 = &\int_{\Omega^e} \rho_0\, \delta u_1 \left[\frac{\partial u_1}{\partial t} + (u_1 \frac{\partial u_1}{\partial x_1} + u_2 \frac{\partial u_1}{\partial x_2} + u_3 \frac{\partial u_1}{\partial x_3}) - f_1\right] d\mathbf{x} \\ &+ \int_{\Omega^e} \left[2\mu \frac{\partial \delta u_1}{\partial x_1}\frac{\partial u_1}{\partial x_1} + \mu \frac{\partial \delta u_1}{\partial x_2}\left(\frac{\partial u_1}{\partial x_2} + \frac{\partial u_2}{\partial x_1}\right) + \mu \frac{\partial \delta u_1}{\partial x_3}\left(\frac{\partial u_1}{\partial x_3} + \frac{\partial u_3}{\partial x_1}\right)\right] d\mathbf{x} \\ &- \oint_{\Gamma^e} \delta u_1\, \mathcal{T}_1\, ds + \int_{\Omega^e} \gamma_e \frac{\partial \delta u_1}{\partial x_1}\left(\frac{\partial u_1}{\partial x_1} + \frac{\partial u_2}{\partial x_2} + \frac{\partial u_3}{\partial x_3}\right) d\mathbf{x} \qquad (4.3-16a)\end{aligned}$$

$$\begin{aligned}0 = &\int_{\Omega^e} \rho_0\, \delta u_2 \left[\frac{\partial u_2}{\partial t} + (u_1 \frac{\partial u_2}{\partial x_1} + u_2 \frac{\partial u_2}{\partial x_2} + u_3 \frac{\partial u_2}{\partial x_3}) - f_2\right] d\mathbf{x} \\ &+ \int_{\Omega^e} \left[2\mu \frac{\partial \delta u_2}{\partial x_2}\frac{\partial u_2}{\partial x_2} + \mu \frac{\partial \delta u_2}{\partial x_1}\left(\frac{\partial u_1}{\partial x_2} + \frac{\partial u_2}{\partial x_1}\right) + \mu \frac{\partial \delta u_2}{\partial x_3}\left(\frac{\partial u_2}{\partial x_3} + \frac{\partial u_3}{\partial x_2}\right)\right] d\mathbf{x} \\ &- \oint_{\Gamma^e} \delta u_2\, \mathcal{T}_2\, ds + \int_{\Omega^e} \gamma_e \frac{\partial \delta u_2}{\partial x_2}\left(\frac{\partial u_1}{\partial x_1} + \frac{\partial u_2}{\partial x_2} + \frac{\partial u_3}{\partial x_3}\right) d\mathbf{x} \qquad (4.3-16b)\end{aligned}$$

$$\begin{aligned}0 = &\int_{\Omega^e} \rho_0\, \delta u_3 \left[\frac{\partial u_3}{\partial t} + (u_1 \frac{\partial u_3}{\partial x_1} + u_2 \frac{\partial u_3}{\partial x_2} + u_3 \frac{\partial u_3}{\partial x_3}) - f_3\right] d\mathbf{x} \\ &+ \int_{\Omega^e} \left[2\mu \frac{\partial \delta u_3}{\partial x_3}\frac{\partial u_3}{\partial x_3} + \mu \frac{\partial \delta u_3}{\partial x_1}\left(\frac{\partial u_1}{\partial x_3} + \frac{\partial u_3}{\partial x_1}\right) + \mu \frac{\partial \delta u_3}{\partial x_2}\left(\frac{\partial u_2}{\partial x_3} + \frac{\partial u_3}{\partial x_2}\right)\right] d\mathbf{x} \\ &- \oint_{\Gamma^e} \delta u_3\, \mathcal{T}_3\, ds + \int_{\Omega^e} \gamma_e \frac{\partial \delta u_3}{\partial x_3}\left(\frac{\partial u_1}{\partial x_1} + \frac{\partial u_2}{\partial x_2} + \frac{\partial u_3}{\partial x_3}\right) d\mathbf{x} \qquad (4.3-16c)\end{aligned}$$

This completes the description of the penalty function method as applied to viscous incompressible flows.

4.3–3 Reduced Integration Penalty Model

The penalty finite element model is obtained from equations (16) by substituting finite element interpolation (4.2–6a) for the velocity field, and $\delta u_i = \mathbf{\Psi}$:

$$\begin{bmatrix} \mathrm{M} & 0 & 0 \\ 0 & \mathrm{M} & 0 \\ 0 & 0 & \mathrm{M} \end{bmatrix} \begin{Bmatrix} \dot{\mathrm{u}}_1 \\ \dot{\mathrm{u}}_2 \\ \dot{\mathrm{u}}_3 \end{Bmatrix} + \begin{bmatrix} \mathrm{C(u)} & 0 & 0 \\ 0 & \mathrm{C(u)} & 0 \\ 0 & 0 & \mathrm{C(u)} \end{bmatrix} \begin{Bmatrix} \mathrm{u}_1 \\ \mathrm{u}_2 \\ \mathrm{u}_3 \end{Bmatrix} +$$

$$\begin{bmatrix} 2\mathrm{K}_{11} + \mathrm{K}_{22} + \mathrm{K}_{33} & \mathrm{K}_{21} & \mathrm{K}_{31} \\ \mathrm{K}_{12} & \mathrm{K}_{11} + 2\mathrm{K}_{22} + \mathrm{K}_{33} & \mathrm{K}_{32} \\ \mathrm{K}_{31} & \mathrm{K}_{32} & \mathrm{K}_{11} + \mathrm{K}_{22} + 2\mathrm{K}_{33} \end{bmatrix} \begin{Bmatrix} \mathrm{u}_1 \\ \mathrm{u}_2 \\ \mathrm{u}_3 \end{Bmatrix}$$

$$+\begin{bmatrix} \hat{\mathbf{K}}_{11} & \hat{\mathbf{K}}_{12} & \hat{\mathbf{K}}_{13} \\ \hat{\mathbf{K}}_{21} & \hat{\mathbf{K}}_{22} & \hat{\mathbf{K}}_{23} \\ \hat{\mathbf{K}}_{31} & \hat{\mathbf{K}}_{32} & \hat{\mathbf{K}}_{33} \end{bmatrix} \begin{Bmatrix} \mathbf{u}_1 \\ \mathbf{u}_2 \\ \mathbf{u}_3 \end{Bmatrix} = \begin{Bmatrix} \mathbf{F}_1 \\ \mathbf{F}_2 \\ \mathbf{F}_3 \end{Bmatrix} \qquad (4.3-18)$$

where $\mathbf{M}, \mathbf{C}(\mathbf{u}), \mathbf{K}_{ij}$, and $\mathbf{F}_i$ are the same as those defined in equation (4.2–13), and

$$\hat{\mathbf{K}}_{ij} = \int_{\Omega^e} \gamma_e \frac{\partial \Psi}{\partial x_i} \frac{\partial \Psi^T}{\partial x_j} \, d\mathbf{x} \qquad (4.3-19)$$

Equation (18) can be expressed symbolically as ($\mathbf{u} = \{\mathbf{u}_1, \mathbf{u}_2, \mathbf{u}_3\}^T$)

$$\bar{\mathbf{M}}\dot{\mathbf{u}} + \left[\bar{\mathbf{C}}(\mathbf{u}) + \bar{\mathbf{K}}^1(\mu) + \bar{\mathbf{K}}^2(\gamma)\right] \mathbf{u} = \bar{\mathbf{F}} \qquad (4.3-20)$$

The actual numerical construction of the penalty terms presented in (19) requires special consideration, the details of which are given in Section 4.5–3.

4.3–4 Consistent Penalty Model

An alternative formulation of the penalty finite element model is known in the literature (see [16,17,20]), and deals more directly with the finite element model than the variational statement. In this formulation, the pressure in the momentum equations (2) is replaced with [three–dimensional version of equation (15)]

$$P = -\gamma_e \left(\frac{\partial u_1}{\partial x_1} + \frac{\partial u_2}{\partial x_2} + \frac{\partial u_3}{\partial x_3} \right) = -\gamma_e \frac{\partial u_i}{\partial x_i} \qquad (4.3-21)$$

and equation (1) is replaced with equation (21). The finite element model of equation (21) is given by

$$\left[\int_{\Omega^e} \Phi \Phi^T dx \right] \mathbf{P} = -\left[\int_{\Omega^e} \gamma_e \Phi \frac{\partial \Psi^T}{\partial x_i} d\mathbf{x} \right] \mathbf{u}_i \qquad (4.3-22)$$

or in terms of a matrix notation

$$\mathbf{M}_p \mathbf{P} = -\gamma_e \mathbf{Q}^T \mathbf{u} \qquad (4.3-23a)$$

or

$$\mathbf{P} = -\gamma_e \mathbf{M}_p{}^{-1} \mathbf{Q}^T \mathbf{u} \qquad (4.3-23b)$$

where $\mathbf{u} = \{\mathbf{u}_1, \mathbf{u}_2, \mathbf{u}_3\}^T$, $\mathbf{Q}^T = (\mathbf{Q}_1, \mathbf{Q}_2, \mathbf{Q}_3)$, and $\mathbf{Q}_i$ and $\mathbf{M}_p$ are given by

$$\mathbf{Q}_i = \int_{\Omega^e} \frac{\partial \Psi}{\partial x_i} \Phi^T \, d\mathbf{x} \; ; \quad \mathbf{M}_p = \int_{\Omega^e} \Phi \Phi^T \, d\mathbf{x} \qquad (4.3-24)$$

When equation (23b) is substituted for the pressure into the finite element model of the momentum equations (4.2–11), we obtain

$$\mathbf{M}\dot{\mathbf{u}} + (\mathbf{C}(\mathbf{u}) + \mathbf{K} + \mathbf{K}_p)\mathbf{u} = \mathbf{F} \tag{4.3-25}$$

with

$$\mathbf{K}_p = \gamma_e \mathbf{Q}\mathbf{M}_p{}^{-1}\mathbf{Q}^T \tag{4.3-26}$$

Equation (4.2–10) is not used in conjunction with equation (25) because it (*i.e.*, the continuity equation) is included in equation (25) through equation (23b).

Equation (25) has the same general form as equation (20). The overall size of the penalty finite element model in equation (20) or (26) is reduced in comparison to the mixed finite element model in equation (4.2–15). Note that to recover the pressure, the inverted form of (23b) is used with a known velocity field. The particular form of the penalty finite element model described here is commonly termed a *consistent penalty model* because it is derived from the discretized form of equation (21). This is in contrast to the reduced penalty model described earlier, which falls into the category known as the *reduced integration penalty (RIP)* methods (see Oden [4,5]). Although the equations of the two models are mathematically identical, there are subtle differences that affect the numerical implementation. The successful numerical implementation of the method described here relies on the ability to efficiently construct the $\mathbf{K}_p$ matrix, *i.e.*, invert $\mathbf{M}_p$ at the element level. This requirement restricts the choices for the basis functions, $\mathbf{\Phi}$, used to represent the pressure [4–10].

4.4 FINITE ELEMENT MODELS OF POROUS FLOW

The derivation of the finite element model for the porous flow problems of Section 1.5 follows the developments of previous sections in a completely analogous manner. For completeness, we present the mixed finite element model of the porous flow equations under isothermal conditions. The penalty finite element model of the equations can also be developed, following the discussion presented in Section 4.3.

The dependent variables are again approximated by the relations in equation (4.2–6). Substituting them into the weak statements of equations (1.5–1) and (1.4–2) results in the following equations:

Continuity:

$$-\left[\int_{\Omega^e} \mathbf{\Phi}\frac{\partial \mathbf{\Psi}^T}{\partial x_i}\,d\mathbf{x}\right]\mathbf{u}_i = 0 \tag{4.4-1}$$

i-th Momentum:

$$\left[\int_{\Omega^e} \frac{\rho_0}{\phi}\mathbf{\Psi}\mathbf{\Psi}^T\,d\mathbf{x}\right]\dot{\mathbf{u}}_i + \left[\int_{\Omega^e} \frac{\rho_0 \hat{c}}{\sqrt{\kappa}}\mathbf{\Psi}\left(\mathbf{\Psi}^T\|\vec{\mathbf{u}}\|\right)\mathbf{\Psi}^T\,d\mathbf{x}\right]\mathbf{u}_i + \left[\int_{\Omega^e} \frac{\mu}{\kappa}\mathbf{\Psi}\mathbf{\Psi}^T\,d\mathbf{x}\right]\dot{\mathbf{u}}_i$$

$$+\left[\int_{\Omega^e} \mu_e\frac{\partial \mathbf{\Psi}}{\partial x_j}\frac{\partial \mathbf{\Psi}^T}{\partial x_j}\,d\mathbf{x}\right]\mathbf{u}_i + \left[\int_{\Omega^e} \mu_e\frac{\partial \mathbf{\Psi}}{\partial x_j}\frac{\partial \mathbf{\Psi}^T}{\partial x_i}\,d\mathbf{x}\right]\mathbf{u}_j - \left[\int_{\Omega^e} \frac{\partial \mathbf{\Psi}}{\partial x_i}\mathbf{\Phi}^T\,d\mathbf{x}\right]\mathbf{P} =$$

$$\left[\int_{\Omega^e} \rho_0 f_i \Psi \, dx\right] + \left\{\oint_{\Gamma^e} \mathcal{T}_i \Psi \, ds\right\} \qquad (4.4-2)$$

where f_i denote the body force components. Equations (1) and (2) can be written symbolically in the matrix form as

Continuity:

$$-\tilde{\mathbf{Q}}^T \mathbf{u} = 0 \qquad (4.4-3)$$

Momentum:

$$\tilde{\mathbf{M}}\dot{\mathbf{u}} + \tilde{\mathbf{C}}(\mathbf{u})\mathbf{u} + \tilde{\mathbf{A}}\mathbf{u} + \tilde{\mathbf{K}}\mathbf{u} - \tilde{\mathbf{Q}}\mathbf{P} = \tilde{\mathbf{F}} \qquad (4.4-4)$$

where $\mathbf{u} = \{\mathbf{u}_1, \mathbf{u}_2, \mathbf{u}_3\}^T$ and

$$\tilde{\mathbf{M}} = \int_{\Omega^e} \frac{\rho_0}{\phi} \Psi \Psi^T \, dx; \;\; \tilde{\mathbf{C}}(\mathbf{u}) = \int_{\Omega^e} \frac{\rho_0 \hat{c}}{\sqrt{\kappa}} \Psi \left(\Psi^T ||\vec{u}||\right) \Psi^T \, dx \qquad (4.4-5)$$

$$\tilde{\mathbf{A}} = \int_{\Omega^e} \frac{\mu_e}{\kappa} \Psi \Psi^T \, dx; \;\; \tilde{\mathbf{K}}_{ij} = \int_{\Omega^e} \mu_e \frac{\partial \Psi}{\partial x_i} \frac{\partial \Psi^T}{\partial x_j} \, dx; \;\; \tilde{\mathbf{Q}}_i = \int_{\Omega^e} \frac{\partial \Psi}{\partial x_i} \Phi^T \, dx \qquad (4.4-6)$$

$$\tilde{\mathbf{F}}_i = \int_{\Omega^e} \rho_0 \Psi f_i dx + \oint_{\Gamma^e} \Psi \mathcal{T}_i ds \qquad (4.4-7)$$

where summation on repeated indices is implied. Equations (3) and (4) can be arranged into a single matrix equation of the same form as given previously for the Navier–Stokes equations [see equations (4.2–14)]:

$$\begin{bmatrix} \tilde{\mathbf{M}} & 0 \\ 0 & 0 \end{bmatrix} \begin{Bmatrix} \dot{\mathbf{u}} \\ \dot{\mathbf{P}} \end{Bmatrix} + \begin{bmatrix} \tilde{\mathbf{C}}(\mathbf{u}) + \tilde{\mathbf{A}} + \tilde{\mathbf{K}} & -\tilde{\mathbf{Q}} \\ -\tilde{\mathbf{Q}}^T & 0 \end{bmatrix} \begin{Bmatrix} \mathbf{u} \\ \mathbf{P} \end{Bmatrix} = \begin{Bmatrix} \tilde{\mathbf{F}} \\ 0 \end{Bmatrix} \qquad (4.4-8)$$

The above equation may take on a somewhat different structure when the various porous flow models are invoked. When the inertia terms are unimportant, then $\tilde{\mathbf{C}} = 0$. For the familiar Darcy flow case ($\hat{c} = 0, \mu_e = 0$), the above equation simplifies to

$$\begin{bmatrix} 0 & 0 \\ 0 & 0 \end{bmatrix} \begin{Bmatrix} \dot{\mathbf{u}} \\ \dot{\mathbf{P}} \end{Bmatrix} + \begin{bmatrix} \tilde{\mathbf{A}} & -\tilde{\mathbf{Q}} \\ -\tilde{\mathbf{Q}}^T & 0 \end{bmatrix} \begin{Bmatrix} \mathbf{u} \\ \mathbf{P} \end{Bmatrix} = \begin{Bmatrix} \tilde{\mathbf{F}} \\ 0 \end{Bmatrix} \qquad (4.4-9)$$

This completes the development of the mixed finite element model of the porous flow equations.

4.5 COMPUTATIONAL CONSIDERATIONS

4.5–1 Properties of the Matrix Equations

It is important to recognize some of the basic features of the matrix equations in (4.2–12), (4.3–18), and (4.4–8) because these characteristics will heavily influence the choice of a solution procedure for the various types of problems studied here. The given matrix equations represent the discrete analogs of the basic conservation equations with each term representing a particular physical process: **M** represents the mass and **C** represents the velocity dependent convective transport term. The viscous diffusion term is represented by **K**, and **Q** represents the pressure gradient operator and $\mathbf{Q}^T$ is the divergence operator. The term **F** contains body forces and surface forces.

An inspection of the structure of the individual matrices in (4.2–13) shows that **M** and **K** are symmetric, while **C** is unsymmetric. This makes $\bar{\mathbf{K}}$ in equation (4.2–15a) unsymmetric. Thus, for general flows, the solution procedure must deal with an unsymmetric system. Conversely, when material properties are constant and flow velocities (*e.g.*, Reynolds number) are sufficiently small, the equations are symmetric and linear.

An additional feature (or difficulty) of the mixed finite element model is the presence of zeroes on the matrix diagonals corresponding to the pressure variables [see equation (4.2–12)]. In formal terms, the equations in (4.2–14), (4.3–20), (4.3–27), and (4.4–8) represent a set of ordinary differential equations in time. The fact that the pressure does not appear explicitly in the continuity equation imparts a number of strong characteristics to the discrete problem. For example, the lack of a time derivative for the pressure [see equation (4.2–14) and (4.4–8)] makes the system time–singular in the pressure and precludes the use of purely explicit time–integration methods. The zero entries on the diagonal cause difficulties in the solution of linear algebraic equations resulting from (4.2–14) or (4.4–8). Direct equation solving methods must use some type of pivoting strategy, while the use of iterative solvers is severely handicapped by poor convergence behavior attributable mainly to the form of the constraint equation. There are a number of other technical problems associated with the incompressibility assumption that directly influence the computational problem. Questions regarding the impossibility of impulsively started flows, proper boundary conditions for derived equations (*e.g.*, vorticity transport equation, pressure Poisson equation), and suitable initial conditions for time–dependent computations are all derived from the special form of the continuity equation. A thorough explanation of many of these topics can be found in the works of Reddy [1–3,12], Oden [4], Oden and Carey [5], Gresho [24–26], and others [6–10].

4.5–2 Choice of Interpolation Functions

In the previous sections we have presented finite element formulations whereby the continuum boundary value problems of interest are reduced to discrete (*i.e.*, finite–dimensional) systems of algebraic equations. In the present section, a discussion of the choice of interpolation functions used in various models is presented.

Of central importance to the development of a finite element model is the choice of particular elements (*i.e.*, interpolation functions) to be included in the element library for the models at hand. For the diffusion problem considered in Section 3.3, a wide variety of element types was presented. The finite element models of conductive heat transfer as well as viscous incompressible flows (see Sections 4.2–4.4) require only the C^0–continuous functions to approximate the field variables. Thus, any of the Lagrange and serendipity family of interpolation functions are admissible for the interpolation of the velocity field in mixed and penalty finite element models.

The choice of interpolation functions used for the pressure variable in the mixed finite element model is further constrained by the special role the pressure plays in incompressible flows; recall that the pressure can be interpreted as a Lagrange multiplier that serves to enforce the incompressibility constraint on the velocity field. From equations (4.2–4) and (4.2–7) it is seen that the approximation function $\mathbf{\Phi}$ used for pressure is the weighting function for the continuity equation (or the incompressibility constraint). It has been shown, both numerically (see Reddy [3] and Taylor and Hood [27]) and theoretically (see Sani *et al.* [14,15]) that, in order to prevent an overconstrained system of discrete equations, the interpolation used for pressure must be at least one order lower than that used for the velocity field (*i.e.*, unequal order interpolation). Further, pressure need not be made continuous across elements because the pressure variable does not constitute a primary variable of the weak form presented in equations (4.2–4) and (4.2–5). Note that the unequal order interpolation criteria can be relaxed for certain "stabilized" formulations such as the methods proposed by Hughes *et al.* [28] and Tezduyar [29], which will not be discussed here.

Convergent finite element approximations of problems with constraints are governed by the ellipticity requirement and the *Ladyzhenskaya–Babuska–Brezzi (LBB) condition* (see Reddy [2, pp. 454–461], Oden [4], Oden and Carey [5], and others [6–10]). The mixed and penalty finite elements used for viscous incompressible fluids must satisfy the LBB condition in order that they yield convergent solutions. It is by no means a simple task to rigorously prove whether every new element developed for the viscous incompressible flows satisfies the LBB condition. The discussion of the LBB condition is beyond the scope of the present study and will not be discussed here.

In the present discussion, two basic finite elements for use in the mixed (also valid for penalty) finite element model have been selected for discussion: a two–dimensional quadrilateral and a three–dimensional hexahedron. The interpolation functions used for the velocity field as well as pressure are identified.

Quadrilateral Element (2–D)

One of the most commonly used elements for two–dimensional flows of viscous incompressible fluids is the nine–node rectangular element shown in Figure 4.5–1. The velocity components and any auxiliary variables that are present are approximated using the biquadratic Lagrange functions given in equation (4.5–1) [also see equation (2.11–7)]. The functions are expressed in terms of the normalized coordinates for the element, (ξ, η), which vary from -1 to $+1$.

Two different pressure approximations are available when the velocities are approximated by quadratic Lagrange functions. The first is a continuous–bilinear approximation, in which the pressure is defined at the corner nodes of the element and is made continuous across element boundaries. The bilinear interpolation functions are defined in equation (4.5–2).

$$\{\Psi^e\} = \frac{1}{4}\begin{Bmatrix} (1-\xi)(1-\eta)(-\xi-\eta-1)+(1-\xi^2)(1-\eta^2) \\ (1+\xi)(1-\eta)(\xi-\eta-1)+(1-\xi^2)(1-\eta^2) \\ (1+\xi)(1+\eta)(\xi+\eta-1)+(1-\xi^2)(1-\eta^2) \\ (1-\xi)(1+\eta)(-\xi+\eta-1)+(1-\xi^2)(1-\eta^2) \\ 2(1-\xi^2)(1-\eta)-(1-\xi^2)(1-\eta^2) \\ 2(1+\xi)(1-\eta^2)-(1-\xi^2)(1-\eta^2) \\ 2(1-\xi^2)(1+\eta)-(1-\xi^2)(1-\eta^2) \\ 2(1-\xi)(1-\eta^2)-(1-\xi^2)(1-\eta^2) \\ 4(1-\xi^2)(1-\eta^2) \end{Bmatrix} \qquad (4.5-1)$$

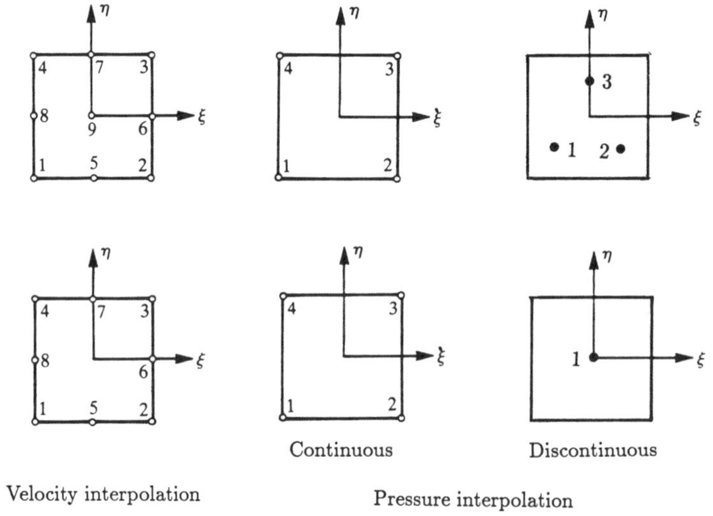

Figure 4.5–1: The quadrilateral elements used for the mixed model.

$$\{\Phi^e\} = \frac{1}{4}\begin{Bmatrix} (1-\xi)(1-\eta) \\ (1+\xi)(1-\eta) \\ (1+\xi)(1+\eta) \\ (1-\xi)(1+\eta) \end{Bmatrix} \qquad (4.5-2)$$

The second pressure approximation involves a discontinuous–linear variation defined on the element by

$$\mathbf{\Phi} = \begin{Bmatrix} 1 \\ x \\ y \end{Bmatrix} \tag{4.5 – 3}$$

Here the unknowns are not nodal point values of the pressure but correspond to the coefficients in $P = a \cdot 1 + b \cdot x + c \cdot y$. In equation (3) the interpolation functions are written in terms of the global coordinates (x, y) for the problem; these functions could also be expressed in terms of the element coordinate system.

When the element interpolation functions are written in terms of the normalized local coordinates (ξ, η), the relationship between the global coordinates (x, y) and the element coordinates (ξ, η) is defined by the transformation (see Section 2.11)

$$x = \mathbf{\Upsilon}^T \mathbf{x} \; ; \qquad y = \mathbf{\Upsilon}^T \mathbf{y} \tag{4.5 – 4}$$

where $\mathbf{\Upsilon}$ is a vector of interpolation functions on the quadrilateral and the $(\mathbf{x}, \mathbf{y})$ are vectors of coordinates, which describe the geometry of the element. This transformation is quite general, and for the nine–node quadrilateral element it allows accurate representation of straight or curved boundaries of a domain. For an isoparametric formulation,we have $\mathbf{\Upsilon} = \mathbf{\Psi}$, where $\mathbf{\Psi}$ is the interpolation functions used to approximate the velocity field. When $\mathbf{\Upsilon} = \mathbf{\Psi}$ is quadratic, and a quadratic interpolation of the element boundary is possible. In a subparametric formulation, we set $\mathbf{\Upsilon} = \mathbf{\Phi}$, and only a linear interpolation of the element boundary is possible.

When the eight–node quadratic element is used to represent the velocity field, a continuous–bilinear pressure approximation may be selected. When a discontinuous pressure variation is utilized with this element, the constant pressure representation over each element must be used. The elements shown in Figure 4.5–1 are known to satisfy the LBB condition and thus give reliable solutions for velocity and pressure fields (see Oden [4], Le Tallec and Ruas [9], and Chapelle and Bathe [10]). Other elements may yield acceptable solutions for the velocity field in some cases but the pressure field is often in error.

Hexahedral Element (3–D)

A standard element used in the analysis of three–dimensional viscous flow problems is the eight–node, hexahedron (brick) shown in Figure 4.5–2. The velocity components are approximated using the trilinear Lagrange functions given in equation (4.5–5)[also see equation (3.3–1)]. The functions are written in terms of the normalized coordinates for the element.

$$\{\Psi^e\} = \frac{1}{8}\begin{Bmatrix} (1-\xi)(1-\eta)(1-\zeta) \\ (1+\xi)(1-\eta)(1-\zeta) \\ (1+\xi)(1+\eta)(1-\zeta) \\ (1-\xi)(1+\eta)(1-\zeta) \\ (1-\xi)(1-\eta)(1+\zeta) \\ (1+\xi)(1-\eta)(1+\zeta) \\ (1+\xi)(1+\eta)(1+\zeta) \\ (1-\xi)(1+\eta)(1+\zeta) \end{Bmatrix} \qquad (4.5-5)$$

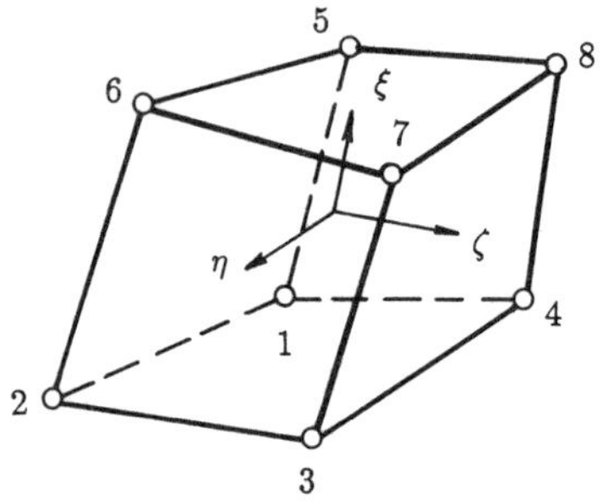

Figure 4.5–2: The eight–node brick element.

The pressure approximation for this element, being one order lower than $\mathbf{\Psi}$, is a constant (and obviously discontinuous between elements),

$$\phi = c \qquad (4.5-6)$$

Since the element shape functions are expressed in the normalized coordinates, the standard parametric transformation is needed

$$x = \mathbf{\Upsilon}^T\mathbf{x} \; ; \qquad y = \mathbf{\Upsilon}^T\mathbf{y} \; ; \qquad z = \mathbf{\Upsilon}^T\mathbf{z} \qquad (4.5-7)$$

For an isoparametric formulation, we take $\mathbf{\Upsilon} = \mathbf{\Psi}$. Note that the eight–node brick element can only have straight edges unless a superparametric representation (*i.e.*,$\mathbf{\Upsilon}$ is set to a triquadratic function) is employed. Superparametric descriptions are rarely used.

4.5–3 Evaluation of Element Matrices in Penalty Models

The numerical evaluation of the coefficient matrices appearing in equation (4.3–18) requires special consideration. This aspect is discussed here by considering the Stokes model for the steady–state case.

For the steady case, equation (4.3–20) is of the form

$$(\mu\bar{\mathbf{K}}^1 + \gamma\bar{\mathbf{K}}^2)\{\mathbf{u}\} = \{\bar{\mathbf{F}}\} \qquad (4.5-8)$$

where $\bar{\mathbf{K}}^1$ is the contribution from the viscous terms and $\bar{\mathbf{K}}^2$ is from the penalty terms, which come from the incompressibility constraint. In theory, as we increase the value of γ, the conservation of mass is satisfied more exactly. However, in practice, for some large value of γ, the contribution from the viscous terms would be negligibly small compared to the penalty terms in the computer. Thus, if $\bar{\mathbf{K}}^2$ is a nonsingular (*i.e.*, invertible) matrix, the solution of equation (4.3–21) for a large value of γ is trivial, $\{\mathbf{u}\} = \{0\}$. While the solution satisfies the continuity equation, it does not satisfy the momentum equations. In this case the discrete problem (21) is said to be overconstrained or "locked". If $\bar{\mathbf{K}}^2$ is singular, then the sum $(\mu\bar{\mathbf{K}}^1 + \gamma\bar{\mathbf{K}}^2)$ is nonsingular (becasue $\bar{\mathbf{K}}^1$ is nonsingular), and a nontrivial solution to the problem can be obtained.

The numerical problem described above is eliminated by proper evaluation of the integrals in $\bar{\mathbf{K}}^1$ and $\bar{\mathbf{K}}^2$. It is found that if the coefficients of $\bar{\mathbf{K}}^2$ (*i.e.*, penalty terms) are evaluated using a numerical integration (see Section 2.12) rule of an order less than that required to integrate them exactly, the finite element equations (4.3–21) give acceptable solutions for the velocity field. This technique of under–integrating the penalty terms is known in the literature as *reduced* (order) *integration*. For example, if a linear rectangular element is used to approximate the velocity field in a two–dimensional problem, the matrix coefficients $\bar{\mathbf{K}}^1$, as well as $\mathbf{M}$ and $\mathbf{C}(\mathbf{u})$, are evaluated using the 2×2 Gauss quadrature, and $\bar{\mathbf{K}}^2$ are evaluated using the one–point (1×1) Gauss quadrature. The one–point quadrature yields a singular $\bar{\mathbf{K}}^2$. Therefore, equation (4.3–21) cannot be inverted, whereas $(\mu\bar{\mathbf{K}}^1 + \gamma\bar{\mathbf{K}}^2)$ is nonsingular and can be inverted (after assembly and imposition of boundary conditions) to obtain a good finite element solution of the original problem. When a quadratic rectangular element is used, the 3×3 Gauss quadrature is used to evaluate $\bar{\mathbf{K}}^1$, $\mathbf{M}$, and $\mathbf{C}(\mathbf{u})$, and the 2×2 Gauss quadrature is used to evaluate $\bar{\mathbf{K}}^2$. Similar comments apply to three–dimensional elements.

The choice of the penalty parameter is largely dictated by the ratio of the magnitude of penalty terms to the viscous and convective terms (or compared to the Reynolds number, R_e), the mesh, and the word length in the computer. The following range of γ is used in computations,

$$\gamma = 10^4 R_e \text{ to } \gamma = 10^{12} R_e \qquad (4.5-9)$$

Concerning the post–computation of pressure in the penalty model, the pressure should be computed, using equation (4.3–21), at integration points corresponding to the reduced Gauss rule. This is equivalent to using an interpolation for pressure that is one order less than the one used for the velocity field. The pressure computed using equation (4.3–21) at the reduced integration points is not always reliable and accurate. The pressures predicted using the linear elements, especially for coarse meshes, are seldom acceptable. Quadratic elements are known to yield more reliable results. In general, triangular elements do not yield stable solutions for pressures. Various techniques have been proposed in the literature to obtain accurate pressure fields (see [14,15,22,23]). An alternative procedure for the post-computation of pressure is discussed in the next section.

4.5–4 Pressure Calculation

As stated earlier, the pressure field computed from equation (4.3–21) in a penalty finite element model is not accurate, especially when solving for flows of fluids with high viscosities (*e. g.*, polymers, extrusion of metals, etc.). An alternative method for computing pressure is to use the Poisson equation (obtained from the momentum equations)

$$-\nabla^2 P = \nabla \cdot (\mathbf{u} \cdot \nabla \mathbf{u}) \tag{4.5-10}$$

Since equation (10) is similar to the heat conduction problem [with Q replaced by the right–hand side expression in equation (10)], it is straightforward to solve it using the finite element method. We will not discuss this method further.

Another alternative and more reliable method of pressure calculation was first proposed by Salonen and Aalto [30] and later used for different problems by Shiojima and Shimazaki [31]. By substituting the known velocity field and/or the shear stresses, the momentum equations can be expressed in terms of the pressure. These equations are solved using the least–squares finite element model (see [23]). Following the developments in Reference 23, a derivation of the pressure computation scheme is presented here.

Since the velocity components u_i are known from the solution of equation (4.3–20) or (4.3–25), their substitution into equation (4.1–7) results in the residual,

$$\mathbf{R} \equiv M_1\hat{\mathbf{e}}_1 + M_2\hat{\mathbf{e}}_2 + M_3\hat{\mathbf{e}}_3 \tag{4.5-11}$$

where M_i is the i–th component of the momentum equation and $\hat{\mathbf{e}}_i$ is the unit vector along the x_i coordinate direction:

$$M_i = \rho_0 \left(\frac{\partial u_i}{\partial t} + u_j \frac{\partial u_i}{\partial x_j} \right) - \frac{\partial}{\partial x_j} \left[-P\delta_{ij} + \mu \left(\frac{\partial u_i}{\partial x_j} + \frac{\partial u_j}{\partial x_i} \right) \right] + \rho_0 f_i \tag{4.5-12}$$

The only unknown in equation (11) is the pressure. We determine P such that

$$\mathbf{R} = 0 \tag{4.5-13}$$

Equation (13) is a vector equation (*i.e.*, there are three equations) in a single unknown, P. Hence, we require the square of the residual vector in equation (13) to be a minimum over an element with respect to P:

$$\frac{\partial}{\partial P}\left(\int_{\Omega^e} \mathbf{R}\cdot\mathbf{R}\, d\mathbf{x}\right) = 0 \tag{4.5-14}$$

This method is known as the *least squares* method. Equation (14) provides the necessary weak statement for the determination of pressure:

$$\int_{\Omega^e} \mathbf{R}\cdot\frac{\partial \mathbf{R}}{\partial P}\, d\mathbf{x} = 0 \tag{4.5-15a}$$

or

$$\int_{\Omega^e} M_i\frac{\partial M_i}{\partial P}\, d\mathbf{x} = 0 \tag{4.5-15b}$$

Substituting $P = \sum_{l=1}^{L} P_l\phi_l$ into equation (15), we obtain

$$\int_{\Omega^e} M_i\frac{\partial \phi_l}{\partial x_i}\, d\mathbf{x} = 0 \tag{4.5-16}$$

where in equation (16), and in the following equations, sum on $i = 1, 2, 3$ is implied. Equation (16) gives the l–th equation of the system of L equations for the nodal values $(P_1, P_2, \cdots, P_L)$. In matrix form, equation (16) has the form

$$\mathbf{K}^*\mathbf{P} = \mathbf{F}^* \tag{4.5-17}$$

where

$$K^*_{lk} = \int_{\Omega^e} \frac{\partial \phi_l}{\partial x_i}\frac{\partial \phi_k}{\partial x_i}\, d\mathbf{x} \tag{4.5-18a}$$

$$F_l^* = \int_{\Omega^e} \frac{\partial \phi_l}{\partial x_i} M_i\, d\mathbf{x} \tag{4.5-18b}$$

The right–hand side vector $\{\mathbf{F}^*\}$ consists of the contributions due to body forces, inertial forces, and viscous forces. The body force terms and inertial force terms are calculated at the nodes in the same manner as they were computed for velocity calculations. The calculation of viscous terms need some special treatment. When linear elements are used, the viscous stresses are constant within the element and their derivatives vanish. The same problem arises in the finite element implementation of equation (10) because it involves second–order derivatives of the velocity field. To overcome this difficulty, the stresses are computed using the standard $2\times2\times2$ Gauss quadrature for the eight–noded (trilinear) brick elements. The stresses at the Gauss points are extrapolated to obtain the contribution of the stresses at the nodes. Such contributions from all surrounding elements are averaged to determine the nodal stresses. The main drawback of this method is that the stresses for the boundary nodes are not as accurate as those inside the domain. However, improvements are possible if the nodal stresses are calculated using an appropriate extrapolation scheme.

4.5–5 Traction Boundary Conditions

Equation (4.2–13) contains the contribution due to the boundary stresses applied to an element (sum on j),

$$\mathbf{F}_i^s = \oint_{\Gamma^e} \Psi T_i \, ds = \oint_{\Gamma^e} \Psi \sigma_{ij} n_j \, ds \tag{4.5 – 19}$$

For example, for two–dimensional problems, equation (19) takes the explicit form,

$$\mathbf{F}_1^s = \mathbf{F}_x^s = \oint_{\Gamma^e} \Psi T_x \, ds = \oint_{\Gamma^e} \Psi \sigma_{xx} n_x \, ds + \oint_{\Gamma^e} \Psi \sigma_{xy} n_y \, ds \tag{4.5 – 20a}$$

$$\mathbf{F}_2^s = \mathbf{F}_y^s = \oint_{\Gamma^e} \Psi T_y \, ds = \oint_{\Gamma^e} \Psi \sigma_{yx} n_x \, ds + \oint_{\Gamma^e} \Psi \sigma_{yy} n_y \, ds \tag{4.5 – 20b}$$

In general, the force vector for each equation contains contributions from both an applied normal stress and a shear stress. Boundary conditions could be specified directly in terms of the quantities shown in equation (20), *i.e.*, the x and y components of the traction vector or the three independent components of the stress tensor. The pressure enters the calculation through the normal stress components: $\sigma_{xx} = -P + \tau_{xx}$, $\sigma_{yy} = -P + \tau_{yy}$, and $\sigma_{zz} = -P + \tau_{zz}$.

As discussed in Section 3.5, the surface integrals shown in equation (20) can be evaluated in terms of the normalized coordinates for an element edge. Using the appropriate definition for the surface element ds,

$$ds = \left[\left(\frac{\partial \hat{\Upsilon}^T}{\partial \hat{s}} \mathbf{x} \right)^2 + \left(\frac{\partial \hat{\Upsilon}^T}{\partial \hat{s}} \mathbf{y} \right)^2 \right]^{\frac{1}{2}} d\hat{s} = \Delta \, d\hat{s} \tag{4.5 – 21a}$$

and the components of the normal,

$$n_x = \frac{1}{\Delta} \frac{\partial \hat{\Upsilon}^T}{\partial \hat{s}} \mathbf{y} \quad ; n_y = -\frac{1}{\Delta} \frac{\partial \hat{\Upsilon}^T}{\partial \hat{s}} \mathbf{x} \tag{4.5 – 21b}$$

the expressions in equation (20) can be written as

$$\mathbf{F_x}^s = \int_{-1}^{+1} \Psi \sigma_{xx} \frac{\partial \hat{\Upsilon}^T}{\partial \hat{s}} \mathbf{y} \, d\hat{s} - \int_{-1}^{+1} \Psi \sigma_{xy} \frac{\partial \hat{\Upsilon}^T}{\partial \hat{s}} \mathbf{x} \, d\hat{s} \tag{4.5 – 22a}$$

$$\mathbf{F_y}^s = \int_{-1}^{+1} \Psi \sigma_{yx} \frac{\partial \hat{\Upsilon}^T}{\partial \hat{s}} \mathbf{y} \, d\hat{s} - \int_{-1}^{+1} \Psi \sigma_{yy} \frac{\partial \hat{\Upsilon}^T}{\partial \hat{s}} \mathbf{x} \, d\hat{s} \tag{4.5 – 22b}$$

Recall that $\hat{\Upsilon}$ is a vector of edge functions for the element, and they may be either linear or quadratic depending on the type of mapping used to describe the geometry of the element. The stress components σ_{ij} in equations (22) are specified along the element boundary. The above derivation may be repeated in a completely analogous manner for the three–dimensional case.

It should be noted that the traction boundary conditions are written in terms of the *total* stress components. The total normal stress components contain pressure as a part. For example, the normal stress $\sigma_{11} = \sigma_{xx}$ is given by [see equation (4.1–3b)]

$$\sigma_{11} = -\hat{P} + 2\mu\left(\frac{\partial u_1}{\partial x_1}\right) \tag{4.5 – 23}$$

In many practical cases, the viscous part of the normal stress is negligibly small and the normal stress is essentially equal to the pressure. When the viscous part is not negligible, the application of a normal stress boundary condition does not distinguish between contributions from the pressure and viscous parts but simply reflects their net effect.

The traction boundary conditions as written in equation (22) imply that the stress components σ_{ij} are known functions of position along the boundary. Though this is the most usual situation, other types of boundary conditions are possible. For example, a slip or "friction" boundary condition relates the shear stress to a change in velocity at the boundary. Along a boundary with $y =$ constant, this condition is written as

$$\sigma_{xy} = \frac{1}{a}(u_x - u_b) \tag{4.5 – 24}$$

where u_x is the fluid velocity at the wall, u_b is the velocity of the boundary, and a is the slip coefficient. If this condition (or its multidimensional generalization) is substituted into equations (24), then a boundary matrix/vector combination is produced since the unknown fluid velocity appears in the equation. This is analogous to the convective heat transfer boundary condition in equation (2.10–11), and is treated in the same manner. Other types of boundary conditions, such as the surface tension condition, may involve other variables, but the basic method of implementation is similar in all cases.

The above description has only considered the case where the applied tractions or stress components are specified directly in the global (x, y, z) coordinate frame. In many situations of practical interest, the stresses will only be known as components normal and tangent to a bounding surface that is not a coordinate surface. It is a fairly straightforward process to rewrite equations (22) so that this situation is properly implemented. The situation is slightly more complex if velocity components normal and tangent to an arbitrary (noncoordinate) surface are to be specified. The standard procedure in this situation is to first transform the momentum equations for the affected boundary nodes to a locally defined coordinate system oriented along the normal and tangent directions. An orthogonal transformation matrix can be constructed to operate on the appropriate element matrix equations. After transformation, the normal and tangential momentum equations can be constrained by the known values of the normal and/or tangential velocity components. After the finite element solution is obtained, the velocity components in the global coordinate frame can be obtained from the normal and tangential components computed on the boundary through a simple definition of the form (for two dimensions),

$$u_x = u_n n_x - u_t n_y \ , \quad u_y = u_n n_y + u_t n_x \tag{4.5 – 25}$$

where (u_n, u_t) denote the normal and tangential components of velocity and (n_x, n_y) are the components of the unit normal vector to the boundary. For incompressible flows, an essential part of this implementation is the definition of the local normal and tangential directions such that a divergence free velocity field is maintained. A complete discussion of this topic can be found in the paper by Engelman *et al.* [32].

4.6 ITERATIVE METHODS FOR NONLINEAR EQUATIONS

4.6–1 General Discussion

The assembled finite element equations associated with (4.2–15), (4.3–20), (4.3–25), and (4.2–8) can be used to solve a very large group of fluid mechanics problems. The solution methods for this system depend significantly on the particular model used, the computational resources available, and the degree of nonlinearity in the system. When the nonisothermal effects discussed in the next chapter are included, the degree of coupling between various physical phenomena adds further complexity to the choice of solution method. The primary difficulty, of course, is the extremes that are embedded within the discrete (and continuum) equations. The flow behavior may be linear or extremely nonlinear; the physical system may be geometrically simple and one–dimensional or complex and multi–dimensional; the flow may have one predominant time and length scale or may have a variety of different scales. Though it is reasonable to believe that there is a "best" algorithm for each combination of characteristics in the equation set, it is impractical to have all of these solution methods available in a single computer code. Rather, a few of the most generally applicable schemes are usually made available with the hope that the characteristics of any particular problem will fall within the bounds of these few techniques. In this section the main focus is placed on methods that have proven to be of value for the most general types of flow problems; little attention is given to methods that only perform well for special types of flows. Also, it must be recognized that even though the algorithms described here are used routinely in both private and commercial software, the development of more effective methods is still an active area of research.

Before proceeding to a discussion of specific algorithms, it is appropriate to outline the common mathematical features of the basic solution techniques. At the outset, it is important to distinguish between the iterative methods used to reduce a set of nonlinear algebraic equations to a linear algebraic system (*e.g.*, the Picard and Newton methods) and the iterative methods used to solve a set of linear algebraic equations (*e.g.*, the Gauss–Seidel iterative method). A linear set of equations can be solved by either a direct (*e.g.*, the Gaussian elimination) or iterative procedure. The present discussion will first consider methods used in the linearization of nonlinear equations. Iterative methods used to solve linear algebraic equations will be discussed in Section 4.7.

Using the symbolic format of (4.2–15), both mixed and penalty finite element models of this chapter can be written as

$$\overline{\mathbf{M}}\dot{\mathbf{U}} + \overline{\mathbf{K}}(\mathbf{U})\mathbf{U} = \overline{\mathbf{F}} \qquad (4.6-1)$$

which represents a set of coupled, nonlinear, differential equations in time. For the time–dependent case, these equations are reduced to nonlinear algebraic equations by means of a time–approximation scheme, as described in Chapters 2 and 3 . The resulting equations are of the form [see equation (2.8–9) or (3.6–5)]

$$\tilde{\mathbf{K}}(\mathbf{U}^*)\mathbf{U}^* = \tilde{\mathbf{F}} \qquad (4.6-2)$$

which must be solved at each time step; the vector $\mathbf{U}^*$ represents the solution variables at some discrete time in the integration interval. For time–independent problems equation (1) is simplified to a form such as (2) by omitting the $\dot{\mathbf{U}}$ term; the $\mathbf{U}^*$ in this case would represent the solution to the steady–state problem. Therefore, regardless of the type of problem under consideration, the solution process for the discrete system must ultimately be concerned with a set of nonlinear algebraic equations.

For convenience, equation (2) can be expressed simply as

$$\mathcal{R}(\mathbf{U}) = 0 \qquad (4.6-3)$$

where it is understood that $\mathcal{R}$ is in general a nonlinear operator and $\mathbf{U}$ is the vector of unknowns. Many of the solution methods of practical interest for (3) can be regarded as variants of the method of successive approximation or functional iteration. Rewrite equation (3) as

$$\mathbf{U} = \mathcal{G}(\mathbf{U}) = \mathbf{U} - \mathcal{R}(\mathbf{U}) \qquad (4.6-4)$$

which is usually termed a *fixed point problem* with $\mathbf{U}$ being the *fixed point* of the operator $\mathcal{G}$ (see Reddy [2]). The form of (4) suggests the following iteration scheme, where the superscript indicates the iteration number,

$$\mathbf{U}^{n+1} = \mathcal{G}(\mathbf{U}^n) = \mathbf{U}^n - \mathcal{R}(\mathbf{U}^n) \qquad (4.6-5)$$

which is often termed the Picard method (or the method of successive substitution). The convergence of the scheme in equation (5) is intimately connected with the contraction mapping principle (see Reddy [2] and Appendix B). In essence, successive substitution can be proven to converge if the operator $\mathcal{G}$ has certain properties (*i.e.*, $\mathcal{G}$ is a *contraction mapping*) and the initial guess for $\mathbf{U}$ is sufficiently "close" to the solution. These properties are relatively difficult to prove for the discrete operators associated with the finite element equations described in this chapter. The true utility of (5) is in its suggestion of an iterative solution procedure which can be generalized to the following form

$$\mathbf{U}^{n+1} = \mathcal{G}(\mathbf{U}^n) = \mathbf{U}^n - \alpha^n \mathbf{A}(\mathbf{U}^n)\mathcal{R}(\mathbf{U}^n) \qquad (4.6-6)$$

where α^n is a constant (that may vary with the iteration number) and $\mathbf{A}(\mathbf{U}^n)$ is a nonsingular matrix which is a function of $\mathbf{U}$. The method specified by (6) provides a recursive method for finding the fixed point of $\mathcal{G}$, *i.e.*, the solution of $\mathcal{R}(\mathbf{U}) = \mathbf{0}$. The success of the algorithm will depend critically on the choices for the parameter α and the matrix $\mathbf{A}$. Note that the product $\mathbf{A}(\mathbf{U}^n)\mathcal{R}(\mathbf{U}^n)$ can be interpreted as a correction vector, $\mathbf{D}(\mathbf{U}^n)$, such that

$$\mathbf{U}^{n+1} = \mathbf{U}^n - \alpha^n \mathbf{A}(\mathbf{U}^n)\mathcal{R}(\mathbf{U}^n) = \mathbf{U}^n - \alpha^n \mathbf{D}(\mathbf{U}^n)$$

where α is now the scalar magnitude of the correction. The basic iterative formula in (6) defines a large class of methods, many of which are of value in a finite element context.

The correction vector $\mathbf{D}(\mathbf{x}^n)$ is often related to a gradient of $\mathcal{R}$ (or more often a gradient of $\mathcal{R}^T\mathcal{R}$ (see Appendix B), and the resulting methods are termed *gradient methods.* If $\mathbf{D}$ can be constructed recursively, then (6) leads to a number of iterative methods, such as the *method of steepest descent* and the *conjugate gradient methods.* These methods are particularly attractive for finite element analyses because the solution of a large matrix system is avoided; in some sense these are "explicit" methods since a global matrix need not be constructed. The major difficulty with these iterative schemes is the task of rationally selecting effective correction vectors. The correction vector $\mathbf{D}$ may also be constructed by defining the matrix $\mathbf{A}$; in this case the correction vector is generated from the solution of a matrix problem and the method is somewhat more "implicit" in nature. Methods in this family of algorithms include the *chord method* and *Newton's method.* These techniques are advantageous because of the many possible definitions for $\mathbf{A}$, though the requirement of a matrix solution is a significant disadvantage.

The discussion to this point has only considered solution methods for the nonlinear algebraic equations of the type in (2) or (3). As outlined above, some of these nonlinear solution techniques (*e.g.*, Newton's method) result in a repetitive linearization of the nonlinear system and the need to solve a linear matrix problem at each iteration. The solution to the matrix problem may be undertaken through a "direct" method, such as Gauss elimination, or through use of an iterative technique, such as a conjugate gradient method or Gauss–Seidel iterative method. Direct methods have the advantage of providing a solution to the matrix problem in a fixed number of operations, although the process generally requires large computer storage and has a large operation count. The iterative techniques may again be related to the fixed point problem and are derived from algorithms for the minimization of a quadratic form (see Appendix B). As indicated previously, the iterative methods have reduced memory requirements and operation counts; the need to condition the matrix to achieve good convergence rates is the biggest drawback to their utilization. For Navier–Stokes equations, the use of iterative matrix solvers is still in the development stages due to problems with matrix conditioning. Further comments on this topic will be reserved for a later chapter.

For general flow problems that are independent of time, the equation set of interest is derived from (1) by setting $\dot{\mathbf{U}} = 0$. That is

$$\overline{\mathbf{K}}(\mathbf{U})\mathbf{U} = \overline{\mathbf{F}} \qquad (4.6-7)$$

where $\mathbf{U}$ is defined as a vector containing all of the unknowns of the problem, $\overline{\mathbf{K}}$ contains the advection and diffusion terms as well as the incompressibility constraint, and $\overline{\mathbf{F}}$ contains all of the applied boundary and volumetric forcing functions. The choice of an iterative method for the nonlinear, algebraic system in (7) is governed by several considerations. The selected method should be capable of providing converged solutions for a wide range of problems with minimal sensitivity to variations in boundary conditions, material properties, and initial data, *i.e.*, the method should have a large radius of convergence (see Appendix B). The rate of convergence should also be reasonably high for reasons of computational economy; the computational work per iterative cycle should also be "small" to increase computational efficiency. No one iterative procedure satisfies all of these for the general span of problems considered here. However, several good methods are available, and they may be employed either separately or in various combinations.

4.6–2 The Picard Method

A particularly simple method derived directly from the fixed point problem is the successive substitution, Picard, or functional iteration method. For the system in (7) this algorithm is given by

$$\overline{\mathbf{K}}(\mathbf{U}^n)\mathbf{U}^{n+1} = \overline{\mathbf{F}}(\mathbf{U}^n) \qquad (4.6-8)$$

where the superscript indicates the iteration number. This algorithm has a reasonably large radius of convergence; the rate of convergence is generally fairly low since it is only a first–order method. An improvement in convergence rate can sometimes be obtained by use of a relaxation formula,

$$\overline{\mathbf{K}}(\mathbf{U}^n)\mathbf{U}^\star = \overline{\mathbf{F}}(\mathbf{U}^n) \qquad (4.6-9a)$$

where

$$\mathbf{U}^{n+1} = \alpha\mathbf{U}^n + (1-\alpha)\mathbf{U}^\star \qquad 0 \leq \alpha < 1 \qquad (4.6-9b)$$

Strongly nonlinear problems often exhibit an oscillatory convergence behavior for successive substitution, in which case the use of (9) would be of significant benefit.

Although the Picard schemes are generally slow to converge, they are often a good choice for the first several iterations in a general solution strategy. The wide radius of convergence allows the Picard method to start with a relatively poor initial solution guess, $\mathbf{U}^0$, and move the solution closer to the true solution and within the radius of convergence of higher–order, more rapidly convergent methods. Note that the starting vector $\mathbf{U}^0$ for the Picard scheme, as well as most other iterative algorithms, will by necessity be selected as a constant or zero vector; the first pass through the Picard method will produce a solution to the linear diffusion (Stokes flow) problem associated with (7). Neither of these solution vectors is particularly "close" to the solution of the nonlinear problem. The availability of a better initial estimate for $\mathbf{U}^0$ can improve dramatically the convergence behavior of the Picard and other solution methods.

4.6–3 Newton's Method

In order to enhance the rate of convergence, a second–order method, such as Newton's method, must be considered. First rewrite equation (7) as

$$\mathbf{R}(\mathbf{U}) = \overline{\mathbf{K}}(\mathbf{U})\mathbf{U} - \overline{\mathbf{F}}(\mathbf{U}) = 0 \qquad (4.6-10)$$

Newton's method is based on a truncated Taylor's series expansion of $\mathbf{R}(\mathbf{U})$ about the known solution $\mathbf{U}^n$:

$$0 = \mathbf{R}(\mathbf{U}^n) + \left.\frac{\partial \mathbf{R}}{\partial \mathbf{U}}\right|_{\mathbf{U}^n} \Delta\mathbf{U} + O(\Delta\mathbf{U})^2 \qquad (4.6-11)$$

where $\Delta\mathbf{U} = (\mathbf{U}^{n+1} - \mathbf{U}^n)$. Omitting the terms of order two and higher, we obtain

$$\mathbf{R}(\mathbf{U}^n) = -\left.\frac{\partial \mathbf{R}}{\partial \mathbf{U}}\right|_{\mathbf{U}^n} (\mathbf{U}^{n+1} - \mathbf{U}^n) \equiv -\mathcal{J}(\mathbf{U}^n)(\mathbf{U}^{n+1} - \mathbf{U}^n) \qquad (4.6-12)$$

where $\mathcal{J}$ is the Jacobian matrix (also known as the *tangent matrix*)

$$\mathcal{J} = \left.\frac{\partial \mathbf{R}}{\partial \mathbf{U}}\right|_{\mathbf{U}^n} \qquad (4.6-13)$$

Solving equation (12) for $\mathbf{U}^{n+1}$, we obtain

$$\mathbf{U}^{n+1} = \mathbf{U}^n - \mathcal{J}^{-1}(\mathbf{U}^n)\mathbf{R}(\mathbf{U}^n) \qquad (4.6-14)$$

Note the correspondence between equation (14) and the general procedure given in equation (6); the use of an α scale factor in (14) is also possible.

To illustrate some of the detail associated with Newton's method, consider the case of a two–dimensional mixed finite element model. In this example, the components of the solution vector $\mathbf{U}^T = [\mathbf{u}_1^T, \mathbf{u}_2^T, \mathbf{P}^T]$ and the components of the vector $\mathbf{R}$ are

$$\begin{aligned}
\mathbf{R}_1 &= \mathbf{C}_1(\mathbf{u}_1)\mathbf{u}_1 + \mathbf{C}_2(\mathbf{u}_2)\mathbf{u}_1 + (2\mathbf{K}_{11} + \mathbf{K}_{22})\mathbf{u}_1 + \mathbf{K}_{12}\mathbf{u}_2 - \mathbf{Q}_1\mathbf{P} - \mathbf{F}_1 \\
\mathbf{R}_2 &= \mathbf{C}_1(\mathbf{u}_1)\mathbf{u}_2 + \mathbf{C}_2(\mathbf{u}_2)\mathbf{u}_2 + \mathbf{K}_{21}\mathbf{u}_1 + (\mathbf{K}_{11} + 2\mathbf{K}_{22})\mathbf{u}_2 - \mathbf{Q}_2\mathbf{P} - \mathbf{F}_2 \\
\mathbf{R}_3 &= -\,\mathbf{Q}_1^T\mathbf{u}_1 - \mathbf{Q}_2^T\mathbf{u}_2
\end{aligned} \qquad (4.6-15)$$

which correspond to the two momentum equations and the incompressibility constraint. The key to Newton's method is the formation of the Jacobian, which in this case is

$$\mathcal{J} = \left[\frac{\partial \mathbf{R}}{\partial \mathbf{U}}\right] = \begin{bmatrix} \frac{\partial \mathbf{R}_1}{\partial \mathbf{u}_1} & \frac{\partial \mathbf{R}_1}{\partial \mathbf{u}_2} & \frac{\partial \mathbf{R}_1}{\partial \mathbf{P}} \\ \frac{\partial \mathbf{R}_2}{\partial \mathbf{u}_1} & \frac{\partial \mathbf{R}_2}{\partial \mathbf{u}_2} & \frac{\partial \mathbf{R}_2}{\partial \mathbf{P}} \\ \frac{\partial \mathbf{R}_3}{\partial \mathbf{u}_1} & \frac{\partial \mathbf{R}_3}{\partial \mathbf{u}_2} & \frac{\partial \mathbf{R}_3}{\partial \mathbf{P}} \end{bmatrix} \qquad (4.6-16)$$

The components of $\mathcal{J}$ can be computed directly from the definition in equation (13). We obtain,

$$\frac{\partial \mathbf{R}_1}{\partial \mathbf{u}_1} = \mathbf{C}_1(\mathbf{u}_1) + \mathbf{C}_1(1)\mathbf{u}_1 + \mathbf{C}_2(\mathbf{u}_2) + 2\mathbf{K}_{11} + \mathbf{K}_{22}$$

$$\frac{\partial \mathbf{R}_1}{\partial \mathbf{u}_2} = \mathbf{C}_2(1)\mathbf{u}_1 + \mathbf{K}_{12}$$

$$\frac{\partial \mathbf{R}_1}{\partial \mathbf{P}} = -\mathbf{Q}_1$$

$$\frac{\partial \mathbf{R}_2}{\partial \mathbf{u}_1} = \mathbf{C}_1(1)\mathbf{u}_2 + \mathbf{K}_{21}$$

$$\frac{\partial \mathbf{R}_2}{\partial \mathbf{u}_2} = \mathbf{C}_1(\mathbf{u}_1) + \mathbf{C}_2(\mathbf{u}_2) + \mathbf{C}_2(1)\mathbf{u}_2 + \mathbf{K}_{11} + 2\mathbf{K}_{22}$$

$$\frac{\partial \mathbf{R}_2}{\partial \mathbf{P}} = -\mathbf{Q}_2$$

$$\frac{\partial \mathbf{R}_3}{\partial \mathbf{u}_1} = -\mathbf{Q}_1{}^T$$

$$\frac{\partial \mathbf{R}_3}{\partial \mathbf{u}_2} = -\mathbf{Q}_2{}^T$$

$$\frac{\partial \mathbf{R}_3}{\partial \mathbf{P}} = 0 \qquad (4.6-17)$$

where (no sum on j)

$$\mathbf{C}_j(\mathbf{u}_j) = \int_{\Omega^e} \rho_0 \mathbf{\Psi}(\mathbf{\Psi}^T \mathbf{u}_j) \frac{\partial \mathbf{\Psi}^T}{\partial x_j} \, d\mathbf{x} \qquad (4.6-18)$$

The components of $\mathcal{J}$ are combinations of the basic element matrices found in the original equations plus a few new terms that arise due to differentiation of terms that are nonlinear in the dependent variables, *e.g.*, the advection matrices. By studying the form of the matrices $\mathbf{C}_j$ given in equation (18), it can be shown that the needed derivatives of these terms can be computed analytically.

All of the nonlinearities found in the components of equation (15) were not explicitly accounted for in the derivatives defining the Jacobian. For example, when the viscosity varies with velocity gradients, the **K** terms are functions of **u** and should appropriately be differentiated with respect to the velocity. The complexity of treating all possible material property and boundary condition variations in this rigorous manner is prohibitive and generally unwarranted. These nonlinearities are usually mild enough that they do not significantly affect the convergence of the Newton algorithm when treated in a "first-order" manner using a Picard or successive substitution procedure. Therefore, a strict Jacobian formulation is normally used only for the highly nonlinear advection terms.

The Newton scheme has a superior (quadratic) rate of convergence compared to the Picard scheme in equation (8). However, Newton's method also has a somewhat smaller radius of convergence (*i.e.*, is more sensitive to the initial solution vector $\mathbf{U}^0$), and therefore is often used in conjunction with a Picard scheme.

4.6–4 Modified and Quasi–Newton Methods

The major drawback to both the successive substitution and Newton algorithms is the computational expense involved in the solution of a large matrix problem at each iteration. One method for avoiding this expense in the Newton scheme is to not compute a new Jacobian at each iteration but instead work with a fixed iteration operator. This procedure, termed a *modified Newton method*, can be expressed as a variant of equation (14),

$$\mathbf{U}^{n+1} = \mathbf{U}^n - \mathcal{J}^{-1}(\mathbf{U}^0)\mathbf{R}(\mathbf{U}^n) \qquad (4.6-19)$$

After the first iteration cycle the method is very inexpensive since it only requires the reduction of the force vector and a back-substitution for each cycle. Unfortunately, the method has a poor rate of convergence and is therefore not recommended.

A method that combines the general efficiency of a modified Newton method and the approximate convergence rate of the full Newton method is the quasi–Newton procedure. Rather than work with a constant or fully updated Jacobian, the quasi–Newton methods seek to approximately update the Jacobian using a simple recursive procedure. The efficiency of the method comes from the ability to update the inverse of the Jacobian, thus saving both the assembly and factorization time of the full method; some additional storage is required to save updates to the Jacobian but this is not a serious disadvantage. A number of variants of the quasi–Newton procedure exist. The most successful scheme is based on the Broyden update [33] as implemented by Engelman *et al.* [34]. This scheme may also be related to the general method in (6).

4.6–5 Continuation Methods

A common failure of all iterative methods described previously is the lack of a sufficiently large radius of convergence. Given an initial estimate of the solution vector, the iterative method may not be able to reach a converged solution, since the initial guess was in some sense "too far away" from the required result. There are two general approaches to the problem of obtaining good initial estimates for a solution vector and both of them involve some type of "tracking" procedure for the solution. The first procedure is simply the method of false transients in which the solution is followed through the use of a time parameter. The ideas associated with dynamic relaxation, which have found use in solid mechanics [35], could also be included with these techniques. Transient algorithms are covered in a following section (see also Section 3.6–3). A second method for circumventing the problems with initial solution vectors, consists of incrementally approaching the final solution through a series of intermediate solutions. These intermediate results may be of physical interest or simply be a means to obtain the required solution. The formal algorithms used to implement such a procedure are termed *continuation (imbedding, incremental loading) methods*, and they can be used in conjunction with any of the previously described iterative procedures [23].

Continuation methods are similar in some respects to Newton's method and can be derived using a similar approach. Assume that the solution to equation (7) depends on a real parameter, λ (*e.g.*, the Reynolds number or magnitude of a boundary condition). Then equation (10) can be written as

$$\mathbf{R}(\mathbf{U},\lambda) = \overline{\mathbf{K}}(\mathbf{U},\lambda)\mathbf{U} - \overline{\mathbf{F}}(\mathbf{U},\lambda) = 0 \qquad (4.6-20)$$

As with Newton's method, if $\mathbf{R}$ has continuous derivatives with respect to $\mathbf{U}$ and is differentiable with respect to λ, then a truncated, two parameter, Taylor's series of $\mathbf{R}(\mathbf{U},\lambda)$ can be used to produce the following algorithm:

$$\left.\frac{\partial \mathbf{R}}{\partial \mathbf{U}}\right|_{\mathbf{U},\lambda} \Delta\mathbf{U} = -\left.\frac{\partial \mathbf{R}}{\partial \lambda}\right|_{\mathbf{U},\lambda} \Delta\lambda$$

or

$$\mathbf{U}^{n+1} = \mathbf{U}^{n} - \mathcal{J}^{-1}(\mathbf{U}^{n},\lambda^{n})\left.\frac{\partial \mathbf{R}}{\partial \lambda}\right|_{\mathbf{U}^{n},\lambda^{n}} \Delta\lambda \qquad (4.6-21)$$

where $\Delta\lambda = \lambda^{n+1} - \lambda^{n}$ and the Jacobian is the same as defined for Newton's method.

The solution scheme in (21) amounts to a piecewise linearization of the solution curve with respect to λ; higher order schemes could be generated. When $\Delta\lambda$ is small, it is probable that the new solution $\mathbf{U}^{n+1}$ is very close to the true solution for λ^{n+1}. However, as $\Delta\lambda$ is made larger, the predicted $\mathbf{U}^{n+1}$ will generally depart from the true solution curve. In such cases, the predicted value may be corrected via any applicable iterative method, *e.g.*, Newton's method. Convergence for this inner iteration should be rapid since the predicted value is relatively close to the true solution.

The continuation procedure outlined here is a very powerful method and may, in fact, be used to investigate the bifurcation and stability characteristics of various flow problems. Extensive developments in this area have been made by Winters and co-workers [36–39]. For difficult problems at large Reynolds numbers, the method is also a good choice. The biggest disadvantage of the algorithm is that there is little guidance for the selection of $\Delta\lambda$ during the solution sequence. Unless specific intermediate solutions are of interest, the sequence of $\Delta\lambda$ should be made as large as possible to ensure minimum computational effort. However, the size of any given

$\Delta\lambda$ is constrained by the radius of convergence requirements of the iterative method (*e.g.*, Newton's method) used to provide the corrected solution at λ^{n+1}. In practice, adaptive continuation has been accomplished mainly through the use of heuristically based $\Delta\lambda$ selection procedures [40]. The theoretical basis for adaptive continuation methods remains an area of research [41,42], especially for applications in solid mechanics.

As a final comment on continuation methods, we note that the algorithm in (21) is a first–order continuation method. A simple and effective "zeroth–order" continuation method has also found fairly widespread use in fluid mechanics and heat transfer problems. In this case, a series of problems are solved at increasing values of the continuation parameter λ. The converged solution at one value of λ^n is used at the starting solution vector at the next higher value of $\lambda^{n+1} = \lambda^n + \Delta\lambda$ in whatever iterative scheme has been selected. The formalism in equation (21) is not used to track the solution curve; some of the bifurcation tracking capabilities are lost in this simplified procedure.

4.6–6 Pressure Correction/Projection Methods

All of the previous iteration methods were based on the idea of a fixed point method and were designed to work with coupled sets of equations, *i.e.*, the velocity and pressure equations were all solved simultaneously. For each iteration method, using either mixed or penalty finite element formulations, the resulting linearized matrix problem was structured such that a direct matrix solution procedure (*e.g.*, Gauss elimination) could be effectively employed. Direct matrix solution methods, however, are decreasingly effective as problem (matrix) size increases due to excessive demands on computer resources. Iterative matrix solutions are significantly less demanding since a global system matrix need not be assembled. Unfortunately, the presence of the incompressibility constraint or penalty matrix, in the matrix system precludes the construction of useful preconditioners and inhibits the convergence and utility of iterative methods. To date, iterative matrix techniques, such as the preconditioned conjugate gradient (PCG) method, have not been successful on coupled Navier–Stokes equations containing a strong incompressibility constraint.

In order to circumvent this difficulty and allow PCG type matrix methods to be used effectively, the basic finite element equation set must be altered. Following ideas developed in the finite difference community, the troublesome incompressibility constraint is removed from the equation set and replaced by a decoupled Poisson equation for the pressure. These types of methods are generally termed pressure correction or pressure projection methods; finite difference methods of this type would be labeled as SIMPLE or SIMPLER methods [43]. Note that this technique is only possible with a mixed finite element formulation; the penalty method does not allow an explicit replacement of the continuity equation.

Consider a simplified component form of the steady, two–dimensional, finite element equations given in (4.2–12)

$$K_{xx}u + K_{xy}v - Q_xP = F_x \qquad (4.6-22a)$$

$$K_{yx}u + K_{yy}v - Q_yP = F_y \qquad (4.6-22b)$$

$$-Q_x^Tu - Q_y^Tv = 0 \qquad (4.6-22c)$$

In writing (22) it is assumed that the nonlinear advection terms have been absorbed in the K matrices. It is now possible to manipulate the equations in (22) to form a consistent, elliptic equation for the pressure. Symbolically, equations (22a) and (22b) are first solved for the u and v velocity components by inverting the appropriate K matrices. The resulting expressions for u and v are then substituted into (22c) which leads to

$$[Q_x^T K_{xx}^{-1} Q_x + Q_y^T K_{yy}^{-1} Q_y]P = -Q_x^T K_{xx}^{-1} f_x - Q_y^T K_{yy}^{-1} f_y \qquad (4.6-23)$$

where $f_x = F_x - K_{xy}v$ and $f_y = F_y - K_{yx}u$. Equation (23) is a discretized form of the pressure Poisson equation that can be used in place of the incompressibility constraint. The system given by (22 a,b) and (23) forms a decoupled system that could (in theory) be solved in an iterative cycle for the unknowns (u, v) and P. Note that (22a) and (22b) can be solved together as a single equation set or individually by component. The practical drawback to the use of (23) is the fact that the inverse K matrices make the pressure matrix impractical to construct; the inverse of a banded matrix is a dense matrix. Fortunately, it has been found by Haroutunian *et al.* [44] and others that a reasonable approximation to the inverse K matrices are sufficient to produce a usable algorithm. In particular, if the diagonal of K is used to approximate K, then the inverse is trivial and the pressure matrix can be readily constructed.

There are a number of methods by which equations (22a), (22b), and (23) can be employed to form a time–independent solution technique. We will only review one possible algorithm here and refer the interested reader to References 44–46. The pressure projection algorithm is defined by the following basic steps that are performed at each iteration until convergence is obtained. An approximation for the pressure is obtained from the pressure Poisson equation

$$[Q_x^T \tilde{K}_{xx}^{-1} Q_x + Q_y^T \tilde{K}_{yy}^{-1} Q_y]P^* = -Q_x^T \tilde{K}_{xx}^{-1} f_x(u^n) - Q_y^T \tilde{K}_{yy}^{-1} f_y(v^n) \qquad (4.6-24)$$

where the f vectors and approximate $\tilde{K}$ matrices are evaluated with velocities from the previous iteration n. Since P^* will generally be a poor approximation to P^{n+1}, a relaxation scheme is often applied to obtain a better estimate for P,

$$P^{n+1} = \alpha P^n + (1-\alpha)P^* \quad 0 \leq \alpha \leq 1 \qquad (4.6-25)$$

With a new value of the pressure (gradient) the momentum equations in (22a,b) can be solved for a velocity field. In component form then

$$K_{xx}u^* = F_x + Q_x P^{n+1} - K_{xy}v^n \qquad (4.6-26)$$

$$K_{yy}v^* = F_y + Q_y P^{n+1} - K_{yx}u^* \qquad (4.6-27)$$

or as a coupled system

$$\begin{bmatrix} K_{xx} & K_{xy} \\ K_{yx} & K_{yy} \end{bmatrix} \begin{Bmatrix} u^* \\ v^* \end{Bmatrix} = \begin{Bmatrix} F_x + Q_x P^{n+1} \\ F_y + Q_y P^{n+1} \end{Bmatrix} \tag{4.6-28}$$

Note that (26)-(28) are generally unsymmetric systems and the left-hand side would be evaluated using u^n and v^n where appropriate. The velocities computed in (26)-(28) are not divergence free. The last step of the algorithm is the projection step where the velocities are forced to satisfy the discretized continuity equation. By rederiving (23) via a slightly different approach, a second pressure Poisson equation provides

$$[Q_x^T \tilde{K}_{xx}^{-1} Q_x + Q_y^T \tilde{K}_{yy}^{-1} Q_y] P^\lambda = -Q_x^T u^* - Q_y^T v^* \tag{4.6-29}$$

The quantity P^λ is actually a Lagrange multiplier that is used to adjust the velocities with the following relations

$$u^{n+1} = u^* + \tilde{K}_{xx}^{-1} Q_x P^\lambda \tag{4.6-30a}$$

$$v^{n+1} = v^* + \tilde{K}_{yy}^{-1} Q_y P^\lambda \tag{4.6-30b}$$

Equations (24)-(30) are used repetitively until convergence is attained. Note that each matrix that requires solution is of a form that is amenable to solution *via* an iterative method such as the conjugate gradient method. Other methods that use a pressure Poisson equation differ mainly in the sequence of steps and how the pressure solution is utilized. The pressure projection method outlined here generally has the best rate of convergence.

4.7 TIME-APPROXIMATION SCHEMES

4.7-1 Preliminary Comments

Here we consider time approximation of the matrix differential equations of the type in (4.6-1):

$$\overline{\mathbf{M}}\dot{\mathbf{U}} + \overline{\mathbf{K}}(\mathbf{U})\mathbf{U} = \overline{\mathbf{F}} \tag{4.7-1}$$

Equation (1) represents a discrete space, continuous time approximation to the original system of partial differential equations. A direct time integration procedure replaces the continuous time derivative with an approximation for the history of the dependent variables over a small portion of the problem time scale. The result is an incremental procedure that advances the solution by discrete steps in time. In constructing such a procedure, questions of numerical stability and accuracy must be considered.

Though explicit integration methods have been used (see [47]) in the solution of (1), preference is usually given to implicit procedures for this class of problems. The explicit methods are plagued by several difficulties, including: (1) the natural implicitness of the pressure in an incompressible flow, (2) severe time step restrictions needed to maintain stability of the integration process, (3) the problems of diagonalizing and inverting $\overline{\mathbf{M}}$ in a cost-effective manner for a variety of element types, and (4) reductions in accuracy due to the diagonalization of $\overline{\mathbf{M}}$. Implicit integration methods, though computationally expensive, are desirable due to their increased stability and the consistent treatment of the pressure.

In the following sections two implicit procedures and one explicit method will be described (see [47,48]). Both of the implicit algorithms make use of a predictor/corrector cycle to improve efficiency and accuracy; both may be used with either a fixed time step or a dynamic time step selection algorithm. The solution of the resulting nonlinear, algebraic system for each time plane is normally obtained by Newton's method. All of these integration methods are direct extensions of the algorithms (*e.g.*, the Crank–Nicolson method) discussed for the heat conduction problem in Chapters 2 and 3.

4.7–2 Forward/Backward Euler Schemes

A first-order integration method that is useful for the equations in (4.6–1) employs a forward Euler scheme as a predictor with the backward Euler method functioning as the corrector step. Omitting the details of the derivation, the application of the explicit, forward Euler formula to equation (4.6–1) yields

$$\overline{\mathbf{M}}\mathbf{U}_p^{n+1} = \overline{\mathbf{M}}\mathbf{U}^n + \Delta t_n \left[\overline{\mathbf{F}}(\mathbf{U}^n) - \overline{\mathbf{K}}(\mathbf{U}^n)\mathbf{U}^n\right] \qquad (4.7-2)$$

This can be written in a form that is more suitable for computation by replacing the bracketed term with a rearranged form of equation (1)

$$\mathbf{U}_p^{n+1} = \mathbf{U}^n + \Delta t_n \dot{\mathbf{U}}^n \qquad (4.7-3)$$

In equations (2) and (3) the superscript on a vector indicates the time plane, the subscript p denotes a predicted value, and $\Delta t_n = t_{n+1} - t_n$ is the time step. By using the form shown in (3) a matrix inversion of $\overline{\mathbf{M}}$ is avoided; the "acceleration" vector $\dot{\mathbf{U}}^n$ is computed from a form of the corrector formula as shown below. Note that since the forward Euler scheme is explicit, it is applied only to the velocity and any auxiliary variables contained in the $\mathbf{U}$ vector; the pressure, being implicit, is not predicted with this formula.

The corrector step of the first-order scheme is provided by the backward Euler (or fully implicit) method. When applied to equation (1) this implicit method yields

$$\overline{\mathbf{M}}\mathbf{U}^{n+1} = \overline{\mathbf{M}}\mathbf{U}^n + \Delta t_n \left[\overline{\mathbf{F}}(\mathbf{U}^{n+1}) - \overline{\mathbf{K}}(\mathbf{U}^{n+1})\mathbf{U}^{n+1}\right] \qquad (4.7-4)$$

or in a form more suitable for computation

$$\left[\frac{1}{\Delta t_n}\overline{\mathbf{M}} + \overline{\mathbf{K}}(\mathbf{U}^{n+1})\right]\mathbf{U}^{n+1} = \frac{1}{\Delta t_n}\overline{\mathbf{M}}\mathbf{U}^n + \overline{\mathbf{F}}(\mathbf{U}^{n+1}) \qquad (4.7-5)$$

The implicit nature of this method is evident from the form of (5). Note that the pressure is included in the vector **U** in (5).

With the discretization of the time derivative, the equation in (5) now represents a set of nonlinear algebraic equations for the solution vector **U** at time t_{n+1}. This is precisely the same type of problem that was encountered in the time-independent case [see equation (4.6–3)] and for which a number of iterative algorithms are available. In practice, only Newton's method is generally used for (4) due to its rapid convergence and the obvious need to limit the number of computations that are repeated at each time plane. As indicated previously, the rate of convergence of Newton's method is greatly increased if the initial estimate of the vector **U** is "close" to the true solution. The solution $\mathbf{U}_p$, predicted from the explicit formula (3) provides this initial guess for the iterative procedure in a cost–effective manner. It should be noted that the use of the predictor step is not mandatory and the backward Euler scheme could be used alone to integrate the equation set. In such a situation, the first estimate of the solution at $n+1$ would be obtained from the solution at the previous time plane and a number of iterations with Newton's method would be required to converge the solution.

4.7–3 Adams-Bashforth/Trapezoid Rule

An integration method that is second-order in time can be developed along the same lines as described above. A second–order equivalent to the forward Euler method is the variable step, Adams-Bashforth predictor given by

$$\mathbf{U}_p^{n+1} = \mathbf{U}^n + \frac{\Delta t_n}{2}\left[\left(2 + \frac{\Delta t_n}{\Delta t_{n-1}}\dot{\mathbf{U}}^n\right) - \left(\frac{\Delta t_n}{\Delta t_{n-1}}\dot{\mathbf{U}}^{n-1}\right)\right] \qquad (4.7-6)$$

where $\Delta t_n = t_{n+1} - t_n$ and $\Delta t_{n-1} = t_n - t_{n-1}$. This formula can be used to predict the solution vector (excluding the pressure) given two "acceleration" vectors from previous time planes; no matrix solution is required.

A compatible corrector formula for use with (6) is available in the form of the trapezoid rule. When applied to equation (1) the trapezoid rule yields the scheme,

$$\left[\frac{2}{\Delta t_n}\overline{\mathbf{M}} + \overline{\mathbf{K}}(\mathbf{U}^{n+1})\right]\mathbf{U}^{n+1} = \frac{2}{\Delta t_n}\overline{\mathbf{M}}\mathbf{U}^n + \overline{\mathbf{M}}\dot{\mathbf{U}}^n + \overline{\mathbf{F}}(\mathbf{U}^{n+1}) \qquad (4.7-7)$$

Equation (7) is again observed to be a nonlinear, algebraic system for the vector $\mathbf{U}^{n+1}$ and can be solved using an iterative procedure such as Newton's method. All of the comments regarding the forward/backward Euler method are also applicable here.

4.7–4 Implicit Integration and Time Step Control

The actual integration process using either of the above two methods can be outlined in a few simple steps that follow closely the heat conduction procedure of Chapters 2 and 3.

At the beginning of each time step it is assumed that all of the required solution and "acceleration" vectors are known and the time increment for the next step has been selected. To advance the solution from time t_n to time t_{n+1} then requires the following steps:

1. A tentative solution vector, $\mathbf{U}_p^{n+1}$, is computed using the predictor equations (3) or (6). The pressure variables are not included in this prediction.

2. The corrector equations (5) or (7) are solved for the "true" solution, $\mathbf{U}^{n+1}$. This involves the iterative solution of (5) or (7) via Newton's method. The predicted values $\mathbf{U}_p^{n+1}$ are used to initialize the Jacobian for the Newton iteration.

3. The "acceleration" vectors are updated using the new solution $\mathbf{U}^{n+1}$ and the "inverted" forms of the corrector formulas. For the first-order method the acceleration is computed from the backward Euler definition

$$\dot{\mathbf{U}}^{n+1} = \frac{1}{\Delta t_n}\left(\mathbf{U}^{n+1} - \mathbf{U}^n\right) \qquad (4.7-8)$$

while the second-order accelerations are derived from the trapezoid rule

$$\dot{\mathbf{U}}^{n+1} = \frac{2}{\Delta t_n}\left(\mathbf{U}^{n+1} - \mathbf{U}^n\right) - \dot{\mathbf{U}}^n \qquad (4.7-9)$$

4. A new integration time step is computed. The time step selection process is based on an analysis of the time truncation errors in the predictor and corrector formulas as described below. If a constant time step is being used, this step is omitted.

5. Return to step 1 for next time increment.

In actual implementation the Newton iteration process in step 2 is not usually carried to absolute convergence. Rather, a one–step Newton correction is employed as advocated in [49]. This procedure is quite efficient and can be very accurate provided the time step is suitably controlled.

The adaptive control of the time step is an added benefit in the use of a predictor/corrector method. The time step estimation formula for the above methods is

$$\Delta t_{n+1} = \Delta t_n \left(b \cdot \frac{\epsilon^t}{d_{n+1}}\right)^m \qquad (4.8-10)$$

where $m = 1/2$, $b = 2$ for the first-order method and $m = 1/3$, $b = 3(1+\Delta t_{n-1}/\Delta t_n)$ for the second-order scheme. The user specified error tolerance for the integration process is ϵ^t, which typically has a value of 0.001. The quantity d_{n+1} is an appropriate norm on the integration error, which is defined as the difference between the predicted solution and the corrected value. Typically, the following norms are used for each velocity variable,

$$d^U_{n+1} = \frac{1}{U_{max}} \left[\sum_{i=1}^{N} \left(U_{i(n+1)} - U^p_{i(n+1)} \right)^2 \right]^{\frac{1}{2}} \qquad (4.8-11)$$

During the integration process a series of new time steps are computed using each of the norms defined in equation (11). The resulting time steps are compared and the smallest is used for the next integration step. This procedure allows the dominant physical process to control the integration algorithm.

4.7–5 Explicit Integration

The primary difficulty in constructing an explicit integration method was first mentioned in Section 4.5–1, and relates to the inherently implicit nature of the pressure. From equations (4.2–15), it is clear that the lack of a coefficient matrix for $\dot{P}$ makes the explicit advancement of this variable impossible. As in the case of the pressure correction methods in Section 4.6–6, this difficulty can be partially circumvented by replacing the standard continuity equation with a pressure Poisson equation. With this substitution, the velocity components can be advanced through time via an explicit integrator; the pressure at each time step must be computed from the implicit Poisson equation.

The explicit procedure that is outlined here was proposed by Gresho *et al.* [47] and follows the general philosophy of the previously described pressure projection method. Consider the simplified, segregated form of the two–dimensional, time–dependent, finite element equations given by (4.2–11)

$$\mathbf{M}\dot{\mathbf{u}} + \mathbf{K}(\mathbf{u})\mathbf{u} - \mathbf{Q}\mathbf{P} = \mathbf{F} \qquad (4.8-12)$$

$$-\mathbf{Q}^T\mathbf{u} = 0 \qquad (4.8-13)$$

where the $\mathbf{K}$ Matrix contains both the diffusive and advective terms and u contains both velocity components. To construct a Poisson equation for the pressure, operate on (12) first with $\mathbf{M}^{-1}$ and then with $\mathbf{Q}^T$; noting from (13) that $\mathbf{Q}^T\dot{\mathbf{u}} = 0$ allows the modified equation (12) to be written as

$$(\mathbf{Q}^T\mathbf{M}^{-1}\mathbf{Q})\mathbf{P} = \mathbf{Q}^T\mathbf{M}^{-1}(\mathbf{F} - \mathbf{K}(\mathbf{u})\mathbf{u}) \qquad (4.8-14)$$

which is an elliptic equation for the pressure P. As in Section 4.6–6 the key to obtaining a viable algorithm is the construction of $\mathbf{M}^{-1}$. If $\mathbf{M}$ is diagonalized or lumped then $\mathbf{M}^{-1}$ is quite simple and the Poisson equation is readily available. Note also that the left–hand side of (14) is a constant matrix and needs only be formed and triangularized once.

The general explicit scheme described by Gresho *et al.* [47] utilized a forward Euler integration method for (12) with a pressure solution obtained from (14). Various numerical procedures were used in conjunction with the basic method to produce a reliable, cost–effective method. Low order (trilinear) velocity elements were used so that M could be diagonalized; one–point Gauss quadrature was used to speed construction of element matrices. So–called "balancing tensor diffusivity" was used to offset the time truncation error associated with the forward Euler scheme. Also, subcycling was introduced for equation (14) in which a new pressure was computed only after several explicit velocity time steps. Though an explicit time integration method can be derived as shown here, its practical implementation is quite complex. The method is only conditionally stable and is still burdened with the solution of at least one matrix problem. More recent work on explicit methods with pressure projection can be found in [48,50].

4.8 SOLUTION OF LINEAR EQUATIONS

4.8–1 General Comments

All of the steady and time–dependent solution methods described in the two previous sections lead ultimately to the point where at least one set (matrix) of algebraic equations must be solved. In general terms, the system

$$\tilde{\mathbf{K}}\mathbf{U}^{*} = \tilde{\mathbf{F}} \tag{4.8-1}$$

must be solved where $\tilde{\mathbf{K}}$ is a banded, sparse matrix that may be either symmetric or unsymmetric depending on the characteristics of the particular flow problem or algorithm. The solution vector $\mathbf{U}^{*}$ may contain several degrees of freedom or a single variable, again depending on the specific algorithm. Other characteristics for $\mathbf{K}$, such as whether or not it is positive definite, and the size of its condition number (ratio of largest to smallest eigenvalues), are also very dependent on the type of finite element model and the particular solution algorithm. However, these characteristics are extremely important, since they strongly influence the type of matrix solution procedure that can be successfully used for any given problem.

A linear set of equations can be solved by either a direct or iterative method. Direct solvers, like the Gauss elimination method, are often used to solve systems of algebraic equations. They provide the solution after a fixed number of steps and are less sensitive to the conditioning of the matrix. However, the main deficiency of the direct solvers is that they require the coefficient matrix to be stored in an ordered format to enhance the matrix band structure and reduce storage requirements. In recent years, the direct solvers have been refined to reduce this deficiency through innovative data management techniques (*e.g.*, frontal solvers, skyline solvers, and others; see Carey and Oden [51]). These improvements enable users to solve moderately large systems of equations efficiently. However, they have been found to be unsuitable for solving very large systems of equations (especially in three–dimensional problems) because they demand out–of–core storage of the equations and hence require large data transfers.

The limitations on CPU time and storage requirements make the use of direct solvers uneconomical even on the present–day high speed and high memory computers, and using direct solvers for complex problems with more than a quarter million equations is impractical. For large systems, iterative methods are more efficient in that they require less storage and CPU time while giving comparable accuracy in the solution. This is due to the fact that the global matrix formation can be avoided, and the major operation is the matrix–vector multiplication as compared to the matrix reduction in direct methods. Another advantage of the iterative methods is that operations can be performed in parallel on an array of processors. The direct solvers can also be parallelized; however, the iterative solvers can be more easily parallelized for massively parallel systems. The demand for numerical solution of complex problems coupled with the cost–ineffectiveness of the direct solvers revived the interest in the use of iterative methods for the solution of systems of equations.

Among the various iterative methods, the Conjugate Gradient (CG) method [52] is most widely used because it is a finite step method (*i.e.*, apart from round–off errors, the solution is achieved in a fixed number of iterations) and it can be used to determine the inverse. However, the number of iterations required depends on the condition number of the coefficient matrix (the condition number of the coefficient matrix is the ratio of the largest to smallest eigenvalue). The convergence of the conjugate gradient method, and iterative methods in general, can be improved by preconditioning and/or scaling the equations [53–55].

The limitations on storage can be overcome by solving the equations at the element level, *i.e.*, use the Gauss–Seidel iteration idea for the set of variables associated with the element. This approach avoids assembly of element matrices to form the global coefficient matrix. This idea of using the element–by–element data structure of the coefficient matrix was first pointed out by Fox and Stanton [56] and Fried [57–59]. The phrase *element–by–element* refers to a particular data structure for finite element techniques wherein information is stored and maintained at the element level rather than assembled into a global data structure [23,60–86]. In this method the matrix–vector multiplications are carried out at the element level and the assembly is carried out on the resultant vector. This idea proves to be very attractive when solving large problems, because the matrix–vector multiplication can be done in parallel on a series of processors. Another advantage of this method is that the resultant savings in storage, compared to direct solvers, allows solution of large problems on small computers. A review of the literature on element–by–element algorithms is presented in [23].

In summary, for iterative solution methods the advantages of the element–by–element data structure over assembling the global coefficient matrix are

1. the need for formation and storage of a global matrix is eliminated, and therefore the total storage and computational costs are low,

2. the amount of storage is independent of the node numbering and mesh topology and depends on the number and type of elements in the mesh, and

3. the element–by–element solution algorithms can be vectorized for efficient use on supercomputers.

The major disadvantage of the element–by–element data structure is the limited number of preconditioners that can be formulated from the unassembled matrices.

4.8–2 Solution Algorithms

In this section, we review three iterative solvers from [23] that are applicable to nonsymmetric, positive definite equation systems that are typical of isothermal flow algorithms. For a comparable direct method, we refer to the *frontal* solution procedure

[87], because it is faster than most direct solvers, it requires less core space as long as active variables can be kept in the core, and it allows for partial pivoting. An additional advantage is that no stringent node numbering scheme is needed, though a judicious element numbering helps to minimize the front width. Details of the frontal method will not be given here as it is a sophisticated implementation of the simple Gauss elimination method which is adequately studied in numerous texts. Implementation issues for the frontal method can be found in [87]. The three iterative solution schemes used here are the Biorthogonal Conjugate Gradient method [54], the Lanczos ORTHORES [54], and the GMRES [80].

The conjugate gradient method for solving a system of equations can be interpreted as the search for the minimum of the energy E of the system. The energy of the system is a minimum when the residual vector $\mathbf{r} = \tilde{\mathbf{F}} - \tilde{\mathbf{K}}\mathbf{U}^*$ vanishes. The algorithm for the biorthogonal conjugate gradient method (also known as two–term form of the steepest descent method, Lanczos/ORTHOMIN) for unsymmetric systems of equations [54,23] is given in Table 4.8–1, and the steps involved in the Lanczos ORTHORES solution algorithm [54,23] are given in Table 4.8–2.

Table 4.8–1: Steps involved in using biorthogonal conjugate gradient method (Lanczos/ORTHOMIN solver).

Repeat the following steps for each nonlinear iteration:

I. *Initial Calculations:*

(1) Form the element stiffness matrix $\tilde{\mathbf{K}}^e$ and force vector $\tilde{\mathbf{F}}^e$.

(2) Apply essential and/or natural boundary conditions, and modify $\tilde{\mathbf{K}}^e$ and $\tilde{\mathbf{F}}^e$.

(3) Store the element matrices in $\bar{\mathbf{A}}$ (whose dimensions are nem, neleq, neleq).†

(4) Store the inverse of the diagonal terms of the global system in $\mathbf{W}$ ($W_{ii} = \sum_{e=1}^{\text{nem}} \tilde{K}_{ii}^{-1}$).

(5) Assemble the global force vector.

II. *Preconditioning:*

Form the preconditioned system of equations

$$\bar{\mathbf{K}}\bar{\mathbf{U}} = \bar{\mathbf{F}}; \quad \bar{\mathbf{K}} = \mathbf{W}^{-1/2}\tilde{\mathbf{K}}\mathbf{W}^{-1/2}\,, \quad \bar{\mathbf{U}} = \mathbf{W}^{1/2}\mathbf{U}^*\,, \quad \bar{\mathbf{F}} = \mathbf{W}^{-1/2}\tilde{\mathbf{F}}$$

III. *Lanczos ORTHOMIN Algorithm:*

---

† nem = number of elements in the finite element mesh, neleq = number of element equations.

(1) For known initial solution vector $\bar{\mathbf{U}}^0$, compute:

$$\mathbf{r}^0 = \bar{\mathbf{F}} - \bar{\mathbf{K}}\bar{\mathbf{U}}^0 , \quad \mathbf{P}^0 = \mathbf{r}^0 , \quad \tilde{\mathbf{r}}^0 = \tilde{\mathbf{P}}^0 = \mathbf{r}^0 ,$$

$$\alpha_0 = 0 , \quad \lambda_0 = \frac{(\mathbf{r}^0, \tilde{\mathbf{r}}^0)}{(\bar{\mathbf{K}}\mathbf{P}^0 , \tilde{\mathbf{r}}^0)} , \quad \bar{\mathbf{U}}^1 = \bar{\mathbf{U}}^0 + \lambda_0 \mathbf{P}^0$$

(2) For each ORTHOMIN iteration $m = 1, 2, 3, \ldots$, compute:§.

$$\lambda_m = \frac{(\mathbf{r}^m, \tilde{\mathbf{r}}^m)}{(\bar{\mathbf{K}}\mathbf{P}^m, \tilde{\mathbf{r}}^m)} , \quad \mathbf{P}^m = \mathbf{r}^m + \alpha_m \mathbf{P}^{m-1}, \quad \tilde{\mathbf{P}}^m = \tilde{\mathbf{r}} + \alpha_m \tilde{\mathbf{P}}^{m-1} ,$$

$$\alpha_m = \frac{(\mathbf{r}^m, \tilde{\mathbf{r}}^m)}{(\mathbf{r}^{m-1}, \tilde{\mathbf{r}}^{m-1})} ,$$

$$\mathbf{r}^{m+1} = \mathbf{r}^m - \lambda_m \bar{\mathbf{K}}\mathbf{P}^m , \quad \tilde{\mathbf{r}}^{m+1} = \tilde{\mathbf{r}}^m - \lambda_m \bar{\mathbf{K}}^T \tilde{\mathbf{P}}^{m-1} , \quad \bar{\mathbf{U}}^{m+1} = \bar{\mathbf{U}}^m + \lambda_m \mathbf{P}^m$$

(3) Convergence criterion: $||\bar{\mathbf{U}}^{m+1}||/||\mathbf{r}^0|| \leq 10^{-6}$

(4) If convergence criterion is satisfied $\bar{U}^* = \mathbf{W}^{-1/2}\bar{\mathbf{U}}^{m+1}$

Table 4.8–2: Steps involved in using Lanczos/ORTHORES solver.

Repeat the following steps for each nonlinear iteration:

I. *Initial Calculations:*

(1) Form the element stiffness matrix $\tilde{\mathbf{K}}^e$ and force vector $\tilde{\mathbf{F}}^e$.

(2) Apply essential and natural boundary conditions, and modify $\tilde{\mathbf{K}}^e$ and $\tilde{\mathbf{F}}^e$.

(3) Store the element matrices in $\bar{\mathbf{A}}$ (whose dimensions are nem, neleq, neleq).

(4) Store the inverse of the diagonal terms of the global system in $\mathbf{W}$ ($W_{ii} = \sum_{e=1}^{\text{nem}} \tilde{K}_{ii}^{-1}$).

(5) Assemble the global force vector.

II. *Preconditioning:*

Form the preconditioned system of equations

$$\bar{\mathbf{K}}\bar{\mathbf{U}} = \bar{\mathbf{F}} , \quad \bar{\mathbf{K}} = \mathbf{W}^{-1/2}\tilde{\mathbf{K}}\mathbf{W}^{-1/2} , \quad \bar{\mathbf{U}} = \mathbf{W}^{1/2}\tilde{\mathbf{U}} , \quad \bar{\mathbf{F}} = \mathbf{W}^{-1/2}\tilde{\mathbf{F}}$$

---

§ $(a, b) = \sum a_i b_i$

III. *Lanczos ORTHORES Algorithm:*

(1) For known initial solution vector $\bar{\mathbf{U}}^0$, compute:

$$\mathbf{r}^0 = \bar{\mathbf{F}} - \bar{\mathbf{K}}\bar{\mathbf{U}}^0$$

(2) For each ORTHORES iteration $m = 0, 1, 2, \ldots$, compute ($\tilde{\mathbf{r}} = \mathbf{r}_0, \lambda^0 = 0$):

$$\lambda^{m+1} = \frac{(\mathbf{r}^m, \tilde{\mathbf{r}}^m)}{(\bar{\mathbf{K}}\mathbf{r}^m, \tilde{\mathbf{r}}^m)}$$

$$\beta^{m+1} = \begin{cases} 1 & ; \text{if } m = 0 \\ \left[1 - \frac{\lambda^{m+1}}{\lambda^m} \frac{(\mathbf{r}^m, \tilde{\mathbf{r}}^m)}{(\mathbf{r}^{m-1}, \tilde{\mathbf{r}}^{m-1})} \frac{1}{\beta_m}\right]^{-1} & ; \text{if } m \geq 1 \end{cases}$$

$$\mathbf{r}^{m+1} = \beta^{m+1}\left(\mathbf{r}^m - \lambda^{m+1}\bar{\mathbf{K}}\mathbf{r}^m\right) + \left(1 - \beta^{m+1}\right)\mathbf{r}^{m-1}$$

$$\tilde{\mathbf{r}}^{m+1} = \beta^{m+1}\left(\tilde{\mathbf{r}}^m - \lambda^{m+1}\bar{\mathbf{K}}\tilde{\mathbf{r}}^m\right) + \left(1 - \beta^{m+1}\right)\tilde{\mathbf{r}}^{m-1}$$

$$\bar{\mathbf{U}}^{m+1} = \beta^{m+1}\left(\bar{\mathbf{U}}^m + \lambda^{m+1}\mathbf{r}^m\right) + \left(1 - \beta^{m+1}\right)\bar{\mathbf{U}}^{m-1}$$

(3) Convergence criterion: $||\bar{\mathbf{U}}^{m+1}||/||\mathbf{r}^0|| \leq 10^{-6}$

(4) If convergence criterion is satisfied $\mathbf{U}^* = \mathbf{W}^{-1/2}\bar{\mathbf{U}}^{m+1}$

The third iterative solver uses the GMRES solution algorithm. For an approximate solution of the form $\bar{\mathbf{U}}_0 + \mathbf{z}$, where $\bar{\mathbf{U}}_0$ is the initial guess and $\mathbf{z}$ is a member of the Krylov space $\mathcal{K}$ of dimension k, the GMRES algorithm determines $\mathbf{z}$ such that $||\ \bar{\mathbf{F}} - \bar{\mathbf{K}}(\bar{\mathbf{U}}_0 + \mathbf{z})\ ||$ is minimized, where $||\ .\ ||$ denotes the L_2–norm. The Krylov space is given by $\mathcal{K} = span\left\{\bar{\mathbf{U}}_0,\ \bar{\mathbf{K}}\bar{\mathbf{U}}_0,\ \bar{\mathbf{K}}^2\bar{\mathbf{U}}_0, \ldots\bar{\mathbf{K}}^{k-1}\bar{\mathbf{U}}_0\right\}$. Therefore, when solving large system of equations, as the value of k increases, the amount of storage required also increases. This drawback can be overcome by employing the GMRES algorithm iteratively by using a smaller value for k and restarting the algorithm after every k steps. The restart version of the GMRES algorithm [80,23] is explained in Table 4.8–3.

The presence of the penalty matrix [$\hat{\mathbf{K}}$ in (4.3–18) and $\mathbf{K}_p$ in (4.3–25)] in the coefficient matrix of the penalty model spoils the condition number. This results in slow convergence when using iterative solvers. However, the convergence of the iterative solvers can be improved by preconditioning the system. In [23], the system of equations is transformed using diagonal scaling matrix (Jacobi/diagonal preconditioning). Accordingly, the system of equations (1) becomes

$$\bar{\mathbf{K}}\bar{\mathbf{U}} = \bar{\mathbf{F}} \qquad (4.8-2)$$

$$\bar{\mathbf{K}} = \mathbf{W}^{-1/2}\tilde{\mathbf{K}}\mathbf{W}^{-1/2};\ \bar{\mathbf{U}} = \mathbf{W}^{1/2}\mathbf{U}^*;\ \bar{\mathbf{F}} = \mathbf{W}^{-1/2}\tilde{\mathbf{F}} \qquad (4.8-3)$$

where, $\mathrm{W}_{ii} = \tilde{K}_{ii}^{-1}$ is a diagonal matrix. During the matrix vector multiplication, the element–by–element data structure is exploited and the multiplications are carried out at the element level and the residuals are then assembled to form the global vector. The diagonal terms are always positive because of the viscous and penalty terms.

Table 4.8–3: Steps involved in using the GMRES solver.

Repeat the following steps for each nonlinear iteration:

I. *Initial Calculations:*

(1) Form the element stiffness matrix $\tilde{\mathbf{K}}^e$ and force vector $\tilde{\mathbf{F}}^e$.

(2) Apply essential and natural boundary conditions, and modify $\tilde{\mathbf{K}}^e$ and $\tilde{\mathbf{F}}^e$.

(3) Store the element matrices in $\bar{\mathbf{A}}$ (whose dimensions are nem, neleq, neleq).

(4) Store the inverse of the diagonal terms of the global system in $\mathbf{W}$ ($W_{ii} = \sum_{e=1}^{\text{nem}} \tilde{K}_{ii}^{-1}$).

(5) Assemble the global force vector.

II. *Preconditioning:*

Form the preconditioned system of equations

$$\bar{\mathbf{K}}\bar{\mathbf{U}} = \bar{\mathbf{F}}\,, \quad \bar{\mathbf{K}} = \mathbf{W}^{-1/2}\tilde{\mathbf{K}}\mathbf{W}^{-1/2}\,, \quad \bar{\mathbf{U}} = \mathbf{W}^{1/2}\tilde{\mathbf{U}}\,, \quad \bar{\mathbf{F}} = \mathbf{W}^{-1/2}\tilde{\mathbf{F}}$$

III. *GMRES Algorithm:*

(1) *Start:* Choose $\bar{\mathbf{U}}^0$ and compute $\mathbf{r}^0 = \bar{\mathbf{F}} - \bar{\mathbf{K}}\bar{\mathbf{U}}^0$ and $\mathbf{v}^1 = \bar{\mathbf{U}}^0 / \| \bar{\mathbf{U}}^0 \|$.

(2) *Iterate:* For $j = 1, 2..., k$ do:

$$h_{i,j} = (\bar{\mathbf{K}}v_j, v_i), \; i = 1, 2, ..., j,$$

$$\hat{v}_{j+1} = \bar{\mathbf{K}}v_j - \textstyle\sum_{i=1}^{j} h_{i,j}v_i,$$

$$h_{j+1,j} = \| \hat{v}_{j+1} \|, \text{ and}$$

$$v_{j+1} = \hat{v}_{j+1}/h_{j+1,j},$$

(3) *Form the approximate solution:*

$\bar{\mathbf{U}}^k = \bar{\mathbf{U}}^0 + \mathbf{V}\mathbf{y}$, where $\mathbf{y}$ minimizes $\| \mathbf{e} - \bar{\mathbf{H}}\mathbf{y} \|, \mathbf{y} \in \mathbf{R}^k$.

(4) *Restart:*

Compute $\mathbf{r}^k = \bar{\mathbf{F}} - \bar{\mathbf{K}}\bar{\mathbf{U}}^k$; check convergence; if satisfied stop, otherwise compute $\bar{\mathbf{U}}^0 := \bar{\mathbf{U}}^k, \mathbf{v}_1 := \bar{\mathbf{U}}^k / \| \bar{\mathbf{U}}^k \|$, and go to step 2.

(5) Convergence criterion: $\|\bar{\mathbf{U}}^{m+1}\| / \|\mathbf{r}^0\| \leq 10^{-6}$

(6) If convergence criterion is satisfied $\mathbf{U}^* = \mathbf{W}^{-1/2}\bar{\mathbf{U}}$

where $\mathbf{V}$ is a $N \times k$ matrix whose columns are l–2 orthonormal basis vectors $\{\mathbf{v}_1, \mathbf{v}_2, \ldots, \mathbf{v}_k\}$, $\mathbf{e} = \{\| \bar{\mathbf{U}}^0 \|, 0, \ldots, 0\}$, and $\bar{\mathbf{H}}$ is the upper $k \times k$ Hessenberg matrix whose entries are the scalars $h_{i,j}$. When using the restart version of the GMRES algorithm, the total number of iteration m can be computed from the number of restarts and the dimension of k.

4.9 POST–PROCESSING

4.9–1 Stress Computation

The results of interest from a flow analysis generally include fluid velocities, forces, and flow patterns. Many of these items are directly available from the finite element results in terms of nodal point quantities; other variables may be derived from these primary quantities. The computation of the viscous stress fields for the fluid flow follow directly from the definition of those quantities.

For a planar two–dimensional geometry the components of the stress tensor arein terms of the pressure and velocity components $u_1 = u_x$ and $u_2 = u_y$:

$$\sigma_{xx} = -P + 2\mu \frac{\partial u_x}{\partial x} \qquad (4.9-1a)$$

$$\sigma_{yy} = -P + 2\mu \frac{\partial u_y}{\partial y} \qquad (4.9-1b)$$

$$\sigma_{xy} = \mu \left(\frac{\partial u_x}{\partial y} + \frac{\partial u_y}{\partial x} \right) \qquad (4.9-1c)$$

Substitution of the finite element approximations for u_i and P into equation (1) yields

$$\sigma_{xx} = -\mathbf{\Phi}^T \mathbf{P} + 2\mu \frac{\partial \mathbf{\Psi}^T}{\partial x} \mathbf{u}_x \qquad (4.9-2a)$$

$$\sigma_{yy} = -\mathbf{\Phi}^T \mathbf{P} + 2\mu \frac{\partial \mathbf{\Psi}^T}{\partial y} \mathbf{u}_y \qquad (4.9-2b)$$

$$\sigma_{xy} = \mu \left(\frac{\partial \mathbf{\Psi}^T}{\partial y} \mathbf{u}_x + \frac{\partial \mathbf{\Psi}^T}{\partial x} \mathbf{u}_y \right) \qquad (4.9-2c)$$

The spatial derivatives of the interpolation functions in equation (2) can be converted to derivatives involving the local element coordinates through use of the coordinate transformation defined in Section 3.3 [see equation (3.3–3)]. When this transformation is invoked the stress components become

$$\sigma_{xx} = -\mathbf{\Phi}^T \mathbf{P} + \frac{\mu}{|\mathbf{J}|}\left[J^*{}_{11}\frac{\partial \mathbf{\Psi}^T}{\partial \xi}\mathbf{u}_x + J^*{}_{12}\frac{\partial \mathbf{\Psi}^T}{\partial \eta}\mathbf{u}_x\right]$$

$$\sigma_{yy} = -\mathbf{\Phi}^T \mathbf{P} + \frac{\mu}{|\mathbf{J}|}\left[J^*{}_{21}\frac{\partial \mathbf{\Psi}^T}{\partial \xi}\mathbf{u}_y + J^*{}_{22}\frac{\partial \mathbf{\Psi}^T}{\partial \eta}\mathbf{u}_y\right]$$

$$\sigma_{xy} = \frac{\mu}{|\mathbf{J}|}\left[J^*{}_{21}\frac{\partial \mathbf{\Psi}^T}{\partial \xi}\mathbf{u}_x + J^*{}_{22}\frac{\partial \mathbf{\Psi}^T}{\partial \eta}\mathbf{u}_x + J^*{}_{11}\frac{\partial \mathbf{\Psi}^T}{\partial \xi}\mathbf{u}_y + J^*{}_{12}\frac{\partial \mathbf{\Psi}^T}{\partial \eta}\mathbf{u}_y\right] \quad (4.9-3)$$

The function Υ that occurs in the Jacobian and its inverse, J^*, is defined as either a quadratic or linear interpolation function, depending on whether the particular element is isoparametric or subparametric. Treatment of the three-dimensional case follows directly from the above procedure.

The expressions in (3) allow the stress components to be evaluated at any point (ξ_0, η_0) within an element for a known element geometry and solution field. Note that since the stresses depend on velocity gradients they are discontinuous between elements.

For an axisymmetric flow, the stress field is computed in the same manner as shown above but with the following definitions for the stress components,

$$\sigma_{rr} = -P + 2\mu\frac{\partial u_r}{\partial r}\ ;\quad \sigma_{zz} = -P + 2\mu\frac{\partial u_z}{\partial z} \quad (4.9-4a)$$

$$\sigma_{\theta\theta} = -P + 2\mu\frac{u_r}{r}\ ;\quad \sigma_{rz} = \mu\left(\frac{\partial u_z}{\partial r} + \frac{\partial u_r}{\partial z}\right) \quad (4.9-4b)$$

For some applications the vorticity field is also of interest and this quantity is computed in the same manner as the fluid stresses. By definition, the vorticity for a two–dimensional flow isgiven by [see equation (4.1–8)]

$$\omega = \frac{\partial u_y}{\partial x} - \frac{\partial u_x}{\partial y} \quad (4.9-5)$$

By using the same methods as outlined for the stress components the vorticity can be evaluated at any point within an element. For three–dimensional flows, the additional components of the vorticity vector may also be computed. However, these quantities are seldom used in practical applications.

4.9–2 Stream Function Computation

A quantity that is often useful in the graphic display of computed flow fields is the stream function. For two–dimensional incompressible flows, the stream function is the remaining nonzero component of a vector potential which satisfies the conservation of mass equation identically. By definition,

$$u_1 = u_x = \frac{\partial \psi}{\partial y}\ ; \qquad u_2 = u_y = -\frac{\partial \psi}{\partial x} \tag{4.9-6}$$

and since the change in the stream function, $\delta\psi$, is an exact differential, then

$$\begin{aligned}\delta\psi &= \int_A^B \mathbf{u}\cdot\hat{\mathbf{n}}\,ds \\ &= \int_A^B (u_x n_x + u_y n_y)\,ds\end{aligned} \tag{4.9-7}$$

with

$$\mathbf{u} = u_x\hat{\mathbf{e}}_x + u_y\hat{\mathbf{e}}_y; \quad \hat{\mathbf{n}} = n_x\hat{\mathbf{e}}_x + n_y\hat{\mathbf{e}}_y \tag{4.9-8}$$

where $\hat{\mathbf{n}}$ is the unit normal to the integration path ds, $\mathbf{u}$ is the velocity vector along the path, and $\hat{\mathbf{e}}_i$ are unit vectors in the coordinate directions.

The calculation of the change in the stream function within a finite element can be carried out using (7) once a suitable integration path AB is identified. In most applications, the integration path is taken along the element boundaries. Consider the typical element boundary shown in Figure 4.9–1 with the following definitions:

$$\begin{aligned} u_x &= \hat{\boldsymbol{\Psi}}^T\mathbf{u}_x\ ; \qquad & \mathbf{u}_y &= \hat{\boldsymbol{\Psi}}^T\mathbf{u}_y \\ x &= \hat{\boldsymbol{\Upsilon}}^T\mathbf{x}\ ; \qquad & y &= \hat{\boldsymbol{\Upsilon}}^T\mathbf{y}\end{aligned} \tag{4.9-9}$$

where $\hat{\boldsymbol{\Psi}}$ and $\hat{\boldsymbol{\Upsilon}}$ are interpolation (edge) functions and $\mathbf{u}_x$, $\mathbf{u}_y$, $\mathbf{x}$, $\mathbf{y}$ are vectors of nodal point velocities and coordinates. The unit normal vector is given by

$$\hat{\mathbf{n}} = \frac{1}{\Delta}\frac{\partial y}{\partial s}\hat{\mathbf{e}}_x - \frac{1}{\Delta}\frac{\partial x}{\partial s}\hat{\mathbf{e}}_y \tag{4.9-10}$$

with ds defined by

$$ds = \left[\left(\frac{\partial x}{\partial \hat{s}}\right)^2 + \left(\frac{\partial y}{\partial \hat{s}}\right)^2\right]^{\frac{1}{2}} d\hat{s} = \Delta\, d\hat{s} \tag{4.9-11a}$$

Using the definitions of equation (9), we obtain

$$ds = \left[\left(\frac{\partial \hat{\boldsymbol{\Upsilon}}^T}{\partial \hat{s}}\mathbf{x}\right)^2 + \left(\frac{\partial \hat{\boldsymbol{\Upsilon}}^T}{\partial \hat{s}}\mathbf{y}\right)^2\right]^{\frac{1}{2}} d\hat{s} = \Delta\, d\hat{s} \tag{4.9-11b}$$

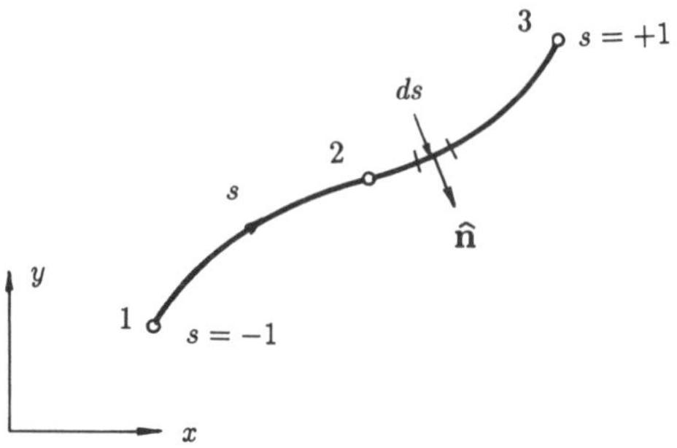

Figure 4.9–1: Definition of the element boundary for stream function computation.

Combining these relations with the definition for $\delta\psi$ produces

$$\delta\psi = \int_{-1}^{+1} \left(\frac{\partial \hat{\boldsymbol{\Upsilon}}^T}{\partial \hat{s}} \mathbf{y} \hat{\boldsymbol{\Psi}}^T \mathbf{u}_x - \frac{\partial \hat{\boldsymbol{\Upsilon}}^T}{\partial \hat{s}} \mathbf{x} \hat{\boldsymbol{\Psi}}^T \mathbf{u}_y \right) d\hat{s} \qquad (4.9-12)$$

The interpolation function definitions were given in Section 4.5; the vector $\hat{\boldsymbol{\Upsilon}}$ can be either linear or quadratic, depending on the shape of the element boundary. The change in the stream function along any element boundary can be computed from equation (12) once the element geometry and velocity fields are specified. Computation of the stream function field for an entire finite element mesh is generated by applying equation (12) along successive element boundaries, starting at a node for which a base value of ψ has been specified.

The calculation of the stream function for axisymmetric flows follows a similar procedure with the appropriate definition for ψ,

$$u_1 = u_r = \frac{1}{r}\frac{\partial \psi}{\partial z}\,; \qquad u_2 = u_z = -\frac{1}{r}\frac{\partial \psi}{\partial r} \qquad (4.9-13)$$

and

$$\mathbf{u} = u_r\ \hat{\mathbf{e}}_r + u_z\ \hat{\mathbf{e}}_z; \quad \hat{\mathbf{n}} = n_r \hat{\mathbf{e}}_r + n_z \hat{\mathbf{e}}_z \qquad (4.9-14)$$

A second approach to computing the stream function for a given velocity field relies on the definition of the vorticity. By substituting the stream function definition in (6) into the vorticity equation (5), the following elliptic partial differential equation is obtained:

$$\omega = -\nabla^2 \psi$$

This equation could be easily cast into a finite element model and solved for ψ since ω is known everywhere in the domain. The main disadvantages to this approach are that boundary conditions for ψ must be supplied for all points on the domain boundary and, more importantly, a separate matrix problem must be solved for each velocity field. The element–by–element procedure is computationally more efficient and is generally the method of choice.

4.9–3 Particle Tracking

Another useful method for visualizing fluid motion involves following the movement of a massless particle that is introduced at a convenient point in the flow. The tracking of particle paths is governed by the set of ordinary differential equations

$$\frac{dx_i}{dt} = \dot{x}_i = u_i \tag{4.9 – 15}$$

Given the initial location of a point, $P(x_i)$, the initial value problem in (15) must be solved to describe the particle trajectory for the known flow field u_i. The integration of (15) is most easily accomplished on an element–by–element basis. Using the standard definitions for an isoparametric mapping of the element geometry and the element approximation for velocity produces

$$\mathbf{J}\begin{Bmatrix} \dot{\eta} \\ \dot{\xi} \\ \dot{\zeta} \end{Bmatrix} = \begin{Bmatrix} \boldsymbol{\Psi}^T(\xi,\eta,\zeta)\mathbf{u}_1 \\ \boldsymbol{\Psi}^T(\xi,\eta,\zeta)\mathbf{u}_2 \\ \boldsymbol{\Psi}^T(\xi,\eta,\zeta)\mathbf{u}_3 \end{Bmatrix} \tag{4.9 – 16}$$

where the Jacobian $\mathbf{J}$ was defined in Section 3.4 and (ξ,η,ζ) are the local element coordinates. For a known starting location (ξ_0,η_0,ζ_0) and velocity field, equation (16) can be integrated by standard methods to yield the particle trajectory across the element. When the particle crosses an element boundary, the procedure is restarted with a new element Jacobian and a new element velocity field.

As shown in (16) the velocity field is independent of time and the integration can be performed with the fixed values of $\mathbf{u}$. Time–dependent flows may also be computed [in which case $\mathbf{u} = \mathbf{u}(t)$]. Conceptually, the method is the same but the numerical implementation now requires interpolation to find the needed particle tracking velocities from the finite element time planes.

It is also possible to generalize this method to consider the motion of real particles in the flow. In this case, a force balance on the particle is required that accounts for drag, gravity effects, or other forces and predicts the particle velocity. The particle velocity is then used in (15) in place of the fluid velocity; the integration procedure is similar to the one shown in (16).

4.10 NUMERICAL EXAMPLES

4.10–1 Preliminary Comments

The true test of any numerical method is its performance in the solution of problems of engineering and scientific interest. In this section, a small sampling of flow problems solved using the finite element models developed herein are presented. No attempt is made to examine the problems in depth but rather the intent is to illustrate the variety and, to some extent, the complexity of problems that can be solved using the finite element method. The examples were solved using the computer codes FEM2DV2 [1], NACHOS II [88], and FIDAP [89]. Program FEM2DV2 contains the reduced integration penalty finite element model for the Stokes problem, NACHOS contains the mixed finite element model for the Navier–Stokes equations, and FIDAP contains both mixed and penalty finite element models of the Navier–Stokes equations. The objective of the first several examples is to evaluate the accuracy of the penalty and mixed finite element models in the light of available analytical or numerical results and to illustrate the effect of the penalty parameter on the accuracy of the solutions.

4.10–2 Fluid Squeezed Between Parallel Plates

Consider the (Stokes) flow of a viscous incompressible material squeezed between two long parallel plates (see Figure 4.10–1a). When the length of the plates is very large compared to both the width of and the distance between the plates, we have a case of plane flow (in the plane formed by the width of and the distance between the plates). Although this is a moving boundary problem, we wish to determine the velocity and pressure fields for a fixed distance between the plates, assuming that a state of plane flow exists. An approximate (analytical) solution to this two-dimensional problem is provided by Nadai [90](also see [1,2]).

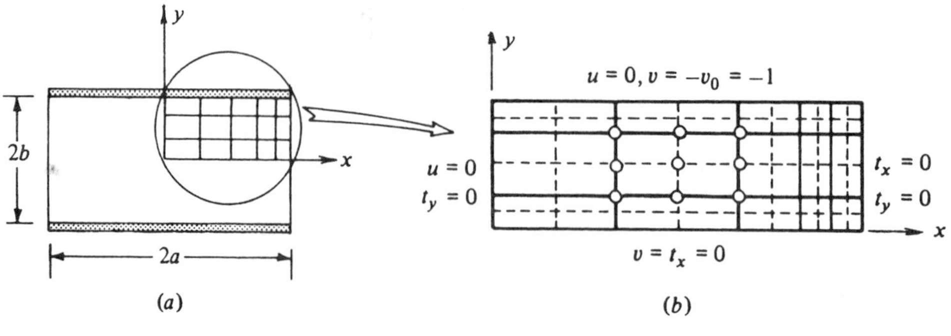

Figure 4.10–1: Schematic and mesh for fluid squeezed between parallel plates.

Let V_0 be the velocity with which the two plates are moving toward each other (*i.e.*, squeezing out the fluid), and let 2b and 2a denote, respectively, the distance between and the width of the plates (see Figure 4.10–1a). Due to the biaxial symmetry present in the problem, it suffices to model only a quadrant of the domain. A 5×3 nonuniform mesh of nine–node quadratic elements is used in the mixed model, and a

10×6 mesh of the four-node linear elements and 5×3 mesh of nine-node quadratic elements are used in the penalty model (see Figure 4.10-1b). The nonuniform mesh, with smaller elements near the free surface (*i.e.*, at $x = a$), is used to approximate accurately the singularity in the shear stress at the point $(a, b) = (6, 2)$. The mesh used for the penalty model has exactly the same number of nodes as the mesh used for the mixed model. There are no specified nonzero secondary variables in the problem. The computer program FEM2DV2 (penalty model) from [1] is used to analyze the problem. The reduced integration rule suggested earlier for the evaluation of penalty terms is used.

The velocities $u_1(x, 0) = u(x, 0)$ obtained with the two models compare well with the analytical solution as shown in Table 4.10-1. The nine-node element gives very good results for both the penalty and mixed models. The influence of the penalty parameter on the accuracy of the solution is clear from the results. Whether the element is linear or quadratic, it is necessary to use a large value of the penalty parameter. For an IBM 3090 computer with double precision arithmetic, a value of $\gamma = 10^{14}$ would be the upper limit beyond which the contribution from the viscous terms is not recognized in comparison to the penalty terms (and the coefficient matrix, being singular, is not invertible).

Table 4.10-1: Comparison of finite element velocity solutions with the analytical solution for fluid squeezed between plates (for $y = 0$).

	$\gamma = 1.0$		$\gamma = 100$		$\gamma = 10^8$		Mixed model	Mixed model	Series
x	**4-node**	**9-node**	**4-node**	**9-node**	**4-node**	**9-node**	**8-node**	**9-node**	**solution**
1	0.0303	0.0310	0.6563	0.6513	0.7576	0.7505	0.7496	0.7497	0.7500
2	0.0677	0.0691	1.3165	1.3062	1.5135	1.4992	1.5038	1.5031	1.5000
3	0.1213	0.1233	1.9911	1.9769	2.2756	2.2557	2.2563	2.2561	2.2500
4	0.2040	0.2061	2.6960	2.6730	3.0541	3.0238	3.0213	3.0203	3.0000
4.5	0.2611	0.2631	3.0718	3.0463	3.4648	3.4307	3.4331	3.4292	3.3750
5.0	0.3297	0.3310	3.4347	3.3956	3.8517	3.8029	3.8249	3.8165	3.7500
5.25	0.3674	0.3684	3.6120	3.5732	4.0441	3.9944	4.0074	3.9893	3.9375
5.5	0.4060	0.4064	3.7388	3.6874	4.1712	4.1085	4.1450	4.1204	4.1250
5.75	0.4438	0.4443	3.8316	3.7924	4.2654	4.2160	4.2188	4.2058	4.3125
6.0	0.4793	0.4797	3.8362	3.7862	4.2549	4.1937	4.2659	4.2364	4.5000

† The three-point Gauss rule for nonpenalty terms and the two-point Gauss rule for penalty terms are used for quadratic elements.

Next, a 12×8 mesh of linear elements and a 6×4 mesh of quadratic elements were used to evaluate the relative accuracies of the rectangular and triangular elements for the penalty model. The meshes are not shown here. The 12×8 mesh of linear triangular elements with full integration of $[K^1]$ and $[K^2]$ and selective integration (*i.e.*, full integration of $[K^1]$ and reduced integration of $[K^2]$) both give the same results for the velocity field. However, in both cases, erroneous results for pressure and stresses are obtained. The 6×4 mesh of quadratic triangular elements with full and selective integrations give the same velocity fields, while the stresses and pressure are predicted to be the same at the same quadrature points. Both the 12 × 8 mesh of linear rectangular elements and the 6 × 4 mesh of nine-node rectangular elements give good results for velocities, pressure, and stresses. The solution accuracy can be increased by using refined meshes.

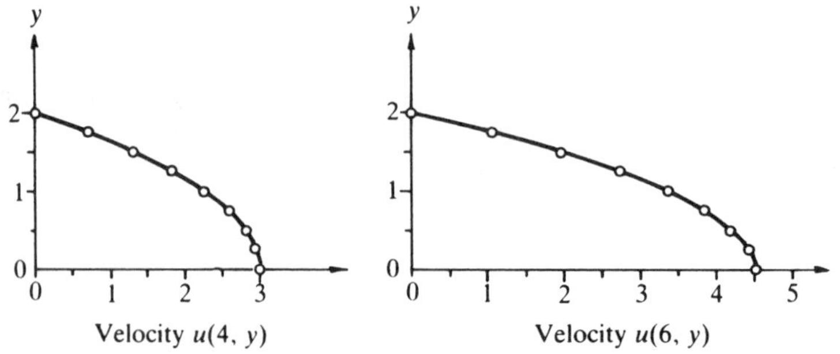

Figure 4.10–2: Velocity fields for fluid sqeezed between parallel plates.

Figure 4.10–2 contains plots of the velocity $u(x, y)$ for $x = 4$ and 6, and Figure 4.10–3 contains plots of pressure $P(x, y)$, for $y =$ constant, computed using the quadratic triangular and nine-node rectangular elements. The pressure in the penalty model was computed using equation (4.3–15) with the 2 × 2 Gauss rule for the quadratic rectangular element and the one-point formula for the quadratic triangular element. If the pressure in the penalty model were computed using the full quadrature rule for rectangular elements, we would obtain erroneous values. The linear triangular element with full as well as reduced integrations gives unstable pressures, while the quadratic triangular element with one- or two-point rules yields good results. In general, the same quadrature rule as that used for the evaluation of the penalty terms in the coefficient matrix must be used to evaluate the pressure, and one should avoid using the linear triangular element in the penalty finite element model.

4.10–3 Flow of a Viscous Lubricant in a Slider Bearing

The slider (or slipper) bearing consists of a short sliding pad moving at a velocity $u = U_o$ relative to a stationary pad inclined at a small angle with respect to the stationary pad, and the small gap between the two pads is filled with a lubricant (see Figure 4.10–4a). Since the ends of the bearing are generally open, the pressure there is atmospheric, P_0. If the upper pad is parallel to the base plate, the pressure everywhere in the gap must be atmospheric (because dP/dx is a constant for flow between parallel plates), and the bearing cannot support any transverse load. If the upper pad is inclined to the base pad, a pressure distribution (in general, a function of x and y) is set up in the gap. For large values of U_o, the pressure generated can be of sufficient magnitude to support heavy loads normal to the base pad.

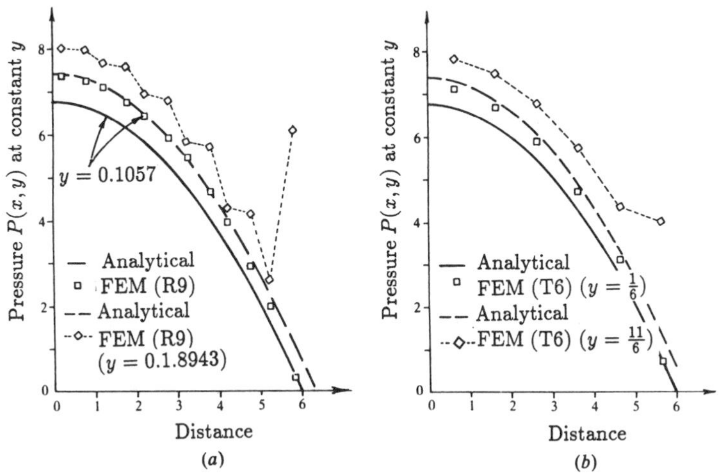

Figure 4.10–3: Pressures for fluid squeezed between parallel plates.

Since the width of the gap and the angle of inclination are generally small, it is assumed that the pressure is not a function of y in developing an analytical solution to the problem. Assuming a two-dimensional flow and a small angle of inclination and neglecting the normal stress gradient (in comparison with the shear stress gradient), the equations governing the flow of the lubricant between the pads can be solved to give (see Schlichting [91] and Reddy [1])

$$u = \left(U_0 - \frac{1}{2\mu}h^2\frac{dP}{dx}\frac{y}{h}\right)\left(1 - \frac{y}{h}\right) \ , \ P = \frac{6\mu U_0 L(h_2 - h)(h - h_1)}{h^2(h_2^2 - h_1^2)} \qquad (4.10-1a)$$

$$\tau_{xy} = \mu\frac{\partial u}{\partial y} = \frac{dP}{dx}\left(y - \frac{h}{2}\right) - \mu\frac{U_o}{h} \qquad (4.10-1b)$$

where

$$h(x) = h_2 + \frac{h_1 - h_2}{L}x \qquad (4.10-1c)$$

In our computations, we choose

$$h_2 = 2h_1 = 8 \times 10^{-4},\ L = 0.36,\ \mu = 8 \times 10^{-4},\ U_0 = 30$$

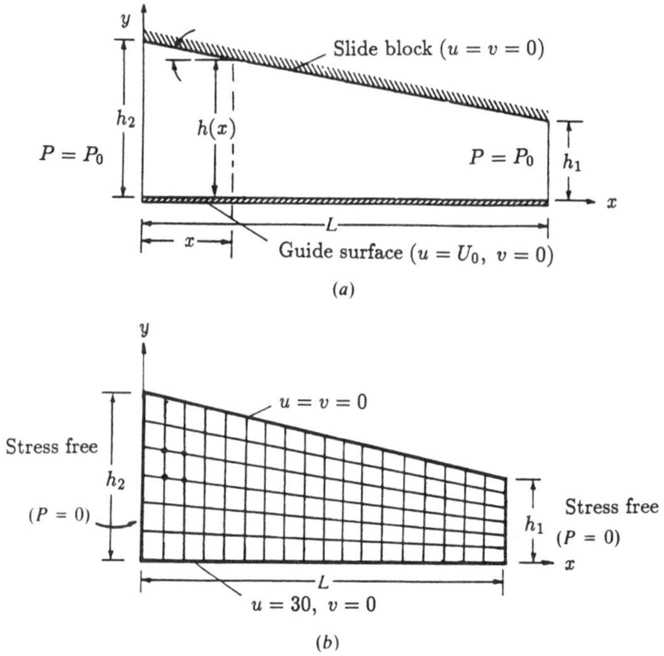

Figure 4.10–4: Schematic and the finite element mesh (Mesh 1) for slider bearing.

First, it should be pointed out that the assumption concerning the pressure not being a function of y is not necessary in the finite element analysis. We use a mesh (Mesh 1) of 18×6 linear quadrilateral elements to analyze the problem. The mesh and boundary conditions are shown in Figure 4.10–4b. Figure 4.10–5 contains plots of the pressure distributions along the length of the bearing for various values of the penalty parameter. The converged pressure is slightly higher than the approximate analytical solution.

To obtain more accurate solutions, a graded mesh (Mesh 2) of 128 linear quadrilateral elements (153 nodes), (equivalently, 32 nine-node quadrilateral elements) is used. Figures 4.10–6a and 4.10–6b contain a comparison of the finite element solutions (for $\gamma = 10^8$) with the analytical solutions for the horizontal velocity, pressure, and shear stress. The pressure was computed at the reduced Gauss points using equation (4.3–15). The results are in excellent agreement with the analytical solutions.

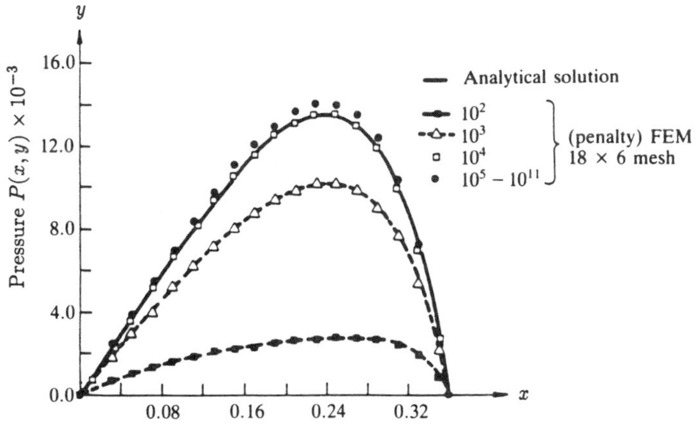

Figure 4.10–5: Pressures obtained with Mesh 1 for slider bearing.

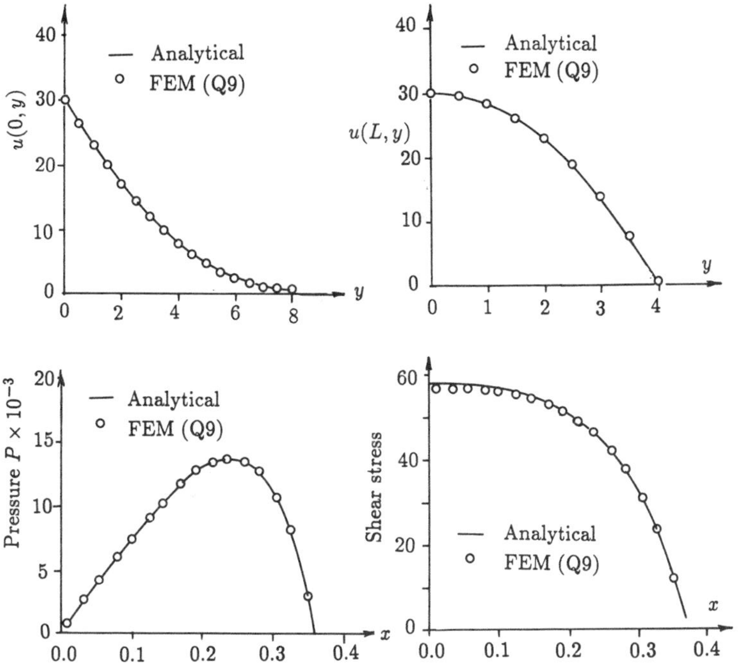

Figure 4.10–6: Velocity field, pressure, and shear stress for the slider bearing.

4.10–4 Wall–Driven 2–D Cavity Flow

Consider the laminar flow of a viscous, incompressible fluid in a square cavity bounded by three motionless walls and a lid moving at a constant velocity in its own plane. The third dimension is assumed to be long enough to have a plane flow. Singularities exists at each corner where the moving lid meets a fixed wall. This example is one which has been extensively studied by analytical, numerical, and experimental methods (see [17,19,27–29,92–100], among others), and it is often used as a benchmark problem to test a new numerical method or formulation. In solving this problem, the mesh used should be such that the boundary layer thickness is resolved. The boundary layer thickness is of the order of $Re^{-\frac{1}{2}}$, where $Re = \rho UL/\mu$ and U is the lid velocity and L is the cavity dimension. Thus the mesh should be graded to have smaller elements near the walls.

Figure 4.10–7 shows the horizontal velocity component along the vertical centerline of the cavity, *i.e.*, plots of $u(0.5, y)$ vs. y, for $Re = 1, 100, 400$, and 10^3. The results were obtained (see Reddy [100]) using a nonuniform, 14 × 14 mesh of linear rectangular elements. Figure 4.10–8 contains the plots of $u(0.5, y)$ vs. y for $Re = 400$ for various meshes. The effect of mesh refinement on the accuracy of the solution is clear.

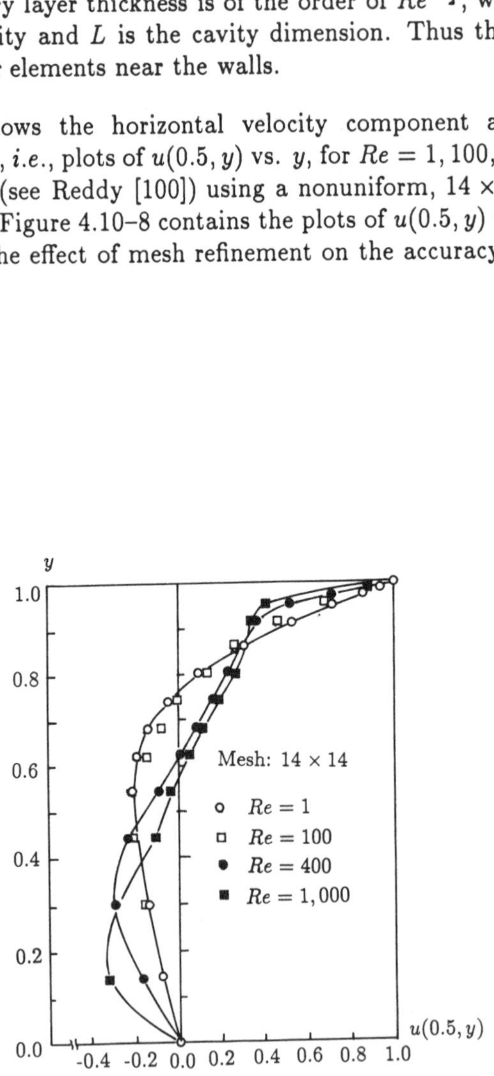

Figure 4.10–7: Horizontal velocity along the vertical centerline of the cavity for various Reynolds numbers (Mesh: 14 × 14).

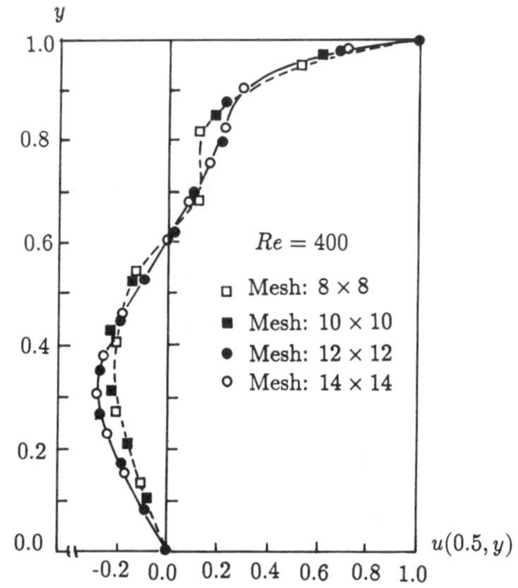

Figure 4.10–8: Horizontal velocity along the vertical centerline of the cavity for various meshes ($Re = 400$).

4.10–5 Wall–Driven 3–D Cavity Flow

The problem considered here is that of a three–dimensional cubical cavity of unit dimension [99–103]. The motion of an incompressible viscous fluid in the cavity is induced by the motion of the lid (see Figure 4.10–9). All other walls of the cavity are assumed to be stationary.

Using the symmetry about the $x = 0$ plane, only half of the domain is modeled using a $4 \times 8 \times 7$ mesh of trilinear (*i.e.*, eight–node brick) elements (see Reddy [103]). The problem is analyzed for $Re = 100$ and 400. Figures 4.10–10 and 4.10–11 contains the plots of velocity components in the midplane of the cavity. The present penalty finite element solution is compared with the finite difference solutions of Dennis *et al.* [101] and Agarwal [102]. In Reference 90, a second–order accurate finite difference method was used, whereas in [102] a third–order accurate finite difference scheme was used. Both used very refined meshes. The present results agree well qualitatively with the results of Agarwal [102], although a very crude mesh was used in the finite element analysis.

4.10–6 Evaluation of the EBE Iterative Solvers

Here we use the wall–driven cavity problem to evaluate the EBE methods discussed in Section 4.8. The entire discussion presented here is taken from [23]. Four nonuniform meshes, Mesh 1: $5 \times 10 \times 10$, Mesh 2: $6 \times 12 \times 12$, Mesh 3: $7 \times 14 \times 14$,

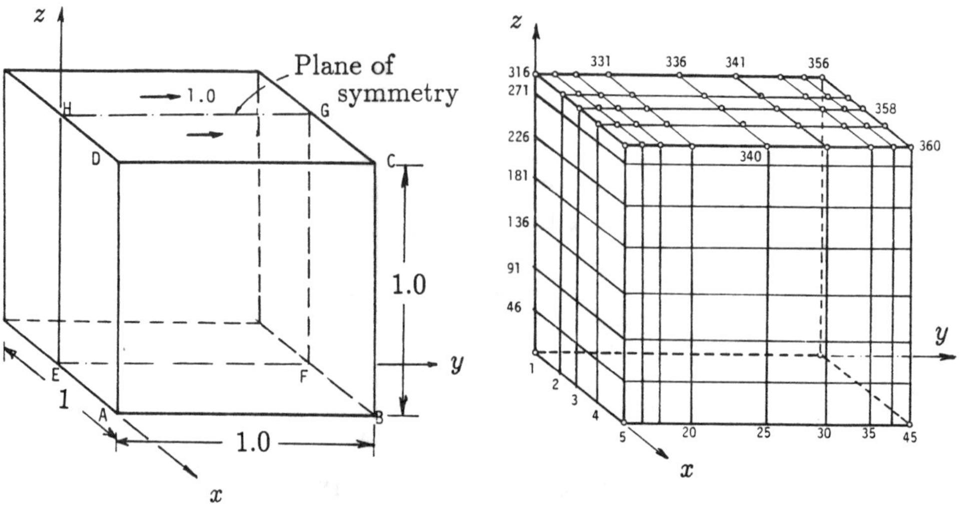

on $x = 0$ (symmetry boundary condition): $u = 0$,

on $x = 0.5$: $u = v = w = 0$,

on $y = 0$ and 1: $u = v = w = 0$,

on $z = 0$: $u = v = w = 0$,

on $z = 1$: $u = w = 0,\ v = 1$.

Figure 4.10–9: Schematic and boundary conditions for a wall-driven cubical cavity.

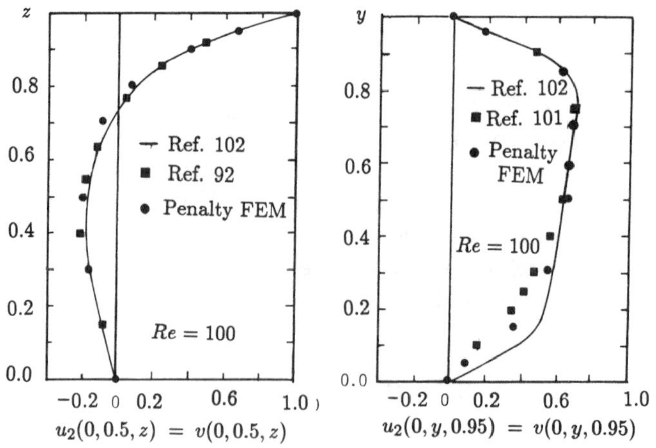

Figure 4.10–10: Velocity fields for a wall-driven cubical cavity, $Re = 100$.

and Mesh 4: 8 × 16 × 16— of trilinear elements were used. The pressure at (0.5, 0.5, 0.0) is taken to be equal to zero, and it is used as an essential boundary condition for the pressure calculation according to equation (4.5–17). The problem was analyzed for *Re* = 10 and 100. For iterative solvers the convergence criterion for velocity and pressure calculations was taken to be $\epsilon_{vel} = 10^{-5}$ and $\epsilon_{pr} = 10^{-4}$ respectively. Figure 4.10–12 shows the finite element discretization of the domain for Mesh 3.

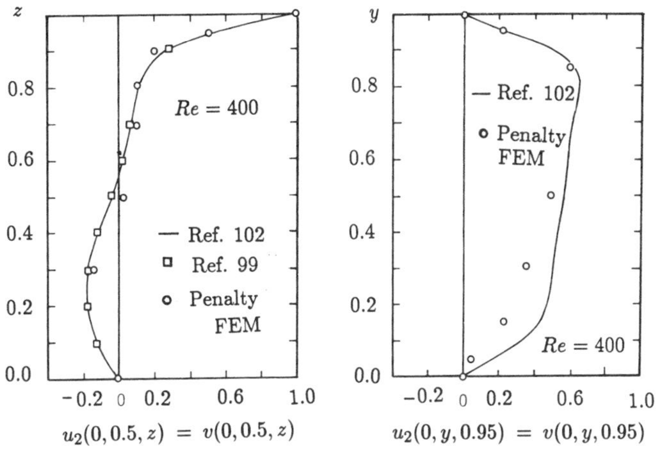

Figure 4.10–11: Velocity fields for a wall-driven cubical cavity, *Re* = 400.

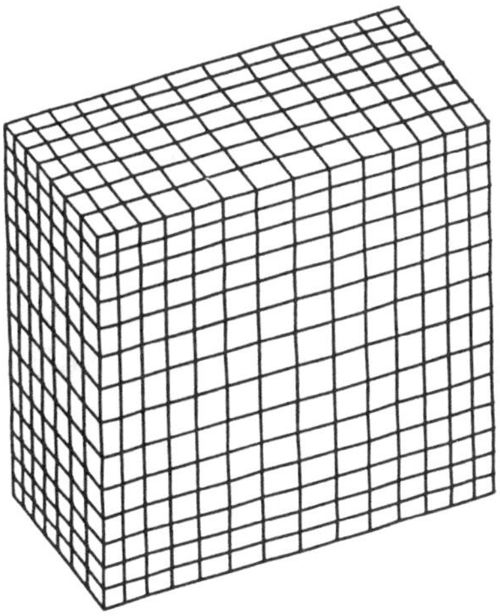

Figure 4.10–12: Finite element mesh (Mesh 3) used for the wall-driven cavity.

Table 4.10–2: Number of iterations for different meshes and solvers.

Mesh	Elements	Equations	GMRES	ORTHOMIN	ORTHORES
1	500	2178	2(1580 + 85)	3(313 + 135 + 1)	3(313 + 137 + 1)
			(19 + 5)	(19 + 9 + 1)	(19 + 9 + 1)
2	864	3549	2(2472 + 137)	3(462 + 243 + 1)	3(463 + 245 + 1)
			(24 + 6)	(24 + 8 + 1)	(24 + 8 + 1)
3	1372	5400	2(3383 + 165)	3(600 + 267 + 1)	3(600 + 269 + 1)
			(35 + 6)	(29 + 9 + 1)	(29 + 9 + 1)
4	2048	7803	2(3811 + 136)	3(789 + 316 + 1)	3(790 + 317 + 1)
			(44 + 6)	(34 + 10 + 1)	(34 + 10 + 1)

It is observed that the convergence rate of the GMRES solver showed no oscillatory behavior but was slow compared to the ORTHOMIN and the ORTHORES solvers. The number of iterations required to arrive at a converged solution for $Re = 10$ for each solver are given in Table 4.10–2. The frontal solver took 3 iterations for each mesh. The number in the parenthesis denotes the number of iterations required to solve the system of equations for each nonlinear iteration. The numbers in parenthesis in the second row for each mesh represent the number of iterations required for pressure calculations. From Table 4.10–2 it is clear that the number of iterations required by the GMRES algorithm is nearly 3.5 times more than those required by the ORTHOMIN or the ORTHORES solvers for the same error tolerances. This reflects in the CPU time required to obtain a converged solution.

Figure 4.10–13 shows the CPU time requirements as a function of the number of elements for different solvers. It is clear that the frontal solver requires more CPU time. This is due to the fact that it is a direct solver and performs an elimination operation which is computationally intensive. Also the convergence of iterative solvers depends to some extent on the initial guess. Once the solution for the linear problem (Stokes) is obtained, it is used as the initial guess for the second iteration and thereon. This results in faster convergence from second the iteration onwards for iterative solvers, as shown in the Table 4.10–2. However, this is not true for the direct solvers because they require the same amount of CPU time for each iteration. Hence the CPU time requirements for direct solvers are enormous. The CPU times for ORTHOMIN were slightly less than those of ORTHORES. This is due to the fact that the GMRES algorithm needed more iterations compared to the ORTHOMIN and the ORTHORES algorithms. A similar trend was observed by [104] for the solution of convection/diffusion problems.

The storage requirements in millions of words for all four solvers for the four meshes are given in Table 4.10–3. From this table it is clear that the storage requirements for ORTHOMIN and ORTHORES are less than those for the GMRES and frontal solvers. The frontal solver requires a very large memory because the equations are stored before backsubstitution. The length of the equations depends on the front width. The front width for each of the meshes is given in parenthesis in Table 4.10–3.

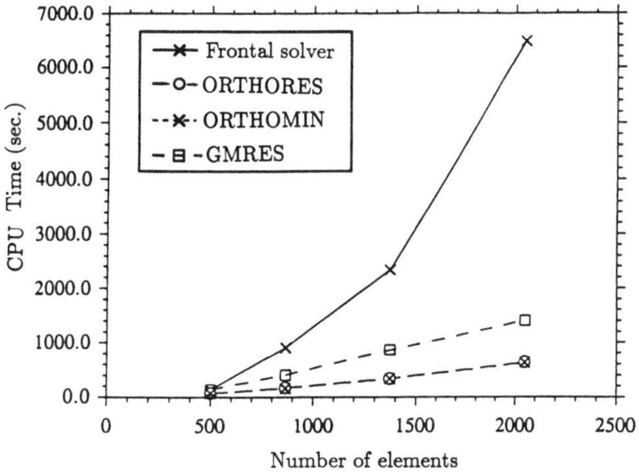

Figure 4.10–13: Comparison of CPU times for different solvers.

Table 4.10–3: Memory requirements (in millions of words) for various methods.

Mesh	Elements	Equations	Frontal	GMRES	ORTHOMIN	ORTHORES
1	500	2178	2.702 (270)	0.822	0.715	0.719
2	864	3549	3.740 (354)	1.257	1.226	1.233
3	1372	5400	7.227 (450)	2.206	1.926	1.947
4	2048	7803	12.146 (558)	3.269	2.878	2.893

The solution for $Re = 100$ was obtained incrementally by using the flow field from the $Re = 10$ solution as the initial solution. The CPU times are given in Table 4.10–4 for Mesh 3 calculations. The difference in the CPU times for the $Re = 100$ and $Re = 10$ solutions gives the time used for the $Re = 100$ solution calculations. From the results shown in Table 4.10–4 it is clear that the iterative solvers converge much faster and need less CPU time because of a good estimate of the initial guess before starting the next Reynolds number flow field calculation.

Table 4.10–4: Comparison of CPU times (in seconds) taken by various methods.

S. no.	Re no.	Frontal	GMRES	ORTHOMIN	ORTHORES
1	10	2345	860	338	341
2	10 and 100	5489	1046	619	626

It is clear that both the ORTHOMIN and the ORTHORES solvers with an element–by–element data structure are superior to the GMRES and frontal solvers.

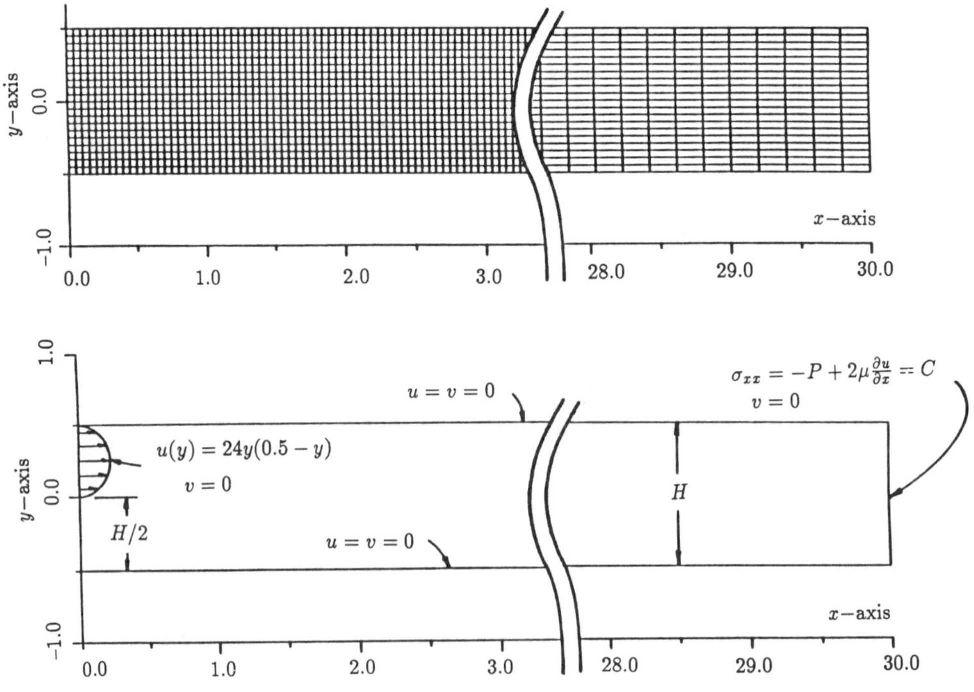

Figure 4.10–14: Schematic and mesh for backward facing step problem.

4.10–7 Backward Facing Step

The laminar flow over a two–dimensional, backward facing step is frequently used as a standard test problem for numerical methods. The problem outlined here was developed as a benchmark for the testing of outflow boundary conditions [105]. The geometry, boundary conditions, and a representative mesh are shown in Figure 4.10–14. The flow problem was solved for a Reynolds number ($Re = U_{avg}H/\nu$) of 800 using zeroth–order continuation and Newton's methods. Mesh refinements from 6×120 to 40×800 elements were used to insure mesh independence of the solution; the velocity field is approximated using the biquadratic interpolation functions, and the linear, discontinuous pressure approximation is used.

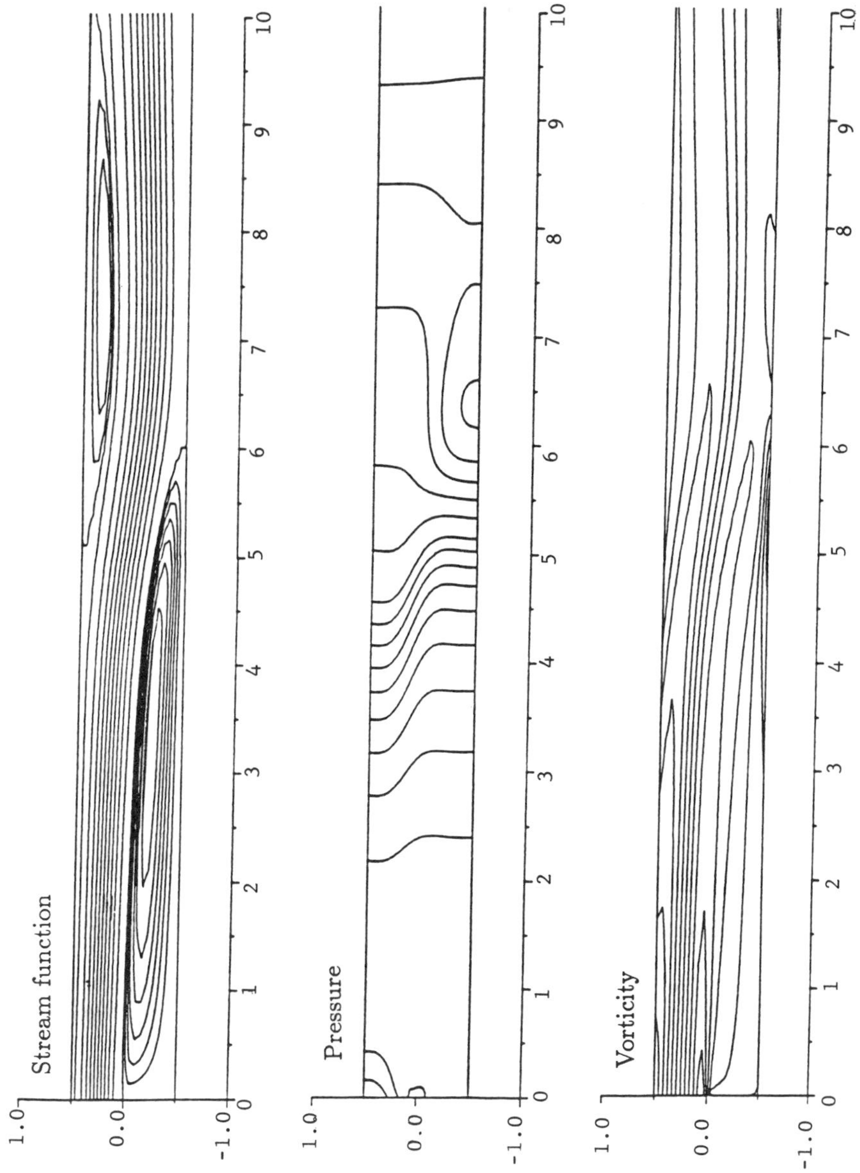

Figure 4.10–15: Contour plots for flow over a backward step; long computational domain.

Figure 4.10–15 contains contour plots of the stream function, pressure, and vorticity obtained on a grid that extended 60 step heights downstream. Two recirculation regions are predicted with their position and intensity corresponding closely to previous numerical solutions. Other simulations were performed on computational domains that were truncated at 30 and 14 step heights downstream. In each case, the traction boundary condition shown in Figure 4.10–14 was specified at outflow. Figure 4.10–16 shows a streamline plot of a result for the shortest domain; the comparison with the long domain result in Figure 4.10–15 is very good, especially considering that the upper wall recirculation zone is truncated by the outflow boundary. A comparison of axial velocities at this location between the solutions obtained with the long and short domains indicates less than a 10% difference. This demonstrates the utility and accuracy of the traction boundary condition for use in short domain internal flow problems.

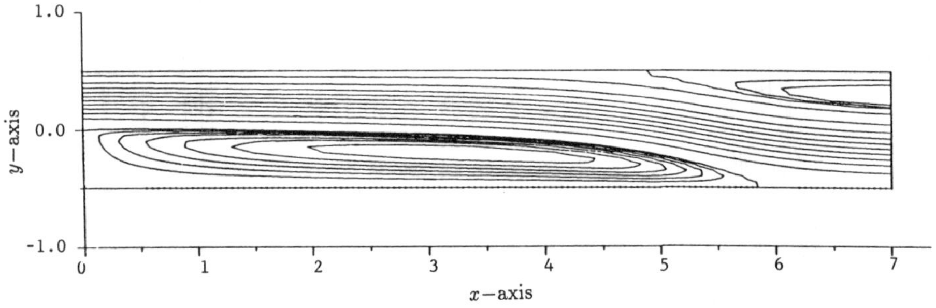

Figure 4.10–16: Contour plots for flow over a backward step; short computational domain.

4.10–8 Flow Past a Submarine

As an example of a large, three-dimensional viscous flow of engineering relevance we consider the flow past an axisymmetric submarine body with an attached fairwater or sail. This simulation was part of a bench mark exercise and consisted of predicting the steady flow past the submarine shape for $Re = 1.2 \times 10^7$. Reference 106 gives details on the entire simulation program.

Shown in Figure 4.10–17 are several views of the surface mesh and surrounding finite element grid. The mesh contained approximately 52,000 eight–node, hexahedral trilinear velocity elements in which the pressure was constant within an element. At the Reynolds number of interest the flow is clearly turbulent and the simulation employed a standard $k - \epsilon$ turbulence model. Wall functions were used to model the viscous sublayer along solid boundaries and provide boundary conditions for the k and ϵ variables. The finite element equations were solved using a pressure correction method as outlined in Section 4.6–6. The individual momentum equations, pressure Poisson equation, and $k - \epsilon$ transport equations were solved sequentially using an early version of the segregated solver developed for the FIDAP code [89]. Further discussion of the segregated solver idea is presented in Section 5.5–3.

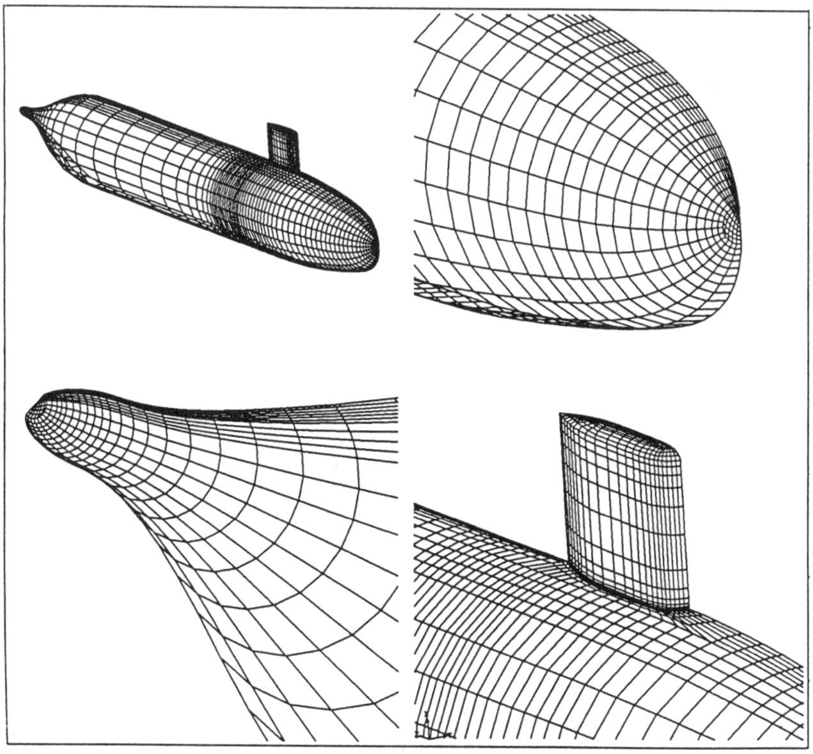

Figure 4.10–17a: Finite element discretization for the surface of the axisymmetric hull with attached fairwater. Clockwise from upper left, the figure illustrates the surface mesh for the complete body and enlarged views of the bow, the hull/fairwater juncture, and the stern tip, respectively.

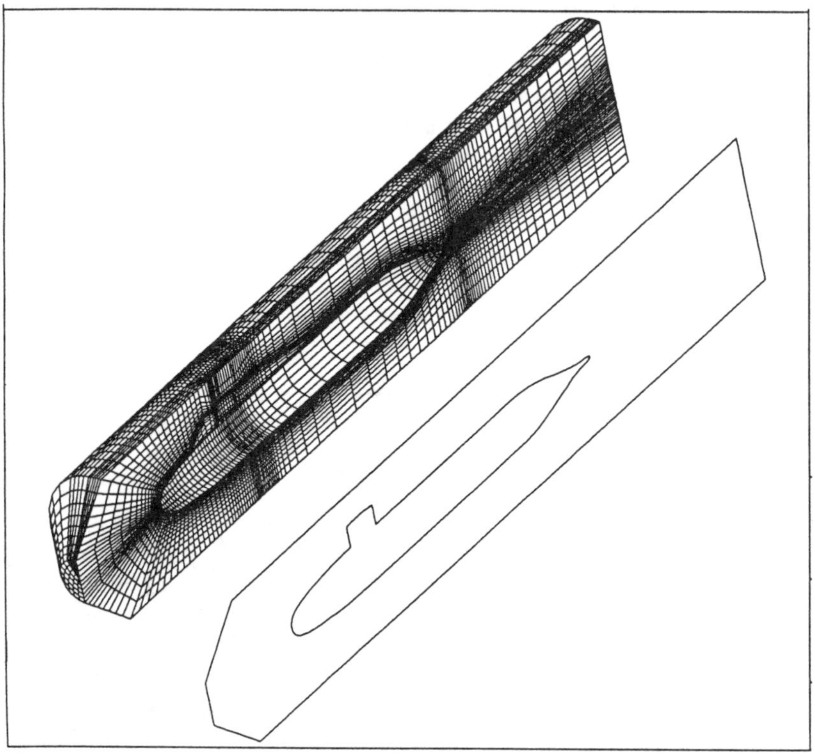

Figure 4.10–17b: Three-dimensional finite element grid used for the numerical computation of steady, turbulent flow ($Re = 1.2 \times 10^7$) past the axisymmetric hull with attached fairwater. This mesh contains 57,523 nodes and 52,295 trilinear brick elements. The lower plot illustrates the relative position of the submarine within the computational domain.

Pressure results for the submarine flow at zero angle of attack are shown as surface contours in Figure 4.10–18. A comparison of this result with an inviscid, potential flow solution showed excellent quantitative agreement. This similarity in results is to be expected since there is virtually no separated flow on the model geometry. One noticeable feature of a viscous flow is the development of secondary flows at internal corners. Shown in Figure 4.10–19 are velocity vectors in a plane normal to the longitudinal axis of the model and located at several stations downstream of the fairwater. The circulation pattern due to the horseshoe vortex that forms at the hull–fairwater junction is clearly visible. Also visible is the vortex emanating from the tip of the fairwater. Further details on this flow and the flow over other hydrodynamic bodies are available in [106].

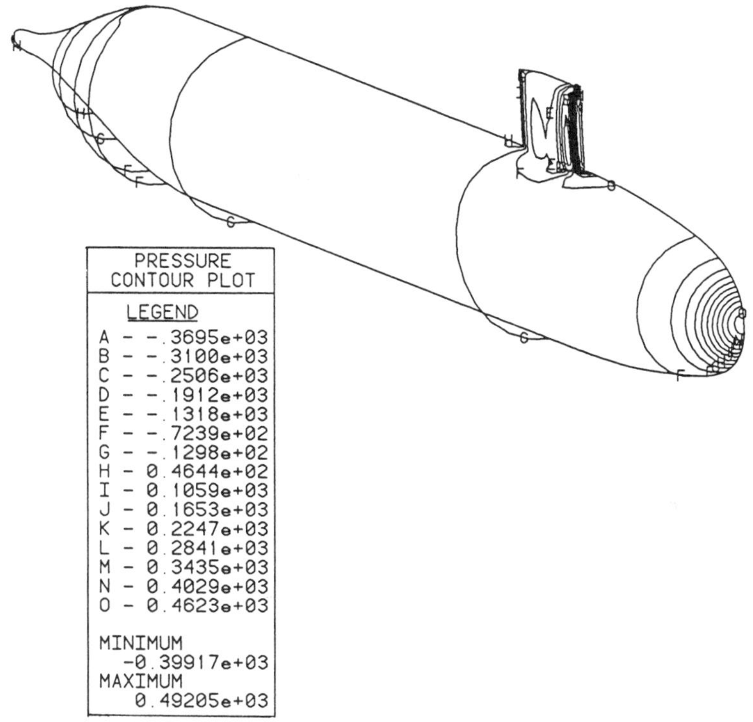

Figure 4.10–18: Pressure contour lines on the surface of the computational domain: 0° attack, 15 contour lines span the pressure extrema.

EXERCISE PROBLEMS

4.1 Consider the vector equations (4.1–1a) and (4.1–2a). Develop the weak statements of the equations in vector form.

Solution:

$$0 = \int_{\Omega^e} Q\,(\nabla \cdot \mathbf{u}) \; d\mathbf{x} \tag{1}$$

and

$$\begin{aligned} 0 &= \int_{\Omega^e} \mathbf{w} \cdot \left[\rho_0 \left(\frac{\partial \mathbf{u}}{\partial t} + \mathbf{u} \cdot \nabla \mathbf{u} \right) - \nabla \cdot \sigma + \rho_0 \mathbf{f} \right] \, d\mathbf{x} \\ &= \int_{\Omega^e} \left[\rho_0 \mathbf{w} \cdot \left(\frac{\partial \mathbf{u}}{\partial t} + \mathbf{u} \cdot \nabla \mathbf{u} \right) + (\nabla \mathbf{w})^T : \sigma + \rho_0 \mathbf{w} \cdot \mathbf{f} \right] \, d\mathbf{x} - \oint_{\Gamma^e} \mathbf{w} \cdot (\hat{\mathbf{n}} \cdot \sigma) ds \end{aligned}$$

where Q and $\mathbf{w}$ are the weight functions. Using equation (4.1–3a), we arrive at the weak form of the momentum equations,

$$0 = \int_{\Omega^e} \left\{ \rho_0 \mathbf{w} \cdot \left(\frac{\partial \mathbf{u}}{\partial t} + \mathbf{u} \cdot \nabla \mathbf{u} \right) + \mu (\nabla \mathbf{w})^T : [\nabla \mathbf{u} + (\nabla \mathbf{u})^T] - (\nabla \cdot \mathbf{w}) P \right.$$
$$\left. + \rho_0 \mathbf{w} \cdot \mathbf{f} \right\} d\mathbf{x} - \oint_{\Gamma^e} \mathbf{w} \cdot \mathcal{T} ds \tag{2}$$

where ':' denotes the "double dot product" ($e.g., \sigma : \mathbf{D} = \sigma_{ij} D_{ji}$). Note that equations (1) and (2) yield equations (4.2–4) and (4.2–5) as special cases when rectangular Cartesian component form of the vectors is used.

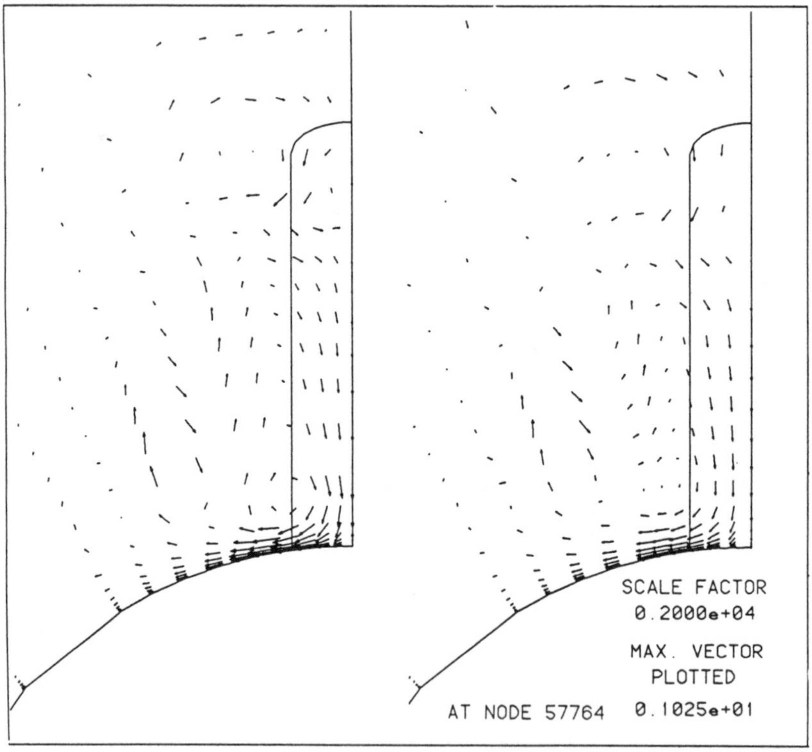

Figure 4.10–19: Velocity vectors depicting the development of the horseshoe vortex which emanates from the hull/fairwater juncture. Velocity vectors in a plane normal to the mean flow direction are illustrated for axial locations corresponding to two (left) and four (right) fairwater chord lengths downstream from the fairwater trailing edge. The mean flow velocity is graphically removed from these illustrations. The fairwater outline is superimposed upon each plot for visual perspective.

4.2 Consider equations (4.2–1) and (4.2–2) in cylindrical coordinates (r, θ, z). For axisymmetric viscous incompressible flows (*i.e.*, the flow field is independent of the θ coordinate), and when the convective (nonlinear) terms are neglected, we have

$$\rho\frac{\partial u}{\partial t} = \frac{1}{r}\frac{\partial}{\partial r}(r\sigma_r) - \frac{\sigma_\theta}{r} + \frac{\partial \sigma_{rz}}{\partial z} + f_r \tag{i}$$

$$\rho\frac{\partial w}{\partial t} = \frac{1}{r}\frac{\partial}{\partial r}(r\sigma_{rz}) + \frac{\partial \sigma_{zz}}{\partial z} + f_z \tag{ii}$$

$$\frac{1}{r}\frac{\partial}{\partial r}(ru) + \frac{\partial w}{\partial z} = 0 \tag{iii}$$

where

$$\sigma_r = -P + 2\mu\frac{\partial u}{\partial r}\ ,\ \sigma_\theta = -P + 2\mu\frac{u}{r}$$

$$\sigma_z = -P + 2\mu\frac{\partial w}{\partial z}\ ,\ \sigma_{rz} = \mu(\frac{\partial u}{\partial z} + \frac{\partial w}{\partial r})$$

Develop the semidiscrete mixed finite element model of the equations.

4.3 Develop the semidiscrete penalty finite element model of the equations in Problem 4.2.

4.4 The equations governing unsteady slow flow of viscous, incompressible fluids in the $x - y$ plane can be expressed in terms of vorticity ω and stream function ψ :

$$\rho\frac{\partial \omega}{\partial t} - \mu\nabla^2\omega = 0$$

$$-\omega - \nabla^2\psi = 0$$

Develop the semidiscrete finite element model of the equations. Discuss the meaning of the secondary variables. Use the α-family of approximation to reduce the ordinary differential equations to algebraic equations.

4.5 Compute the tangent coefficient matrix for the penalty finite element model in equation (4.3–18).

REFERENCES

1. J. N. Reddy, *An Introduction to the Finite Element Method*, Second Edition, McGraw–Hill, New York (1993).

2. J. N. Reddy, *Applied Functional Analysis and Variational Methods in Engineering*, McGraw–Hill, New York (1986); reprinted by Krieger Publishing, Melbourne, Florida (1991).

3. J. N. Reddy, "On Penalty Function Methods in the Finite Element Analysis of Flow Problems," *International Journal of Numerical Methods in Fluids*, **2**, 151–171 (1982).

4. J. T. Oden, "RIP Methods for Stokesian Flows," in R. H. Gallagher, O. C. Zienkiewicz, J. T. Oden, and D. Norrie (eds.), *Finite Element Method in Flow Problems*, Vol. IV, John Wiley & Sons, London (1982).

5. J. T. Oden and G. F. Carey: *Finite Elements, Mathematical Aspects*, Vol. IV, Prentice–Hall, Englewood Cliffs, New Jersey (1983).

6. F. Brezzi and M. Fortin, *Mixed and Hybrid Finite Element Methods*, Springer–Verlag, Berlin (1991).

7. F. Brezzi and K. J. Bathe, "The inf–sup Condition, Equivalent Forms and Applications," in *Reliability of Methods for Engineering Analysis*, K. J. Bathe and D. R. J. Owen (eds.), Pineridge Press, Swansea, U.K. (1986).

8. D. S. Malkus and E. T. Olsen, "Obtaining Error Estimates for Optimally Constrained Incompressible Finite Elements," *Computer Methods in Applied Mechanics and Engineering*, **45**, 331–353 (1984).

9. P. Le Tallac and V. Ruas, "On the Convergence of the Bilinear–Velocity Constant–Pressure Finite Element Method in Viscous Flow," *Computer Methods in Applied Mechanics and Engineering*, **54**, 235–243 (1986).

10. D. Chapelle and K. J. Bathe, "The inf–sup Test," *Computers & Structures*, **47**(4/5), 537–545 (1993).

11. J. N. Reddy, "On the Accuracy and Existence of Solutions to Primitive Variable Models of Viscous Incompressible Fluids," *International Journal of Engineering Science*, **16**, 921-929 (1978).

12. J. N. Reddy, "On the Finite Element Method with Penalty for Incompressible Fluid Flow Problems," in *The Mathematics of Finite Elements and Applications III*, J. R. Whiteman (ed.), Academic Press, New York (1979).

13. R. Temam, *Theory and Numerical Analysis of the Navier-Stokes Equations*, North-Holland, Amsterdam (1977).

14. R. L. Sani, P. M. Gresho, R. L. Lee, and D. F. Griffiths, "The Cause and Cure (?) of the Spurious Pressures Generated by Certain FEM Solutions of the Incompressible Navier-Stokes Equations: Part 1," *International Journal of Numerical Methods in Fluids*, **1**, 17–43 (1981).

15. R. L. Sani, P. M. Gresho, R. L. Lee, and D. F. Griffiths, "The Causeand Cure (?) of the Spurious Pressures Generated by Certain FEM Solutions of the Incompressible Navier-Stokes Equations: Part 2," *International Journal of Numerical Methods in Fluids*, **1**, 171–204 (1981).

16. T. J. R. Hughes, R. L. Taylor, and J. F. Levy, "A Finite Element Method for Incompressible Viscous Flows," in *Proceedings of the Second International Symposium on Finite Element Methods in Flow Problems*, S. Margherita, Italy (1976).

17. T. J. R. Hughes, W. K. Liu, and A. Brooks, "Review of Finite Element Analysis of Incompressible Viscous Flows by Penalty Function Formulation," *Journal of Computational Physics,* **30**, 1–60 (1979).

18. M. Bercovier and M. Engelman, "A Finite Element for the Numerical Solution of Viscous Incompressible Flows," *Journal of Computational Physics,* **30**, 181–201 (1979).

19. M. Engelman, R. Sani, P. M. Gresho, and M. Bercovier, "Consistent vs. Reduced Integration Penalty Methods for Incompressible Media Using Several Old and New Elements," *International Journal of Numerical Methods in Fluids,* **2**, 25–42 (1982).

20. R. S. Marshall, J. C. Heinrich, and O. C. Zienkiewicz, "Natural Convection in a Square Enclosure by a Finite Element, Penalty Function Method, Using Primitive Fluid Variables," *Numerical Heat Transfer*, **1**, 315–330 (1978).

21. J. C. Heinrich and M. Strada, "Penalty Finite Element Analysis of Natural Convection at High Rayleigh Numbers," in *Finite Elements in Fluids, Vol. 4*, R. H. Gallagher, D. Norrie, J. T. Oden and O. C. Zienkiewicz (eds.), John Wiley & Sons, New York (1982).

22. M. P. Reddy and J. N. Reddy, "Finite–Element Analysis of Flows of Non–Newtonian Fluids in Three–Dimensional Enclosures," *International Journal of Non–Linear Mechanics*, **27**, 9–26 (1992).

23. M. P. Reddy, J. N. Reddy, and H. U. Akay, "Penalty Finite Element Analysis of Incompressible Flows Using Element by Element Solution Algorithms," *Computer Methods in Applied Mechanics and Engineering,* **100**, 169–205 (1992).

24. P. M. Gresho, "Some Current CFD Issues Relevant to the Incompressible Navier-Stokes Equations," *Computer Methods in Applied Mechanics and Engineering,* **87**, 201–252 (1991).

25. P. M. Gresho, "Incompressible Fluid Dynamics: Some Fundamental Formulation Issues," *Annual Review of Fluid Mechanics*, **23**, 413–453 (1991).

26. P. M. Gresho, "Some Interesting Issues in Incompressible Fluid Dynamics, Both in the Continuum and in Numerical Simulation," in *Advances in Applied Mechanics,* J. W. Hutchinson and T. Y. Wu (eds.), **28**, 46–140 (1992).

27. C. Taylor and P. Hood, "A Numerical Solution of the Navier-Stokes Equations Using FEM Techniques," *Computers and Fluids*, **1**, 73–100 (1973).

28. T. J. R. Hughes, L. P. Franca, and M. Balestra, "A New Finite Element Formulation for Computational Fluid Dynamics, V. Circumventing the Babuska–Brezzi Condition: A Stable Petrov-Galerkin Formulation for the Stokes Problem Accommodating Equal–Order Interpolations," *Computer Methods in Applied*

Mechanics and Engineering, **59**, 85–99 (1986).

29. T. E. Tezduyar, "Stabilized Finite Element Formulations for Incompressible Flow Computations," in *Advances in Applied Mechanics,* J. W. Hutchinson and T. Y. Wu (eds.), **28**, 1–44 (1992).

30. E. M. Salonen and J. Aalto, "A Pressure Determination Scheme," *Proceedings of the Fifth International Conference on Numerical Methods in Laminar and Turbulent Flow,* C. Taylor, M. D. Olson, P. M. Gresho, and W. G. Habashi (eds.), Pineridge Press, Swansea, U.K. (1985).

31. T. Shiojima and Y. Shimazaki, "A Pressure–Smoothing Scheme for Incompressible Flow Problems," *International Journal of Numerical Methods in Fluids,* **9**, 557–567 (1989).

32. M. Engelman, R. L. Sani, and P. M. Gresho, "The Implementation of Normal and/or Tangential Boundary Conditions in Finite Element Codes for Incompressible Fluid Flow," *International Journal of Numerical Methods in Fluids,* **2**, 225–238 (1982).

33. G. C. Broyden, "A Class of Methods for Solving Nonlinear Simultaneous Equations," *Mathematics of Computation,* **19**, 577–593 (1965).

34. M. S. Engelman, G. Strang, and K. J. Bathe, "The Application of Quasi-Newton Methods in Fluid Mechanics," *International Journal for Numerical Methods in Engineering,* **17**, 707–718 (1981).

35. P. Underwood, "Dynamic Relaxation," in *Computational Methods for Transient Analysis,* T. Belytschko and T. J. R. Hughes (eds.), North-Holland, Amsterdam, 245–265 (1983).

36. C. P. Jackson and K. H. Winters, "A Finite Element Study of the Bernard Problem Using Parameter–Stepping and Bifurcation Search," *International Journal of Numerical Methods in Fluids,* **4**, 127–145 (1984).

37. K. H. Winters, K. A. Cliffe, and C. P. Jackson, "The Prediction of Instabilities Using Bifurcation Theory," in *Numerical Methods for Transient and Coupled Problems,* R. W. Lewis *et al.* (eds.), John Wiley & Sons, New York, 179–198 (1987).

38. K. H. Winters and R. O. Jack, "Anomalous Convection at Low Prandtl Number," *Communications in Applied Numerical Methods,* **5**, 401–404 (1989).

39. K. Winters, "Bifurcation and Stability: A Computational Approach," *Computational Physics Communications,* **65**, 299–309 (1991).

40. W. F. Schmidt, "Adaptive Step Size Selection for Use with the Continuation Method," *International Journal for Numerical Methods in Engineering,* **12**, 677–694 (1978).

41. W. C. Rheinboldt, "On the Solution of Some Nonlinear Equations Arising in the Application of Finite Element Methods," in *Proceedings of the Conference on The Mathematics of Finite Elements and Applications,* J. R. Whiteman (ed.),

Brunel University, Academic Press, London (1975).

42. W. C. Rheinboldt, "Numerical Analysis of Continuation Methods for Nonlinear Structural Problems," *Computers and Structures*, **13**, 103–113 (1981).

43. S. V. Patankar, *Numerical Heat Transfer and Fluid Flow,* McGraw–Hill, New York (1980).

44. V. Haroutunian, M. S. Engelman, and I. Hasbani, "Segregated Finite Element Algorithms for the Numerical Solution of Large–Scale Incompressible Flow Problems," *International Journal for Numerical Methods in Fluids,* **17**, 323–348 (1993).

45. A. C. Benim and W. Zinser, "A Segregated Formulation of Navier–Stokes Equations with Finite Elements," *Computer Methods in Applied Mechanics and Engineering,* **57**, 223–237 (1986).

46. J. G. Rice and R. J. Schnipke, "An Equal–Order Velocity–Pressure Formulation that Does Not Exhibit Spurious Pressure Modes," *Computer Methods in Applied Mechanics and Engineering,* **58**, 135–149 (1986).

47. P. M. Gresho, S. T. Chan, R. L. Lee, and C. D. Upson, "A Modified Finite Element Method for Solving the Time–Dependent, Incompressible Navier–Stokes Equations. Part 1: Theory," *International Journal of Numerical Methods in Fluids,* **4**, 557–598 (1984).

48. P. M. Gresho and S.T. Chan, "On the Theory of Semi–Implicit Projection Methods for Viscous Incompressible Flow and Its Implementation Via a Finite Element Method That Introduces a Nearly Consistent Mass Matrix, Part 1–Theory," *Internatinal Journal for Numerical Methods in Fluids,* **11**, 587–620 (1990).

49. P. M. Gresho, R. L. Lee, S. T. Chan, and R. L. Sani, "Solution of the Time–Dependent Incompressible Navier-Stokes and Boussinesq Equations Using the Galerkin Finite Element Method," in *Proc. IUTAM Symposium on Approximation Methods for Navier–Stokes Problems*, Paderborn, W. Germany, Springer-Verlag, Berlin (1979).

50. P. M. Gresho and S. T. Chan, "On the Theory of Semi–Implicit Projection Methods for Viscous Incompressible Flow and Its Implementation Via a Finite Element Method that Introduces a Nearly Consistent Mass Matrix, Part 2–Implementation," *Internatinal Journal for Numerical Methods in Fluids,* **11**, 621–659 (1990).

51. G. F. Carey and J. T. Oden, *Finite Elements: Computational Aspects,* Prentice–Hall, Englewood Cliffs, New Jersey (1984).

52. M. R. Hestenes and E. L. Stiefel, "Methods of Conjugate Gradients for Solving Linear Systems," *National Bureau of Standards Journal of Research,* **49**, 409–436 (1952).

53. G. H. Golub and C. F. Van Loan, *Matrix Computations, 2nd ed.,* The Johns Hopkins University Press, Baltimore, Maryland (1989).

54. K. C. Jea and D. M. Young, "On the Simplification of Generalized Conjugate-Gradient Methods for Nonsymmetrizable Linear Systems," *Linear Algebra Applications*, **52**, 399–417 (1983).

55. D. J. Evans, "Use of Preconditioning in Iterative Methods for Solving Linear Equations with Symmetric Positive Definite Matrices," *Computer Journal*, **4**, 73–78 (1961).

56. R. L. Fox and E. L. Stanton, "Developments in Structural Analysis by Direct Energy Minimization," *AIAA Journal*, **6**, 1036–1042 (1968).

57. I. Fried, "Gradient Methods for Finite Element Eigenproblems," *AIAA Journal*, **7**, 739–741 (1969).

58. I. Fried, "More on Generalized Iterative Methods in Finite Element Analysis," *AIAA Journal*, **7**, 565–567 (1969).

59. I. Fried, "A Gradient Computational Procedure for the Solution of Large Problems Arising from the Finite Element Discretization Method," *International Journal for Numerical Methods in Engineering*, **2**, 477–494 (1970).

60. A. J. Wathen, "An Analysis of Some Element-By-Element Techniques," *Computer Methods in Applied Mechanics and Engineering*, **74**, 271–287 (1989).

61. T. J. R. Hughes, I. Levit, and J. Winget, "An Element-By-Element Implicit Algorithm for Heat Conduction," *Journal of the Engineering Mechanics Division, ASCE*, **109**(2), 576–585 (1983).

62. M. Ortiz, P. M. Pinsky, and R. L. Taylor, "Unconditionally stable Element-by-Element Algorithms for Dynamic Problems," *Computer Methods in Applied Mechanics and Engineering*, **36**, 223–239 (1983)

63. T. J. R. Hughes, I. Levit, and J. Winget, "An Element-By-Element Solution Algorithm for Problems of Structural and Solid Mechanics," *Computer Methods in Applied Mechanics and Engineering*, **36**, 241–254 (1983).

64. J. M. Winget and T. J. R. Hughes, "Solution Algorithms for Nonlinear Transient Heat Conduction Analysis Employing Element-By-Element Iterative Strategies," *Computer Methods in Applied Mechanics and Engineering*, **52**, 711–815 (1985).

65. T. J. R. Hughes, R. M. Ferencz, and J. O. Hallquist, "Large-Scale Vectorized Implicit Calculations in Solid Mechanics on a Cray X-MP/48 Utilizing EBE Preconditioned Conjugate Gradients," *Computer Methods in Applied Mechanics and Engineering*, **61**, 215–248 (1987).

66. B. Jiang and G. F. Carey, "Subcritical Flow Computation Using an Element-By-Element Conjugate Gradient Method," *Proceedings of the Fifth International Symposium on Finite Elements and Flow Problems*, University of Texas at Austin, 103–106 (1984).

67. G. F. Carey and B. Jiang, "Element-By-Element Linear and Nonlinear Solution Schemes," *Communications in Applied Numerical Methods*, **2**, 145–153 (1986).

68. G. F. Carey, "Parallelism in Finite Element Modeling," *Communications in Applied Numerical Methods,* **2**, 281–287 (1986).

69. B. Jiang and G. F. Carey, "Adaptive Refinement for Least–Squares Finite Elements with Element–By–Element Conjugate Gradient Solution," *International Journal for Numerical Methods in Engineering,* **24**, 569–580 (1987).

70. J. L. Hayes and P. Devloo, "An Element–By–Element Block Iterative Method for Large Non–Linear Problems," *In Innovative Methods for Nonlinear Behavior,* W. K. Liu *et al.* (eds.), Pineridge Press, Swansea, U. K., 51–62 (1985).

71. A. L. G. A. Coutinho *et al.*, "On the Application of an Element–By–Element Lanczos Solver to Large Offshore Structural Engineering Problems," *Computers & Structures,* **27**(1), 27–37 (1987).

72. E. K. Buratynski, "An Element–By–Element Method for Heat Conduction CAE Including Composite Problems," *International Journal for Numerical Methods in Engineering,* **26**, 199–215 (1988).

73. R. B. King and V. Sonnad, "Implementation of an Element–by–Element Solution Algorithm for the Finite Element Method on a Coarse–Grained Parallel Computer," *Computer Methods in Applied Mechanics and Engineering,* **65**, 47–59 (1987).

74. I. Levit, "Element By Element Solvers of Order N," *Computers & Structures,* **27**(3), 357–360 (1987).

75. A. de La Bourdinnaye, "The Element By Element Method as a Preconditioner for Linear Systems Coming from Finite Element Models," *International Journal of Supercomputer Applications,* **3**, 60–68 (1989).

76. K. V. G. Prakhya, "Some Conjugate Gradient Methods for Symmetric and Nonsymmetric Systems," *Communications in Applied Numerical Methods,* **4**, 531–539 (1988).

77. G. F. Carey, E. Barragy, R. Mclay, and M. Sharma, "Element–by–Element Vector and Parallel Computations," *Communications in Applied Numerical Methods,* **4**, 299–307 (1988).

78. L. J. Hayes and P. Devloo, "A Vectorized Version of a Sparse Matrix–Vector Multiply," *International Journal for Numerical Methods in Engineering,* **23**, 1043–1056 (1986).

79. F. Shakib, T. J. R. Hughes, and Z. Johan, "A Multi–Element Group Preconditioned GMRES Algorithm for Nonsymmetric Systems Arising in Finite Element Analysis," *Computer Methods in Applied Mechanics and Engineering,* **75**, 415–456 (1989).

80. Y. Saad and M. H. Schultz, "GMRES: A Generalized Minimal Residual Algorithm for Solving Nonsymmetric Linear Systems," *SIAM Journal of Scientific and Statistical Computations,* **7**(3), 856–869 (1986).

81. F. Angeleri, V. Sonnad, and K. J. Bathe, "Studies of Finite Element Procedures—

An Evaluation of Preconditioned Iterative Solvers," *Computers & Structures,* **32**(3/4), 671–677 (1989).

82. T. E. Tezduyar and J. Liou, "Element-by-Element and Implicit-Explicit Finite Element Formulations for Computational Fluid Dynamics," *First International Symposium on Domain Decomposition Methods for Partial Differential Equations,* R. Glowinski, G. H. Golub, G. A. Meurant, and J. Periaux (eds.), SIAM, Philadelphia, Pennsylvania, 281–300 (1988).

83. T. E. Tezduyar and J. Liou, "Grouped Element-By-Element Iteration Schemes for Incompressible Flow Computations," *Computational Physics Communications,* **53**, 441–453 (1989).

84. T. E. Tezduyar and J. Liou, "Parallel Incompressible Flow Computations with the Grouped Element-By-Element Method," *Proceedings of the Parallel CFD'90, Implementations and Results Using Parallel Computers,* May 1990, Indianapolis, Indiana (1990).

85. J. Liou and T. E. Tezduyar, "A Clustered Element-By-Element Iteration Method for Finite Element Computations," University of Minnesota Supercomputer Institute Research Report, 90/116 (1990).

86. D. S. Kershaw, "The Incomplete Choleski Conjugate Gradient Method for Iterative Solution of System of Linear Equations," *Journal of Computational Physics,* **26**, 43–65 (1978).

87. P. Hood, "Frontal Solution Program for Unsymmetric Matrices," *International Journal for Numerical Methods in Engineering,* **10**, 379–399 (1976); also see **10**, 1055 (1976) for a correction.

88. D. K. Gartling, "NACHOS II - A Finite Element Computer Program for Incompressible Flow Problems," Sandia National Laboratories Report, SAND86-1816 and SAND86-1817, Albuquerque, New Mexico (1987).

89. FIDAP Manual, Version 5.0, Fluid Dynamics International, Evanston, Illinois (1990).

90. A. Nadai, *Theory of Flow and Fracture of Solids, Vol. II,* McGraw-Hill, New York (1963).

91. H. Schlichting, *Boundary-Layer Theory* (translated by J. Kestin), Seventh Edition, McGraw-Hill, New York (1979).

92. O. R. Burggraf, "Analytical and Numerical Studies of the Structure of Steady Separated Flows," *Journal of Fluid Mechanics,* **24**(1), 113–151 (1966).

93. F. Pan and A. Acrivos, "Steady Flow in Rectangular Cavities," *Journal of Fluid Mechanics,* **28**, 643 (1967).

94. R. T. Cheng, "Numerical Solution of the Navier–Stokes Equations by the Finite Element Method," *Physics of Fluids,* **15** (12), 2098–2105 (1972).

95. G. D. Mallison and G. de Vahl Davis, "The Method of False Transient for the Solution of Coupled Elliptic Equations", *Journal of Computational Physics,* **12**, 435–461 (1967).

96. G. de Vahl Davis and G. D. Mallison, "An Evaluation of Upwind and Central Difference Approximations by a Study of Recirculating Flow," *Computers and Fluids*, **4**, 29–43 (1976).

97. T. J. R. Hughes, "A Simple Scheme for Developing Upwind Finite Elements," *International Journal for Numerical Methods in Engineering,* **12**, 1359–1365 (1978).

98. M. Kawahara *et al.*, " Steady and Unsteady Finite Element Analysis of Incompressible Viscous Fluids," *International Journal for Numerical Methods in Engineering,* **10**, 437–456 (1976).

99. H. Takami and K. Kuwahara, "Numerical Study of Three–Dimensional Flow Within a Cubic Cavity," *Journal of Physical Society of Japan,* **37**, 1695–1698 (1974).

100. J. N. Reddy, "Penalty Finite Element Methods for the Solution of Advection and Free Convection Flows,", *Finite Element Methods in Engineering,* A. P. Kabaila and V. A. Pulmano (eds.), University of New South Wales, Sydney, Australia, 583–598 (1979).

101. S. C. R. Dennis, D. B. Ingham, and R. N. Cook, "Finite Difference Methods for Calculating Steady Incompressible Flows in Three Dimensions," *Journal of Computational Physics,* **33**, 325–339 (1979).

102. R. K. Agarwal, "A Third–Order Accurate Upwind Scheme for Navier–Stokes Solutions in Three Dimensions," McDonnell Douglas Research Laboratory, Report No. MDRL 81–20, St. Louis, Missouri (1981).

103. J. N. Reddy, "Penalty–Finite–Element Analysis of 3–D Navier–Stokes Equations," *Computer Methods in Applied Mechanics and Engineering,* **35**, 87–106 (1982).

104. H. P. Langtangen and A. Tevito, "A Numerical Comparison of Conjugate Gradient–Like Methods," *Communications in Applied Numerical Methods,* **34**, 793–798 (1988).

105. D. K. Gartling, "A Test Problem for Outflow Boundary Conditions—Flow Over a Backward-Facing Step," *International Journal of Numerical Methods in Fluids,* **11**, 953–967 (1990).

106. R. C. Givler, D. K. Gartling, M. S. Engelman, and V. Haroutunian, "Navier–Stokes Simulations of Flow Past Three–Dimensional Submarine Models," *Computer Methods in Applied Mechanics and Engineering,* **87**, 175–200 (1991).

Chapter Five

CONVECTIVE HEAT TRANSFER

5.1 INTRODUCTION

5.1–1 Background

The general problem area of convective heat transfer is a particularly challenging one since it represents a situation involving inherently coupled problems, *i.e.*, problems involving multiple physical phenomena. Indeed, the subject itself represents a joining of two classical areas of applied mechanics — fluid mechanics and heat transfer.

Convection problems are generally divided into two major categories depending primarily on the forces that are responsible for the fluid motion. In *forced convection*, the fluid motion is due to the application of pressure or viscous forces on the fluid boundary. *Free* or *natural convection* problems are characterized by the fluid motion which is produced by temperature–induced buoyancy forces. This distinction between the types of convection is not always possible because the two types of driving forces (surface and volume) may appear in varying degrees and combinations.

Since the convective heat transfer process depends directly on the motion of a fluid medium, the Navier–Stokes equations as well as the advection–diffusion equation (*i.e.*, energy equation; see Chapter 2 for a review) that describes thermal energy transport must be considered in the development of a computational scheme. These equations consist of a set of coupled partial differential equations in terms of the velocity components, temperature, and pressure.

In this chapter, finite element models based on the weak formulation of the governing equations of viscous incompressible fluids in the presence of buoyancy forces and inertial effects are developed. The finite element model development is completely analogous to those developed for heat conduction in Chapter 3 and viscous incompressible flows in Chapter 4. Therefore, detailed discussions of the actual model development will not be repeated here.

5.1–2 Governing Equations

The laws describing the flow of Newtonian fluids were presented in Chapter 1, and they are summarized here for ready reference. The equations are written for a fluid region Ω_f using a Cartesian coordinate system x_i in an Eulerian reference frame, with the subscript (or index) $i = 1, 2, 3$ (or 1 and 2 for two–dimensional problems), and the usual summation convention on repeated indices is used [see equations (1.4–6)–(1.4–9)]; time is denoted by t. An extended form of the Boussinesq approximation is used, which allows the fluid properties to be functions of the thermodynamic state (*e.g.*, pressure and temperature) and the density ρ to vary with temperature T according to the relation

$$\rho = \rho_0[1 - \beta(T - T_0)] \tag{5.1-1}$$

where β is the coefficient of thermal expansion and the subscript zero indicates a reference condition. The variation of density as given in (1) is permitted only in the description of the body force; the density in all other situations is assumed to be that of the reference state, ρ_0. The governing equations of convective heat transfer are summarized below.

Conservation of Mass:

$$\frac{\partial u_i}{\partial x_i} = 0 \tag{5.1-2}$$

Conservation of Momentum:

$$\rho_0\left(\frac{\partial u_i}{\partial t} + u_j\frac{\partial u_i}{\partial x_j}\right) - \frac{\partial}{\partial x_j}\left[-P\delta_{ij} + \mu\left(\frac{\partial u_i}{\partial x_j} + \frac{\partial u_j}{\partial x_i}\right)\right] + \rho_0 g_i\beta(T - T_0) = 0 \tag{5.1-3}$$

Conservation of Energy:

$$\rho_0 C_v\left(\frac{\partial T}{\partial t} + u_j\frac{\partial T}{\partial x_j}\right) - \frac{\partial}{\partial x_i}\left(k_{ij}\frac{\partial T}{\partial x_j}\right) - Q - \Phi = 0 \tag{5.1-4a}$$

where u_i denote the velocity components, P is a modified pressure defined by $P = \hat{P}\rho_0 g_i x_i$, ρ_0 is the density, g_i are the gravitational force components, T is the temperature, C_v is the specific heat of the fluid at constant volume, Q is the rate of internal heat generation, μ is the shear viscosity of the fluid, k_{ij} are the components of the thermal conductivity tensor, and Φ is the viscous dissipation in the fluid,

$$\Phi = 2\mu D_{ij}D_{ij} \; ; \quad D_{ij} = \frac{1}{2}\left(\frac{\partial u_i}{\partial x_j} + \frac{\partial u_j}{\partial x_i}\right) \tag{5.1-4b}$$

Recall that equation (2) is also known as the continuity equation or divergence–free condition on the velocity field and equations (3) are known as the Navier–Stokes equations.

The boundary conditions are given by

$$u_i = f_i^u(s,t) \qquad \text{on } \Gamma_u \tag{5.1-5a}$$

$$\mathcal{T}_i \equiv \sigma_{ij}(s,t)n_j(s) = f_i^{\mathcal{T}}(s,t) \qquad \text{on } \Gamma_{\mathcal{T}} \tag{5.1-5b}$$

for the fluid mechanics part of the problem, and

$$T = \hat{T}(s,t) \qquad on\ \Gamma_T \tag{5.1-6a}$$

$$\left(k_{ij}\frac{\partial T}{\partial x_j}\right) n_i + q_c + q_r = q(s,t) \qquad on\ \Gamma_q \tag{5.1-6b}$$

for the heat transfer part of the problem. Here σ_{ij} denote the components of the total stress tensor (*i.e.*, viscous and hydrostatic),

$$\sigma_{ij} = \mu\left(\frac{\partial u_i}{\partial x_j} + \frac{\partial u_j}{\partial x_i}\right) - P\delta_{ij} \tag{5.1-7}$$

We will not explicitly consider the energy equation for solid regions in this chapter as it was considered in detail in Chapter 3. However, it should be recognized that the inclusion of solid body conduction regions in any convection problem (*i.e.*, the so-called conjugate problem) is a relatively straightforward procedure. Also, as mentioned in Chapter 1, convection problems may require the addition of auxiliary variables to describe all of the physical phenomena of interest. We will not consider this case explicitly since the formulation and solution methods for transport equations follow exactly the treatment of the energy equation. Note, however, that the inclusion of auxiliary transport equations may significantly increase the magnitude of the computational effort.

There are five partial differential equations expressed in terms of five unknowns (u_1, u_2, u_3, P, T) for a three-dimensional problem. Following the developments of Chapters 3 and 4, we shall consider two finite element models of equations (1)–(3). The first one is the *velocity-pressure-temperature model* or *mixed model*. The second model is the *penalty-finite element model*. These models were discussed in Chapter 4 for isothermal flows. Extension of these models to convective heat transfer problems is straightforward, as will be shown in this chapter.

5.2 MIXED FINITE ELEMENT MODEL

Let us denote expressions on the left side of the equalities in (2)–(4) by $f_1, \mathbf{f}_2$, and f_3, respectively. As discussed in Section 4.2, the weighted-integral statements of the three equations over a typical element Ω^e are given by

$$\int_{\Omega^e} w_1 f_1 \, d\mathbf{x} = 0$$

$$\int_{\Omega^e} \mathbf{w}_2 \cdot \mathbf{f}_2 \, d\mathbf{x} = 0$$

$$\int_{\Omega^e} w_3 f_3 \, d\mathbf{x} = 0 \tag{5.2-1}$$

where $(w_1, \mathbf{w}_2, w_3)$ are the weight functions, which will be equated to the interpolation functions used for $(P, \mathbf{u}, T)$, respectively.It should be recalled from Chapters 3 and 4 that the weighted-integral statements in equation (1) are reduced to weak statements by integrating the viscous and diffusion parts. All other terms are kept as they are.

Suppose that the dependent variables (T, u_i, P) are approximated by expansions of the form,

$$T(\mathbf{x},t) = \sum_{m=1}^{M} \theta_m(\mathbf{x})T_m(t) = \mathbf{\Theta}^T\mathbf{T} \qquad (5.2-2a)$$

$$u_i(\mathbf{x},t) = \sum_{n=1}^{N} \psi_n(\mathbf{x})u_i^n(t) = \mathbf{\Psi}^T\mathbf{u}_i \qquad (5.2-2b)$$

$$P(\mathbf{x},t) = \sum_{l=1}^{L} \phi_l(\mathbf{x})P_l(t) = \mathbf{\Phi}^T\mathbf{P} \qquad (5.2-2c)$$

where $\mathbf{\Theta}$, $\mathbf{\Psi}$, and $\mathbf{\Phi}$ are vectors of interpolation (or shape) functions, $\mathbf{T}$, $\mathbf{u}_i$, and $\mathbf{P}$ are vectors of nodal values of temperature, velocity components, and pressure, respectively, and superscript $(\cdot)^T$ denotes a transpose of the enclosed vector or matrix. The weight functions $(w_1, \mathbf{w_2}, w_3)$ have the following correspondence (see Reddy [1] for further details),

$$w_1 \approx \phi_l, \; \mathbf{w}_2 \approx \psi_n, \; w_3 \approx \theta_m \qquad (5.2-3)$$

Substitution of equations (2) and (3) into the weak forms associated with equation (1) results in the following finite element equations:

Continuity:

$$-\left[\int_{\Omega^e} \mathbf{\Phi}\frac{\partial \mathbf{\Psi}^T}{\partial x_i}\, d\mathbf{x}\right]\mathbf{u}_i = 0 \qquad (5.2-4)$$

i-th Momentum:

$$\left[\int_{\Omega^e} \rho_0 \mathbf{\Psi}\mathbf{\Psi}^T\, d\mathbf{x}\right]\dot{\mathbf{u}}_i + \left[\int_{\Omega^e} \rho_0 \mathbf{\Psi}(\mathbf{\Psi}^T\mathbf{u}_j)\frac{\partial \mathbf{\Psi}^T}{\partial x_j}\, d\mathbf{x}\right]\mathbf{u}_i + \left[\int_{\Omega^e} \mu\frac{\partial \mathbf{\Psi}}{\partial x_j}\frac{\partial \mathbf{\Psi}^T}{\partial x_j}\, d\mathbf{x}\right]\mathbf{u}_i$$

$$+\left[\int_{\Omega^e} \mu\frac{\partial \mathbf{\Psi}}{\partial x_j}\frac{\partial \mathbf{\Psi}^T}{\partial x_i}\, d\mathbf{x}\right]\mathbf{u}_j - \left[\int_{\Omega^e} \frac{\partial \mathbf{\Psi}}{\partial x_i}\mathbf{\Phi}^T\, d\mathbf{x}\right]\mathbf{P} =$$

$$-\left[\int_{\Omega^e} \rho_0 g_i \beta \mathbf{\Psi}\mathbf{\Theta}^T\, d\mathbf{x}\right]\mathbf{T} + \left\{\int_{\Omega^e} \rho_0 g_i \beta T_0 \mathbf{\Psi}\, d\mathbf{x}\right\} + \left\{\oint_{\Gamma^e} \mathbf{\Psi}T_i\, ds\right\} \qquad (5.2-5)$$

Energy:

$$\left[\int_{\Omega^e} \rho_0 C \Theta \Theta^T \, dx\right] \dot{\mathbf{T}} + \left[\int_{\Omega^e} \rho_0 C_v \Theta (\Psi^T \mathbf{u}_j) \frac{\partial \Theta^T}{\partial x_j} \, dx\right] \mathbf{T} + \left[\int_{\Omega^e} k_{ij} \frac{\partial \Theta}{\partial x_i} \frac{\partial \Theta^T}{\partial x_j} \, d\mathbf{x}\right] \mathbf{T}$$

$$= \left\{\int_{\Omega^e} \Theta Q \, d\mathbf{x}\right\} + \left\{\int_{\Omega^e} \Theta \Phi \, d\mathbf{x}\right\} + \left\{\oint_{\Gamma^e} \Theta q \, ds\right\} \qquad (5.2-6)$$

or

$$\begin{bmatrix} \mathbf{M} & 0 & 0 & 0 \\ 0 & \mathbf{M} & 0 & 0 \\ 0 & 0 & \mathbf{M} & 0 \\ 0 & 0 & 0 & 0 \end{bmatrix} \begin{Bmatrix} \dot{\mathbf{u}}_1 \\ \dot{\mathbf{u}}_2 \\ \dot{\mathbf{u}}_3 \\ \dot{\mathbf{P}} \end{Bmatrix} + \begin{bmatrix} \mathbf{C}(\mathbf{u}) & 0 & 0 & 0 \\ 0 & \mathbf{C}(\mathbf{u}) & 0 & 0 \\ 0 & 0 & \mathbf{C}(\mathbf{u}) & 0 \\ 0 & 0 & 0 & 0 \end{bmatrix} \begin{Bmatrix} \mathbf{u}_1 \\ \mathbf{u}_2 \\ \mathbf{u}_3 \\ \mathbf{P} \end{Bmatrix} +$$

$$\begin{bmatrix} \hat{\mathbf{K}}_{11} & \mathbf{K}_{21} & \mathbf{K}_{31} & -\mathbf{Q}_1 \\ \mathbf{K}_{12} & \hat{\mathbf{K}}_{22} & \mathbf{K}_{32} & -\mathbf{Q}_2 \\ \mathbf{K}_{13} & \mathbf{K}_{23} & \hat{\mathbf{K}}_{33} & -\mathbf{Q}_3 \\ -\mathbf{Q}_1^T & -\mathbf{Q}_2^T & -\mathbf{Q}_3^T & 0 \end{bmatrix} \begin{Bmatrix} \mathbf{u}_1 \\ \mathbf{u}_2 \\ \mathbf{u}_3 \\ \mathbf{P} \end{Bmatrix} = \begin{Bmatrix} \mathbf{F}_1(\mathbf{T}) \\ \mathbf{F}_2(\mathbf{T}) \\ \mathbf{F}_3(\mathbf{T}) \\ 0 \end{Bmatrix} \qquad (5.2-7)$$

$$[\mathbf{N}]\left\{\dot{\mathbf{T}}\right\} + [\mathbf{D}(\mathbf{u})]\{\mathbf{T}\} + [\mathbf{L}]\{\mathbf{T}\} = \{\mathbf{G}(\mathbf{T})\} \qquad (5.2-8)$$

The coefficient matrices shown in equations (7) and (8) are defined by

$$\begin{aligned} \hat{\mathbf{K}}_{11} &= 2\mathbf{K}_{11} + \mathbf{K}_{22} + \mathbf{K}_{33} \\ \hat{\mathbf{K}}_{22} &= \mathbf{K}_{11} + 2\mathbf{K}_{22} + \mathbf{K}_{33} \\ \hat{\mathbf{K}}_{33} &= \mathbf{K}_{11} + \mathbf{K}_{22} + 2\mathbf{K}_{33} \end{aligned} \qquad (5.2-9a)$$

$$\begin{aligned} \mathbf{M} &= \int_{\Omega^e} \rho_0 \Psi \Psi^T \, dx \; ; \quad \mathbf{C}(\mathbf{u}) = \int_{\Omega^e} \rho_0 \Psi (\Psi^T \mathbf{u}_j) \frac{\partial \Psi^T}{\partial x_j} \, d\mathbf{x} \\ \mathbf{K}_{ij} &= \int_{\Omega^e} \mu \frac{\partial \Psi}{\partial x_i} \frac{\partial \Psi^T}{\partial x_j} \, d\mathbf{x} \; ; \quad \mathbf{Q}_i = \int_{\Omega^e} \frac{\partial \Psi}{\partial x_i} \Phi^T \, d\mathbf{x} \\ \mathbf{F}_i(\mathbf{T}) &= -\int_{\Omega^e} \rho_0 g_i \beta \Psi \Psi^T \mathbf{T} \, d\mathbf{x} + \int_{\Omega^e} \rho_0 g_i \beta T_0 \Psi \, d\mathbf{x} + \oint_{\Gamma^e} \Psi T_i \, ds \\ \mathbf{D}(\mathbf{u}) &= \int_{\Omega^e} \rho_0 C \Psi (\Psi^T \mathbf{u}_j) \frac{\partial \Psi^T}{\partial x_j} \, d\mathbf{x} \\ \mathbf{N} &= \int_{\Omega^e} \rho_0 C_v \Theta \Theta^T \, dx \; ; \quad \mathbf{L} = \int_{\Omega^e} k \frac{\partial \Theta}{\partial x_i} \frac{\partial \Theta^T}{\partial x_i} \, d\mathbf{x} \\ \mathbf{G} &= \int_{\Omega^e} \Theta Q \, d\mathbf{x} + \int_{\Omega^e} \Theta \Phi \, d\mathbf{x} + \oint_{\Gamma^e} \Theta q \, ds \end{aligned} \qquad (5.2-9b)$$

where summation on repeated indices is implied. The above equations can be written symbolically as

Continuity:

$$-\mathbf{Q}^T\mathbf{u} = 0 \tag{5.2-10}$$

Momentum:

$$\mathbf{M}\dot{\mathbf{u}} + \mathbf{C}\mathbf{u} + \mathbf{K}\mathbf{u} - \mathbf{Q}\mathbf{P} + \mathbf{B}\mathbf{T} = \mathbf{F} \tag{5.2-11}$$

Energy:

$$\mathbf{N}\dot{\mathbf{T}} + \mathbf{D}\mathbf{T} + \mathbf{L}\mathbf{T} = \mathbf{G} \tag{5.2-12}$$

where the superposed dot represents a time derivative and $\mathbf{u}^T = \{\mathbf{u}_1^T, \mathbf{u}_2^T, \mathbf{u}_3^T\}$. In writing equation (11) the buoyancy term $\mathbf{B}$ has been separated from the general force vector $\mathbf{F}$. The expression $\mathbf{BT}$ has the meaning

$$\mathbf{BT} = \begin{bmatrix} \mathbf{B}_1 & 0 & 0 \\ 0 & \mathbf{B}_2 & 0 \\ 0 & 0 & \mathbf{B}_3 \end{bmatrix} \begin{Bmatrix} \mathbf{T} \\ \mathbf{T} \\ \mathbf{T} \end{Bmatrix} \tag{5.2-13a}$$

where

$$\mathbf{B}_i = \int_{\Omega^e} \rho_0 \beta g_i \mathbf{\Psi}^T \mathbf{\Theta}\, d\mathbf{x} \tag{5.2-13b}$$

Note that the solid body conduction equation of Chapter 3 is obtained from (12) by neglecting the transport term $\mathbf{D}$.

Equations (10)–(12) can be combined into a single matrix equation

$$\begin{bmatrix} \mathbf{M} & 0 & 0 \\ 0 & 0 & 0 \\ 0 & 0 & \mathbf{N} \end{bmatrix} \begin{Bmatrix} \dot{\mathbf{u}} \\ \dot{\mathbf{P}} \\ \dot{\mathbf{T}} \end{Bmatrix} + \begin{bmatrix} \mathbf{C}(\mathbf{u}) + \mathbf{K}(\mathbf{u},\mathbf{T}) & -\mathbf{Q} & \mathbf{B}(\mathbf{T}) \\ -\mathbf{Q}^T & 0 & 0 \\ 0 & 0 & \mathbf{D}(\mathbf{u}) + \mathbf{L}(\mathbf{T}) \end{bmatrix} \begin{Bmatrix} \mathbf{u} \\ \mathbf{P} \\ \mathbf{T} \end{Bmatrix} = \begin{Bmatrix} \mathbf{F}(\mathbf{T}) \\ 0 \\ \mathbf{G}(\mathbf{T},\mathbf{u}) \end{Bmatrix} \tag{5.2-14}$$

In a more symbolic format, equation (14) can be expressed as

$$\bar{\mathbf{M}}\dot{\mathbf{U}} + \bar{\mathbf{K}}\mathbf{U} = \bar{\mathbf{F}} \tag{5.2-15a}$$

where

$$\mathbf{U}^T = \{\mathbf{u}_1^T, \mathbf{u}_2^T, \mathbf{u}_3^T, \mathbf{P}^T, \mathbf{T}^T\} \qquad (5.2-15b)$$

Note that the general form of equation (15) is the same as the nonlinear diffusion equation (3.6–3). Therefore, the time–approximations discussed in Chapter 3 are readily applicable to the ordinary differential equations in (15).

The three sets of interpolation functions used in (2) should be of the Lagrange type, *i.e.*, derived by interpolating only the values of the functions—and not their derivatives. There are three different finite elements associated with the three field variables (u_i, P, T). Although, in principle, one can use different interpolation functions for the velocity field and temperature, it is practical to use the same type of interpolation for the two field variables (T, u_i). The choice of interpolation functions used for the pressure variable (see Section 4.5–2) in the mixed finite element model is constrained by the special role the pressure plays in incompressible flows. In order to prevent an overconstrained system of discrete equations, the interpolation used for pressure must be at least one order lower than that used for the velocity field (*i.e.*, unequal interpolation). Further, pressure need not be made continuous across elements because the pressure variable does not constitute a primary variable of the mixed model.

For two–dimensional flows of viscous incompressible fluids, often the Lagrange quadratic interpolation is used for the velocity components. Two different pressure approximations are available when the velocities are approximated by quadratic Lagrange functions. The first is a continuous–bilinear approximation, in which the pressure is defined at the corner nodes of the element and is made continuous across element boundaries. The second pressure approximation involves a discontinuous–linear variation defined on the element by

$$\mathbf{\Phi} = \begin{Bmatrix} 1 \\ x \\ y \end{Bmatrix} \qquad (5.2-16)$$

Here the unknowns are not nodal point values of the pressure but correspond to the coefficients in $P = a \cdot 1 + b \cdot x + c \cdot y$. A standard element used in the analysis of three–dimensional viscous flow problems is the eight–node, hexahedron (brick) element, where the velocity components are approximated using the trilinear Lagrange functions and the pressure is a constant (and obviously discontinuous between elements).

5.3 PENALTY FINITE ELEMENT MODEL

5.3–1 Preliminary Comments

Recall that the velocity field of viscous incompressible flows must satisfy the momentum equations (5.1–3) and in addition the continuity equation (5.1–2). Equation (5.1–2) is treated as a constraint on the velocity field. In the penalty function method, the constrained problem is reformulated as an unconstrained problem. The discussion presented in Section 4.3 is applicable here with the body force components

f_i in equations (4.3–16) replaced with the buoyancy forces, $g_i\beta(T-T_0)$. The weak statements associated with the penalty formulation of convective heat transfer are given by

Continuity and Momentum:

$$\begin{aligned}0 = &\int_{\Omega^e} \rho_0\, \delta u_1 \left[\frac{\partial u_1}{\partial t} + (u_1\frac{\partial u_1}{\partial x_1} + u_2\frac{\partial u_1}{\partial x_2} + u_3\frac{\partial u_1}{\partial x_3}) - g_1\beta(T-T_0)\right] d\mathbf{x} \\ &+ \int_{\Omega^e} \left[2\mu\frac{\partial \delta u_1}{\partial x_1}\frac{\partial u_1}{\partial x_1} + \mu\frac{\partial \delta u_1}{\partial x_2}\left(\frac{\partial u_1}{\partial x_2} + \frac{\partial u_2}{\partial x_1}\right) + \mu\frac{\partial \delta u_1}{\partial x_3}\left(\frac{\partial u_1}{\partial x_3} + \frac{\partial u_3}{\partial x_1}\right)\right] d\mathbf{x} \\ &- \oint_{\Gamma^e} \delta u_1\, T_1\, ds + \int_{\Omega^e} \gamma_e\, \frac{\partial \delta u_1}{\partial x_1}\left(\frac{\partial u_1}{\partial x_1} + \frac{\partial u_2}{\partial x_2} + \frac{\partial u_3}{\partial x_3}\right) d\mathbf{x} \qquad (5.3-1a)\end{aligned}$$

$$\begin{aligned}0 = &\int_{\Omega^e} \rho_0\, \delta u_2 \left[\frac{\partial u_2}{\partial t} + (u_1\frac{\partial u_2}{\partial x_1} + u_2\frac{\partial u_2}{\partial x_2} + u_3\frac{\partial u_2}{\partial x_3}) - g_2\beta(T-T_0)\right] d\mathbf{x} \\ &+ \int_{\Omega^e} \left[2\mu\frac{\partial \delta u_2}{\partial x_2}\frac{\partial u_2}{\partial x_2} + \mu\frac{\partial \delta u_2}{\partial x_1}\left(\frac{\partial u_1}{\partial x_2} + \frac{\partial u_2}{\partial x_1}\right) + \mu\frac{\partial \delta u_2}{\partial x_3}\left(\frac{\partial u_2}{\partial x_3} + \frac{\partial u_3}{\partial x_2}\right)\right] d\mathbf{x} \\ &- \oint_{\Gamma^e} \delta u_2\, T_2\, ds + \int_{\Omega^e} \gamma_e\, \frac{\partial \delta u_2}{\partial x_2}\left(\frac{\partial u_1}{\partial x_1} + \frac{\partial u_2}{\partial x_2} + \frac{\partial u_3}{\partial x_3}\right) d\mathbf{x} \qquad (5.3-1b)\end{aligned}$$

$$\begin{aligned}0 = &\int_{\Omega^e} \rho_0\, \delta u_3 \left[\frac{\partial u_3}{\partial t} + (u_1\frac{\partial u_3}{\partial x_1} + u_2\frac{\partial u_3}{\partial x_2} + u_3\frac{\partial u_3}{\partial x_3}) - g_3\beta(T-T_0)\right] d\mathbf{x} \\ &+ \int_{\Omega^e} \left[2\mu\frac{\partial \delta u_3}{\partial x_3}\frac{\partial u_3}{\partial x_3} + \mu\frac{\partial \delta u_3}{\partial x_1}\left(\frac{\partial u_1}{\partial x_3} + \frac{\partial u_3}{\partial x_1}\right) + \mu\frac{\partial \delta u_3}{\partial x_2}\left(\frac{\partial u_2}{\partial x_3} + \frac{\partial u_3}{\partial x_2}\right)\right] d\mathbf{x} \\ &- \oint_{\Gamma^e} \delta u_3\, T_3\, ds + \int_{\Omega^e} \gamma_e\, \frac{\partial \delta u_3}{\partial x_3}\left(\frac{\partial u_1}{\partial x_1} + \frac{\partial u_2}{\partial x_2} + \frac{\partial u_3}{\partial x_3}\right) d\mathbf{x} \qquad (5.3-1c)\end{aligned}$$

Energy:

$$\begin{aligned}0 = &\int_{\Omega^e} \rho_0 C_v\, \delta T \left[\frac{\partial T}{\partial t} + (u_1\frac{\partial T}{\partial x_1} + u_2\frac{\partial T}{\partial x_2} + u_3\frac{\partial T}{\partial x_3}) - Q\right] d\mathbf{x} \\ &+ \int_{\Omega^e} k_{ij}\frac{\partial \delta T}{\partial x_i}\frac{\partial T}{\partial x_j} d\mathbf{x} - \oint_{\Gamma^e} \delta T\, q\, ds \qquad (5.3-2)\end{aligned}$$

where γ_e is the penalty parameter. We note that the pressure does not appear explicitly in the weak forms (1) and (2), although it is a part of the boundary stresses, T_i [see equations (5.1–5b) and (5.1–7)]. An approximation for the pressure can be post–computed from the relation [see equation (4.4–21)]

$$P = -\gamma_e\left(\frac{\partial u_1}{\partial x_1} + \frac{\partial u_2}{\partial x_2} + \frac{\partial u_3}{\partial x_3}\right) = -\gamma_e\frac{\partial u_i}{\partial x_i} \qquad (5.3-3)$$

5.3–2 Reduced Integration Penalty Model

The penalty finite element model is obtained from equation (1) by substituting finite element interpolation (5.2–2a) for the velocity field and $\delta u_i = \Psi^T$:

$$\begin{bmatrix} \mathbf{M} & 0 & 0 \\ 0 & \mathbf{M} & 0 \\ 0 & 0 & \mathbf{M} \end{bmatrix} \begin{Bmatrix} \dot{\mathbf{u}}_1 \\ \dot{\mathbf{u}}_2 \\ \dot{\mathbf{u}}_3 \end{Bmatrix} + \begin{bmatrix} \mathbf{C}(\mathbf{u}) & 0 & 0 \\ 0 & \mathbf{C}(\mathbf{u}) & 0 \\ 0 & 0 & \mathbf{C}(\mathbf{u}) \end{bmatrix} \begin{Bmatrix} \mathbf{u}_1 \\ \mathbf{u}_2 \\ \mathbf{u}_3 \end{Bmatrix} +$$

$$\begin{bmatrix} 2\mathbf{K}_{11} + \mathbf{K}_{22} + \mathbf{K}_{33} & \mathbf{K}_{21} & \mathbf{K}_{31} \\ \mathbf{K}_{12} & \mathbf{K}_{11} + 2\mathbf{K}_{22} + \mathbf{K}_{33} & \mathbf{K}_{32} \\ \mathbf{K}_{13} & \mathbf{K}_{23} & \mathbf{K}_{11} + \mathbf{K}_{22} + 2\mathbf{K}_{33} \end{bmatrix} \begin{Bmatrix} \mathbf{u}_1 \\ \mathbf{u}_2 \\ \mathbf{u}_3 \end{Bmatrix}$$

$$+ \begin{bmatrix} \hat{\mathbf{K}}_{11} & \hat{\mathbf{K}}_{12} & \hat{\mathbf{K}}_{13} \\ \hat{\mathbf{K}}_{21} & \hat{\mathbf{K}}_{22} & \hat{\mathbf{K}}_{23} \\ \hat{\mathbf{K}}_{31} & \hat{\mathbf{K}}_{32} & \hat{\mathbf{K}}_{33} \end{bmatrix} \begin{Bmatrix} \mathbf{u}_1 \\ \mathbf{u}_2 \\ \mathbf{u}_3 \end{Bmatrix} = \begin{Bmatrix} \mathbf{F}_1 \\ \mathbf{F}_2 \\ \mathbf{F}_3 \end{Bmatrix} \qquad (5.3-4)$$

where $\mathbf{M}, \mathbf{C}(\mathbf{u}), \mathbf{K}_{ij}$, and $\mathbf{F}_i$ are the same as those defined in equation (5.2–9), and

$$\hat{\mathbf{K}}_{ij} = \int_{\Omega^e} \gamma_e \frac{\partial \Psi}{\partial x_i} \frac{\partial \Psi^T}{\partial x_j} \, d\mathbf{x} \qquad (5.3-5)$$

The finite element model of the energy equation, equation (5.2–12), remains unchanged.

The comments made in Section 4.5–3 concerning the evaluation of the integrals in $\hat{\mathbf{K}}$ of equation (4) should be recalled here. These coefficients (*i.e.*, penalty terms) should be evaluated using a numerical integration rule of an order less than that required to integrate them exactly. For example, if a linear rectangular element is used to approximate the velocity field for two–dimensional problems, all matrix coefficients except the penalty terms should be evaluated using the 2×2 Gauss quadrature, and the penalty terms should be evaluated using the one–point (1×1) Gauss quadrature. When a quadratic rectangular element is used, the 3×3 Gauss quadrature is used to evaluate the non–penalty terms, and the 2×2 Gauss quadrature is used to evaluate the penalty terms. Similar comments apply to three–dimensional elements.

Concerning the post–computation of pressure in the penalty model, in general, the pressure computed from equation (3) at the integration points is not always reliable and accurate, and one is advised to use either the Poisson equation for pressure [see equation (4.5–10)] or the scheme suggested in Section 4.5–4 [see equation (4.5–17)].

5.3–3 Consistent Penalty Model

In this model, the finite element model of equation (3) is constructed first,

$$\left[\int_{\Omega^e} \Phi \Phi^T \, dx \right] \mathbf{P} = - \left[\int_{\Omega^e} \gamma_e \Phi \frac{\partial \Psi^T}{\partial x_i} \, d\mathbf{x} \right] \mathbf{u}_i \qquad (5.3-6a)$$

or in matrix notation

$$\mathbf{M}_p\mathbf{P} = -\gamma_e \mathbf{Q}^T\mathbf{u} \qquad (5.3-6b)$$

Inverting equation (6b) for $\mathbf{P}$, we obtain

$$\mathbf{P} = -\gamma_e {\mathbf{M}_p}^{-1}\mathbf{Q}^T\mathbf{u} \qquad (5.3-7)$$

where $\mathbf{u}^T = \{\mathbf{u}_1^T, \mathbf{u}_2^T, \mathbf{u}_3^T\}$, $\mathbf{Q}^T = \{\mathbf{Q}_1^T, \mathbf{Q}_2^T, \mathbf{Q}_3^T\}$, and $\mathbf{Q}_i$ and $\mathbf{M}_p$ are given by

$$\mathbf{Q}_i = \int_{\Omega^e} \frac{\partial \boldsymbol{\Psi}}{\partial x_i}\boldsymbol{\Phi}^T\, d\mathbf{x}\,; \quad \mathbf{M}_p = \int_{\Omega^e} \boldsymbol{\Phi}\boldsymbol{\Phi}^T\, d\mathbf{x} \qquad (5.3-8)$$

Next, equation (7) is substituted for the pressure into the finite element model (5.2–11) to obtain

$$\mathbf{M}\dot{\mathbf{u}} + (\mathbf{C}(\mathbf{u}) + \mathbf{K} + \mathbf{K}_p)\mathbf{u} = \mathbf{F} \qquad (5.3-9a)$$

with

$$\mathbf{K}_p = \gamma_e \mathbf{Q}{\mathbf{M}_p}^{-1}\mathbf{Q}^T \qquad (5.3-9b)$$

It should be noted that reduced integration is not used to evaluate $\mathbf{K}_p$. For straight–sided elements, one can show the equivalence of matrix $\hat{\mathbf{K}}$ in equation (4) and the matrix $\mathbf{K}_p$. The post–computation of the pressure P using equation (7) in the consistent penalty method yields very good results.

5.4 FINITE ELEMENT MODELS OF POROUS FLOW

The derivation of the finite element model for the porous flow problem follows the developments of Section 4.4 in a completely analogous manner. For completeness, we record the mixed finite element model including the buoyancy term:

Continuity:

$$-\left[\int_{\Omega^e} \boldsymbol{\Phi}\frac{\partial \boldsymbol{\Psi}^T}{\partial x_i}\, d\mathbf{x}\right]\mathbf{u}_i = 0 \qquad (5.4-1)$$

Momentum:

$$\left[\int_{\Omega^e} \frac{\rho_0}{\phi}\boldsymbol{\Psi}\boldsymbol{\Psi}^T\, d\mathbf{x}\right]\dot{\mathbf{u}}_i + \left[\int_{\Omega^e} \frac{\rho_0\hat{c}}{\sqrt{\kappa}}\boldsymbol{\Psi}(\boldsymbol{\Psi}^T||\vec{\mathbf{u}}||)\boldsymbol{\Psi}^T\, d\mathbf{x}\right]\mathbf{u}_i + \left[\int_{\Omega^e} \frac{\mu}{\kappa}\boldsymbol{\Psi}\boldsymbol{\Psi}^T\, d\mathbf{x}\right]\dot{\mathbf{u}}_i$$

$$+\left[\int_{\Omega^e} \mu_e\frac{\partial \boldsymbol{\Psi}}{\partial x_j}\frac{\partial \boldsymbol{\Psi}^T}{\partial x_j}\, d\mathbf{x}\right]\mathbf{u}_i + \left[\int_{\Omega^e} \mu_e\frac{\partial \boldsymbol{\Psi}}{\partial x_j}\frac{\partial \boldsymbol{\Psi}^T}{\partial x_i}\, d\mathbf{x}\right]\mathbf{u}_j - \left[\int_{\Omega^e} \frac{\partial \boldsymbol{\Psi}}{\partial x_i}\boldsymbol{\Phi}^T\, d\mathbf{x}\right]\mathbf{P} =$$

$$-\left[\int_{\Omega^e} \rho_0 g_i \beta \Psi \Theta d\mathbf{x}\right]\mathbf{T} + \left\{\int_{\Omega^e} \rho_0 g_i \beta T_0 \Psi \, d\mathbf{x}\right\} + \left\{\oint_{\Gamma^e} \Psi T_i ds\right\} \qquad (5.4-2)$$

Energy:

$$\left[\int_{\Omega^e} (\rho_0 C)_e \Theta\Theta^T \, d\mathbf{x}\right]\dot{\mathbf{T}} + \left[\int_{\Omega^e} \rho_0 C \Theta(\Psi^T \mathbf{u_j}) \frac{\partial \Theta^T}{\partial x_j} \, d\mathbf{x}\right]\mathbf{T} +$$

$$\left[\int_{\Omega^e} k_{ij} \frac{\partial \Theta}{\partial x_i} \frac{\partial \Theta^T}{\partial x_j} \, d\mathbf{x}\right]\mathbf{T} = \left\{\int_{\Omega^e} \Theta Q \, d\mathbf{x}\right\} + \left\{\oint_{\Gamma^e} \Theta q_n \, ds\right\} \qquad (5.4-3)$$

Note that the dissipation term is omitted from the energy equation. The above equations can be written symbolically in the matrix form as

Continuity:

$$-\tilde{\mathbf{Q}}^T \mathbf{u} = 0 \qquad (5.4-4)$$

Momentum:

$$\tilde{\mathbf{M}}\dot{\mathbf{u}} + \tilde{\mathbf{C}}\mathbf{u} + \tilde{\mathbf{K}}\mathbf{u} - \tilde{\mathbf{Q}}\mathbf{P} + \tilde{\mathbf{B}}\mathbf{T} = \tilde{\mathbf{F}} \qquad (5.4-5)$$

Energy:

$$\tilde{\mathbf{N}}\dot{\mathbf{T}} + \tilde{\mathbf{D}}\mathbf{T} + \tilde{\mathbf{L}}\mathbf{T} = \tilde{\mathbf{G}} \qquad (5.4-6)$$

where $\mathbf{u}^T = \{\mathbf{u}_1^T, \mathbf{u}_2^T, \mathbf{u}_3^T\}$ and

$$\tilde{\mathbf{M}} = \int_{\Omega^e} \frac{\rho_0}{\phi} \Psi\Psi^T \, d\mathbf{x}; \quad \tilde{\mathbf{C}}(\mathbf{u}) = \int_{\Omega^e} \frac{\rho_0 \hat{c}}{\sqrt{\kappa}} \Psi\Psi^T ||\vec{\mathbf{u}}||\Psi^T \, d\mathbf{x}$$

$$\tilde{\mathbf{A}} = \int_{\Omega^e} \frac{\mu}{\kappa} \Psi\Psi^T \, d\mathbf{x}; \quad \tilde{\mathbf{K}}_{ij} = \int_{\Omega^e} \mu_e \frac{\partial \Psi}{\partial x_i} \frac{\partial \Psi^T}{\partial x_j} \, d\mathbf{x}; \quad \tilde{\mathbf{Q}}_i = \int_{\Omega^e} \frac{\partial \Theta}{\partial x_i} \Phi^T \, d\mathbf{x}$$

$$\tilde{\mathbf{F}}_i(\mathbf{T}) = -\int_{\Omega^e} \rho_0 g_i \beta \Psi \Theta d\mathbf{x} + \int_{\Omega^e} \rho_0 g_i \beta T_0 \Psi \, d\mathbf{x} + \oint_{\Gamma^e} \Psi T_i ds$$

$$\tilde{\mathbf{D}}(\mathbf{u}) = \int_{\Omega^e} \rho_0 C \Theta(\Psi^T \mathbf{u}_j) \frac{\partial \Theta^T}{\partial x_j} \, d\mathbf{x}$$

$$\tilde{\mathbf{N}} = \int_{\Omega^e} (\rho_0 C)_e \Theta\Theta^T \, d\mathbf{x}; \quad \tilde{\mathbf{L}} = \int_{\Omega^e} k_e \frac{\partial \Theta}{\partial x_i} \frac{\partial \Theta^T}{\partial x_i} \, d\mathbf{x}$$

$$\tilde{\mathbf{G}} = \int_{\Omega^e} \Theta Q \, d\mathbf{x} + \oint_{\Gamma^e} \Theta q_n \, ds \qquad (5.4-7)$$

where summation on repeated indices is implied. The porous flow equations (5)–(7) can be arranged into a single matrix equation of the same form as given previously for the flow equations:

$$\begin{bmatrix} \tilde{M} & 0 & 0 \\ 0 & 0 & 0 \\ 0 & 0 & \tilde{N} \end{bmatrix} \begin{Bmatrix} \dot{u} \\ \dot{P} \\ \dot{T} \end{Bmatrix} + \begin{bmatrix} \tilde{C}(u) + \tilde{A} + \tilde{K} & -\tilde{Q} & \tilde{B}(T) \\ -\tilde{Q}^T & 0 & 0 \\ 0 & 0 & \tilde{D}(u) + \tilde{L}(T) \end{bmatrix} \begin{Bmatrix} u \\ P \\ T \end{Bmatrix} = \begin{Bmatrix} \tilde{F}(T) \\ 0 \\ \tilde{G}(T, u) \end{Bmatrix} \qquad (5.4-8)$$

5.5 SOLUTION METHODS

5.5–1 General Discussion

The finite element models developed in Sections 5.2–5.4 describe a very large group of heat transfer and fluid mechanics problems ranging in complexity from simple solid body conduction through isothermal flows, forced and free convection, to mixed mode convection/conduction. The solution methods used depend on the model, computational resources, the nonlinearity of the system, and the strength of the coupling between equations (see Section 4.6).

The solution procedures described in Section 4.6 were all variants of the basic fixed point method (see Appendix B) and were generally designed to operate concurrently on all of the equations present in a finite element model. The strong coupling between equations (*e.g.*, momentum and energy) that is characteristic of convective heat transfer problems makes these combined equation methods optimal from the standpoint of convergence rate. The disadvantage, of course, is that a very large and computationally expensive matrix problem must be treated at each iteration. The requirement to perform larger (more elements and higher dimensionality) and more complex (physical phenomena) simulations has reached the point where the usual direct matrix methods for combined equations are prohibitively expensive. A natural choice to make the matrix problem more affordable (while retaining the standard fixed point schemes) is to switch from direct, Gauss elimination type methods to the iterative matrix methods, such as the preconditioned conjugate gradient (PCG) method. Unfortunately, the development of iterative methods for combined equation sets has been severely handicapped by the lack of good preconditioner techniques that can adequately treat the dominating effect of the incompressibility constraint. To make progress with current iterative methods, the combined equation approach must be sacrificed for alternative formulations of the discrete equations.

In the following sections we first review Newton's method for convective heat transfer problems, a combined equation technique that is heavily used for two–dimensional problems and some smaller three–dimensional applications. Various segregated solution methods are then reviewed. Throughout this section we will only consider the time–independent equations. From the discussion in Section 4.6 it should be recognized that any of the standard time integration procedures can be applied to the convective heat transfer equations. The resulting nonlinear algebraic equations for each time step are then analogous to the steady state problem, and they can be treated with the algorithms described in the following sections.

5.5–2 Newton's Method

The application of Newton's method to the combined equations for convective heat transfer (5.2–14) or (5.4–8) follows directly from the isothermal case described in Section 4.6–3. The iterative procedure is given by

$$\mathbf{U}^{n+1} = \mathbf{U}^n - \mathcal{J}^{-1}(\mathbf{U}^n)\mathbf{R}(\mathbf{U}^n) \qquad (5.5-1)$$

where the vector of unknowns is $\mathbf{U}^T = \{\mathbf{u}_1^T, \mathbf{u}_2^T, \mathbf{u}_3^T, \mathbf{P}^T, \mathbf{T}^T\}$ for the mixed finite element model and $\mathbf{U}^T = \{\mathbf{u}_1^T, \mathbf{u}_2^T, \mathbf{u}_3^T, \mathbf{T}^T\}$ for the penalty finite element model. The definition of the Jacobian $\mathcal{J}$ is expanded to include the variations in the equations with respect to the vector of nodal temperatures, $\mathbf{T}$. Thus for the two–dimensional mixed finite element model we have

$$\mathcal{J} = \frac{\partial \mathbf{R}}{\partial \mathbf{U}} = \begin{bmatrix} \frac{\partial \mathbf{R}_1}{\partial \mathbf{u}_1} & \frac{\partial \mathbf{R}_1}{\partial \mathbf{u}_2} & \frac{\partial \mathbf{R}_1}{\partial \mathbf{P}} & \frac{\partial \mathbf{R}_1}{\partial \mathbf{T}} \\ \frac{\partial \mathbf{R}_2}{\partial \mathbf{u}_1} & \frac{\partial \mathbf{R}_2}{\partial \mathbf{u}_2} & \frac{\partial \mathbf{R}_2}{\partial \mathbf{P}} & \frac{\partial \mathbf{R}_2}{\partial \mathbf{T}} \\ \frac{\partial \mathbf{R}_3}{\partial \mathbf{u}_1} & \frac{\partial \mathbf{R}_3}{\partial \mathbf{u}_2} & \frac{\partial \mathbf{R}_3}{\partial \mathbf{P}} & \frac{\partial \mathbf{R}_3}{\partial \mathbf{T}} \\ \frac{\partial \mathbf{R}_4}{\partial \mathbf{u}_1} & \frac{\partial \mathbf{R}_4}{\partial \mathbf{u}_2} & \frac{\partial \mathbf{R}_4}{\partial \mathbf{P}} & \frac{\partial \mathbf{R}_4}{\partial \mathbf{T}} \end{bmatrix} \qquad (5.5-2)$$

where

$$\begin{aligned} \mathbf{R}_1 &= \mathbf{C}_1(\mathbf{u}_1)\mathbf{u}_1 + \mathbf{C}_2(\mathbf{u}_2)\mathbf{u}_1 + (2\mathbf{K}_{11} + \mathbf{K}_{22})\mathbf{u}_1 + \mathbf{K}_{12}\mathbf{u}_2 - \mathbf{Q}_1\mathbf{P} + \mathbf{B}_1\mathbf{T} - \mathbf{F}_1 \\ \mathbf{R}_2 &= \mathbf{C}_1(\mathbf{u}_1)\mathbf{u}_2 + \mathbf{C}_2(\mathbf{u}_2)\mathbf{u}_2 + \mathbf{K}_{21}\mathbf{u}_1 + (\mathbf{K}_{11} + 2\mathbf{K}_{22})\mathbf{u}_2 - \mathbf{Q}_2\mathbf{P} + \mathbf{B}_2\mathbf{T} - \mathbf{F}_2 \\ \mathbf{R}_3 &= -\mathbf{Q}_1^T\mathbf{u}_1 - \mathbf{Q}_2^T\mathbf{u}_2 \\ \mathbf{R}_4 &= \mathbf{D}_1(\mathbf{u}_1)\mathbf{T} + \mathbf{D}_2(\mathbf{u}_2)\mathbf{T} + (\mathbf{L}_{11} + \mathbf{L}_{22})\mathbf{T} - \mathbf{G} \end{aligned} \qquad (5.5-3)$$

which correspond to the two momentum equations, the incompressibility constraint, and the energy equation. The matrices appearing in equation (3) are defined in equation (5.2–9). The components of $\mathcal{J}$ are computed directly from the definitions in (3) and yield the following values:

$$\begin{aligned} &\frac{\partial \mathbf{R}_1}{\partial \mathbf{u}_1} = \mathbf{C}_1(\mathbf{u}_1) + \mathbf{C}_2(\mathbf{u}_2) + \mathbf{C}_1(1)\mathbf{u}_1 + 2\mathbf{K}_{11} + \mathbf{K}_{22} \\ &\frac{\partial \mathbf{R}_1}{\partial \mathbf{u}_2} = \mathbf{C}_2(1)\mathbf{u}_1 + \mathbf{K}_{21}\ ; \quad \frac{\partial \mathbf{R}_1}{\partial \mathbf{P}} = -\mathbf{Q}_1\ ; \quad \frac{\partial \mathbf{R}_1}{\partial \mathbf{T}} = \mathbf{B}_1 \\ &\frac{\partial \mathbf{R}_2}{\partial \mathbf{u}_1} = \mathbf{C}_1(1)\mathbf{u}_2 + \mathbf{K}_{12}\ ; \quad \frac{\partial \mathbf{R}_2}{\partial \mathbf{P}} = -\mathbf{Q}_2 \\ &\frac{\partial \mathbf{R}_2}{\partial \mathbf{u}_2} = \mathbf{C}_1(\mathbf{u}_1) + \mathbf{C}_2(\mathbf{u}_2) + \mathbf{C}_2(1)\mathbf{u}_2 + \mathbf{K}_{11} + 2\mathbf{K}_{22} \end{aligned}$$

$$\frac{\partial \mathbf{R}_2}{\partial \mathbf{T}} = \mathbf{B}_2 \; ; \; \frac{\partial \mathbf{R}_3}{\partial \mathbf{u}_1} = -\mathbf{Q}_1{}^T \; ; \; \frac{\partial \mathbf{R}_3}{\partial \mathbf{u}_2} = -\mathbf{Q}_2{}^T \; ; \; \frac{\partial \mathbf{R}_3}{\partial \mathbf{P}} = 0$$

$$\frac{\partial \mathbf{R}_3}{\partial \mathbf{T}} = 0 \; ; \; \frac{\partial \mathbf{R}_4}{\partial \mathbf{u}_1} = \mathbf{D}_1(1)\mathbf{T} \; ; \; \frac{\partial \mathbf{R}_4}{\partial \mathbf{u}_2} = \mathbf{D}_2(1)\mathbf{T} \; ; \; \frac{\partial \mathbf{R}_4}{\partial \mathbf{P}} = 0$$

$$\frac{\partial \mathbf{R}_4}{\partial \mathbf{T}} = \mathbf{D}_1(\mathbf{u}_1) + \mathbf{D}_2(\mathbf{u}_2) + \mathbf{L}_{11} + \mathbf{L}_{22} \qquad (5.5-4)$$

The three–dimensional formulation follows directly from the above definitions.

The advantages and disadvantages of Newton's method as outlined in Section 4.6–3 for the isothermal case hold equally for the convective heat transfer system. Also, the other variations of the fixed point schemes, such as Picard iteration, modified and quasi–Newton methods, and the various continuation techniques can be directly extended to the nonisothermal situation with comparable effectiveness.

5.5–3 Segregated Equation Methods

Segregated equation methods are not specifically solution algorithms but rather an approach to decomposing (5.2–14) into a series of smaller problems that can be solved efficiently using direct matrix methods or are more amenable to PCG type methods. For convective heat transfer problems (or other transport problems) a particularly simple segregated or cyclic procedure is given by the following for the $(n+1)$ iteration step:

Step 1 - (Energy Equation)

$$\mathbf{D}(\mathbf{u}^n, \mathbf{T}^n)\mathbf{T}^{n+1} + \mathbf{L}(\mathbf{T}^n)\mathbf{T}^{n+1} = \mathbf{G}(\mathbf{u}^n, \mathbf{T}^n) \qquad (5.5-5)$$

Step 2 - (Continuity and Momentum Equations)

$$-\mathbf{Q}^T \mathbf{u}^{n+1} = 0 \qquad (5.5-6)$$

$$\mathbf{C}(\mathbf{u}^n)\mathbf{u}^{n+1} + \mathbf{K}(\mathbf{T}^{n+1})\mathbf{u}^{n+1} - \mathbf{Q}\mathbf{P}^{n+1} = \mathbf{F}(\mathbf{T}^{n+1}) \qquad (5.5-7)$$

The energy equation is typically solved first in each cycle, especially when considering free convection–dominated flows since the buoyancy term drives the fluid motion; the buoyancy term is included in $\mathbf{F}$ in equation (6).

The procedure given in equations (5)–(7), termed a cyclic Picard method, has some advantages over a combined equation method in terms of computer storage and execution time per iteration. Unfortunately, as with most variants of successive substitution, the procedure has a relatively slow rate of convergence; the convergence rate is further reduced due to the lagging of the coefficients induced by the splitting of the equations. This method has been used successfully for convection problems of moderate nonlinearity by several investigators [2–5].

Though the cyclic procedure described above offers some reduction in computational requirements, the combined momentum and continuity equations, especially in three dimensions, can still represent a very large matrix problem.

Following the idea of segregating equations to its logical extreme, [6] the system in (5.2–14) can be decomposed into the following series of equations:

$$(\mathbf{C_1}(\mathbf{u}) + \mathbf{K_1})\mathbf{u_1} - \mathbf{Q_1}\mathbf{P} = \mathbf{F_1} \qquad (5.5-8a)$$

$$(\mathbf{C_2}(\mathbf{u}) + \mathbf{K_2})\mathbf{u_2} - \mathbf{Q_2}\mathbf{P} = \mathbf{F_2} \qquad (5.5-8b)$$

$$\mathbf{Q_1}^T\mathbf{u_1} + \mathbf{Q_2}^T\mathbf{u_2} = 0 \qquad (5.5-8c)$$

$$(\mathbf{D}(\mathbf{u}) + \mathbf{L})\mathbf{T} = \mathbf{G} \qquad (5.5-8d)$$

for the case of two-dimensional, nonisothermal flow; the three–dimensional case follows the same procedure as does the inclusion of additional transport equations. The $(\mathbf{C}, \mathbf{D})$, and $(\mathbf{K}, \mathbf{L})$ matrices, represent the advection and diffusion operators, respectively. A segregated solution approach to equations (8) would solve each individual momentum equation, the continuity equation, and the energy equation for the appropriate variable in a cyclic procedure similar to the one shown in equations (5)–(7). As each variable is updated from a new solution, it is used where appropriate in the coefficients of the remaining equations, resulting in another type of cyclic Picard scheme. By reducing (5.2–14) to the series of equations in (8), a significant reduction in size and cost of the matrix problem is realized. The penalty to be paid is the slower convergence rate of the cyclic procedure. However, for very large problems there is a cross–over point for these competing effects and the segregated approach is clearly superior to standard combined equation methods.

A significant complication in the use of equations (8) arises due to the fact that there is no equation in the set from which the pressure may be directly obtained. The solution to this dilemma involves the construction of a pressure Poisson equation. Solving equations (8a) and (8b) symbolically for the velocities $\mathbf{u}_1$ and $\mathbf{u}_2$ and substituting this result into the incompressibility condition (8c) produces a consistent Poisson equation for the pressure

$$\left[\mathbf{Q}_1^{\mathrm{T}}(\mathbf{C_1}(\mathbf{u}) + \mathbf{K_1})^{-1}\mathbf{Q_1} + \mathbf{Q}_2^{\mathrm{T}}(\mathbf{C_2}(\mathbf{u}) + \mathbf{K_2})^{-1}\mathbf{Q_2}\right] \mathbf{P} = \mathbf{F} \qquad (5.5-9)$$

where the explicit form of $\mathbf{F}$ is not important in this discussion. Equation (9), when used in place of the continuity equation, provides a complete set of equations for all of the unknowns. A number of possible variations exist for the sequential solution of equations (8) and (9), each of which produces a sightly different algorithm. In general, the Poisson equation (9) provides an initial estimate of the pressure field which is used in the pressure gradients of the momentum equations; velocity estimates are then obtained from each momentum equation in turn. Since the velocity field is not divergence free at this point, a second application of the Poisson equation is required; this solution supplies a pressure correction that allows the velocity fields to be adjusted to satisfy incompressibility. This general algorithm is analogous to the family of SIMPLE methods developed by Patankar for finite difference methods [7]. Details regarding several variants of the method can be found in the work of Haroutunian *et al.* [6], who were the original developers of this technique for finite element applications. Note that another significant benefit in the usage of the segregated approach with the pressure Poisson equation is the applicability of iterative matrix solution methods for the individual equations in the sequence. The removal of the incompressibility constraint allows useful preconditioners to be found for each equation in the system. The combination of preconditioned, conjugate gradient (PCG) matrix methods with a segregated equation approach provides the most effective solution method presently available for large scale applications.

5.6 CONVECTION WITH CHANGE OF PHASE

There are two major difficulties in simulating melting and solidification change of phase for convection heat transfer problems — accounting for the latent heat effect and the application of fluid boundary conditions on the phase boundary. The latent heat problem has been detailed in the heat conduction section and its solution via the enthalpy method carries over directly to the convection problem. The boundary condition difficulty, however, is unique to the convection problem.

As part of the formulation of the phase change problem in Section 1.8, it was noted that on the phase boundary the fluid velocity was assumed to be zero [equation (1.8–1)]. This is a particularly difficult condition to apply in standard finite element formulations for two reasons. The location of the phase boundary is not known *a priori*, and even if it were known, the boundary does not generally coincide with a set of element edges (nodes) to which an essential boundary condition can be applied. One method to circumvent these difficulties is to explicitly track the motion of the phase boundary and allow the finite element mesh to adjust to this motion. Such tracking or remeshing schemes allow the phase boundary to always remain aligned with element edges and thus simplify the application of the no–slip boundary condition. However, these procedures are not without their shortcomings. They are expensive, complex to implement for general applications, and are presently limited in application to simple phase boundary motions and shapes.

A second, more approximate, method for imposing the no-slip boundary condition relies on the use of a temperature–dependent fluid viscosity and the allowed variation of material properties within each finite element. Depending on the type of variable coefficient implementation that is used in the construction of element matrices, a variable viscosity is evaluated at the numerical quadrature points or nodal points within an element (see Section 3.3 for details). During each cycle of the solution process the temperature at the viscosity evaluation point is compared with the solidus (liquidus) temperature of the material. If the temperature is above the liquidus, the appropriate fluid viscosity is assigned at the evaluation point; for temperatures below the solidus temperature, the evaluation point viscosity is assigned a "large" value to "solidify" the fluid. Typically, the "solid viscosity" is set at four to five orders of magnitude larger than the normal fluid viscosity; the viscosity function is usually designed to take on a realistic, or at least convenient, variation between the liquidus and solidus temperatures (*i.e.*, through the "mushy" zone).

The viscosity evaluation method proposed here allows the phase boundary computation to be maintained on a subelement basis since a given element may contain both fluid and solid phases. However, this procedure is also similar to a smearing technique since a distinct phase boundary is never identified. This is in keeping with the spirit and implementation of the latent heat procedures described previously. The utility of this approach is demonstrated in a subsequent section and in [8].

5.7 CONVECTION WITH ENCLOSURE RADIATION

The inclusion of enclosure radiation effects in convective heat transfer problems involves the direct application of the methods from Section 3.8. By assumption, the fluid in the enclosure is nonparticipating and the radiative exchange therefore only influences the flux and temperature on the fluid boundaries. For purposes of computation the radiation problem is decoupled from the convection/conduction part of the problem and simply provides flux boundary conditions at each iteration or time step of the simulation.

5.8 POST–COMPUTATION OF HEAT FLUX

The results of interest from a convective heat transfer analysis generally include fluid velocities, flow patterns, temperature distributions, and heat fluxes. Most of these items are directly available from the finite element results in terms of nodal point quantities. The heat flux (or flux of an auxiliary variable), however, being a derivative of the temperature, requires an auxiliary computation. Details of a method for performing this operation were given in Section 3.3. The simplest procedure for computing the components of the heat flux vector is based on the element level shape function for the temperature. That is, since

$$T(x_i,t) = \Theta^T(x_i)\mathbf{T}(t) \qquad (5.8-1)$$

then the flux components are

$$q_i(x_i,t) = -k\frac{\partial \Theta^T}{\partial x_i}\mathbf{T}(t) \qquad (5.8-2)$$

The relation in (2) is valid for any element and allows the flux to be evaluated at any time and at any point within the element. Note that in evaluating the derivative of Θ the element geometry must be considered for mapped or parametric elements [see equation (3.3–3)]. Maximum accuracy in the shape function derivative is obtained by evaluating the derivative at the element numerical integration points (typically, the 2×2 Gauss points in a quadrilateral element) [9]. Since fluxes are normally required on element or domain boundaries, the integration point values are extrapolated to the nodes [10] and the discontinuous element level fluxes averaged between adjacent elements. Other evaluation points within an element have been employed, such as the points midway between nodes on the element boundary, or the Gauss points on the element boundary. All of these methods provide results of acceptable accuracy provided the computational mesh is sufficiently refined. Note that a common feature of these procedures is the need to compute a unit normal on an element boundary if a total flux is to be derived. Methods for determining unique and consistent normals to element boundaries are discussed by Engleman *et al.* [11]. It is also important to recall that the above methods were described for the case of a conductive heat flux as would be computed for a solid boundary or a no flow–through surface. The heat flux computation for an open flow boundary would have to include terms describing the advective transport of energy.

A second, less familiar, approach for computing flux quantities relies on the consistent application of the weighted residual method. This idea was proposed by

Larock and Herrmann [12] and Marshall *et al.* [13], and refined by Gresho *et al.* [14]. In essence, a weighted residual equation is derived that relates the boundary flux to local heat sources and computed temperature gradients. The consistent flux equation requires a matrix solution, which for many cases can be simplified by the use of a "lumping" procedure. Though somewhat more accurate than the simpler basis function derivative method, this procedure requires more computation, especially for the advection–diffusion equation. For details of this procedure, see [14].

5.9 NUMERICAL EXAMPLES

5.9–1 Preliminary Comments

The numerical examples selected for this section are intended to illustrate the performance of the finite element models presented for convective heat transfer problems. Again, the problems are not examined in full engineering detail, but are sufficiently complete to demonstrate the flexibility and capability of the recommended algorithms. All of the examples were solved using the codes NACHOS II [15] and FIDAP [16] with the mixed finite element model, unless otherwise stated. For numerical examples of natural convection problems solved by the penalty finite element models, the reader may consult the review paper by Pelletier *et al.* [17].

5.9–2 Concentric Tube Flow

The first example we consider is the forced convection/conduction problem of flow in a concentric tube. The problem schematic is shown in Figure 5.9–1 along with the finite element mesh of eight–noded elements and boundary conditions. The fluid in the inner tube is a medium–weight oil and the counterflowing cooling fluid is water. The central tube is made of copper and the outside concentric tube is made of steel. All material properties are assumed to be constant.

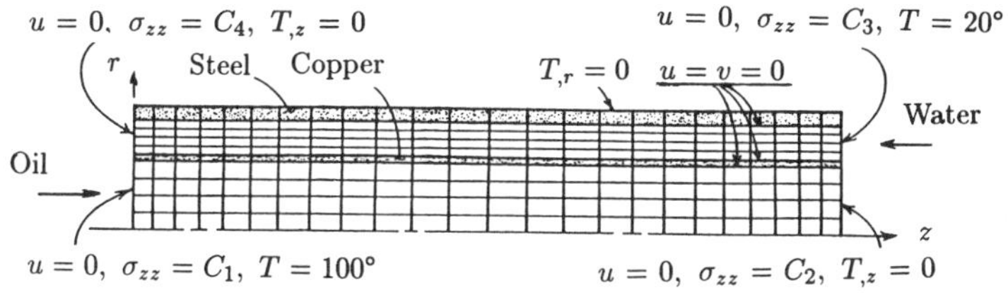

Figure 5.9–1: Schematic for concentric tube flow.

As is typical for internal flow problems, flux type boundary conditions are used at the outflow boundary for both fluid regions to allow a smooth solution field without an excessive tube length. Also, in this case a specified traction boundary condition was specified at the inflow boundary; the difference in the normal tractions across the tube length produce the driving force for the fluid. Note that the nonlinear terms are identically zero in the fluid and thermal equations in this case, and therefore the problem can be solved without iteration.

The solution to the hydrodynamic part of the problem consists of a fully developed Poiseuille flow for the oil and flow in a concentric annulus for the water. Both computed profiles are essentially exact for this model. Three cases were computed in which the oil Peclet number ($Pe = \mu C_p/k$) was varied by changing the pressure drop imposed on the tube. A typical isotherm plot is shown in Figure 5.9–2. The computed temperature profiles along the inside of the copper tube are shown in Figure 5.9–3. Further details on this analysis and other conjugate tube flow problems are available in [18].

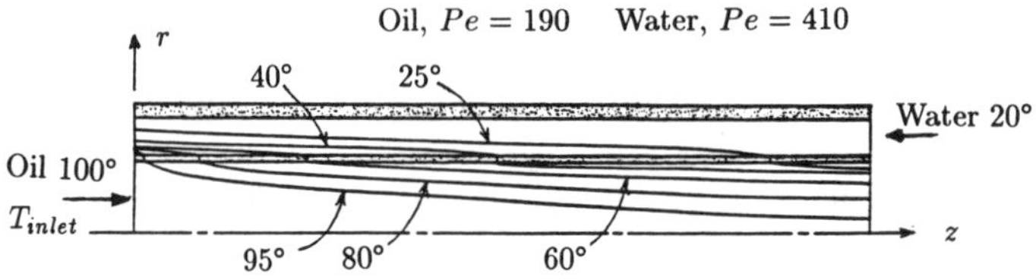

Figure 5.9–2: Isotherms for the concentric tube flow problem.

5.9–3 Tube Flow with Change of Phase

Consider the problem of a heated fluid flowing through a cold tube, the temperature of which is below the fluid solidification temperature. The problem geometry is shown in Figure 5.9–4. The fluid within the tube is initially quiescent with a uniform temperature equal to the inlet temperature. Fluid motion is initiated by imposing a specified pressure drop on the tube. The methods described in Sections 3.3 and 4.7 were used to model the latent heat release and the immobilization of the fluid at the phase boundary [8].

The initial temperature response of the fluid is due solely to conduction at the tube wall. The large initial heat loss to the tube cools the fluid below the solidus temperature, producing a solid crust on the tube wall. As the fluid velocity increases, convection along the length of the tube becomes important.

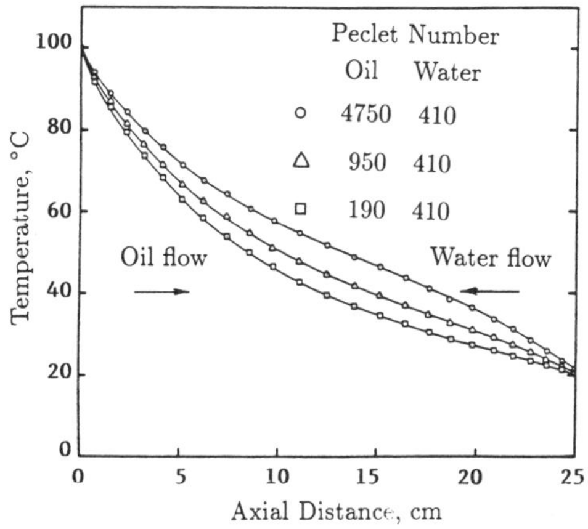

Figure 5.9–3: Axial temperature profiles for the concentric tube flow problem.

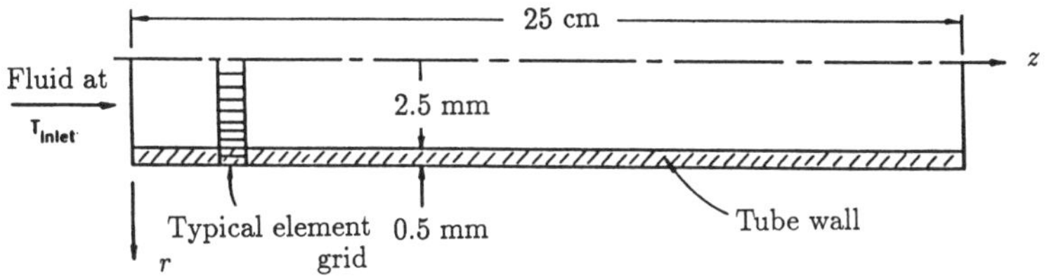

5.9–4: Schematic for tube flow with change of phase.

Figure 5.9–5 shows the result of a computation in which the tube eventually becomes obstructed due to the crust formation. Each frame shows the phase change isotherm which corresponds to the phase boundary. Figures 5.9–6 and 5.9–7 show velocity and temperature profiles at a station four tube diameters from the inlet. The velocity profiles show the effects of initial fluid acceleration, reduction in tube diameter due to crust growth, and finally the reduction in flow rate due to increasing viscosity.

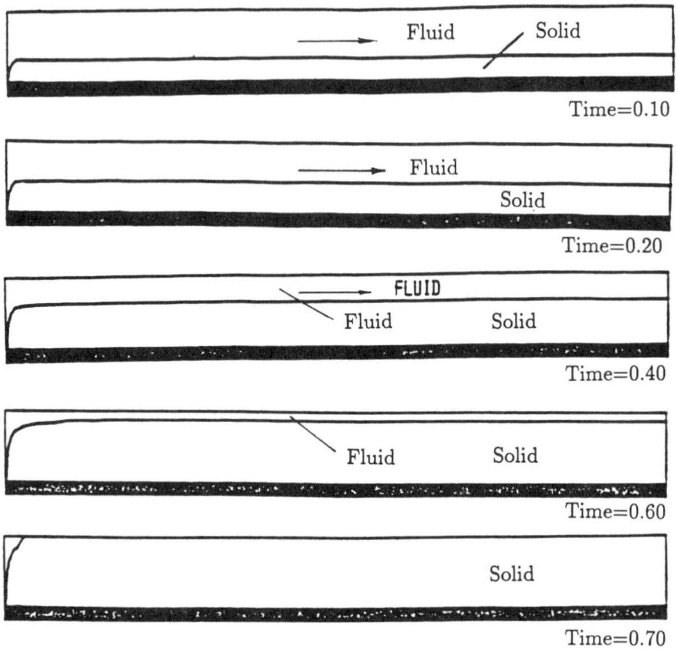

Figure 5.9–5: Motion of phase boundary for impulsively started tube flow.

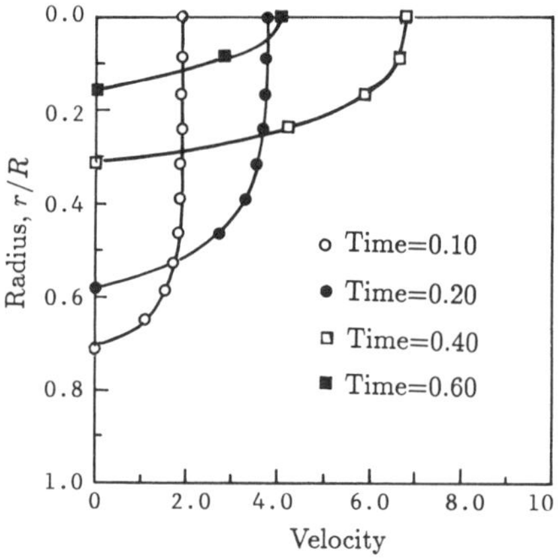

Figure 5.9–6: Velocity profiles for tube flow, $z/R = 8$.

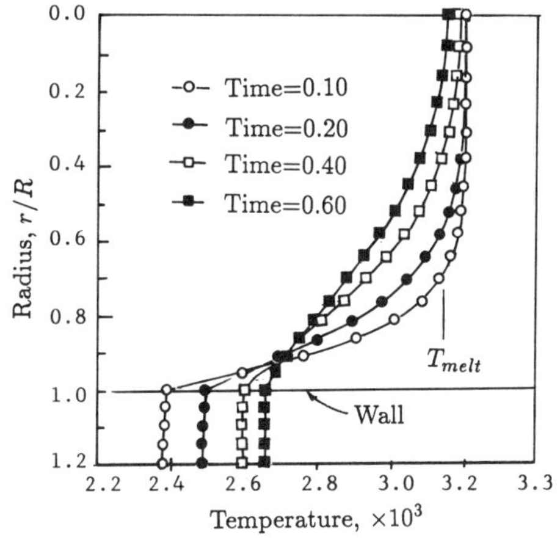

Figure 5.9–7: Temperature profiles for tube flow, $z/R = 8$.

5.9–4 Heated Cavity

A standard bench mark problem for natural convection algorithms consists of a square, closed cavity with insulated top and bottom faces and differentially heated vertical boundaries. For the present case, the contained fluid is a gas with constant material properties. A schematic of the problem with the boundary conditions and nondimensional parameters is shown in Figure 5.9–8. Since the velocity field is specified on all boundaries of the domain, the pressure field is determined only to within a constant value; the pressure level within the cavity is set by specifying a single pressure value at one node on the cavity boundary. A typical mesh consisting of eight-node elements is also shown in Figure 5.9–8.

The cavity problem was initially solved using the cyclic Picard scheme employed in the earliest versions of the NACHOS code. More recent solutions have been obtained using Newton's method for the combined equation set and zeroth order continuation through the Rayleigh number ($Ra = \beta g(T_H - T_C)D^3/\nu\kappa$) range of interest. Results for this problem in the form of streamline and isotherm plots are shown in Figures 5.9–9 and 5.9–10 for Rayleigh numbers of $Ra = 10^4$ and 10^6, respectively ($Pr = \nu/\kappa = 0.71$). For the lower Rayleigh number, the flow is relatively weak and the thermal field is only slightly perturbed from a conduction solution. At the higher Rayleigh number, the flow field develops a considerable structure while the thermal field becomes vertically stratified in the core of the cavity with high heat flux regions along the vertical boundaries. As required by the form of the boundary conditions and boundary value problem, the computed solutions are exactly centro–symmetric as seen in the contour plots. A detailed comparison of the present solutions with a large number of other numerical simulations is contained in [19].

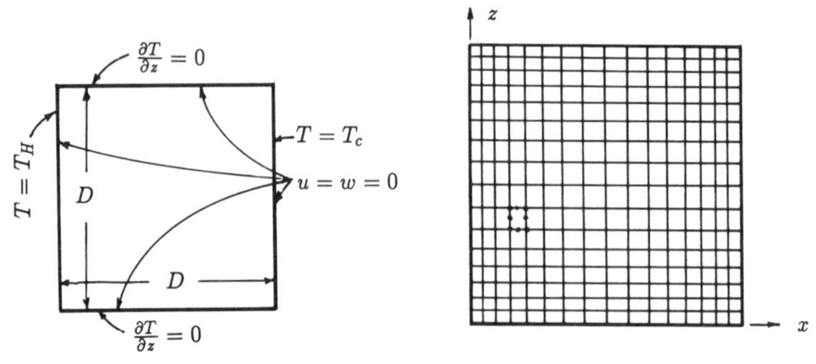

Figure 5.9–8: Schematic for natural convection in a cavity.

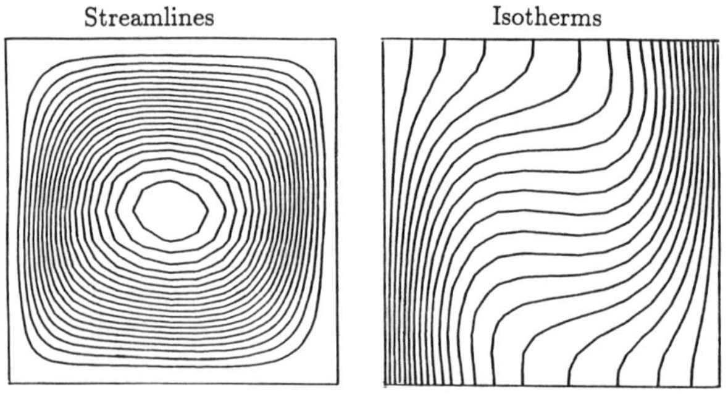

Figure 5.9–9: Streamlines and isotherms for natural convection in a cavity, $Ra = 10^4$.

5.9–5 Solar Receiver

The schematic shown in Figure 5.9–11 represents a cross section of an annular solar receiver tube surrounded by an eccentrically located glass envelope. The inner tube carries a heat transfer fluid that is heated by a flux that varies with position around the tube; the incident flux is due to solar energy being concentrated on the tube by a parabolic trough collector. The glass envelope provides a shield to reduce the forced convection (wind) heat loss from the collector tube. Of interest in this problem is the prediction of heat loss from the inner tube due to natural convection in the annular space. Parameters such as inner tube temperature distribution, working fluid in the annulus, annular gap, and eccentricity of the envelope were varied to assess their importance on overall heat loss.

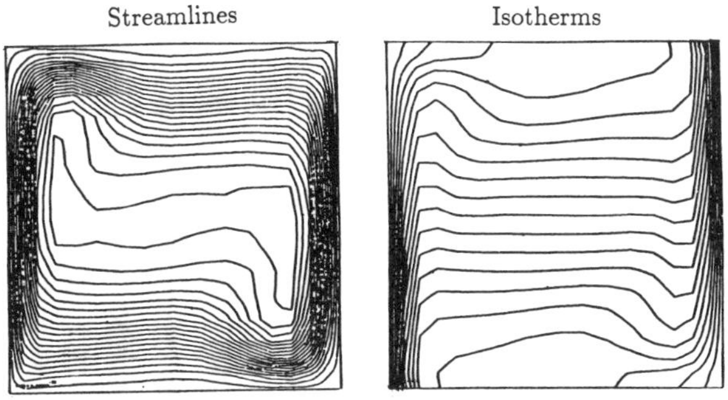

Figure 5.9–10 Streamlines and isotherms for natural convection in a cavity, $Ra = 10^6$.

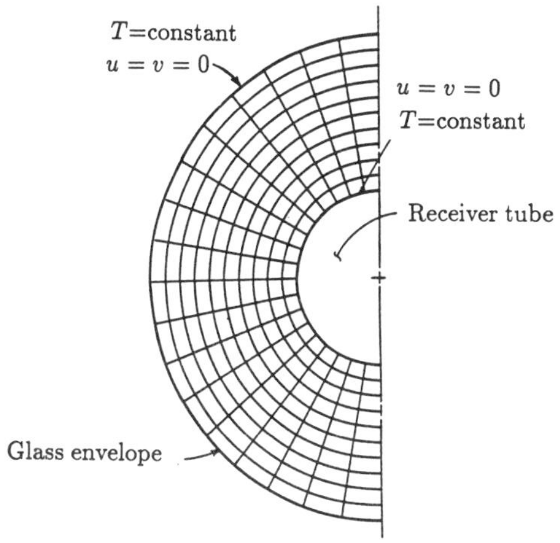

Figure 5.9–11: Schematic for the annular solar receiver.

Figures 5.9–12 through 5.9–15 contain streamline and isotherm plots for an air-filled annulus for various temperature and geometric configurations. The flow pattern and heat flux distribution is quite sensitive to variations in these parameters even though the Rayleigh number is the same for all cases. The computed heat transfer results for these cases are compared with experimental data in Figure 5.9–16.

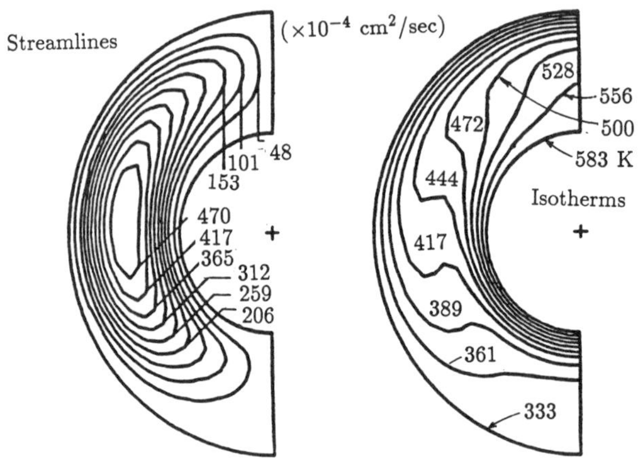

Figure 5.9–12: Streamline and isotherm plots for the solar receiver; uniform wall temperature, $Ra = 1.2 \times 10^4$.

5.9–6 Tube Bundle

The natural convection heat transfer from three tubes contained within a larger diameter conduit is modeled with the finite element grids shown in Figure 5.9–17. The geometry is symmetric, allowing the simulation to be performed on one–half of the conduit domain. The interior tubes have constant, but different wall temperatures; the exterior of the conduit is held at a temperature that is lower than the interior tubes. The two meshes shown in Figure 5.9–17 represent two very different mesh generation techniques. The more structured mesh was developed in three sections using a standard mapping technique. The completely unstructured quadrilateral mesh was produced using a new method called "paving" [20] which works with the entire region at one time. The only input to the paving algorithm, other than a boundary description, is an element size along each boundary.

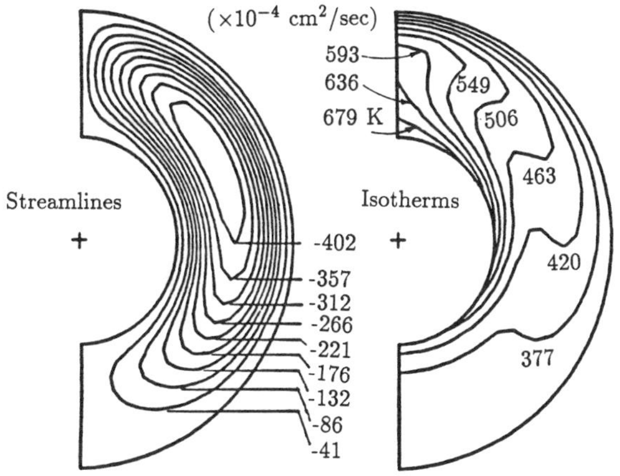

Figure 5.9–13: Streamline and isotherm plots for the solar receiver; asymmetric wall temperature, hot on top, $Ra = 1.2 \times 10^4$.

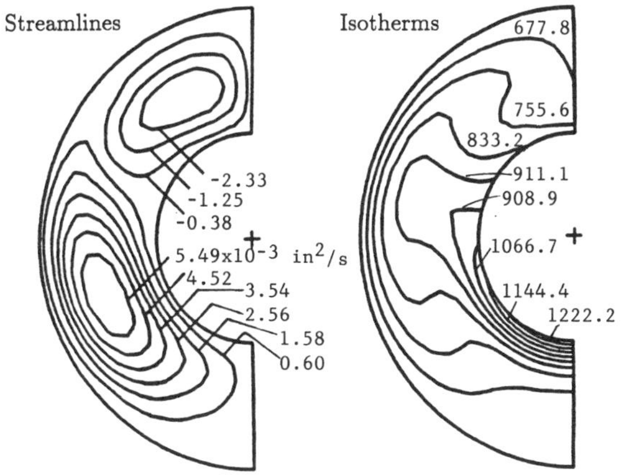

Figure 5.9–14: Streamline and isotherm plots for the solar receiver; uniform wall temperature, hot on bottom, $Ra = 1.2 \times 10^4$.

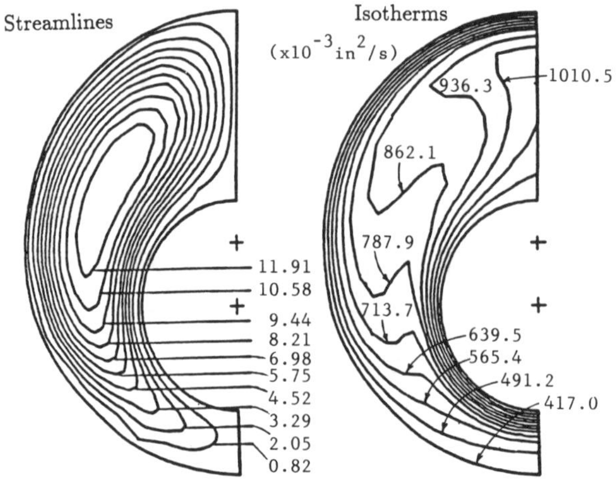

Figure 5.9–15: Streamline and isotherm plots for the solar receiver; uniform wall temperature, eccentric geometry, $Ra = 1.2 \times 10^4$.

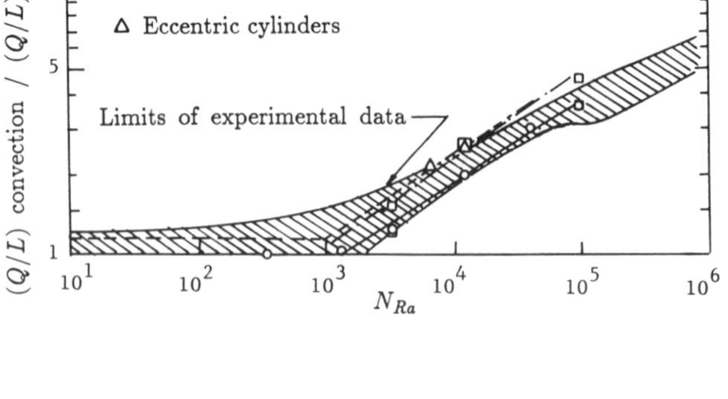

Figure 5.9–16: Heat loss by natural convection in annular spaces of a solar receiver. Computed points are shown as symbols.

Figure 5.9–17: Finite element meshes for tube bundle geometry.

Isotherm and streamline plots for two cases are shown in Figures 5.9–18 and 5.9–19. Figure 5.9–18 illustrates the flow produced when a small temperature difference ($\Delta T_{max} = 10$) is maintained between the hottest interior tube and the exterior boundary. The second case shows the result when the temperature difference from the first case is doubled (*i.e.*, $\Delta T_{max} = 20$). Solutions for both cases were obtained using a solution strategy with several Picard iterations followed by a Newton procedure. The computed flow fields show a characteristic plume rising from the lower cylinder with an overall circulation within the larger conduit.

5.9–7 Volumetrically Heated Fluid

This example consists of a volumetrically heated fluid contained in an isothermal circular cylindrical container. The problem and computational mesh are shown in Figure 5.9–20. At time zero, the initially quiescent, isothermal fluid is heated volumetrically at a uniform rate. The transient solution is obtained by an implicit integration scheme using the trapezoid rule with quasilinearization.

The series of streamline and isotherm plots in Figure 5.9–21 shows the development of the flow at a moderate modified Grashof number, $Gr = g\beta l^5 Q/\nu^2 k =$

2×10^5. A time history of the maximum stream function shown in Figure 5.9–22 shows the overshoot that occurs during the initial transient. The stream function vs. time plot of Figure 5.9–23 shows the interaction between the two cells at a slightly higher Grashof number, $Gr = 4 \times 10^5$. A further series of plots in Figures 5.9–24a and 5.9–24b illustrate the additional dynamics present in the flow at Grashof number, $Gr = 4 \times 10^5$. The development of a secondary cell that undergoes a damped oscillation is quite evident. The development of secondary cells is quite common in volumetrically heated fluids. Further details of this analysis are available in [21].

5.9–8 Porous/Fluid Layer

A simple problem that illustrates the use of the saturated porous media formulation in conjunction with an incompressible, viscous flow is shown in Figure 5.9–25. A rectangular geometry is divided vertically with a porous material occupying the right half of the cavity. The top and bottom boundaries of the cavity are insulated while the vertical boundaries are held at uniform but different temperatures. The steady natural convection problem in the cavity was modeled using a Brinkman model for flow in the porous medium. Newton's method was used to solve the flow problem.

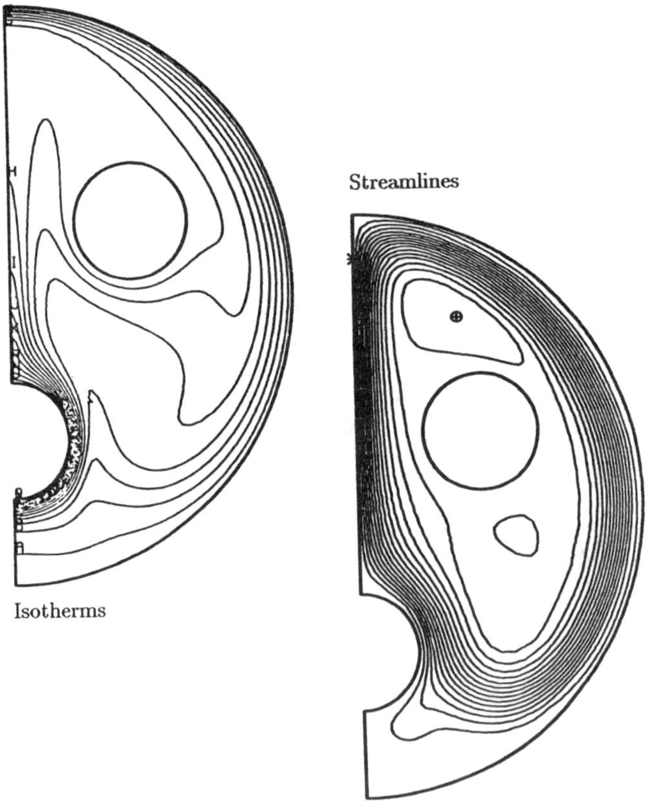

5.9–18: Contour plots for natural convection in a tube bundle, $\Delta T_{max} = 10$.

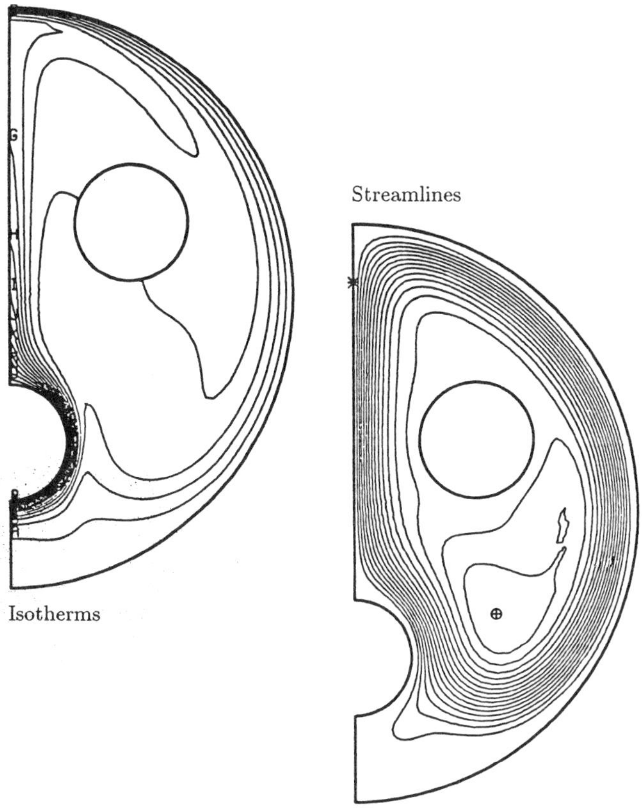

5.9–19: Contour plots for natural convection in a tube bundle, $\Delta T_{max} = 20$.

For small values of the Darcy number (or permeability), $Da = \sqrt{k}/L$, the dominant fluid motion is confined to the fluid layer with little circulation occurring in the porous layer. At higher Darcy numbers, the fluid motion spans the cavity though the velocities in the porous layer are still much smaller than in the bulk fluid. Representative stream function and isotherm plots are shown in Figures 5.8–26 and 5.8–27 for extremes in the Darcy number. Further discussion on the modeling of conjugate problems and the use of various porous/layer/bulk fluid interface conditions can be found in [22].

5.9–9 Curing of an Epoxy

This example demonstrates the use of both the phase change procedures and the auxiliary equation formulation to predict the gelation behavior of an epoxy. The geometry for the simulation is a simple rectangular crucible that initially contains a well–stirred mixture of epoxy resin and curing agent (see Figure 5.9–28). The crucible is held in a constant–temperature oven during the course of the curing process. Since the curing reaction is exothermic, the volumetric heat addition to the fluid is given by

$$Q = \rho \Delta H \frac{D\alpha}{Dt} = \rho \left(C_p^p - C_p^r \right) T \frac{D\alpha}{Dt} \tag{5.9-1}$$

where ΔH is the change in enthalpy of the reacting material due to its change in state, and the superscripts p and r denote the products and reactants, respectively. The variable α is an internal state variable that describes the extent of the gelation reaction; α can be interpreted as the ratio of the mass of reacted material (per unit volume) to the total mass of material (per unit volume). Therefore, α varies between 0 (unreacted) to 1 (fully reacted) with $\alpha = 0.6$ indicating the point of gelation (solidification). The time evolution of the extent of reaction variable, α, is provided by an advection–diffusion equation

$$\frac{\partial \alpha}{\partial t} + u_j \frac{\partial \alpha}{\partial x_j} = \frac{\partial}{\partial x_j} \left(D \frac{\partial \alpha}{\partial x_j} \right) + R \tag{5.9-2}$$

where R is the reaction rate. For this particular material the reaction rate is often assumed to be of second-order with Arrhenius rate constants

$$R = \left[A_1 exp(-E_1/\mathcal{R}T) + \alpha A_2 exp(-E_2/\mathcal{R}T) \right] (1 - \alpha) \tag{5.9-3}$$

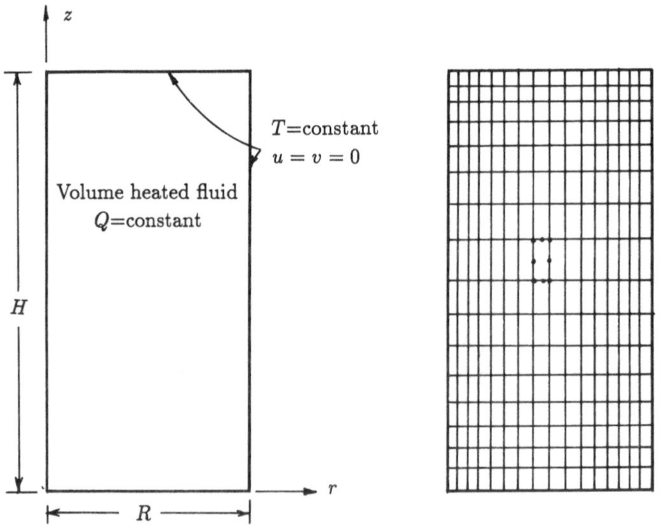

5.9–20: Schematic for volumetrically heated fluid in a circular cylinder.

where A_1, A_2 are pre–exponential factors, E_1, E_2 are activation energies, and $\mathcal{R}$ is the gas constant.

The above equations must be coupled to the standard nonisothermal, Navier–Stokes equations to describe the entire curing process. In addition, material properties, which vary with the extent of the reaction and the temperature must be specified. The present problem was solved using the adaptive time step version of the trapezoid rule; the finite element approximations used quadratic functions for the velocity, temperature, and extent of the reaction and a linear, discontinuous pressure approximation. Further details on the model and analysis are described in [23].

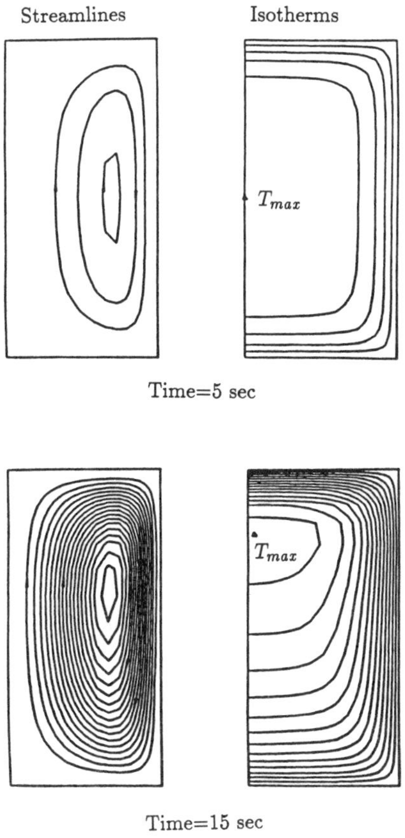

Figure 5.9–21a: Streamlines and isotherms for volumetrically heated fluid ($Gr = 2 \times 10^5$).

Shown in Figure 5.9–29 is a time history plot of the adaptive time step employed for the analysis. It is quite evident that the size of the time step reflects the changing complexity of the flow problem. Figures 5.9–30 through 5.9–33 show contour plots of the field variables at four times during the curing process. As with most volumetrically heated fluids, multiple convection cells are predicted during the low heat release part of the process. When the exothermic reaction begins to accelerate, gelation occurs first at the top of the crucible and proceeds downward as a planar front. A comparison of the velocity of the gel front with the limited experimental data available showed good agreement, considering the uncertainty in material properties.

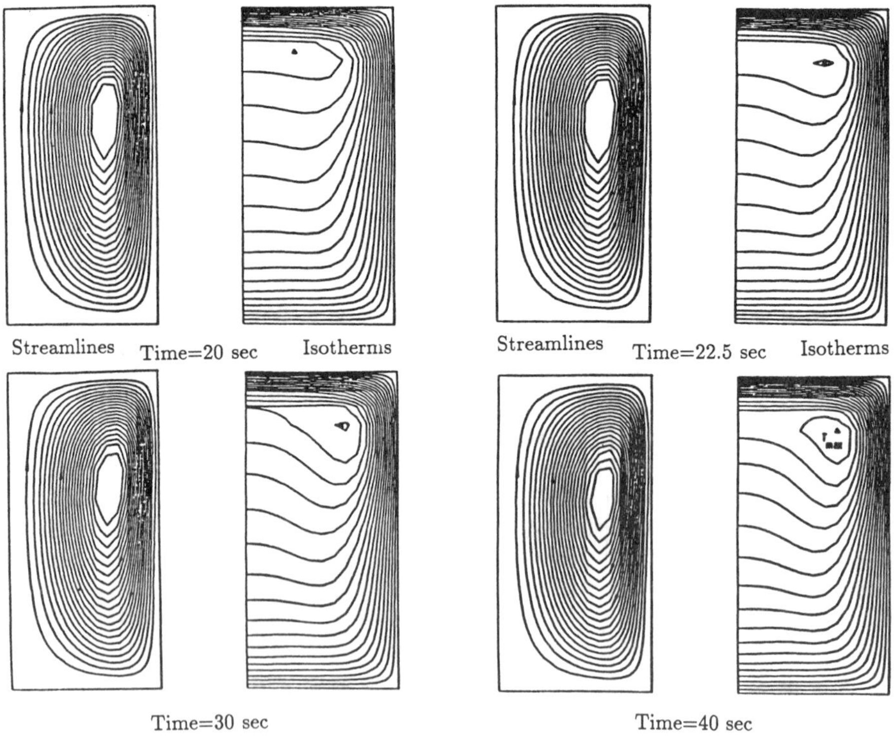

Figure 5.9–21b: Streamlines and isotherms for volumetrically heated fluid ($Gr = 2 \times 10^5$).

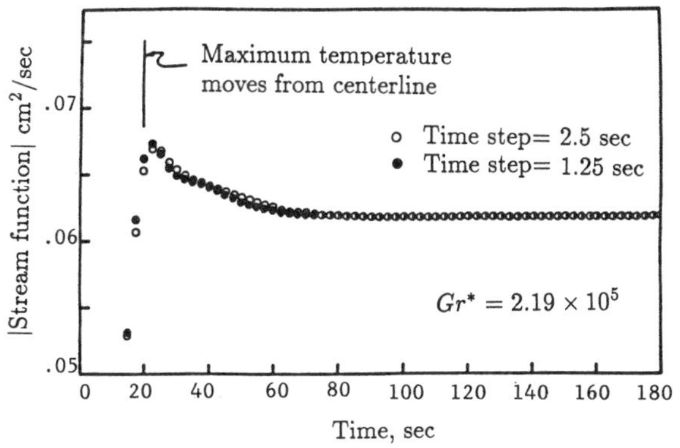

Figure 5.9–22: Stream function versus time, volumetrically heated fluid, $Gr = 2 \times 10^5$.

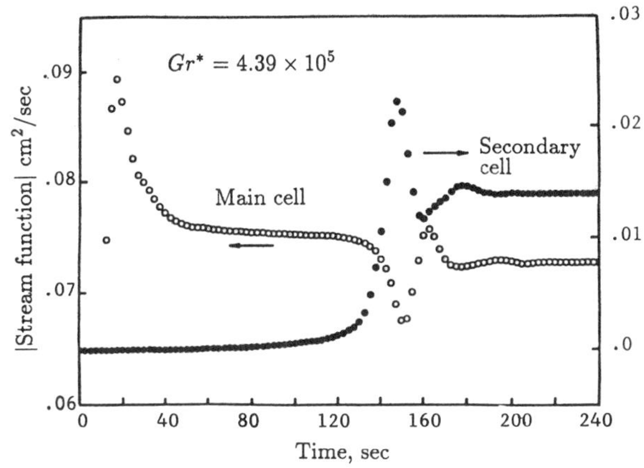

Figure 5.9–23: Stream function versus time, volumetrically heated fluid, $Gr = 4 \times 10^5$.

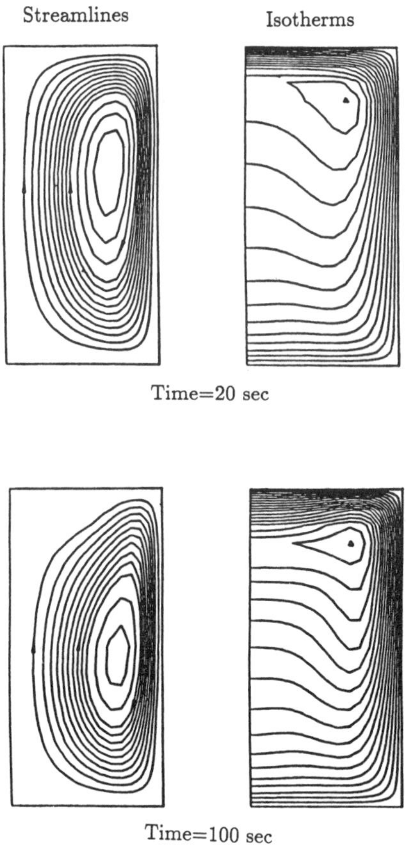

Figure 5.9–24a: Streamlines and isotherms for volumetrically heated fluid ($Gr = 4 \times 10^5$).

REFERENCES

1. J. N. Reddy, *An Introduction to the Finite Element Method*, Second Edition, McGraw–Hill, New York (1993).

2. D. K. Gartling, "Convective Heat Transfer Analysis by the Finite Element Method," *Computer Methods in Applied Mechanics and Engineering,* **12**, 365–382 (1977).

3. B. Tabarrok and R. C. Lin, "Finite Element Analysis of Free Convection Flows," *International Journal of Heat and Mass Transfer*, **20**, 953–960 (1977).

4. J. N. Reddy and A. Satake, "A Comparison of Various Finite Element Models of Natural Convection in Enclosures," *Journal of Heat Transfer*, **102**, 659–666 (1980).

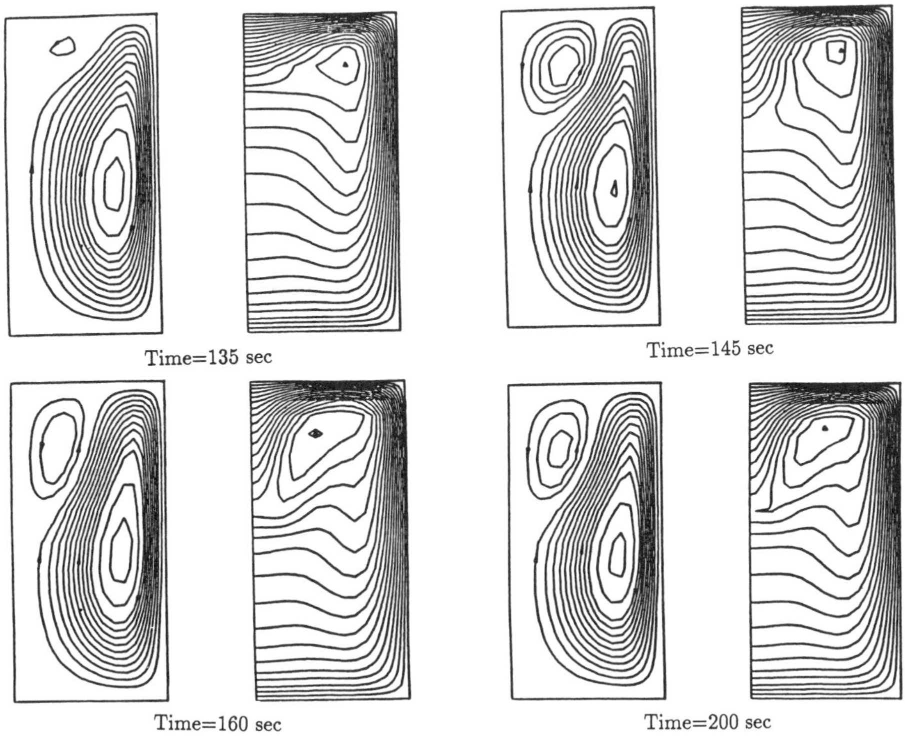

Figure 5.9–24b: Streamlines and isotherms for volumetrically heated fluid ($Gr = 4 \times 10^5$).

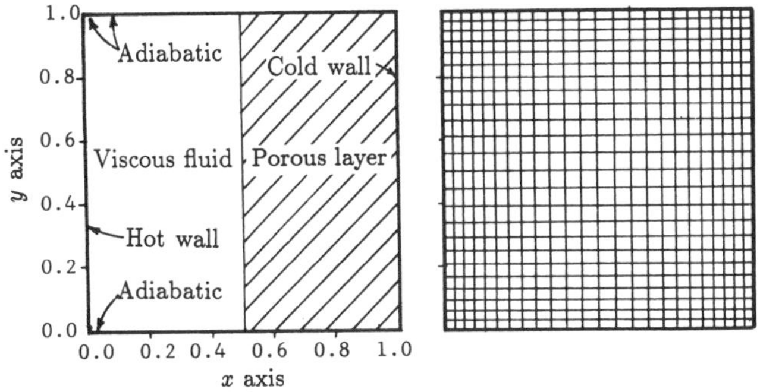

Figure 5.9–25: Schematic and mesh for a conjugate, porous/fluid layer problem.

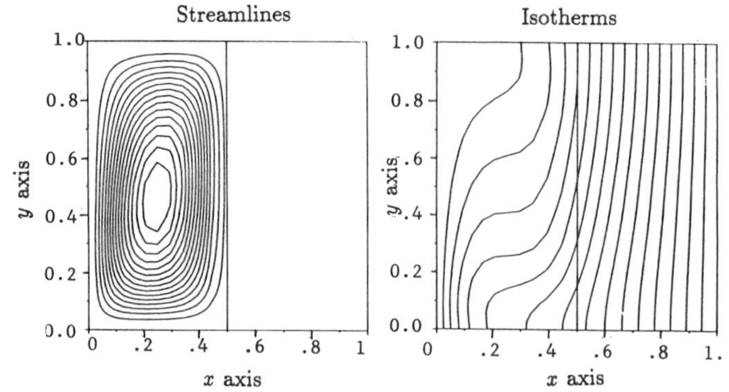

Figure 5.9–26: Contour plots for a conjugate, porous/fluid layer, $Ra = 10^5, Da = 10^{-5}$.

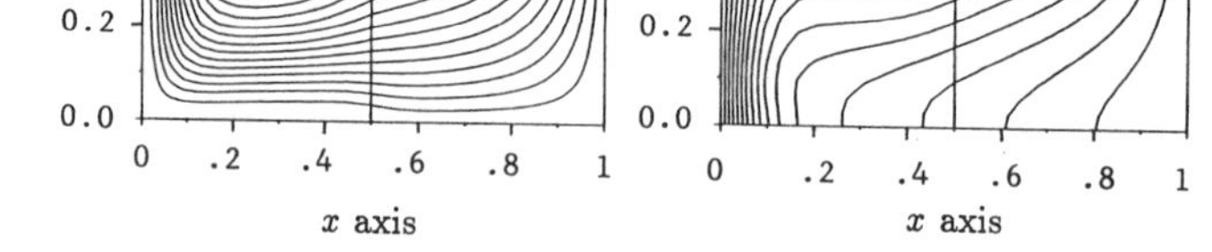

Figure 5.9–27: Contour plots for a conjugate, porous/fluid layer, $Ra = 10^5, Da = 10^{-3}$.

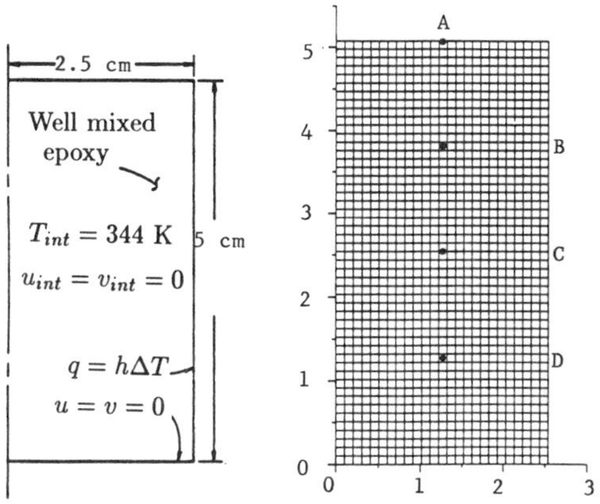

Figure 5.9–28: Schematic and mesh for epoxy curing simulation.

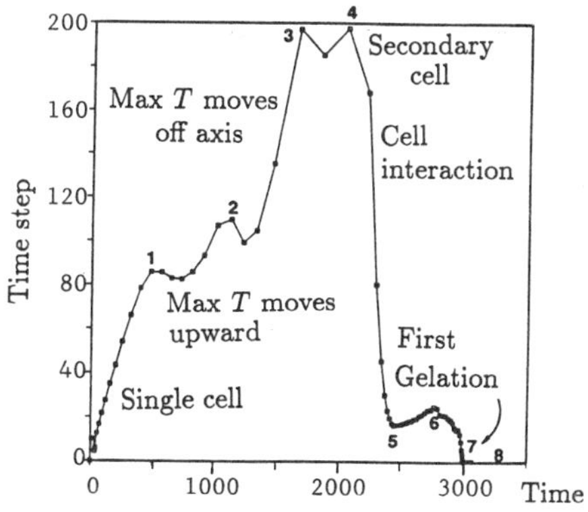

Figure 5.9–29: Time step history for epoxy curing simulation.

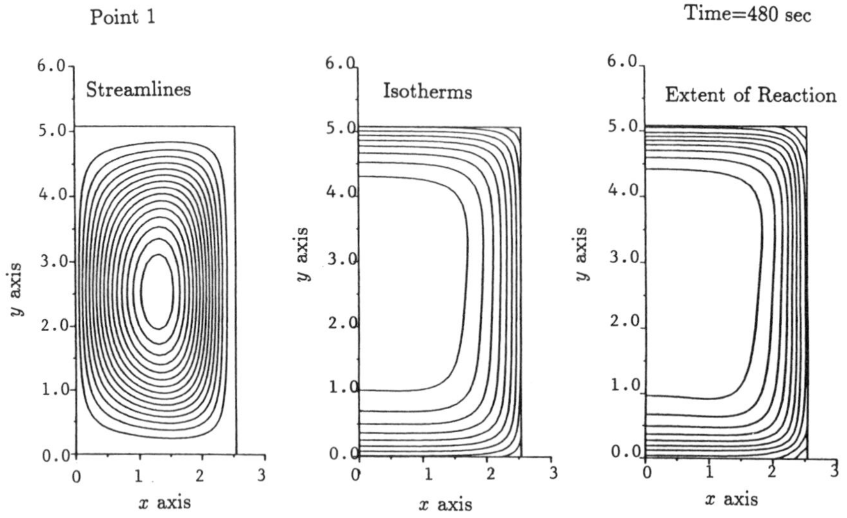

Figure 5.9–30: Contour plots for epoxy curing simulation, time = 480 sec.

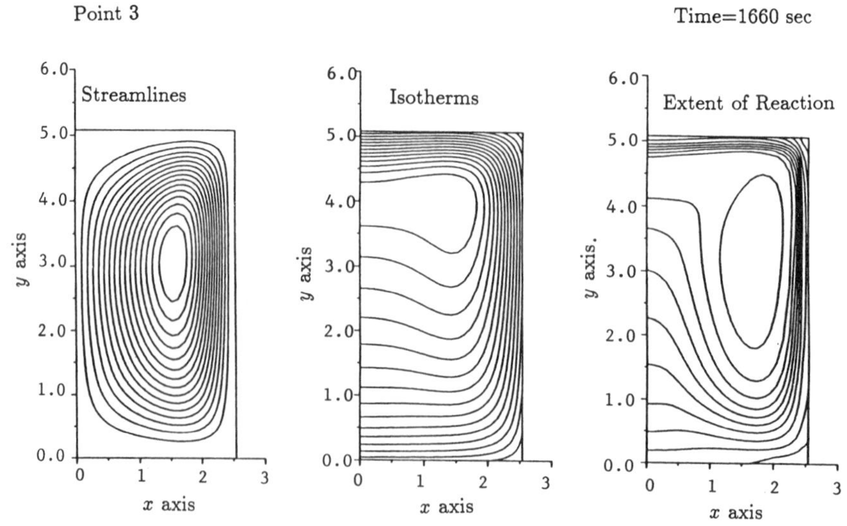

Figure 5.9–31: Contour plots for epoxy curing simulation, time = 1660 sec.

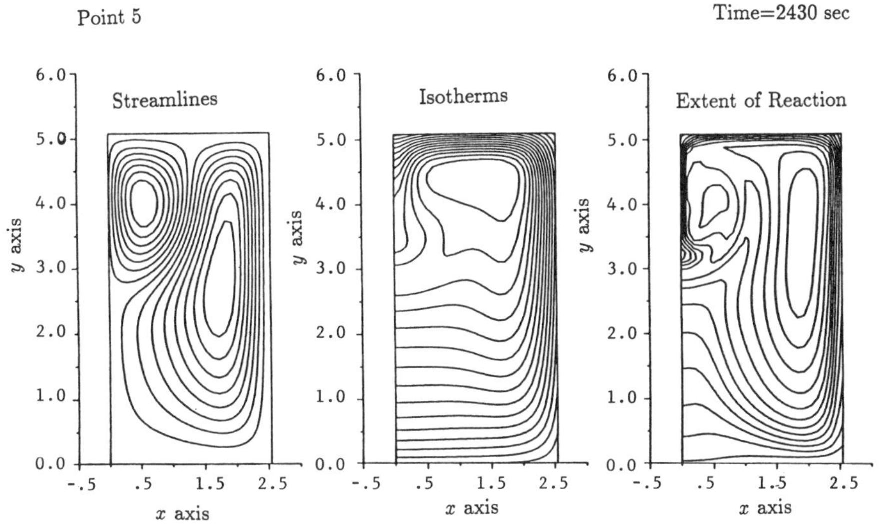

Figure 5.9–32: Contour plots for epoxy curing simulation, time = 2340 sec.

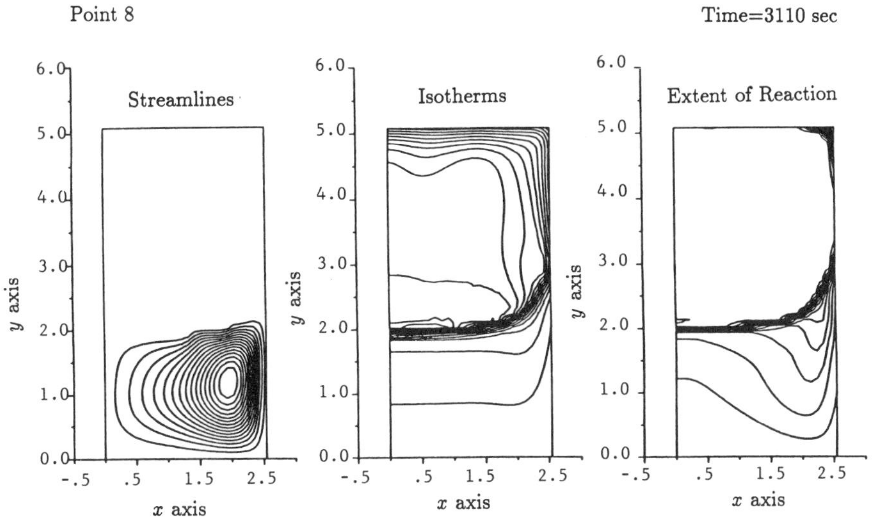

Figure 5.9–33: Contour plots for epoxy curing simulation, time =3110 sec.

5. W. N. R. Stevens, "Finite Element, Stream Function-Vorticity Solution of Steady Laminar Natural Convection," *International Journal for Numerical Methods in Fluids,* **2**, 349–366 (1982).

6. V. Haroutunian, M. S. Engelman, and I. Hasbani, "Segregated Finite Element Algorithms for the Numerical Solution of Large-Scale Incompressible Flow Problems," *International Journal for Numerical Methods in Fluids,* **17**, 323–348 (1993).

7. S. V. Patankar, *Numerical Heat Trasfer and Fluid Flow,* McGraw–Hill, New York (1980).

8. D. K. Gartling, "Finite Element Analysis of Convective Heat Transfer Problems with Phase Change," *Computer Methods in Fluids,* K. Morgan, C. Taylor, and C. A. Brebbia (eds.), Pentech Press, London, U. K. (1974).

9. D. J. Naylor, "Stresses in Nearly Incompressible Materials by Finite Elements with Application to the Calculation of Excess Pore Pressures," *International Journal for Numerical Methods in Engineering,* **8**, 443–460 (1974).

10. E. Hinton, F. C. Scott, and R. E. Ricketts, "Local Least Squares Stress Smoothing for Parabolic Isoparametric Elements," *International Journal for Numerical Methods in Engineering,* **9**, 235–256 (1975).

11. M. Engelman, R. L. Sani, and P. M. Gresho, "The Implementation of Normal and/or Tangential Boundary Conditions in Finite Element Codes for Incompressible Fluid Flow," *International Journal for Numerical Methods in Fluids,* **2**, 225–238 (1982).

12. B. E. Larock and L. R. Herrmann, "Improved Flux Prediction Using Low Order Finite Elements," in *Proc. First International Conference on Finite Elements in Water Resources,* Gray and Pinder (eds.), Pentech Press, London, U. K. (1977).

13. R. S. Marshall, J. C. Heinrich and O. C. Zienkiewicz, "Natural Convection in a Square Enclosure by a Finite Element, Penalty Function Method, Using Primitive Fluid Variables," *Numerical Heat Transfer,* **1**, 315–330 (1978).

14. P. M. Gresho, R. L. Lee, and R. L. Sani, "The Consistent Method for Computing Derived Boundary Quantities When the Galerkin FEM is Used to Solve Thermal and/or Fluids Problems," in *Proc. Second International Conference on Numerical Methods in Thermal Problems,* Venice, Italy (1981).

15. D. K. Gartling, "NACHOS II – A Finite Element Computer Program for Incompressible Flow Problems," Sandia National Laboratories Report, SAND86–1816 and SAND86–1817, Albuquerque, New Mexico (1978).

16. FIDAP Manual, Version 5.0, Fluid Dynamics International, Evanston, Illinois (1990).

17. D. H. Pelletier, J. N. Reddy, and J. A. Schetz, "Some Recent Developments and Trends in Finite Element Computational Natural Convection," in *Annual Review of Numerical Fluid Mechanics and Heat Transfer,* Vol. 2, C. L. Tien and T. C. Chawla (eds.), Hemisphere, New York, 39–85 (1989).

18. D. K. Gartling, "Convective Heat Transfer Analysis by the Finite Element Method," *Computer Methods in Applied Mechanics and Engineering,* **12**, 365–382 (1977).

19. G. deVahl Davis and I. P. Jones, "Natural Convection in a Square Cavity: A Comparison Exercise," *International Journal for Numerical Methods in Fluids,* **3**, 227–248 (1983).

20. T. D. Blacker and M. B. Stephenson, "Paving – A New Approach to Automated Quadrilateral Mesh Generation," Sandia National Laboratories Report, SAND90–0249, Albuquerque, New Mexico (1990).

21. D. K. Gartling, "A Finite Element Analysis of Volumetrically Heated Fluids in an Axisymmetric Enclosure," in *Finite Elements in Fluids, Vol. 4*, R. H. Gallagher, D. Norrie, J. T. Oden, and O. C. Zienkiewicz (eds.), John Wiley & Sons, New York (1982).

22. D. K. Gartling, "A Finite Element Formulation for Incompressible, Conjugate, Viscous/Porous Flow Problems," in *Proc. International Conference on Computational Methods in Flow Analysis, Vol. 1*, H. Niki and M. Kawahara (eds.), Okayama, Japan, 619–626 (1988).

23. D. K. Gartling, "The Numerical Simulation of Viscous Flow with Change of Phase and Chemical Reaction," in *Proc. International Conference on Computational Engineering Science, Vol 2*, S. Atluri and G. Yagawa (eds.), Atlanta, Georgia (1988).

Chapter Six

INELASTIC FLUIDS

6.1 INTRODUCTION

In Chapters 4 and 5 we studied the finite element models of *Newtonian fluids* (*i.e.*, fluids whose constitutive behavior is linear). Fluids that are not described by the Newtonian constitutive relations are commonly encountered in a wide variety of industrial processes. For example, such materials include motor oils, high molecular weight liquids, slurries, pastes, and other complex mixtures. The processing and transport of such fluids are central problems in the chemical, food, plastics, petroleum, and polymer industries.

Non–Newtonian behavior manifests itself in a number of different ways. Most such fluids exhibit a shear rate dependent viscosity, with "shear thinning" (*i.e.*, decreasing viscosity with increasing shear rate) being the most prevalent behavior. Other phenomena associated with the elasticity and memory of the fluid, such as recoil, are also observed in many situations. Differences in the normal stress components occur in many flows and lead to such well–known effects as rod climbing, known as the *Weissenberg effect*, and the curvature of the free surface in an open channel flow. A comprehensive list and discussion of these and other non–Newtonian effects is given in the book by Bird *et al.* [1].

For the present discussion non–Newtonian fluids can conveniently be separated into two distinct categories: (1) inelastic fluids or fluids without memory, and (2) viscoelastic fluids, in which memory effects are significant. The distinction is an important one from both a physical and computational point of view. Basically, inelastic fluids can be viewed as generalizations (in some sense) of the Newtonian fluid. The viscosity function for such materials depends on the rate of deformation of the fluid and thus allows "shear thinning" effects to be modeled. For numerical computations, inelastic fluids can be treated using minor extensions to the standard finite element models developed for Newtonian fluids. Viscoelastic fluids, on the other hand, represent a significant departure from the Newtonian limit in terms of both physical behavior and computational complexity. The primary difficulty here is the "memory" of the fluid; the motion of a material element depends not only on the present stress state, but also on the deformation history of the material element. This history dependence leads to very complex constitutive equations and the need for special computational procedures. It is the purpose of the present chapter to study some aspects of the finite element simulation of inelastic fluids. The finite element simulation of viscoelastic fluids will be discussed in Chapter 7.

Following this introduction, a brief outline of the continuum equations, boundary conditions, and pertinent constitutive equations of inelastic fluids will be presented, followed by a description of their finite element models. The emphasis will be on the treatment of the material nonlinearities inherent in these types of fluids. A section on the simulation of free surface flows is also presented, since many non-Newtonian (and Newtonian) applications include this type of geometric complication. Several numerical examples will also be presented.

The approach taken in the following sections will be very pragmatic with regard to constitutive relations for non-Newtonian fluids. No attempt will be made to theoretically justify the models used or describe in detail their positive or negative aspects. For readers interested in these questions or other topics in rheology, a number of references can be consulted, including the textbooks by Bird *et al.* [1], Lodge [2], and Walters [3]. The book by Crochet *et al.* [4] covers applications of finite difference and finite element methods to non-Newtonian fluids. The text by Pearson [5] treats a variety of theoretical and practical topics regarding the analysis of industrial flow problems.

6.2 GOVERNING EQUATIONS

For completeness and ready reference, the basic equations of motion are reviewed once again. The section concludes with a discussion of representative inelastic, non-Newtonian constitutive relations.

6.2–1 Conservation Equations

The equations resulting from conservation of mass, momentum, and energy for flows of viscous, incompressible, inelastic fluids consist of the standard continuity equation, the Navier–Stokes equations, and the energy equation. For most flows of interest the assumptions of incompressibility and laminar flow are easily justified. Utilizing standard indicial notation, where repeated subscripts imply summation, the governing equations in a rectangular Cartesian coordinate system can be summarized as follows:

Conservation of Mass:

$$\frac{\partial u_i}{\partial x_i} = 0 \qquad (6.2-1)$$

Conservation of Momentum:

$$\rho\left(\frac{\partial u_i}{\partial t} + u_j\frac{\partial u_i}{\partial x_j}\right) = \frac{\partial \sigma_{ij}}{\partial x_j} + \rho g_i \qquad (6.2-2)$$

Conservation of Energy:

$$\rho C\left(\frac{\partial T}{\partial t} + u_j\frac{\partial T}{\partial x_j}\right) = -\frac{\partial q_i}{\partial x_i} + Q + \Phi \qquad (6.2-3)$$

The constitutive equations for the total stress, σ_{ij}, and heat flux, q_i, are given by

$$\sigma_{ij} = -P\delta_{ij} + 2\mu(D_{ij})D_{ij} \qquad (6.2-4)$$

and

$$q_i = -k\frac{\partial T}{\partial x_i} \tag{6.2-5}$$

where equation (4) is written for an isotropic, inelastic, non–Newtonian or generalized Newtonian fluid [1]. Equation (5) is the well known Fourier heat conduction law. The flow kinematics are given by

$$D_{ij} = \frac{1}{2}(L_{ij} + L_{ji}) \ ; \quad L_{ij} = \frac{\partial u_i}{\partial x_j} \tag{6.2-6}$$

where D_{ij} and L_{ij} are the rate of deformation and velocity gradient tensors, respectively. The coefficient μ in equation (4) is a scalar viscosity function which, in general, depends on the rate of deformation tensor in a manner that will be made clear in a subsequent section. The function Φ in equation (3) represents the viscous dissipation in the fluid, and it enters the energy equation as a volume source of thermal energy. The dissipation depends on the velocity gradients and is given by

$$\Phi = 2\mu D_{ij}D_{ij} \tag{6.2-7}$$

The remaining terms in equations (1)–(5) are the same as defined in earlier chapters for Newtonian flows.

Though the above equation set is quite general and describes most practical inelastic flow problems, several simplifications to the equation system are often considered. Many processing operations involving non–Newtonian fluids occur at very low Reynolds numbers, where $Re = \rho U_{ref} D/\mu$ and U_{ref} is a reference velocity and D is a characteristic dimension in the flow problem. For such cases the nonlinear terms in the momentum equation may be neglected. Despite this simplification, the momentum equations are still nonlinear due to the variation of the viscosity function with the rate of deformation tensor. There are many instances when viscous dissipation can also be neglected, in which case the energy equation is simplified. Various other simplifications involving the energy equation are possible, depending on whether buoyancy effects or temperature–dependent material properties are considered important. Such possibilities are treated in the same manner as for a Newtonian fluid and thus will not be considered further in the present chapter.

6.2–2 Boundary Conditions

Equations (1)–(7) represent, in general, a coupled problem that requires boundary conditions on both the fluid motion and the energy transport. The necessary boundary conditions are of the standard type and consist of specified velocities or tractions for the momentum equations and specified temperature or heat flux for the energy equation. Symbolically the boundary conditions for the momentum equation are

$$u_i = f_i^u(s,t) \qquad \text{on } \Gamma_u \tag{6.2-8a}$$

$$\mathcal{T}_i = \sigma_{ij}(s,t)n_j(s) = f_i^{\mathcal{T}}(s,t) \qquad \text{on } \Gamma_{\mathcal{T}} \tag{6.2-8b}$$

where s is the coordinate along the boundary, t is the time, n_i is the outward unit normal to the boundary, and Γ_f is the total boundary enclosing the fluid domain, Ω_f, with $\Gamma_f = \Gamma_u \cup \Gamma_T$. Note that the conditions written in equation (8) are in component form, *i.e.*, there is a condition on each component of the velocity and stress vectors.

The thermal boundary conditions, for the nonisothermal case, are given by

$$T = \hat{T}(s,t) \qquad \text{on } \Gamma_T \tag{6.2-9a}$$

$$\left(k\frac{\partial T}{\partial x_i}\right) n_i + q_c + q_r = q(s,t) \qquad \text{on } \Gamma_q \tag{6.2-9b}$$

where Γ_{ht} is the total boundary enclosing the heat transfer region and $\Gamma_{ht} = \Gamma_T \cup \Gamma_q$. Also, q_c and q_r refer to the convective and radiative components given by

$$q_c = h_c(s,T,t)(T - T_c) \tag{6.2-10a}$$

$$q_r = h_r(s,T,t)(T - T_r) \tag{6.2-10b}$$

where h_c and h_r are the convective and radiative heat transfer coefficients, and T_c and T_r are the reference (or sink) temperatures for convective and radiative transfer.

The above boundary conditions apply to most standard situations where the fluid is contained by fixed boundaries or enters/leaves the domain, Ω_f. One other type of boundary (or interface) condition that requires further discussion concerns conditions along a free surface between two fluids. Many non–Newtonian (and Newtonian) flows involve situations where the fluid (liquid) is extruded, spun, drawn or flows in a sheet or jet, such that a free surface interface exists in the problem domain. In most cases, one of the fluids is a gas and its motion relative to the other fluid is neglected. The case of an interface between two immiscible liquids can also be included in this formulation, but will not be considered here (see Section 1.7–1). A typical free surface geometry is shown in Figure 6.2–1.

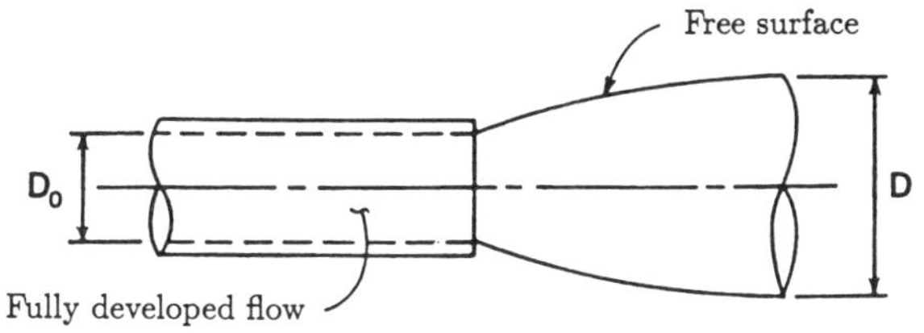

Figure 6.2–1: Typical free surface flow.

The general boundary conditions on a free surface require that the velocity and stress are continuous across the interface. There is also a kinematic condition that requires the interface to remain an interface. For the simplified case in which one fluid is a gas, the continuity condition on the velocity is removed since motion in the gas is ignored. The gas is assumed to have a uniform pressure, P_a, and shear stresses (*i.e.*, drag forces) are neglected. The free surface boundary conditions for this case then are

$$\sigma_{ij} n_j = \gamma \left(\frac{1}{R_1} + \frac{1}{R_2} \right) n_i - P_a n_i \qquad (6.2-11)$$

for the stress continuity condition and

$$\frac{\partial F}{\partial t} + u_j \frac{\partial F}{\partial x_j} = 0 \qquad (6.2-12)$$

for the kinematic condition. In equations (11) and (12), γ is the surface tension, R_i are the principal radii of curvature for the free surface, and $F(x_i, t) = 0$ represents the equation for the free surface. In this formulation, surface tension gradient (*i.e.*, thermocapillary) effects have been neglected. For nonisothermal free surface flows, appropriate (standard) thermal boundary conditions must also be applied at the free surface.

In general, the function F, which describes the location of the free surface, is not known *a priori* and must, therefore, be located as part of the solution to the boundary value problem. For time–dependent free surface flows, such as the filling of a cavity or mold with a viscous fluid, equation (12) provides an evolution equation for the free surface location. In the case of a time–independent free surface flow, equation (12) simplifies to

$$u_i n_i = 0 \qquad (6.2-13)$$

which states that the free surface is a stream surface (or streamline in two dimensions). Note that in the steady flow case there is no explicit equation for the free surface; equation (13) only provides another condition to apply along the free surface. The numerical treatment of these geometric nonlinearities will be considered in a later section.

6.2–3 Constitutive Equations

The form of the constitutive equation for a generalized Newtonian fluid was given above as

$$\sigma_{ij} = -P\delta_{ij} + 2\mu(D_{ij})D_{ij} \qquad (6.2-14)$$

where σ_{ij} are the components of the total stress tensor, P is the pressure, δ_{ij} is the Kronecker delta (or components of the unit tensor), and D_{ij} are the components of the rate of deformation tensor. Of interest in the present section are particular forms for the *deviatoric stress* or *extra stress* components defined by

$$\tau_{ij} = 2\mu(D_{ij})D_{ij} \tag{6.2-15}$$

The viscosity for non–Newtonian fluids is found to depend on the rate of deformation tensor (see [1,6]):

$$\mu = \mu(D_{ij}) = \mu(I_1, I_2, I_3) \tag{6.2-16}$$

where the I_i are the *invariants* of D_{ij}, defined by

$$I_1 = tr(\mathbf{D}) = \sum_i D_{ii} \tag{6.2-17a}$$

$$I_2 = \frac{1}{2}tr(\mathbf{D}^2) = \frac{1}{2}\sum_i\sum_j D_{ij}D_{ji} \tag{6.2-17b}$$

$$I_3 = \frac{1}{3}tr(\mathbf{D}^3) = \frac{1}{3}\sum_i\sum_j\sum_k D_{ij}D_{jk}D_{ki} \tag{6.2-17c}$$

where tr denotes the trace. For an incompressible fluid, $I_1 = \nabla \cdot u = 0$. Also, there is no theoretical or experimental evidence to suggest that the viscosity depends on I_3; thus, the dependence on the third invariant is eliminated. Equation (15) then reduces to

$$\mu = \mu(D_{ij}) = \mu(I_2) \tag{6.2-18}$$

for a generalized Newtonian fluid. The viscosity can also depend on the thermodynamic state of the fluid, which for incompressible fluids usually implies a dependence only on the temperature.

Though equation (18) gives the general functional form for the viscosity function, experimental observation and a limited theoretical base must be used to provide specific models for non–Newtonian viscosities. A variety of models have been proposed and correlated with experimental data (*e.g.*, [1]). Several of the most useful and popular models are cataloged below.

Power–Law Model

The simplest and most familiar non–Newtonian viscosity model is the power–law model which has the form

$$\mu = K I_2^{(n-1)/2} \tag{6.2-19}$$

where n and K are parameters, which are perhaps functions of temperature, termed the *power law index* and *consistency*, respectively. One of the most common features of many non–Newtonian fluids is the "power law" decrease in the apparent viscosity with increasing shear (deformation) rate as modeled by equation (19). Such fluids, with an index $n < 1$ are termed *shear thinning* or *pseudoplastic*. A few materials are *shear thickening* or *dilatant* and have an index $n > 1$. The Newtonian viscosity function is obtained with $n = 1$. The admissible range of the index is bounded below by zero due to stability considerations.

When considering nonisothermal flows, the following empirical relations for n and K have proved useful

$$n = n_0 + B\left(\frac{T - T_0}{T_0}\right) \qquad (6.2-20)$$

$$K = K_0 exp(-A[T - T_0]/T_0) \qquad (6.2-21)$$

where subscript zero indicates a reference condition and A and B are constants for each fluid.

Carreau Model

A major deficiency in the power–law model is that it fails to predict upper and lower limiting viscosities for extreme values of the deformation rate, I_2. This problem is alleviated in the multiple parameter Carreau model, which is of the form

$$\mu = \mu_\infty + (\mu_0 - \mu_\infty)\left(1 + [\lambda I_2]^2\right)^{(n-1)/2} \qquad (6.2-22)$$

In equation (22), μ_0 and μ_∞ are the initial and infinite shear rate viscosities, respectively, and λ is a time constant. The remaining parameters were defined previously.

To illustrate the differences between the power–law and Carreau models, a plot of $\log \mu$ versus $\log I_2$ is shown in Figure 6.2–2 for several examples of each model. Like the power–law model, the Carreau viscosity is seen to have a "power law" region, which is bounded on either end by regions of constant viscosity.

Bingham Model

The Bingham fluid differs from most other fluids in that it can sustain an applied stress without fluid motion occurring. The fluid possesses a yield stress, τ_0, such that when the applied stresses are below τ_0 no motion occurs; when the applied stresses exceed τ_0 the material flows, with the viscous stresses being proportional to the excess of the stress over the yield condition. Typically, the constitutive equation after yield is taken to be Newtonian (Bingham model), though other forms such as a power–law equation (Herschel–Buckley model) are possible.

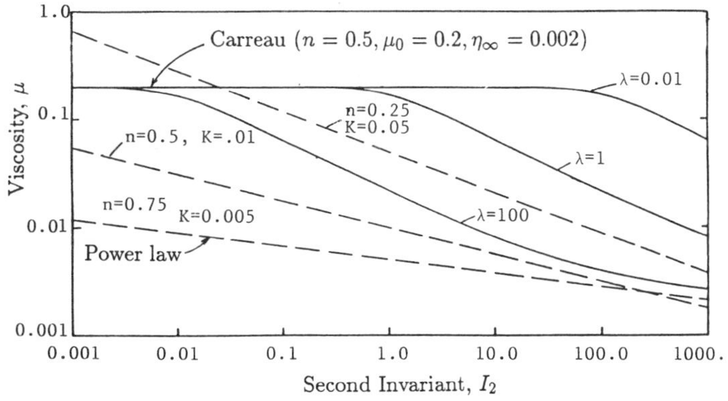

Figure 6.2–2: Viscosity functions for the power–law and Carreau models.

In a general form, the Bingham model is expressed as [7,8]

$$\tau_{ij} = \left(\frac{\tau_0}{\sqrt{I_2}} + 2\mu\right) D_{ij} \qquad \text{when } \frac{1}{2}tr(\tau^2) \geq \tau_0^2 \qquad (6.2-23)$$

$$D_{ij} = 0 \qquad \text{when } \frac{1}{2}tr(\tau^2) < \tau_0^2 \qquad (6.2-24)$$

From equation (23) the apparent viscosity of the material beyond the yield point is $\left(\tau_0/\sqrt{I_2} + 2\mu\right)$. For a Herschel–Buckley fluid the μ in (23) is given by equation (19). The inequalities in equations (23–24) describe a von Mises yield criterion. The implementation of the Bingham model into a computational procedure requires that (23–24) be modified slightly due to the condition $D_{ij} = 0$. This point will be covered in detail in a later section.

The above models serve to illustrate some of the typical viscosity functions that are available to model inelastic, non–Newtonian fluids. A more extensive list is available in [1].

6.3 FINITE ELEMENT MODELS

6.3–1 Introduction

The finite element formulation of non–Newtonian flows follows very closely the formulations developed for Newtonian flow problems. Therefore, in the present section only a brief overview of the general formulation will be given, with more attention focused on those aspects that are unique to the non–Newtonian problem. For a detailed description of the finite element models of viscous incompressible flow problems, the reader is referred to Chapter 4.

As in the case of Newtonian fluids, there are several different formulations that may be used to construct the finite element models of a non–Newtonian fluid. The equations given in Section 6.2–1 are in terms of the velocity, pressure, and temperature as the dependent variables. However, equations using the stream function and vorticity or the stream function alone could also be employed. The primary argument in favor of the use of primitive variables (*i.e.*, velocity, pressure, and temperature) for non–Newtonian flows comes from the free surface problem. The free surface boundary conditions given in (6.2–11) contain the pressure explicitly, and therefore can be more conveniently imposed in a primitive variable formulation. Also, as stated earlier, the use of the stream function–vorticity or stream function formulations present difficulties in the imposition of boundary conditions, and the stream function formulation requires higher–order interpolation (*i.e.*, C^1–continuity).

Here, we develop finite element models based on the primitive variables. The mixed and penalty models described in Chapter 4 are presented here for inelastic fluids.

6.3–2 Mixed Model

The weak form of equations (6.2–1) through (6.2–3) can be developed from their Galerkin integrals, as explained in Chapter 5 (see Section 5.2). The velocity components are approximated by Lagrange interpolation functions, one order higher than those used for the pressure. Suppose that the dependent variables (u_i, P, T) are approximated by expansions of the form

$$T(\mathbf{x},t) = \sum_{n=1}^{N} \psi_n(\mathbf{x}) T_n(t) = \mathbf{\Psi}^T \mathbf{T} \qquad (6.3-1a)$$

$$u_i(\mathbf{x},t) = \sum_{m=1}^{M} \psi_m(\mathbf{x}) u_i^m(t) = \mathbf{\Psi}^T \mathbf{u}_i \qquad (6.3-1b)$$

$$P(\mathbf{x},t) = \sum_{l=1}^{L} \phi_l(\mathbf{x}) P_l(t) = \mathbf{\Phi}^T \mathbf{P} \qquad (6.3-1c)$$

where $\mathbf{\Psi}$ and $\mathbf{\Phi}$ are vectors of interpolation (or shape) functions, $\mathbf{T}$, $\mathbf{u}_i$, and $\mathbf{P}$ are (column) vectors of nodal values of temperature, velocity components, and pressure, respectively, and superscript $(\cdot)^T$ denotes a transpose of the enclosed vector or matrix. Note that the standard practice of interpolating the temperature and velocity variables with the same shape functions has been employed here. Substitution of equations (1a)–(1c) into the weak forms of equations (6.2–1)–(6.2–4) [*i.e.*, equations (5.2–1)] results in the following set of nonlinear algebraic equations [see equations (5.2–4) through (5.2–6) for details]:

Continuity:

$$-\mathbf{Q}^{\mathrm{T}} \mathbf{u} = 0 \qquad (6.3-2)$$

Momentum:

$$\mathbf{M\dot{u} + Cu + K(\mu)u - QP + BT = F} \qquad (6.3-3)$$

Energy:

$$\mathbf{N\dot{T} + DT + LT = G} \qquad (6.3-4)$$

where the superposed dot represents a time derivative and $\mathbf{u}^T = \{\mathbf{u}_1^T, \mathbf{u}_2^T, \mathbf{u}_3^T\}$. This set of equations is virtually identical to those used in Newtonian problems except for the dependence of the viscous diffusion term **K** on the velocity, *i.e.*, the rate of deformation (because of viscosity's dependence on the rate of deformation tensor), and possibly temperature.

For the three–dimensional case, equations (2)–(4) have the following explicit form [the continuity equation (2) and momentum equations (3) are combined into one]:

$$\begin{bmatrix} \mathrm{M} & 0 & 0 & 0 \\ 0 & \mathrm{M} & 0 & 0 \\ 0 & 0 & \mathrm{M} & 0 \\ 0 & 0 & 0 & 0 \end{bmatrix} \begin{Bmatrix} \dot{\mathrm{u}}_1 \\ \dot{\mathrm{u}}_2 \\ \dot{\mathrm{u}}_3 \\ \dot{\mathrm{P}} \end{Bmatrix} + \begin{bmatrix} \mathrm{C(u)} & 0 & 0 & 0 \\ 0 & \mathrm{C(u)} & 0 & 0 \\ 0 & 0 & \mathrm{C(u)} & 0 \\ 0 & 0 & 0 & 0 \end{bmatrix} \begin{Bmatrix} \mathrm{u}_1 \\ \mathrm{u}_2 \\ \mathrm{u}_3 \\ \mathrm{P} \end{Bmatrix} +$$

$$\begin{bmatrix} \hat{\mathrm{K}}_{11} & \mathrm{K}_{21} & \mathrm{K}_{31} & -\mathrm{Q}_1 \\ \mathrm{K}_{12} & \hat{\mathrm{K}}_{22} & \mathrm{K}_{32} & -\mathrm{Q}_2 \\ \mathrm{K}_{13} & \mathrm{K}_{23} & \hat{\mathrm{K}}_{33} & -\mathrm{Q}_3 \\ -\mathrm{Q}_1^T & -\mathrm{Q}_2^T & -\mathrm{Q}_3^T & 0 \end{bmatrix} \begin{Bmatrix} \mathrm{u}_1 \\ \mathrm{u}_2 \\ \mathrm{u}_3 \\ \mathrm{P} \end{Bmatrix} = \begin{Bmatrix} \mathrm{F}_1(\mathrm{T}) \\ \mathrm{F}_2(\mathrm{T}) \\ \mathrm{F}_3(\mathrm{T}) \\ 0 \end{Bmatrix} \qquad (6.3-5)$$

$$[\mathrm{N}]\left\{\dot{\mathrm{T}}\right\} + [\mathrm{D(u)}]\{\mathrm{T}\} + [\mathrm{L}]\{\mathrm{T}\} = \{\mathrm{G(T)}\} \qquad (6.3-6)$$

The coefficient matrices shown in equations (5) and (6) are defined by [from equations (6.2–12)]

$$\begin{aligned} \hat{\mathrm{K}}_{11} &= 2\mathrm{K}_{11} + \mathrm{K}_{22} + \mathrm{K}_{33} \\ \hat{\mathrm{K}}_{22} &= \mathrm{K}_{11} + 2\mathrm{K}_{22} + \mathrm{K}_{33} \\ \hat{\mathrm{K}}_{33} &= \mathrm{K}_{11} + \mathrm{K}_{22} + 2\mathrm{K}_{33} \end{aligned} \qquad (6.3-7a)$$

$$\mathrm{M} = \int_{\Omega^e} \rho_0 \boldsymbol{\Psi}\boldsymbol{\Psi}^T \, d\mathrm{x} \; ; \quad \mathrm{C(u)} = \int_{\Omega^e} \rho_0 \boldsymbol{\Psi}(\boldsymbol{\Psi}^T \mathrm{u}_j)\frac{\partial \boldsymbol{\Psi}^T}{\partial x_j} \, d\mathrm{x}$$

$$\mathrm{K}_{ij} = \int_{\Omega^e} \mu \frac{\partial \boldsymbol{\Psi}}{\partial x_i}\frac{\partial \boldsymbol{\Psi}^T}{\partial x_j} \, d\mathrm{x} \; ; \quad \mathrm{Q}_i = \int_{\Omega^e} \frac{\partial \boldsymbol{\Psi}}{\partial x_i} \boldsymbol{\Phi}^T \, d\mathrm{x}$$

$$\mathrm{F}_i(\mathrm{T}) = -\int_{\Omega^e} \rho_0 g_i \beta \boldsymbol{\Psi}\boldsymbol{\Phi}^T \, d\mathrm{x} + \int_{\Omega^e} \rho_0 g_i \beta T_0 \boldsymbol{\Psi} \, d\mathrm{x} + \oint_{\Gamma^e} \boldsymbol{\Psi} T_i \, ds$$

$$\mathbf{D}(\mathbf{u}) = \int_{\Omega^e} \rho_0 C \boldsymbol{\Phi}(\boldsymbol{\Psi}^T \mathbf{u}_j) \frac{\partial \boldsymbol{\Phi}^T}{\partial x_j}\, d\mathbf{x}$$

$$\mathbf{N} = \int_{\Omega^e} \rho_0 C \boldsymbol{\Psi}\boldsymbol{\Psi}^T\, d\mathbf{x}\ ; \quad \mathbf{L} = \int_{\Omega^e} k \frac{\partial \boldsymbol{\Psi}}{\partial x_i} \frac{\partial \boldsymbol{\Psi}^T}{\partial x_i}\, d\mathbf{x}$$

$$\mathbf{G} = \int_{\Omega^e} \boldsymbol{\Psi} Q\, d\mathbf{x} + \int_{\Omega^e} 2\mu \boldsymbol{\Psi}\Phi\, d\mathbf{x} + \oint_{\Gamma^e} \boldsymbol{\Psi} q_n\, ds \qquad (6.3-7b)$$

where summation on repeated indices is implied. Note that the finite element model is nonlinear because of the nonlinearity in the convective terms as well as the viscosity. In addition, the conductivity k can be a function of temperature. Equations (2)–(4) can be combined into a single matrix equation

$$\begin{bmatrix} \mathbf{M} & 0 & 0 \\ 0 & 0 & 0 \\ 0 & 0 & \mathbf{N} \end{bmatrix} \begin{Bmatrix} \dot{\mathbf{u}} \\ \dot{\mathbf{P}} \\ \dot{\mathbf{T}} \end{Bmatrix} + \begin{bmatrix} \mathbf{C}(\mathbf{u}) + \mathbf{K}(\mathbf{u},\mathbf{T},\mu) & -\mathbf{Q} & \mathbf{B}(\mathbf{T}) \\ -\mathbf{Q}^T & 0 & 0 \\ 0 & 0 & \mathbf{D}(\mathbf{u}) + \mathbf{L}(\mathbf{T}) \end{bmatrix} \begin{Bmatrix} \mathbf{u} \\ \mathbf{P} \\ \mathbf{T} \end{Bmatrix}$$

$$= \begin{Bmatrix} \mathbf{F}(\mathbf{T}) \\ 0 \\ \mathbf{G}(\mathbf{T},\mathbf{u}) \end{Bmatrix} \qquad (6.3-8)$$

or in a more symbolic format as

$$\bar{\mathbf{M}}\dot{\mathbf{U}} + \bar{\mathbf{K}}(\mathbf{u},\mathbf{T},\mu)\mathbf{U} = \bar{\mathbf{F}} \qquad (6.3-9)$$

where

$$\mathbf{U}^T = \{\mathbf{u}_1^T, \mathbf{u}_2^T, \mathbf{u}_3^T, \mathbf{P}^T, \mathbf{T}^T\} \qquad (6.3-10)$$

This completes the development of the mixed finite element model for the inelastic case.

6.3–3 Penalty Model

In the penalty function method, the continuity equation is treated as a constraint (see Section 5.3) and the problem is reformulated as an unconstrained problem. The pressure, which is the Lagrange multiplier, does not appear explicitly as a dependent variable in the formulation, although it is a part of the boundary stresses [see equations (5.3–5)]. In two dimensions, an approximation for the pressure can be post-computed from the relation

$$P = -\gamma_e \left(\frac{\partial u_1}{\partial x_1} + \frac{\partial u_2}{\partial x_2}\right) \qquad (6.3-11)$$

where γ_e is the penalty parameter. The penalty finite element model is given by

$$\begin{bmatrix} \mathbf{M} & 0 \\ 0 & \mathbf{M} \end{bmatrix} \begin{Bmatrix} \dot{\mathbf{u}}_1 \\ \dot{\mathbf{u}}_2 \end{Bmatrix} + \begin{bmatrix} \mathbf{C(u)} & 0 \\ 0 & \mathbf{C(u)} \end{bmatrix} \begin{Bmatrix} \mathbf{u}_1 \\ \mathbf{u}_2 \end{Bmatrix} +$$

$$\left(\begin{bmatrix} 2\mathbf{K}_{11} + \mathbf{K}_{22} & \mathbf{K}_{21} \\ \mathbf{K}_{12} & \mathbf{K}_{11} + 2\mathbf{K}_{22} \end{bmatrix} + \begin{bmatrix} \hat{\mathbf{K}}_{11} & \hat{\mathbf{K}}_{21} \\ \hat{\mathbf{K}}_{12} & \hat{\mathbf{K}}_{22} \end{bmatrix} \right) \begin{Bmatrix} \mathbf{u}_1 \\ \mathbf{u}_2 \end{Bmatrix} = \begin{Bmatrix} \mathbf{F}_1 \\ \mathbf{F}_2 \end{Bmatrix} \qquad (6.3-12)$$

where $\mathbf{M}, \mathbf{C(u)}, \mathbf{K}_{ij}$ (which depend on the viscosity) and $\mathbf{F}_i$ are the same as those defined in equation (7), and

$$\hat{\mathbf{K}}_{ij} = \int_{\Omega^e} \gamma_e \frac{\partial \Psi}{\partial x_i} \frac{\partial \Psi^T}{\partial x_j} \, d\mathbf{x} \qquad (6.3-13)$$

The energy equation remains unchanged as in equation (6). In matrix form, equations (13) and (6) can be expressed as

$$\bar{\mathbf{M}}\dot{\mathbf{U}} + \bar{\mathbf{K}}(\mathbf{u}, \mathbf{T}, \mu)\mathbf{U} = \bar{\mathbf{F}} \qquad (6.3-14)$$

where

$$\mathbf{U}^T = \{\mathbf{u}_1^T, \mathbf{u}_2^T, \mathbf{u}_3^T, \mathbf{T}^T\} \qquad (6.3-15)$$

This completes the discussion of the penalty finite element model for the inelastic case.

6.3–4 Matrix Evaluations

The appearance of a nonlinear diffusion term, $\mathbf{K}$, in equations (9) and (14) influences two basic aspects of the finite element analysis, namely, the matrix evaluation and solution algorithms. Matrix evaluation methods will be discussed first.

The techniques used for the construction of element level matrices that contain variable coefficients have generally fallen into one of two categories — the reconstruction methods and the hypermatrix methods. The differences in these methods and their application to non–Newtonian formulations are best explained via a specific example. Consider the construction of a particular component of the $\mathbf{K}$ matrix in equation (9), for example,

$$\mathbf{K}_{11} = \mathbf{K}_{\mathbf{xx}} = \int_{\Omega_e} \frac{\partial \Psi}{\partial x} \mu(I_2, T) \frac{\partial \Psi^T}{\partial x} \, d\mathbf{x} \qquad (6.3-16)$$

where Ψ is the vector of shape functions used for the velocity field. The viscosity function is shown in its most general form with a dependence on I_2 and the temperature; the invariant I_2 depends on velocity gradients [see equation (6.2–17)].

For most finite element applications the integration in equations (7) and (16) is performed via a numerical quadrature. Such a computation requires that each function in the integrand be evaluated at each integration point within the element. Therefore, the evaluation of **K** for a non-Newtonian fluid requires that the velocity gradients and perhaps temperature be evaluated at integration points, using the values from the latest available solution $(\bar{u}_m, \bar{T}_n)$ at time t_s:

$$T(\mathbf{x}, t_s) = \sum_{n=1}^{N} \theta_n(\mathbf{x}) \bar{T}_n(t_s) \tag{6.3-17a}$$

$$u_i(\mathbf{x}, t_s) = \sum_{m=1}^{M} \psi_m(\mathbf{x}) \bar{u}_i^m(t_s) \tag{6.3-17b}$$

In many standard finite element programs the above quadrature procedure is carried out each time the element matrix is required. As a result of the iterative methods used to solve the nonlinear equations, the matrix formation operations may be required many times, especially for time-dependent problem, resulting in a large computational cost.

An alternative to the matrix reconstruction method, the hypermatrix approach may be used. This method reduces some of the computational cost but at the cost of some additional storage and I/O in the computer program. Since the viscosity is a function of position in the element (due to its functional dependence on the velocity field and temperature), it is natural to interpolate the viscosity in the same way as a dependent variable. Thus, let the viscosity be represented by

$$\mu = \mathbf{\Psi}^T \hat{\mu} \tag{6.3-18}$$

where $\mathbf{\Psi}$ is a vector of shape functions and $\hat{\mu}$ is a vector of nodal point viscosity values. Substitution of (18) in equation (16) yields

$$\mathbf{K}_{\mathbf{xx}} = \int_{\Omega_e} \frac{\partial \mathbf{\Psi}}{\partial x} (\mathbf{\Psi}^T \hat{\mu}) \frac{\partial \mathbf{\Psi}^T}{\partial x} \, d\mathbf{x} \tag{6.3-19}$$

Since the $\hat{\mu}$ are the nodal values, the integral in equation (19) can be constructed once and stored as a three-dimensional array $\mathbf{K}(NI, NJ, NK)$, where each index corresponds to one of the shape functions in (19). This matrix is called the *hypermatrix*. A product of the hypermatrix with a known vector of nodal point viscosities produces the required element matrix without repeated numerical quadrature. This technique has been used successfully (see [9,10]) for the nonlinear advection terms in the momentum equations [*i.e.*, the $\mathbf{C}(\mathbf{u})\mathbf{u}$ term in equation (7)] as well as for material property variations (see [11]). It is most effective when the variable coefficient is a nodal point quantity or depends on a nodal point quantity, since this permits shape function evaluations to be avoided. Unfortunately, this is not the case for the non-Newtonian viscosity which depends on velocity gradients. It is well known that the most accurate points within a quadrilateral element at which to evaluate derivatives are the 2 × 2 Gauss integration points (see [12]). Thus, for maximum accuracy in the viscosity evaluation the invariant I_2 should be evaluated at the Gauss points. However, the extrapolation of Gauss point values of I_2 to the nodes by standard methods (see [13,14]) in order to use equation (19) has not proven to be a viable technique for most non–Newtonian viscosity models. In general, predictions of I_2 at the nodes via extrapolation are very inaccurate and lead to poor overall accuracy or nonconvergence of the solution method. As an alternative to extrapolation an averaging method is recommended in which the Gauss point values of I_2 are averaged over the element and a single value used to evaluate the viscosity at the nodes.

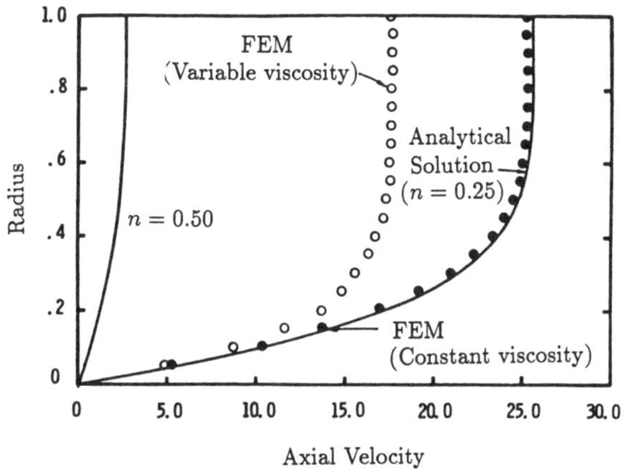

Figure 6.3–1: Velocity profile for a power fluid in a tube; various methods for viscosity computation.

Figure 6.3–1 shows velocity profiles obtained with both the extrapolation and averaging procedures described above. The results are for a power–law fluid in a cylindrical tube. The inaccuracies generated by extrapolation are on the order of 30% while the averaging procedure yields results within a few percent of the analytical result. The averaging procedure in conjunction with equation (19) has been used successfully in other problems with a variety of viscosity models (see [15,16]).

To conclude this section it should be pointed out that the choice of a reconstruction or hypermatrix procedure is dependent on the type of computing resources available and the structure of the finite element software. Both methods are equivalent and effective. The relative costs of CPU and I/O time will heavily influence the selection.

6.4 SOLUTION METHODS

The solution procedure selected for the non–Newtonian problem given in equation (6.3–1) must be capable of treating several different types of nonlinearity. For nonzero Reynolds number flows the nonlinear advection terms are significant and in many cases may dominate the problem. The non–Newtonian fluid introduces a second type of nonlinear behavior while the presence of unknown free surface boundaries can introduce a geometric nonlinearity. In the present section only the first two types of phenomena will be considered; free surface solutions are covered in a subsequent section. Also, the discussion will focus on steady isothermal problems since many of the procedures and results obtained for this type of problem can be extended in a straightforward manner to more complex situations. The solution methods described in Chapters 4 and 5 are also applicable here with only minor alterations.

For the steady–state case, equation (6.3–1) takes the form

$$\mathbf{C}(\mathbf{u})\mathbf{u} + \mathbf{K}(\mu, \mathbf{u}) - \mathbf{Q}\mathbf{P} = \mathbf{F} \qquad (6.4-1a)$$

$$-\mathbf{Q}^{\mathbf{T}}\mathbf{u} = 0 \qquad (6.4-1b)$$

or, as a single matrix equation we have

$$\hat{\mathbf{K}}(\mathbf{U})\mathbf{U} = \hat{\mathbf{F}} \qquad (6.4-2)$$

where the vector $\mathbf{U}$ now contains the velocity components $(\mathbf{u_1}, \mathbf{u_2}, \mathbf{u_3})$ and the pressure $\mathbf{P}$. The dependence of μ on $\mathbf{u}$ has been absorbed into $\hat{\mathbf{K}}$. There are a wide variety of iterative algorithms that can be applied to equation (2), and we will consider only two such schemes in detail here.

The simplest method is Picard's method (also known as successive substitution or functional iteration), which can be written in the following form

$$\hat{\mathbf{K}}(\mathbf{U}^n)\mathbf{U}^{n+1} = \hat{\mathbf{F}}^n \qquad (6.4-3)$$

where the superscript n indicates the iteration level. Equation (3) describes a particularly simple procedure in which the nonlinear coefficients in the problem are evaluated using dependent variable data from the previous iteration. The method has a linear rate of convergence but works for a relatively large range of problems. For shear thinning materials it is observed that as the power law index, n [see equation (6.2–20) for example] decreases, the rate of convergence decreases markedly (see [4]).

A more sophisticated algorithm with a higher rate of convergence is Newton's method. For the nonlinear equation in (2) Newton's method may be written as

$$\mathbf{J}(\mathbf{U}^n)\left[\mathbf{U}^{n+1} - \mathbf{U}^n\right] = -\hat{\mathbf{K}}(\mathbf{U}^n)\mathbf{U}^n + \hat{\mathbf{F}}^n \qquad (6.4-4)$$

where **J** is the Jacobian defined by

$$\mathbf{J}(\mathbf{U}^n) = \frac{\partial}{\partial \mathbf{U}} \left[\hat{\mathbf{K}}(\mathbf{U})\mathbf{U} - \hat{\mathbf{F}} \right] \Big|_{\mathbf{U}^n} \tag{6.4-5}$$

Newton's method is the standard solution procedure for Newtonian problems since it is well–suited to the quadratic nonlinearity occurring in the advection terms of the momentum equations. However, experience has shown that this procedure, as written in equation (6.3–29), does not perform well for many types of generalized Newtonian fluids (see [17]). In particular, Newton's method does not work well for viscosity models with shear thinning. It is therefore recommended that the non–Newtonian behavior in equation (4) be treated using a Picard method; the advection terms, if present, should be treated using Newton's method.

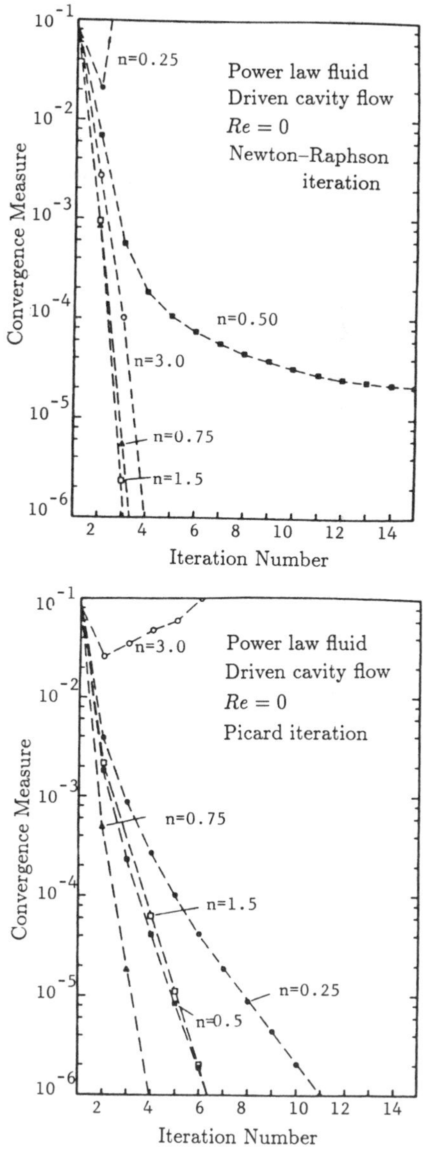

Figure 6.4–1: Iterative convergence for Picard and Newton methods.

To illustrate the behavior of the Picard and Newton algorithms, a simple creeping flow problem ($Re = 0$) was solved using a power–law viscosity model. The problem consists of the ubiquitous driven cavity flow in which a fluid is contained in a square cavity three sides of which are stationary; the fourth side of the cavity moves at unit velocity in its own plane. Shown in Figure 6.4–1 are plots of the convergence measure (relative norm on the change in the solution between iterations) versus iteration number for both algorithms and several values of the power–law index. The Picard scheme converges for all values of the index below unity (shear thinning) but diverges for shear thickening fluids. The Newton method performs in the opposite way, with rapid convergence for shear thickening and divergence for smaller values of the index.

Other methods of solution of equation (4) are possible though the Picard and Newton methods represent the behavioral limits for most algorithms. Tanner and Milthorpe [17] have used a combination of a Picard and Newton method successfully for several nonlinear viscosity models. Also, Engelman [18] has found the quasi–Newton or variable metric method to be very cost effective for non–Newtonian flows.

The solution procedures outlined above are applicable to all of the common inelastic, non–Newtonian models. However, the Bingham model [equation (6.2–24)] deserves some further comments due to its unique method of implementation. For a typical velocity–based finite element method, the exact Bingham constitutive equation is quite difficult to use, mainly as a result of the requirement that no motion take place below the yield condition. To circumvent this difficulty it has proved useful to employ an approximate Bingham equation given by

$$\tau_{ij} = \left(\frac{\tau_0 (1 - \mu/\mu_r)}{\sqrt{I_2}} + 2\mu \right) D_{ij} \qquad \text{when} \quad \frac{1}{2} tr(\tau^2) \geq \tau_0^2 \qquad (6.4-6a)$$

$$\tau_{ij} = 2\mu_r D_{ij} \qquad \text{when} \quad \frac{1}{2} tr(\tau^2) < \tau_0^2 \qquad (6.4-6b)$$

where μ_r is a pre-yield viscosity and $\mu/\mu_r \ll 1$. A plot of the constitutive model given by (6) is shown in Figure 6.4–2. Unlike the true Bingham material, the fluid described by equation (6) can undergo deformation below the yield point though the magnitude of the motion can be made arbitrarily small (*i.e.*, approach the Bingham model) by increasing μ_r relative to μ. Experience has shown that when $\mu/\mu_r \sim .01$ or less then the solution is virtually independent of μ_r and thus approximates the Bingham model with excellent accuracy [16,17].

Iterative convergence of Bingham fluids using a Picard method [see equation (3)] can be very slow since the apparent viscosity after yield is essentially like that of a power–law fluid with index $n = 0$. To accelerate the convergence a particular form of extrapolation can be used. Note first that for the present model the yield stress, τ_0, corresponds to a value of the rate of deformation (invariant) given by $\sqrt{I_{2_0}} = \tau_0/(\mu_r - \mu)$ as seen in Figure 6.4–2. At the start of the iterative procedure the fluid is usually assumed to have a viscosity, μ_r, from which an initial velocity field and values of I_2 may be computed. For those elements which have yielded $(\sqrt{I_2} \geq \sqrt{I_{2_0}})$ an extrapolation at constant stress should be used to estimate a new value of the apparent viscosity. Failure to perform this extrapolation can double or triple the number of iterations needed to reach convergence. Based on Figure 6.4–2 (dashed arrow) it is possible to provide an analytical expression for the projected value of I_2 needed for the viscosity evaluation. That is

$$\sqrt{I_2} = \frac{2\mu_r\sqrt{I_2^{int}} - \tau_0}{2\mu} \qquad (6.4-7)$$

where the superscript *int* indicates the value of I_2 computed initially using a viscosity of μ_r. The above procedure has been used in a variety of steady–state problems with good success [15,16] and is readily extended to time–dependent flows.

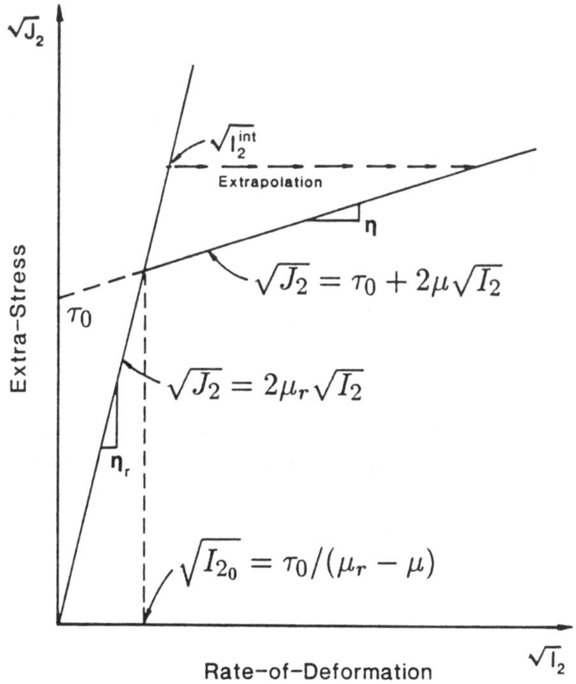

Figure 6.4–2: Bi–viscosity model for computations involving Bingham fluids.

6.5 FREE SURFACE PROBLEMS

6.5–1 Preliminary Comments

As noted previously, many non–Newtonian (and Newtonian) problems of practical interest contain a second complicating feature in the form of a free surface boundary. Boundary conditions for the free surface problem were given in Section 6.2–2 and were seen to be somewhat different for the steady–state flow problem and the case of a dynamic free surface. Likewise, the solution algorithms for steady and time–dependent free surface flows tend to be quite different since the boundary interface conditions dictate the numerical procedure. In this section we provide an overview of the types of procedures used for both static (quasi–static) free surfaces and advancing front type problems.

6.5–2 Time–Independent Free Surfaces

Steady flow problems with a free surface boundary are somewhat easier than the dynamic case since only a single location and shape of the free surface needs to be computed. For some geometries, such as extrusions and jets with a simple cross section, the free surface shapes are relatively simple and the required algorithms are straightforward and easily implemented. When the free surface shape can be very complex, such as in the various coating flows, the numerical algorithms are necessarily more sophisticated. In the following, the discussion is limited to two–dimensional problems to simplify the notation and equations; the extension to three dimensions is conceptually the same though the details are more complex.

The appropriate boundary conditions for a free surface involve the balance of forces in the directions normal and tangent to the surface. Neglecting drag effects from the surrounding medium, these conditions are given by equation (6.2–10) and are written for the two–dimensional case as

$$\sigma_{nn} = \gamma \left(\frac{1}{R} \right) - P_a \qquad (6.5-1)$$

$$\sigma_{ns} = 0 \qquad (6.5-2)$$

where γ is the surface tension, R is the principal radius of curvature of the surface, and P_a is the ambient pressure. Note that equation (1) is written for a planar problem; an axisymmetric problem would require consideration of a second radius of curvature. In addition to the boundary conditions (1) and (2), a kinematic condition on the surface is required; the free surface is a streamline and thus

$$u_i n_i = 0 \qquad (6.5-3)$$

where n is the outward normal. For many applications surface tension effects are negligible. Also, without loss of generality the ambient pressure may be set to zero. Thus, the appropriate boundary conditions in many cases correspond to a vanishing of the normal and tangential stresses on the free surface.

The zero stress conditions encountered for the zero surface tension case are precisely the conditions that are enforced by the "natural" boundary conditions arising in the Galerkin finite element form of the momentum equations. With these conditions automatically enforced, it remains to find a method for locating the position of the free surface. For simple free surface shapes a procedure proposed by Tanner *et al.* [19] and used by others (see [20,21]) can be employed.

The computation starts with an assumed free surface shape and a given finite element mesh. The appropriate boundary conditions are applied and the problem solved for a velocity and pressure field. In general, the computed velocity field will not satisfy the kinematic constraint in equation (13) along the assumed free surface boundary. Referring to Figure 6.5–1, the location of the free surface is updated using the following relation

$$\frac{d}{dx}f(x) = \frac{v}{u} \qquad (6.5-4)$$

where u, v are the velocity components on the surface. Equation (4) can be integrated numerically (*e.g.*, Simpson's rule) to produce a new surface shape $f(x)$. The success of the integration is dependent on knowing at least one (fixed) point on the free surface. With a new free surface location a new finite element mesh can be constructed and the above procedure is repeated until convergence is obtained. For problems with material and/or convective nonlinearities, the free surface iteration is normally intertwined with the other iterative procedures.

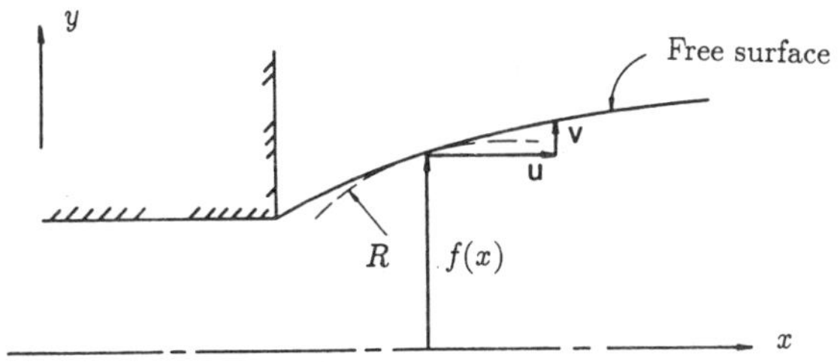

Figure 6.5–1: Nomenclature for free surface boundary computations.

When surface tension is not negligible, the procedure outlined above may still be used but with minor modifications. Instead of applying the zero normal stress boundary condition, a normal force due to the surface tension must be applied at each iteration. Again referring to Figure 6.5–1, the radius of curvature is given by

$$R = \frac{\left[1 + \left(\frac{df}{dx}\right)^2\right]^{3/2}}{\left(\frac{d^2f}{dx^2}\right)} \qquad (6.5-5)$$

which may be computed at any iteration since $f(x)$ is then known. Equations (1) and (5) allow the normal stress at the free surface to be computed and applied at each cycle of the free surface iteration. The procedure outlined above has been used to study a number of extrusion and jet flow problems with good success. However, for applications where the shape of the free surface is rather complex and the indicated procedure is not adequate, more sophisticated algorithms must be employed.

An alternative approach, developed by Saito and Scriven [22] and Scriven and Kistler [23], employs a finite element parameterization in which location parameters (degrees of freedom) for the free surface become unknowns in the general problem formulation. Though the details of implementation are fairly involved, the general idea can be obtained from the sketch in Figure 6.5–2. As indicated by the figure, nodes on the free surface are free to move along fixed lines, called spines, such that their location relative to a fixed base point is determined by a single nodal parameter, h_i. The coordinates for the free surface nodes are given by

$$x_i^{fs} = \alpha_x^i h_i + x_i^{bl} \tag{6.5-6a}$$

$$y_i^{fs} = \alpha_y^i h_i + y_i^{bl} \tag{6.5-6b}$$

where α^i are components of the direction vector along the spine, and x_i^{bl} and y_i^{bl} are coordinates for the base point. Note that in an actual implementation, nodes for all the elements between the free surface and the base line would also be parameterized such that node spacing ratios would be maintained with movement of the free surface. To complete the formulation for the free surface, it is observed that the surface is composed of a series of element edges with each edge containing two or three nodes, depending on the type of finite element being employed. The isoparametric description of the element edge then has

$$x = \mathbf{\Upsilon}^{\mathbf{T}}\mathbf{x}^{\mathbf{fs}} \tag{6.5-7a}$$

$$y = \mathbf{\Upsilon}^{\mathbf{T}}\mathbf{x}^{\mathbf{fs}} \tag{6.5-7b}$$

where $\mathbf{x}^{\mathbf{fs}}, \mathbf{y}^{\mathbf{fs}}$ are computed from (6) and $\mathbf{\Upsilon}$ is an appropriate linear or quadratic edge (interpolation) function.

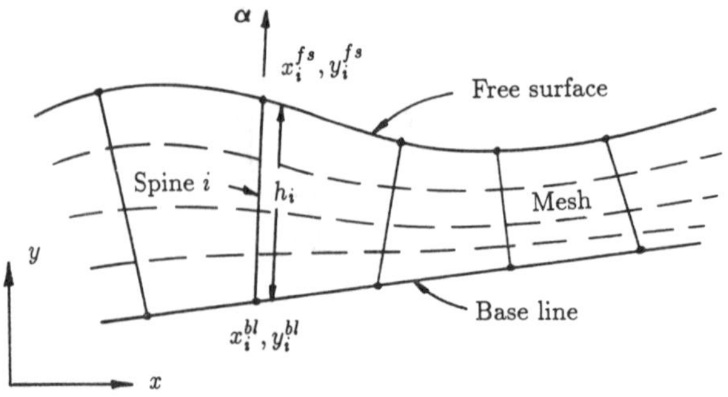

Figure 6.5–2: Nomenclature for spines used in free surface description.

With the above relations, the standard finite element discretization of the momentum and continuity equations can be implemented. The proper accounting must be made in the boundary integrals for the free surface stresses acting on the parameterized element edges. A final equation is also needed since the h parameters are now unknowns in the global equation set. The last equation needed is the kinematic condition (3) which can be written in a weighted residual form as

$$\int_{\Gamma_{fs}} \Upsilon u_n d\Gamma = \int_{\Gamma_{fs}} \Upsilon(u_x n_x + u_y n_y) d\Gamma = 0 \qquad (6.5-8)$$

where the weighting function is the interpolation function defining the free surface [see equation (7)]. With the proper definition for the velocities in the element and the mass consistent normals at the surface [24], the surface integrals in equation (8) can be evaluated and assembled to provide an equation for each boundary node. Analogous with the matrix equations (6.3–5) and (6.3–6), the matrix problem for a free surface flow is then given by

$$\mathbf{C}(\mathbf{u})\mathbf{u} + \mathbf{K}(\mathbf{u})\mathbf{u} - \mathbf{Q}\mathbf{P} + \mathbf{B}\mathbf{h} = \mathbf{F} \qquad (6.5-9a)$$

$$-\mathbf{Q}^T\mathbf{u} = 0 \qquad (6.5-9b)$$

$$\mathbf{K}_{fs}\mathbf{u} = 0 \qquad (6.5-9c)$$

where $\mathbf{h}$ is the global vector of free surface parameters, $\mathbf{B}$ corresponds to the surface stress conditions, and $\mathbf{K}_{fs}$ is the kinematic constraint from (8).

The solution to (9) proceeds in the usual way, with the use of Newton's method being the preferred algorithm. Other fixed point schemes such as the Picard method are not employed here since at each iteration the velocities would have to be interpolated onto the new nodal point locations defined by new $\mathbf{h}$ values. However, a Newton procedure avoids this difficulty if the full Jacobian is used, *i.e.*, the variations of all equations with respect to the free surface parameters $\mathbf{h}$ are constructed. The construction of these new Jacobian entries is not trivial since both the integrands and the limits of integration depend on $\mathbf{h}$. The usual methods for isoparametric mapping of the elements can be employed to simplify this task and make the Newton algorithm usable.

The method of spines is a formal method of embedding the kinematic free surface condition directly into the equations for the system; the method given in equation (4) uses the kinematic condition outside the basic equations as a control on the (free surface) iteration procedure. Though the spine algorithm is more general and can handle some problems of great complexity, it is not without its drawbacks. The fact that nodes are constrained to a "one-dimensional" motion along the spine limits severely the ability of the mesh to conform to a complex surface shape; re–entrant corners are particularly difficult to describe with spines. In essence, the initial mesh design and spine placement must anticipate the final free surface shape. The method is practical in some two–dimensional applications and a limited number of three–dimensional flows. To avoid some of these difficulties, work has progressed on changing the way the mesh is parameterized but retaining most of the other solution features of the spine approach. Christodoulou and Scriven [25] have proposed using body–fitted coordinates and elliptic mesh generation to parametrize the entire mesh. This appears to be more flexible and useful than spines and is extendable to three–dimensions. Elliptic mesh generation does, however, have its limitations. This area continues to be one of heavy research activity because of its importance and difficulty. A recent review of free surface algorithms can be found in [26].

6.5–3 Time–Dependent Free Surfaces

The methods outlined in the previous section can, in some cases, be extended to handle time–dependent free surface problems. For situations where the amplitude of the free surface motion is relatively small or the motion is almost one–dimensional, the previous methods would be suitable. However, of particular interest in the time domain are problems involving the filling of cavities, such as in molding or casting applications. In this case, the quasi–Lagrangian methods based on mesh updating are not very applicable due to the very large and complex motions of the advancing fluid front; a simpler, Eulerian method is required.

The Volume of Fluid (VOF) method for dynamic free surfaces uses a fixed Eulerian mesh and employs the kinematic condition in (6.2–11) to describe the location of the fluid surface. The hyperbolic equation,

$$\frac{\partial F}{\partial t} + u_j \frac{\partial F}{\partial x_j} = 0 \qquad (6.5-10)$$

describes the motion of the free surface F due to the fluid velocity u_j. In a finite element framework, equation (10) can be discretized using a Galerkin weighted–residual formulation to produce an advection equation for the nodal values of F. The function F varies between zero and unity, with zero indicating the absence of fluid and unity denoting that fluid is present. Being hyperbolic, equation (10) requires only initial values of F at the inlet regions for the problem. As the function F is computed at nodes throughout the mesh, the relevant fluid and thermal equations are solved on an ever–expanding set of grid points indicated by the condition $F = 1$.

In standard implementations, equation (10) is advanced using an explicit time integration method, while the fluid mechanics and heat transfer parts of the problem are solved implicitly. This is done mainly for computational efficiency though the ability to subcycle the kinematic equation does offer some accuracy advantages. With an explicit formulation, the F equation has a CFL (Courant–Friedrich–Levy) limit on the allowable time step. Again, this is not a severe limitation since subcycling is available. At the free surface (usually indicated by $F = 0.5$) the usual stress boundary conditions are applied; the standard choice is for a traction free surface as surface tension effects are somewhat difficult to incorporate in this formulation.

As with all front tracking methods of this type, the fidelity of the free surface shape and the details of the flow at the surface are not as good as with a method that locally fits (meshes) the free surface. However, this algorithm is capable of simulating the overall behavior of complex filling flows as demonstrated in the next section.

6.6 NUMERICAL EXAMPLES

6.6–1 Preliminary Comments

In this section a few example problems are described to demonstrate some of the procedures discussed in previous sections. The first examples consider inelastic, non–Newtonian flows and the second group illustrates a couple of free surface applications. Most of the examples presented here were obtained using the mixed finite element model contained in the NACHOS II code [9], unless otherwise stated. Use of the penalty finite element models for power–law fluids can be found in References 27–29. For further examples, the literature cited in the text should be consulted.

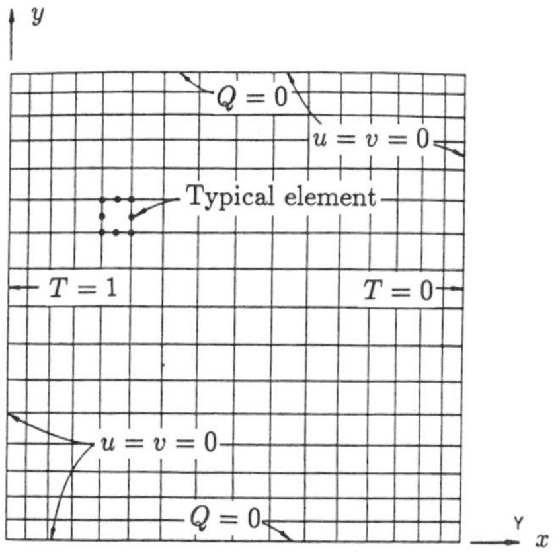

Figure 6.6–1: Schematic of natural convection in a cavity.

6.6–2 Buoyancy Driven Flow in a Cavity

The first example consists of a planar, rectangular cavity that contains a power law fluid as shown in Figure 6.6–1. The vertical walls of the cavity are held at uniform but different temperatures, while the horizontal boundaries are insulated. The differentially heated walls produce a buoyancy driven flow in the cavity, the details of which are dependent on the power law index. The computational boundary conditions for this problem consist of specified temperatures on the vertical walls, zero fluxes on the horizontal walls, and zero velocities on all four boundaries. The pressure level was set by specifying a pressure value at one point on the boundary of the cavity.

Shown in Figures 6.6–2 and 6.6–3 are contour plots for the temperature and stream function obtained for three values of the power–law index. The $n = 1.0$ case corresponds to a Newtonian fluid at a Rayleigh number of $Ra = 1 \times 10^4$ while the other two cases are for shear thinning fluids at the same nominal Rayleigh number. The effect of the shear thinning viscosity is evident as the fluid motion becomes more vigorous and complex for the same basic driving force. The illustrated cases were solved using a Picard iteration scheme starting from a zero initial solution field and required approximately 10 to 12 iterations to reach convergence.

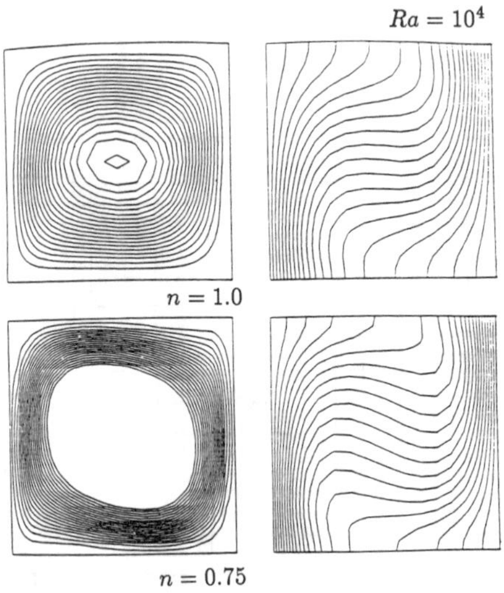

Figure 6.6–2: Streamlines and isotherms, natural convection in a cavity, power law fluid, $n = 0.75$.

6.6–3 Driven Cavity Flow

An isothermal driven cavity flow is considered as a second example. A Bingham fluid is contained in a planar cavity, three sides of which are fixed, with the fourth side moving at unit velocity in its own plane. For purposes of computation, the Bingham parameters are taken to be $\mu = 1.0$ and $\mu_r = 1000.0$ with τ_0 being varied over the range $0 \leq \tau_0 \leq 0.50$. The nominal Reynolds number for the problem is $Re = 100$. The computational mesh for this problem is the same as shown in Figure 6.6–1. Again, a Picard scheme was used to solve the problem with the extrapolation procedure of Section 6.3–2 also being employed.

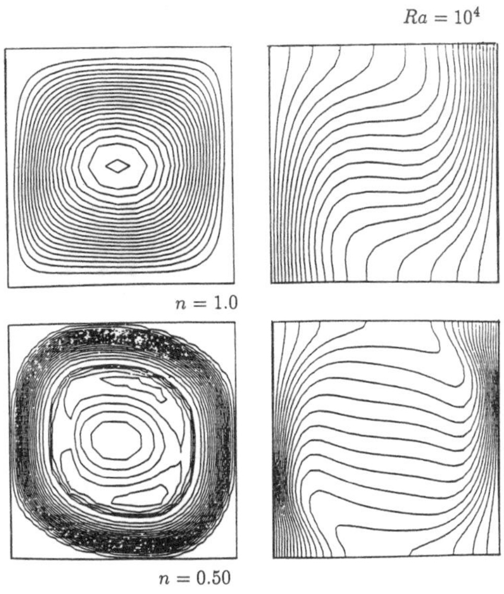

Figure 6.6–3: Streamlines and isotherms, natural convection in a cavity, power law fluid, $n = 0.50$.

Figure 6.6–4 contains contour plots of the stream function at four different values of the yield stress. The $\tau_0 = 0$ case is a Newtonian fluid. With increasing yield stress, the motion in the cavity is reduced and relegated to a small area near the moving lid of the cavity. This effect is seen more clearly in the velocity profiles shown in Figure 6.6–5. Note also that, as shown in Figure 6.6–5, the velocity profiles show a "break" near the vortex center. This is due to the relative magnitudes of the deformation rate above and below the vortex center. Actually, the break point should occur at zero velocity but is offset slightly due to the use of a constant viscosity within each element and a relatively coarse mesh.

6.6–4 Squeeze Film Flow

The third example considers the flow of a plastic (Bingham) fluid in a squeeze film apparatus. This type of device is often used to measure the yield parameters for lubricating greases and other "stiff" materials. The geometry and a typical mesh are shown in Figure 6.6–6. Two circular plates are separated by a fluid layer of initial height h. At time zero the upper plate moves downward with a fixed velocity. The motion of the plate is slow enough that a creeping flow assumption may be used and a series of steady–state problems examined with h as a parameter.

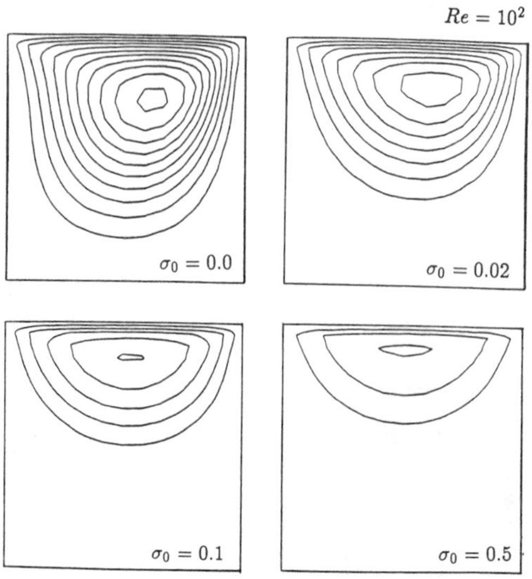

Figure 6.6–4: Streamlines for driven cavity flow; Bingham fluid; various yield stresses.

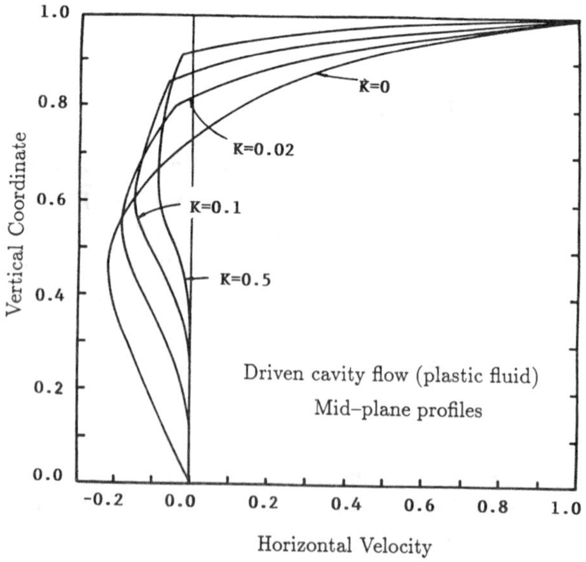

Figure 6.6–5: Mid-plane velocity profiles for driven cavity flow; Bingham fluid.

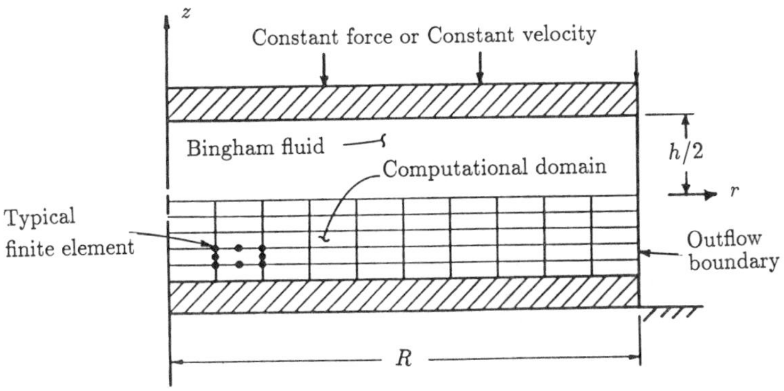

Figure 6.6–6: Schematic of squeeze film flow.

Boundary conditions for this simulation consisted of a specified nonzero vertical velocity and a zero horizontal velocity along the top surface of the domain. Along the vertical centerline the horizontal velocity is set to zero and the shear stress (natural boundary condition) vanishes due to symmetry. The horizontal symmetry plane has a zero vertical velocity and a zero shear stress. The outflow boundary assumes a zero vertical velocity (parallel flow assumption) and a constant normal traction. A Picard iteration scheme with extrapolation was used as a solution procedure; up to 12 iterations were required to obtain convergence.

Radial velocity profiles for two plate separations are shown in Figure 6.6–7, where they are compared to an analytical solution developed in [16]. The fluid near the centerplane of the device flows as an unyielded plug while near the plate a more viscous boundary layer–like flow occurs. The demarcation line between yielded and unyielded material is illustrated in Figure 6.6–8. The agreement with the analytical solution is quite good.

6.6–5 Crystal Growth From the Melt

As an illustration of the use of spines to compute a free surface shape, the problem of crystal growth from a melt using the Czochralski process is considered. This example is a brief excerpt from the extensive study performed by Sackinger [30] and Sackinger and *et al.* [31]. Figure 6.6–9 shows a schematic of the axisymmetric crystal growth model. There are two free surfaces in this problem: the melt/gas interface denoted by H_1 and the melt/crystal interface marked by H_0. Both surfaces are parameterized with a spine technique and the location of each boundary is computed in a fully coupled manner with the momentum, continuity, and energy equations for each material region. Note that the radius of the crystal is also an unknown and was computed as part of the solution. This simulation also included the effect of Lorentz forces on the melt, rotation of the crucible and/or the crystal, and thermocapillary forces on the free surface H_1. The steady form of this problem is well suited to the spine technique, since the movement of the nodes on each free surface is essentially one–dimensional. Note that the spines associated with H_0 and H_1 are permitted to slide along the r coordinate as a result of variations in the crystal radius, R.

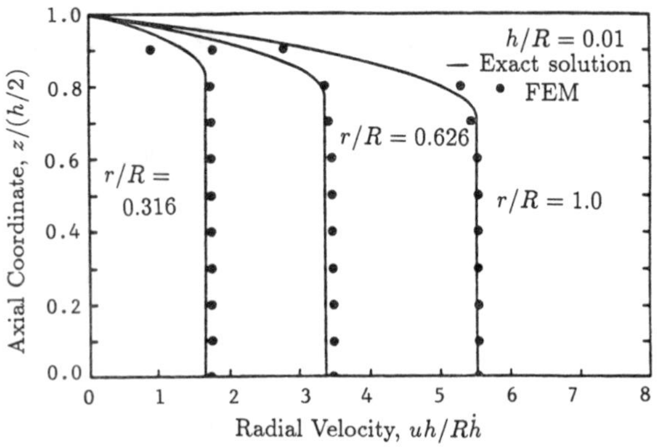

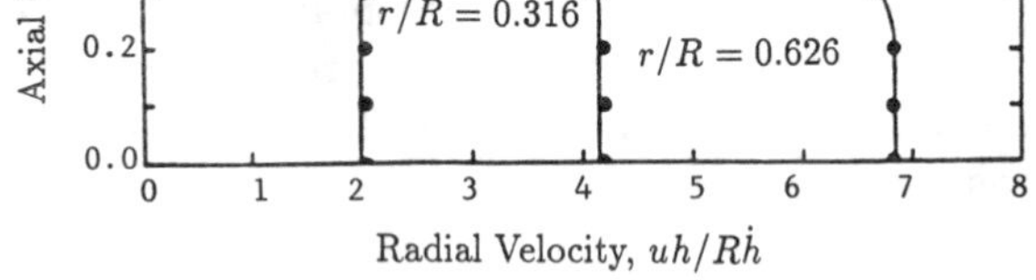

Figure 6.6–7: Radial velocity profiles for the squeeze film flow problem; Bingham fluid.

Figure 6.6–10 contains typical finite element meshes of nine–node, quadrilateral elements. Contour plots for the streamlines and isotherms for two cases with very different melt pool volumes are shown in Figures 6.6–11 and 6.6–12. The free surface shapes are quite different for these two cases, as are the temperature and flow fields.

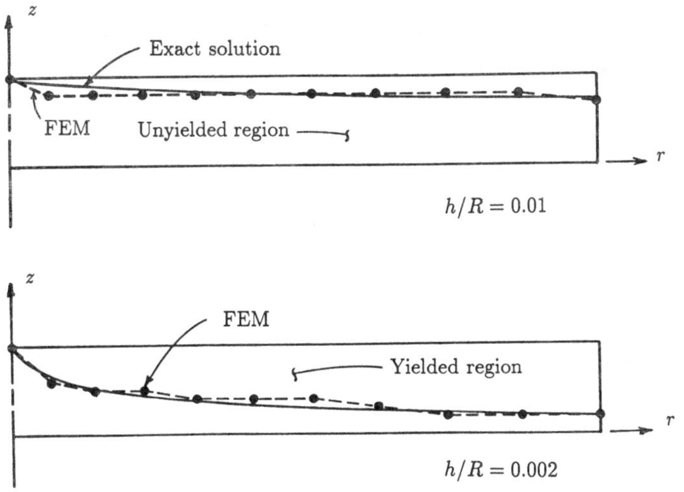

Figure 6.6–8: Plots of the yield surface for the squeeze film flow problem; Bingham fluid.

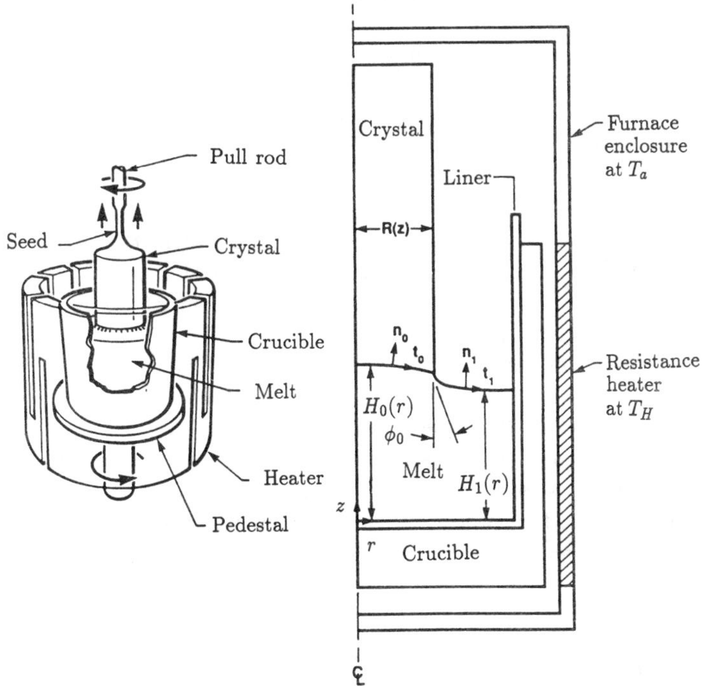

Figure 6.6–9: Schematic of the crystal growth problem.

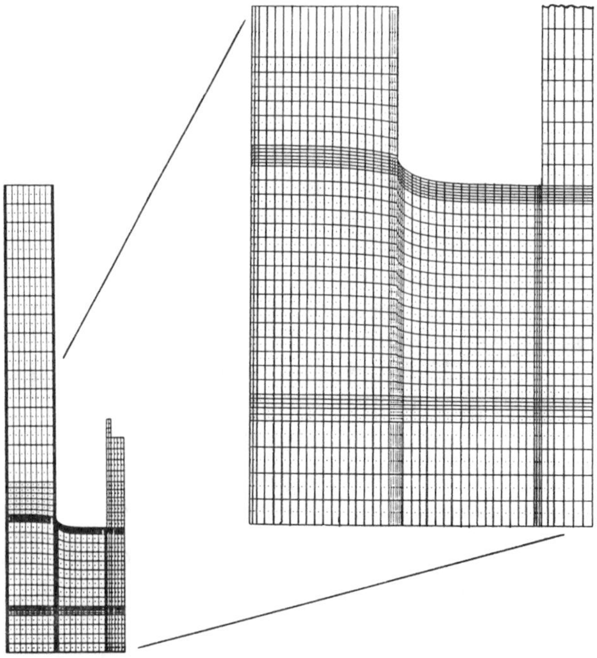

Figure 6.6–10: Typical finite element meshes for the free surface problem.

6.6–6 Mold Filling

The filling of a cavity with a liquid presents a particularly difficult challenge for a computational algorithm. The present example considers the problem of filling a flat plate mold with a liquid metal in an experimental investment casting setup. This simulation was performed by R. C. Givler (personal communication) using the commercial finite element code, ProCAST [32]. The free surface algorithm in ProCAST is a refined version of the VOF method described in Section 6.5–3.

Figure 6.6–13 contains the mesh used for the filling simulation; the mold is not shown in this view. Liquid metal (stainless steel) enters the computational domain through the circular cross section at the top of the edge gate; the fluid temperature and pressure are specified at inflow. For this particular simulation the heat transfer problem is also considered in which the metal loses energy to the mold and the mold is cooled via radiation to the surroundings.

Figures 6.6–14 and 6.6–15 show the pressure fields at two times during the filling process as viewed from both the top and bottom of the model. It is quite noticeable that the molten metal runs along the bottom of the mold in a film that is less than the channel thickness. Filling of the mold is completed when the fluid reaches the far end of the mold, is turned toward the inlet, and fills the upper part of the channel.

Note that the gate thickness is less than the channel thickness, so this behavior is not unexpected.

$T_H = 1835.13$ K

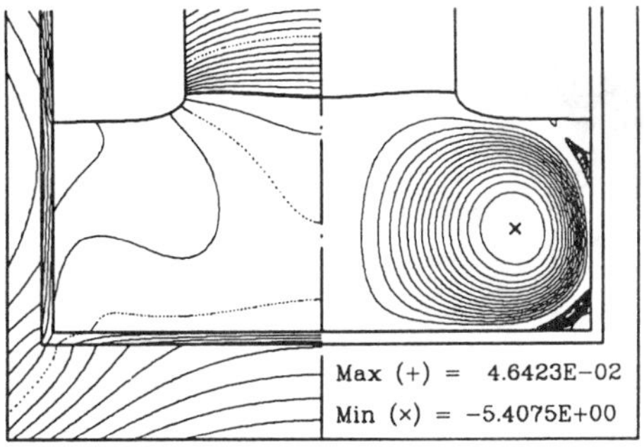

$V_m = 2.5$ (974 cm^3), $Gr^* = 1.022 \times 10^7$,
$Re^* = 2287$, $\Delta H_0 = 0.92$ mm

Figure 6.6–11: Isotherms and streamlines for the crystal growth problem; melt volume = 2.5, $Gr = 1.022 \times 10^7$.

$T_H = 1970.71$ K

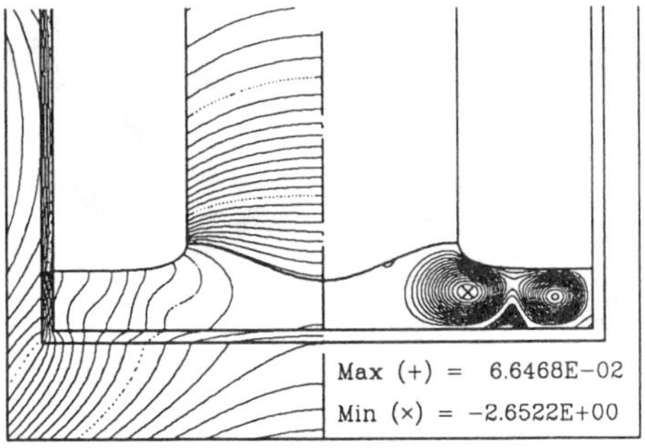

$V_m = 0.8$ (292 cm^3), $Gr^* = 1.992 \times 10^7$
$Re^* = 3180$, $\Delta H_0 = 9.84$ mm

Figure 6.6–12: Isotherms and streamlines for the crystal growth problem; melt volume = 0.8, $Gr = 1.992 \times 10^7$.

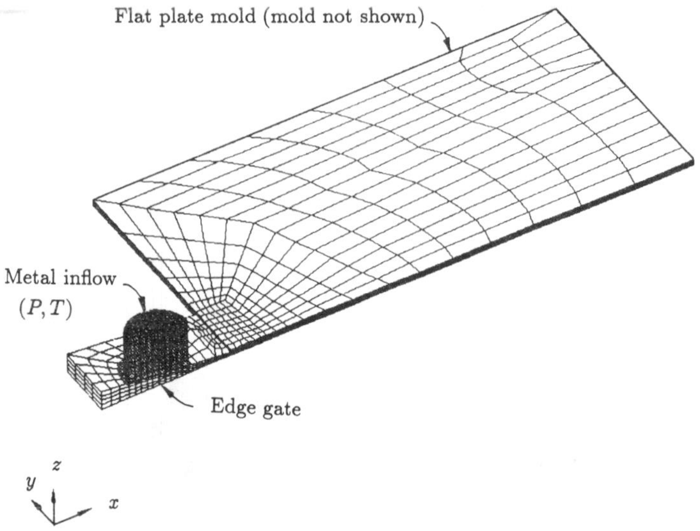

Figure 6.6–13: Schematic and mesh for mold filling.

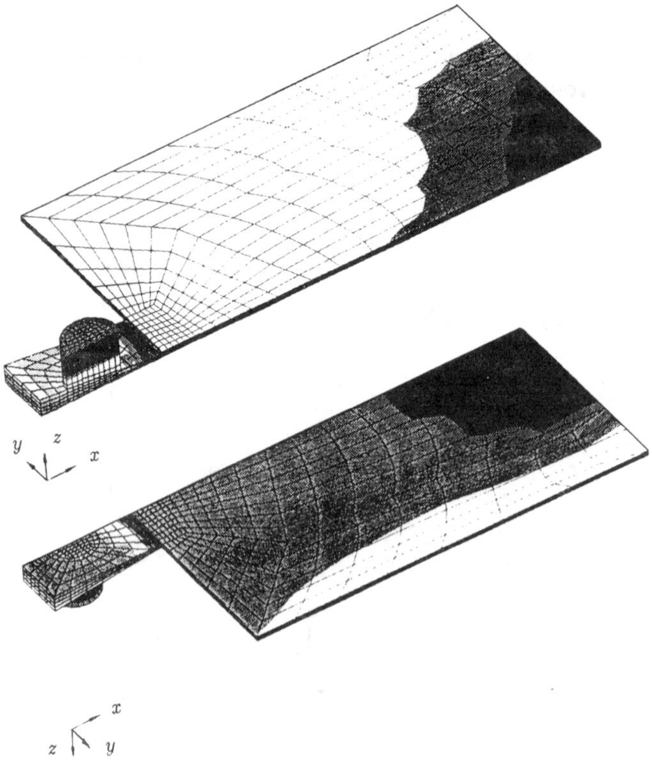

Figure 6.6–14: Top and bottom views of pressure field and free surface during filling; time = 1.45.

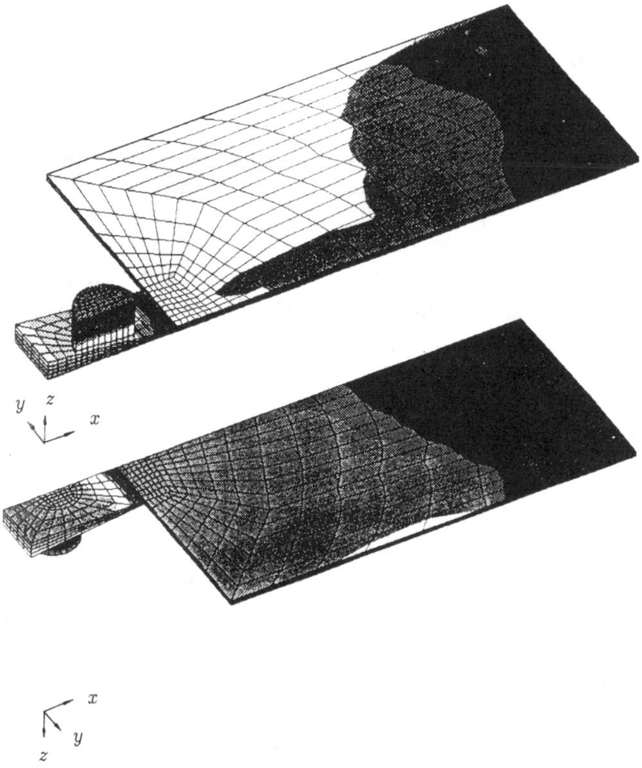

Figure 6.6–15: Top and bottom views of pressure field and free surface during filling; time = 2.09.

REFERENCES

1. R. B. Bird, R. C. Armstrong, and O. Hassager, *Dynamics of Polymeric Liquids, Vol. 1: Fluid Mechanics*, Second Edition, John Wiley & Sons, New York (1971).

2. A. S. Lodge, *Elastic Liquids*, Academic Press, New York (1964).

3. K. Walters, *Rheometry*, Chapman and Hall, London, U. K. (1975).

4. M. J. Crochet, A. R. Davies, and K. Walters, *Numerical Simulation of Non-Newtonian Flow*, Elsevier, Amsterdam, The Netherlands (1984).

5. J. R. A. Pearson, *Mechanics of Polymer Processing*, Elsevier, London, U. K. (1985).

6. R. B. Bird, W. E. Stewart, and E. N. Lightfoot, *Transport Phenomena*, John Wiley & Sons, New York (1960).

7. J. G. Oldroyd, "A Rational Formulation of the Equations of Plastic Flow for a Bingham Solid," *Proceedings of the Cambridge Philosophical Society*, **43**, 100–105 (1947).

8. W. Prager, *Introduction to the Mechanics of Continua*, Ginn and Company, Boston, Massachusetts (1961).

9. D. K. Gartling, "NACHOS II–A Finite Element Computer Program for Incompressible Flow Problems," Sandia National Laboratories Report, SAND86–1816, Albuquerque, New Mexico (1987).

10. R. L. Lee and P. M. Gresho, "Development of a Three–Dimensional Model of the Atmospheric Boundary Layer Using the Finite Element Method," *Lawrence Livermore National Laboratory Report, UCRL–52366*, Livermore, California (1977).

11. D. K. Gartling, "Finite Element Analysis of Convective Heat Transfer Problems with Change of Phase," in *Computer Methods in Fluids*, K. Morgan, C. Taylor and C. A. Brebbia, (eds.), Pentech Press, London, U.K., 257–284 (1980).

12. D. J. Naylor, "Stresses in Nearly Incompressible Materials by Finite Elements with Application to the Calculation of Excess Pore Pressures," *International Journal for Numerical Methods in Engineering*, **8**, 443–460 (1974).

13. E. Hinton and J. S. Campbell, "Local and Global Smoothing of Discontinuous Finite Element Functions Using a Least Squares Method," *International Journal for Numerical Methods in Engineering*, **8**, 461–480 (1974).

14. E. Hinton, F. C. Scott, and R. E. Ricketts, "Local Least Squares Stress Smoothing for Parabolic Isoparametric Elements," *International Journal for Numerical Methods in Engineering*, **9**, 235–238 (1975).

15. D. K. Gartling, "The Numerical Simulation of Plastic Fluids," in *Proceedings of Third International Conference on Numerical Methods in Laminar and Turbulent Flow*, C. Taylor, C. Johnson, and I. Smith (eds.), Pineridge Press, Swansea, U.K., 669–679 (1983).

16. D. K. Gartling and N. Phan Thien, "A Numerical Simulation of a Plastic Fluid in a Parallel-Plate Plastometer," *Journal of Non–Newtonian Fluid Mechanics*, **14**, 347–360 (1984).

17. R. I. Tanner and J. F. Milthorpe, "Numerical Simulation of the Flow of Fluids with Yield Stresses," in *Proceedings of Third International Conference on Numerical Methods in Laminar and Turbulent Flow*, C. Taylor, C. Johnson and I. Smith (eds.), Pineridge Press, Swansea, U.K., 680–690 (1983).

18. M. S. Engelman, "Quasi–Newton Methods in Fluid Dynamics," in *The Mathematics of Finite Elements and Applications, IV*, J. R. Whiteman (ed.), Academic Press, London, U.K., 479–487 (1982).

19. R. I. Tanner, R. E. Nickell, and R. W. Bilger, "Finite Element Methods for the Solution of Some Incompressible Non-Newtonian Fluid Mechanics Problems with Free Surfaces," *Computer Methods in Applied Mechanics and Engineering,*

6, 155–174 (1975).

20. K. R. Reddy and R. I. Tanner, "Finite Element Solution of Viscous Jet Flows with Surface Tension," *Computers and Fluids*, **6**, 83–91 (1978).

21. O. C. Zienkiewicz, P. C. Jain, and E. Onate, "Flow of Solids During Forming and Extrusion: Some Aspects of Numerical Solutions," *International Journal of Solids and Structures*, **14**, 15–38 (1978).

22. H. Saito and L. E. Scriven, "Study of Coating Flow by the Finite Element Method," *Journal of Computational Physics*, **42**, 53–76 (1981).

23. L. E. Scriven and S. F. Kistler, "Coating Flow Theory by Finite Element and Asymptotic Analysis of the Navier-Stokes System," in *Finite Element Flow Analysis*, T. Kawai (ed.), University of Tokyo Press, Tokyo, Japan, 503–510 (1982).

24. M. Engelman, R. L. Sani, and P. M. Gresho, "The Implementation of Normal and/or Tangential Boundary Conditions in Finite Element Codes for Incompressible Fluid Flow," *International Journal for Numerical Methods in Fluids*, **2**, 225–238 (1982).

25. K. N. Christodoulou and L. E. Scriven, "Discretization of Free Surface Flows and Other Moving Boundary Problems," *Journal of Computational Physics*, **99**, 39–55 (1992).

26. P. M. Schweizer and S. F. Kistler (eds.), *Liquid Film Coating, Scientific Principles and Their Technical Implications*, Van Nostrand–Reinhold, New York (1994).

27. M. Iga and J. N. Reddy, "Penalty Finite Element Analysis of Free Surface Flows of Power–Law Fluids," *International Journal of Non-Linear Mechanics*, **24**(5), 383–399 (1989).

28. M. P. Reddy and J. N. Reddy, "Finite Element Analysis of Flows of Non–Newtonian Fluids in Three–Dimensional Enclosures," *International Journal of Non-Linear Mechanics*, **27**(1), 9–26 (1992).

29. M. P. Reddy and J. N. Reddy, "Numerical Simulation of Forming Processes Using a Coupled Fluid Flow and Heat Transfer Model," *International Journal for Numerical Methods in Engineering*, **35**, 807–833 (1992).

30. P. A. Sackinger, "Flows and Transitions During Solidification: Simulation of Hydrodynamics, Heat Transfer and Free Boundaries in Czochralski Growth," Ph.D. Dissertation, Dept. of Chem. Engr., MIT, Cambridge, Massachusetts (1989).

31. P. A. Sackinger, R. A. Brown, and J. J. Derby, "A Finite Element Method for Analysis of Fluid Flow, Heat Transfer and Free Interfaces in Czochralski Crystal Growth," *International Journal for Numerical Methods in Fluids*, **9**, 453–492 (1989).

32. ProCAST User's Manual, Version 2.0, *UES, Inc.*, Dayton, Ohio (1992).

Chapter Seven

VISCOELASTIC FLUIDS

7.1 INTRODUCTION

As noted in the introduction to Chapter 6, non–Newtonian behavior has many facets. In Chapter 6 we dealt with the modeling of "shear rate" dependent viscosity behavior but ignored the problems associated with fluid elasticity and memory. In this chapter we will focus on viscoelastic fluids and their numerical simulation by the finite element method.

The following discussion will be concerned with the so–called simple fluid with fading memory [1–4]. No attempt will be made to describe the general theory of such materials as this topic is well outside the scope of the present text. Rather, we will simply state the standard balance laws in a convenient form and then describe some of the constitutive equations that have been found useful for numerical experimentation and computation.

7.2 CONSERVATION EQUATIONS

The analytical description of the motion of a continuous medium is based on conservation of mass, momentum, and energy, and the associated equations of state and constitutive relations. The present development will be limited to laminar flows of incompressible, viscous isotropic fluids. The fluid motion is assumed to be isothermal to avoid the discussion of the energy equation, as it was discussed in detail in Chapters 2, 5, and 6. We also assume that, for simplicity, the flows are two–dimensional. Of course, the extension of the present discussion to three dimensions is conceptually straightforward, although in practice the subject is sufficiently complex to have received relatively little attention.

The equations of interest for two–dimensional flows are written in the Cartesian coordinate system (x_1, x_2) using the Eulerian description:

Conservation of Mass:

$$\frac{\partial u_i}{\partial x_i} = 0 \qquad (7.2-1)$$

Conservation of Momentum:

$$\rho \left(\frac{\partial u_i}{\partial t} + u_j \frac{\partial u_i}{\partial x_j} \right) = \frac{\partial \sigma_{ij}}{\partial x_j} + \rho g_i \qquad (7.2-2)$$

where u_i denotes the i–th component of velocity, σ_{ij} the components of the total (Cauchy) stress tensor, ρ the density, g_i the component of body force, and t denotes time. The standard indicial notation is used, with summation on repeated indices.

The field equations in (1) and (2) are to be solved in conjunction with constitutive equations of the fluid and boundary and initial conditions of the flow problem. The constitutive equations for viscoelastic fluids is the major focus of this chapter, and will be discussed in Section 7.3. Since the boundary conditions depend on the dependent variables used, they are discussed subsequent to the discussion of the constitutive relations.

7.3 CONSTITUTIVE EQUATIONS

7.3–1 Introduction

For an incompressible fluid the total stress is given by

$$\sigma_{ij} = -P\delta_{ij} + \tau_{ij} \tag{7.3-1}$$

where P is a pressure and τ_{ij} is the extra–stress. It should be noted that due to the particular forms that will be used to describe the extra–stress, the pressure defined in equation (1) will not, in general, be the same as a Newtonian pressure or hydrostatic pressure. That is, the extra–stress may not be traceless (*i.e.*, $\tau_{ii} \neq 0$), in which case P is not the usual mean normal stress found in Newtonian fluid mechanics [5,6].

The central issue for the constitutive description of a viscoelastic fluid is the choice of an equation that relates the extra–stress in (1) to the flow kinematics. For the general class of materials called simple fluids, such a relationship can be established in functional form where the current extra–stress is related to the history of deformation in the fluid. Typically, such a relation is expressed as

$$\tau_{ij} = \mathcal{F}_{ij}[G_{kl}(s),\ 0 < s < \infty] \tag{7.3-2}$$

where $\mathcal{F}_{ij}$ is a tensor–valued functional, G_{kl} is a finite deformation tensor (related to the Cauchy–Green tensor), and $s = t - t'$ is the time lapse from time t' to the present time, t. Unfortunately, the generality of a functional form, such as (2), does not produce usable constitutive equations for general flow problems; the solution of practical problems requires that approximate forms of (2) be developed.

Since there are numerous ways to approximate the general functional given in (2), a wide variety of constitutive equations have developed out of the simple fluid theory. To date, none of these approximate constitutive equations are universally applicable; that is, no single constitutive equation is capable of predicting all of the observed behavior in elastic fluids. In essence, the present state–of–the–art mandates that a spectrum of constitutive equations be considered for use, with the specific choice being dictated by the ability of a given model to predict the dominant, non–Newtonian effects expected in a particular application.

To further complicate the situation, approximate constitutive equations can be of several different forms. For certain limiting flow conditions, such as "small" strains, "slow" flows, or "slightly" elastic fluids, the functional in (2) can be approximated by an expansion in terms of a small parameter. Such a formulation leads to a set of "hierarchical" equations [1,2] for the extra–stress in terms of the rate of deformation tensor and its various derivatives. In these models the extra–stress is given explicitly in terms of the flow kinematics which, in principle, allows equations (7.2–1), (7.2–2), and (7.3–1) to be written in terms of only the velocity and pressure, as in a Newtonian flow. Though convenient in form, these hierarchical formulas are of very limited value in general flows due to the "smallness" assumptions used in their derivation. Some computational work has been done using these types of models [7–9], however, due to their limited applicability they will not be given further consideration here.

The two remaining major categories of approximate constitutive relations include the integral and differential models. The integral model represents the extra–stress in terms of an integral over past time of the fluid deformation history. For a differential model the extra–stress is determined from a differential (evolution) equation that relates the stress and stress rate to the flow kinematics. The choice between a differential and integral formulation is a crucial one with regard to numerical simulation since the computational algorithms for each type are very different. In the present book we will emphasize the differential form since the majority of the current numerical work involves this formulation. References to work with integral models will also be given.

7.3–2 Differential Models

The differential constitutive equations to be considered here are implicit, rate–type models, generally associated with the names of Oldroyd, Maxwell, and Jeffrey. Before proceeding to the description of specific models it is important to list a number of definitions. For an Eulerian reference frame the material time derivative (or convected derivative) of a symmetric, second–order tensor can be defined in several ways, all of which are frame invariant. Let φ_{ij} denote the Cartesian components of a second-order tensor. Then the *upper-convected* (or codeformational) derivative is defined by

$$\bar{\varphi}_{ij} = \frac{\partial \varphi_{ij}}{\partial t} + u_k \frac{\partial \varphi_{ij}}{\partial x_k} - L_{ik}\varphi_{kj} - L_{jk}\varphi_{ki} \qquad (7.3-3a)$$

and the *lower–convected* derivative is defined as

$$\tilde{\varphi}_{ij} = \frac{\partial \varphi_{ij}}{\partial t} + u_k \frac{\partial \varphi_{ij}}{\partial x_k} + L_{ki}\varphi_{kj} + L_{kj}\varphi_{ki} \qquad (7.3-3b)$$

where u_k are the components of the velocity vector and L_{ij} are the components of the velocity gradient tensor $\mathbf{L}$, defined by

$$\mathbf{L} = \nabla \mathbf{u}, \quad \text{or} \quad L_{ij} = \frac{\partial u_j}{\partial x_i} \qquad (7.3-4)$$

Since both equations (3a) and (3b) are admissible convected derivatives, their linear combination is also admissible:

$$\hat{\varphi}_{ij} = \left(1 - \frac{a}{2}\right) \bar{\varphi}_{ij} + \left(\frac{a}{2}\right) \tilde{\varphi}_{ij} \tag{7.3-5}$$

Equation (5) is a general convected derivative which reduces to (3a) for $a = 0$ and (3b) for $a = 2$. When $a = 1$ [average of (3a) and (3b)] the convected derivative in (5) is termed a *corotational* or *Jaumann derivative.*

Using τ as the extra–stress tensor with the Cartesian components denoted by τ_{ij}, the upper–convected derivative of the stress components is given by

$$\bar{\tau}_{ij} = \frac{\partial \tau_{ij}}{\partial t} + u_k \frac{\partial \tau_{ij}}{\partial x_k} - L_{ik}\tau_{kj} - L_{jk}\tau_{ki} \tag{7.3-6a}$$

and the lower–convected derivative is

$$\tilde{\tau}_{ij} = \frac{\partial \tau_{ij}}{\partial t} + u_k \frac{\partial \tau_{ij}}{\partial x_k} + L_{ki}\tau_{kj} + L_{kj}\tau_{ki} \tag{7.3-6b}$$

The corotational derivative is

$$\hat{\tau}_{ij} = \left(1 - \frac{a}{2}\right) \bar{\tau}_{ij} + \left(\frac{a}{2}\right) \tilde{\tau}_{ij} \tag{7.3-7}$$

All of these derivatives have been used in various differential constitutive equations. The selection of one type of derivative over another is usually based on the physical plausability of the resulting constitutive equation and the matching of experimental data to the model for simple (viscometric) flows.

The simplest differential constitutive models are the upper- and lower–convected Maxwell fluids, which are defined by the following equations:

Upper–Convected Maxwell Fluid:

$$\tau_{ij} + \lambda \bar{\tau}_{ij} = 2\mu_0 D_{ij} \tag{7.3-8a}$$

Lower–Convected Maxwell Fluid:

$$\tau_{ij} + \lambda \tilde{\tau}_{ij} = 2\mu_0 D_{ij} \tag{7.3-8b}$$

where λ is a relaxation time for the fluid, μ_0 is a viscosity, and D_{ij} are the components of the rate of deformation (or strain rate) tensor defined by

$$\mathbf{D} = \frac{1}{2}\left[\nabla \mathbf{u} + (\nabla \mathbf{u})^T\right], \text{ or } D_{ij} = \frac{1}{2}(L_{ij} + L_{ji}) \tag{7.3-9}$$

The upper–convected Maxwell model (8a) has been used extensively in testing numerical algorithms; the lower–convected and corotational forms of the Maxwell fluid predict physically unrealistic behavior and are not generally used.

By employing the general convected derivative (7) in a Maxwell–like model the following constitutive equation is produced:

Johnson–Segalman Model:

$$\tau_{ij} + \lambda \hat{\tau}_{ij} = 2\mu_0 D_{ij} \tag{7.3-10}$$

which is the Johnson–Segalman fluid [10]. By slightly modifying equation (10) to include a variable coefficient for τ_{ij}, the Phan Thien–Tanner model [11] is derived.

Phan Thien–Tanner Model:

$$Y(\tau)\tau_{ij} + \lambda \hat{\tau}_{ij} = 2\mu_0 D_{ij} \tag{7.3-11a}$$

where

$$Y(\tau) = 1 + \frac{\epsilon\lambda}{\mu_0}\, tr(\tau) \tag{7.3-11b}$$

and ϵ is a constant. This equation is somewhat better than (10) in representing actual material behavior.

The Johnson–Segalman and Phan Thien–Tanner models suffer a common defect in that, for a monotonically increasing shear rate, there is a region where the shear stress decreases — a physically unrealistic behavior. To correct this anomaly the constitutive equations can be altered using the following procedure. Let the extra–stress be decomposed into two partial stresses, τ^s_{ij} and τ^p_{ij} such that

$$\tau_{ij} = \tau^s_{ij} + \tau^p_{ij} \tag{7.3-12}$$

In equation (12) τ^s_{ij} is a purely viscous stress and τ^p_{ij} is a viscoelastic component. The decomposition used here is sometimes thought of in terms of a combination of a Newtonian solvent (superscript s) and a polymer additive (superscript p). Using the Johnson–Segalman fluid as an example, then

$$\tau^s_{ij} = 2\mu^s_0\, D_{ij}$$

$$\tau^p_{ij} + \lambda \hat{\tau}^p_{ij} = 2\mu^p_0\, \hat{D}_{ij} \tag{7.3-13}$$

It is possible to eliminate the partial stresses in (12) and (13) to produce a new constitutive relation of the following form

$$\tau_{ij} + \lambda \hat{\tau}_{ij} = 2\mu \left(D_{ij} + \lambda' \hat{D}_{ij} \right) \tag{7.3-14}$$

with $\mu = (\mu_0^s + \mu_0^p)$ and $\lambda' = \lambda\mu_0^s/\mu$, where λ' is a retardation time. The constitutive equation in (14) is recognized as a type of Oldroyd fluid — it is, in fact, a reduced form of the eight–constant Oldroyd model [1,12]. For particular choices of the convected derivative in equation (14), specific models can be generated. When $a = 0$ ($\hat{\tau}_{ij}$ becomes $\bar{\tau}_{ij}$) then (14) becomes the Oldroyd B fluid; the case $a = 2$ ($\hat{\tau}_{ij}$ becomes $\tilde{\tau}_{ij}$) produces the Oldroyd A fluid. In order to ensure a monotonically increasing shear stress the inequality $\mu_0^s \geq \mu_0^p/8$ must be satisfied. The stress decomposition employed above can also be used with the Phan Thien–Tanner model to produce a correct shear stress behavior.

In all of the above constitutive equations the material parameters, λ and μ_0, were assumed to be constants. For some constitutive equations the constancy of these parameters leads to material (or viscometric) functions that do not accurately represent the behavior of real elastic fluids. For example, the shear viscosity predicted by a Maxwell fluid is a constant, when in fact viscoelastic fluids normally exhibit a shear thinning behavior. This situation can be remedied to some degree by allowing the parameters λ and μ_0 to be functions of the invariants of the rate of deformation tensor as was done for the generalized Newtonian fluid (see Section 6.2–2). Using the upper–convected Maxwell fluid as an example, then

$$\tau_{ij} + \lambda(I_2)\bar{\tau}_{ij} = 2\mu(I_2)D_{ij} \qquad (7.3-15)$$

where I_2 is the second invariant of the strain rate tensor $\mathbf{D}$ [see equation (9)], $I_2 = 1/2(D_{ij}D_{ij})$. The constitutive equation in (15) is usually termed a White–Metzner model [13]. White–Metzner forms of other differential models, such as the Oldroyd fluids, have also been developed and used in various situations.

The above list of differential models is by no means exhaustive though it does include many of the constitutive equations that have been used in computational work. Though the equations have been stated without derivation or motivation, they all rest on a reasonable theoretical basis. Some of the equations (*e.g.*, Oldroyd and Maxwell) were developed as generalizations of simple, linear viscoelastic models. Linear viscoelasticity can be formally developed from the previously cited "hierarchical" equations or informally from mechanical analogies and heuristic arguments [12]. Other differential models, such as the Johnson–Segalman and Phan Thien–Tanner fluids, were derived using statistical mechanics ideas and a conceptual (network) model for the microstructure of a viscoelastic fluid. Due to their approximate nature, all of the above models are limited in their ability to represent true viscoelastic behavior; the more complex models provide a reasonable qualitative description of many observed phenomena. A catalog of the strengths, weaknesses, and limitations of these and other differential models can be found in [1,6,12,14].

7.3–3 Integral Models

An approximate integral model for a viscoelastic fluid represents the extra–stress in terms of an integral over the past history of the fluid deformation. A general form for a single integral model can be expressed as (see [15])

$$\tau_{ij} = \int_{-\infty}^{t} 2m(t-t')H_{ij}(t,t')dt' \qquad (7.3-16)$$

where t is the current time, m is a scalar memory function (or relaxation kernel), and H_{ij} is a nonlinear deformation measure (tensor) between the past time, t', and t.

There are many possible forms for both the memory function and the deformation measure. Normally the memory function is a decreasing function of the time lapse $s = t - t'$. Typical of such a function is the exponential given by

$$m(t-t') = m(s) = \frac{\mu_0}{\lambda^2}\, e^{-s/\lambda} \qquad (7.3-17)$$

where the parameters μ_0, λ, and s were defined previously. Like the choice of a convected derivative in a differential model, the selection of a deformation measure for use in equation (16) is somewhat arbitrary. One particular form that has received some attention is given by

$$H_{ij} = \phi_1(I, II)C_{ij}^{-1} + \phi_2(I, II)C_{ij} \qquad (7.3-18)$$

In equation (18) C_{ij} is the Cauchy–Green deformation tensor, C_{ij}^{-1} is its inverse, called the Finger tensor, and the ϕ_i are scalar functions of the invariants of the deformation tensors, *i.e.*, $I = tr(C_{ij}^{-1})$ and $II = tr(C_{ij})$. The indicated deformation tensor is defined by (see Malvern [16])

$$C_{ij} = \frac{\partial \chi_m}{\partial x_i}\frac{\partial \chi_m}{\partial x_j} \qquad (7.3-19)$$

where χ_m is the location at time t' of a fluid particle that is now at position x_i and time t. The form of the deformation measure in equation (18) is still quite general, though specific choices for the functions ϕ_i and the memory function m lead to several well–known constitutive models. Among these are the Kaye–BKZ fluid [6,12,14] and the Lodge rubber–like liquid [17].

As a specific example of an integral model of the type given by the previous equations we consider the Maxwell fluid. Setting $\phi_1 = 1$ and $\phi_2 = 0$ in equation (18) and using the memory function shown in (17) allows a constitutive equation of the following form

$$\tau_{ij} = \frac{\mu_0}{\lambda^2}\int_{-\infty}^{t} exp\left[-(t-t')/\lambda\right]\left[C_{ij}^{-1}(t') - \delta_{ij}\right] dt' \qquad (7.3-20)$$

In writing equation (20), the Finger deformation tensor in (18) has been replaced with the more usual Finger strain tensor (*e.g.*, [14–16]). The above constitutive equation is an integral equivalent to the upper–convected Maxwell model shown in differential form in equation (8a). Note that in this case the extra–stress is given in an explicit form, though its evaluation requires that the strain history be known for each fluid particle. Also, it is important to emphasize that though the Maxwell fluid has both a differential and integral form, this is not generally true for other constitutive equations.

Since the emphasis here is on differential models we will not dwell on other specific forms of integral constitutive equations. A good source for further discussion of such models is the textbook by Bird *et al.* [12]. It is appropriate to note in closing this section that some of the integral models are very good at reproducing realistic viscoelastic behavior [6,14]. However, this improvement in modeling accuracy is offset to a large degree by the difficulties in using such models in general computational schemes. Further comments on computational procedures for integral models will be reserved for a later section.

7.3–4 Boundary Conditions

The equations in (7.2–1) and (7.2–2) are to be solved in conjunction with an appropriate constitutive relation as discussed in Section 7.3. The set of boundary conditions for the problem at hand consist of either specified velocities or tractions on the boundary of the fluid domain, with one condition needed for each component of velocity. However, there is also an additional set of conditions for the viscoelastic problem due to the "memory" of the fluid. The extra condition in reality, is an initial condition on the components of the fluid extra–stress, though it is often implemented as a boundary condition, especially for flow–through type boundaries. For time–dependent problems the initial fluid stresses must be given at all points in the problem domain. All problems that contain an inflow boundary require the specification of a strain (stress) history of the fluid crossing the boundary. Completely confined flows do not require any additional data beyond the initial stress-state. Additional comments on boundary conditions can be found in [18].

7.4 FINITE ELEMENT MODEL OF DIFFERENTIAL FORM

7.4–1 Preliminary Comments

In this section we will outline typical finite element procedures for the solution of viscoelastic flow problems. At the outset it is important to realize that a standard, well–established computational procedure has not yet evolved for this particular class of flow problems. Numerous different formulations have been proposed [4], but none has proved adequate for the solution of flows with highly elastic fluids. Further comments on the unresolved general problems in this area will be given in Section 7.6.

The wide variety of possible constitutive equations and finite element formulations makes detailed explanation of many specific algorithms impossible in this limited text. Therefore, the discussion will be focused on a single, mixed finite element formulation for a particular differential constitutive equation. A brief outline of typical procedures for integral models will also be presented.

7.4–2 Summary of Governing Equations

The governing equations of interest are summarized below in rectangular Cartesian form.

Conservation of Mass:

$$\frac{\partial u_i}{\partial x_i} = 0 \qquad (7.4-1)$$

Conservation of Momentum:

$$\rho\left(\frac{\partial u_i}{\partial t}+u_j\frac{\partial u_i}{\partial x_j}\right)=\frac{\partial \sigma_{ij}}{\partial x_j} \tag{7.4-2}$$

Constitutive Equations:

$$\sigma_{ij}=-P\delta_{ij}+\tau_{ij} \tag{7.4-3}$$

$$\tau_{ij}=\tau_{ij}^s+\tau_{ij}^p \tag{7.4-4}$$

$$\tau_{ij}^s=2\mu_0^s D_{ij} \tag{7.4-5}$$

$$\tau_{ij}^p+\lambda\bar{\tau}_{ij}^p=2\mu_0^p D_{ij} \tag{7.4-6}$$

where in equation (2) the body force term has been neglected for simplicity. The constitutive equation employed is the Oldroyd B model.

The above set of equations can be combined in various ways, each of which leads to a slightly different finite element procedure. For example, combining equations (3)–(5) and substituting for σ_{ij} in the momentum balance (2) leads to the following set of equations:

$$\frac{\partial u_i}{\partial x_i}=0 \tag{7.4-7}$$

$$\rho\left(\frac{\partial u_i}{\partial t}+u_j\frac{\partial u_i}{\partial x_j}\right)=-\frac{\partial P}{\partial x_j}\delta_{ij}+\frac{\partial}{\partial x_j}\left(2\mu_0^s D_{ij}\right)+\frac{\partial}{\partial x_j}\left(\tau_{ij}^p\right) \tag{7.4-8}$$

$$\tau_{ij}^p+\lambda\bar{\tau}_{ij}^p=2\mu_0^p D_{ij} \tag{7.4-9}$$

The above equation set is a natural extension of work on a Newtonian fluid, and was the first one considered by workers in viscoelastic flows [19–21]. However, as pointed out by Crochet *et al.* [1], the mixed finite element formulation associated with (7)–(9) has some restrictions with regard to the choice of approximating functions for the extra–stress variables.

Another possible formulation was proposed by Chang *et al.* [22] and is the one that will be discussed here. The implicit constitutive equation (6) can be rearranged and substituted into (8). The result is a set of equations of the following form

$$\frac{\partial u_i}{\partial x_i}=0 \tag{7.4-10}$$

$$\rho\left(\frac{\partial u_i}{\partial t}+u_j\frac{\partial u_i}{\partial x_j}\right)=-\frac{\partial P}{\partial x_j}\delta_{ij}+\frac{\partial}{\partial x_j}\left(2\mu D_{ij}\right)-\frac{\partial}{\partial x_j}\left(\lambda\bar{\tau}_{ij}^p\right) \tag{7.4-11}$$

$$\tau_{ij}^p+\lambda\bar{\tau}_{ij}^p=2\mu_0^p D_{ij} \tag{7.4-12}$$

where $\mu=(\mu_0^s+\mu_0^p)$, D_{ij} is defined in (7.3–9), and the stress rate, $\bar{\tau}_{ij}$, is given in (7.3–6a). The inclusion of the stress rate in the momentum equation appears to be of some benefit for numerical computations [1].

For two-dimensional geometries the equations in (10)–(12) provide six relations for the six unknowns, u_1, u_2, P, τ_{11}^p, τ_{22}^p, and τ_{12}^p. In an axisymmetric geometry the system would increase to seven equations with the circumferential (hoop) stress, τ_{33}^p, as the seventh unknown. The distinguishing feature of this system, as compared to a Newtonian problem, is the implicit nature of the constitutive equation that forces the extra-stresses to remain as dependent variables.

7.4–3 Finite Element Model

Application of the finite element procedure to the partial differential equations in (10)–(12) follows the standard format. The velocity components, pressure, and extra-stress components are approximated by expansions of the form

$$u_i = \sum_{m=1}^{M} \Psi_m u_i^m = \mathbf{\Psi}^T \mathbf{u}_i \qquad (7.4-13a)$$

$$P = \sum_{n=1}^{N} \Phi_n P_n = \mathbf{\Phi}^T \mathbf{P} \qquad (7.4-13b)$$

$$\tau_{ij} = \sum_{k=1}^{K} \Pi^k \tau_{ij}^k = \mathbf{\Pi}^T \boldsymbol{\tau}_{ij} \qquad (7.4-13c)$$

where $\mathbf{\Phi}$, $\mathbf{\Psi}$, and $\mathbf{\Pi}$ are vectors of basis functions, $\mathbf{u}_i$, $\mathbf{P}$, and τ_{ij} are vectors of nodal unknowns, and M, N, and K indicate the number of nodes at which the various unknowns are defined. The superscript p for the extra-stress has been omitted for clarity. Using the finite element approximations (13) in standard weighted-integral statements (*i.e.*, weak forms) of the system in (10)–(12) (see Chapters 5 and 6), the following system of finite element equations can be obtained:

Momentum:

$$\left[\int_{\Omega^e} \rho_0 \mathbf{\Phi}\mathbf{\Phi}^T \mathbf{u}_j \frac{\partial \mathbf{\Phi}^T}{\partial x_j}\, d\mathbf{x}\right]\mathbf{u}_i + \left[\int_{\Omega^e} \mu \frac{\partial \mathbf{\Phi}}{\partial x_j}\frac{\partial \mathbf{\Phi}^T}{\partial x_j}\, d\mathbf{x}\right]\mathbf{u}_i + \left[\int_{\Omega^e} \mu \frac{\partial \mathbf{\Phi}}{\partial x_j}\frac{\partial \mathbf{\Phi}^T}{\partial x_i}\, d\mathbf{x}\right]\mathbf{u}_j$$

$$+\left[\int_{\Omega^e} -\frac{\partial \mathbf{\Phi}}{\partial x_i}\mathbf{\Psi}^T\, d\mathbf{x}\right]\mathbf{P} - \left[\int_{\Omega^e} \lambda \frac{\partial \mathbf{\Phi}}{\partial x_j}\bar{\tau}_{ij}\, d\mathbf{x}\right] = \left[\oint_{\Gamma^e} \mathbf{\Phi}\tau_{ij} n_j\, ds\right] \qquad (7.4-14)$$

Continuity:

$$\left[\int_{\Omega^e} -\mathbf{\Psi}\frac{\partial \mathbf{\Phi}^T}{\partial x_i}\, d\mathbf{x}\right]\mathbf{u}_i = 0 \qquad (7.4-15)$$

Constitutive:

$$\left[\int_{\Omega^e} \Pi\ \Pi^T\, dx\right]\tau_{ij}+\left[\int_{\Omega^e} \Pi\lambda\bar{\tau}_{ij}\, dx\right]-\left[\int_{\Omega^e} \mu_0^p\Pi\frac{\partial\Phi^T}{\partial x_j}\, dx\right]u_i-\left[\int_{\Omega^e} \mu_0^p\Pi\frac{\partial\Phi^T}{\partial x_i}\, dx\right]u_j=0 \qquad (7.4-16)$$

where the steady–state form of the equations has been considered. In arriving at the above equations, the Green–Gauss or divergence theorem has been used to reduce the second-order diffusion terms to first-order terms plus a boundary integral. The appearance of the boundary integral corresponds to the "natural" boundary conditions for the problem. Also, the stress rate has not been written out due to its complexity; the two–dimensional components of the stress rate are written below in terms of the finite element functions,

$$\bar{\tau}_{11}=\Phi^T u_1\frac{\partial\Pi^T}{\partial x_1}\tau_{11}+\Phi^T u_2\frac{\partial\Pi^T}{\partial x_2}\tau_{11}-2\frac{\partial\Phi^T}{\partial x_1}u_1\Pi^T\tau_{11}-2\frac{\partial\Phi^T}{\partial x_2}u_1\Pi^T\tau_{12} \qquad (7.4-17)$$

$$\bar{\tau}_{22}=\Phi^T u_1\frac{\partial\Pi^T}{\partial x_1}\tau_{22}+\Phi^T u_2\frac{\partial\Pi^T}{\partial x_2}\tau_{22}-2\frac{\partial\Phi^T}{\partial x_1}u_2\Pi^T\tau_{12}-2\frac{\partial\Phi^T}{\partial x_2}u_2\Pi^T\tau_{22} \qquad (7.4-18)$$

$$\bar{\tau}_{12}=\Phi^T u_1\frac{\partial\Pi^T}{\partial x_1}\tau_{12}+\Phi^T u_2\frac{\partial\Pi^T}{\partial x_2}\tau_{12}-\frac{\partial\Phi^T}{\partial x_1}u_2\Pi^T\tau_{11}-\frac{\partial\Phi^T}{\partial x_2}u_1\Pi^T\tau_{22} - \left(\frac{\partial\Phi^T}{\partial x_1}u_1+\frac{\partial\Phi^T}{\partial x_2}u_2\right)\Pi^T\tau_{12} \qquad (7.4-19)$$

Equations (17)–(19) can be substituted directly into (14) and (16) to produce the complete form of the finite element equations.

Once the form of the interpolation functions Φ, Ψ, and Π is specified (*i.e.*, a particular element is selected), and the geometry of the element is known (*i.e.*, x_i) then the integrals in (14)–(16) may be evaluated to produce the required coefficient matrices. The integrals are evaluated via a numerical quadrature procedure, as discussed in Chapters 2 and 3. The discrete system is given by the following matrix equations (it is common to write the momentum and continuity equations as a single system):

$$\begin{bmatrix} C_i(u_i) & 0 & 0 \\ 0 & C_i(u_i) & 0 \\ 0 & 0 & 0 \end{bmatrix}\begin{Bmatrix} u_1 \\ u_2 \\ P \end{Bmatrix}+\begin{bmatrix} 2K_{11}+K_{22} & K_{21} & Q_1 \\ K_{12} & K_{11}+2K_{22} & Q_2 \\ Q_1^T & Q_2^T & 0 \end{bmatrix}\begin{Bmatrix} u_1 \\ u_2 \\ P \end{Bmatrix}$$
$$+\begin{bmatrix} R_1^{11}(u_1,u_2) & R_1^{22}(u_1,u_2) & R_1^{12}(u_1,u_2) \\ R_2^{11}(u_1,u_2) & R_2^{22}(u_1,u_2) & R_2^{12}(u_1,u_2) \\ 0 & 0 & 0 \end{bmatrix}\begin{Bmatrix} \tau_{11} \\ \tau_{22} \\ \tau_{12} \end{Bmatrix}=\begin{Bmatrix} F_1 \\ F_2 \\ 0 \end{Bmatrix} \qquad (7.4-20)$$

and for the constitutive equation

$$\begin{bmatrix} N & 0 & 0 \\ 0 & N & 0 \\ 0 & 0 & N \end{bmatrix}\begin{Bmatrix} \tau_{11} \\ \tau_{22} \\ \tau_{12} \end{Bmatrix}-\begin{bmatrix} 2D_1(u_1) & 0 & 2D_2(u_1) \\ 0 & 2D_2(u_2) & 2D_1(u_2) \\ D_1(u_2) & D_2(u_1) & D_i(u_i) \end{bmatrix}\begin{Bmatrix} \tau_{11} \\ \tau_{22} \\ \tau_{12} \end{Bmatrix}+$$

$$\begin{bmatrix} \mathbf{C}_i^*(\mathbf{u}_i) & 0 & 0 \\ 0 & \mathbf{C}_i^*(\mathbf{u}_i) & 0 \\ 0 & 0 & \mathbf{C}_i^*(\mathbf{u}_i) \end{bmatrix} \begin{Bmatrix} \tau_{11} \\ \tau_{22} \\ \tau_{12} \end{Bmatrix} - \begin{bmatrix} 2\mathbf{L}_1 & 0 & 0 \\ 0 & 2\mathbf{L}_2 & 0 \\ \mathbf{L}_1 & \mathbf{L}_2 & 0 \end{bmatrix} \begin{Bmatrix} \mathbf{u}_1 \\ \mathbf{u}_2 \\ \mathbf{P} \end{Bmatrix} = \begin{Bmatrix} 0 \\ 0 \\ 0 \end{Bmatrix} \tag{7.4-21}$$

where sum on repeated indices is implied. The coefficient matrices shown in equations (20) and (21) are defined by

$$\mathbf{C}_i(\mathbf{u}_j) = \int_{\Omega^e} \rho_0 \mathbf{\Phi}\mathbf{\Phi}^T \mathbf{u}_j \frac{\partial \mathbf{\Phi}^T}{\partial x_i}\, d\mathbf{x}$$

$$\mathbf{K}_{ij} = \int_{\Omega^e} \mu \frac{\partial \mathbf{\Phi}}{\partial x_i} \frac{\partial \mathbf{\Phi}^T}{\partial x_j}\, d\mathbf{x}$$

$$\mathbf{Q}_i = -\int_{\Omega^e} \frac{\partial \mathbf{\Phi}}{\partial x_i} \mathbf{\Psi}^T\, d\mathbf{x}$$

$$\mathbf{F}_i = \oint_{\Gamma^e} \mathbf{\Phi} \tau_{ij} n_j\, ds$$

$$\mathbf{N} = \int_{\Omega^e} \mathbf{\Pi}\, \mathbf{\Pi}^T\, d\mathbf{x}$$

$$\mathbf{C}_i^*(\mathbf{u}_j) = \int_{\Omega^e} \lambda \mathbf{\Pi} \mathbf{\Phi}^T \mathbf{u}_j \frac{\partial \mathbf{\Pi}^T}{\partial x_i}\, d\mathbf{x}$$

$$\mathbf{D}_i(\mathbf{u}_j) = \int_{\Omega^e} \lambda \mathbf{\Pi} \frac{\partial \mathbf{\Phi}^T}{\partial x_i} \mathbf{u}_j \mathbf{\Pi}^T\, d\mathbf{x}$$

$$\mathbf{L}_i = \int_{\Omega^e} \mu_0^p \mathbf{\Pi} \frac{\partial \mathbf{\Phi}^T}{\partial x_i}\, d\mathbf{x} \tag{7.4-22}$$

and the matrices $\mathbf{R}_k^{ij}$ are defined by

$$\begin{aligned}
\mathbf{R}_1^{11} &= -\mathbf{S}_{11}(\mathbf{u}_1) - \mathbf{S}_{12}(\mathbf{u}_2) + 2\mathbf{T}_{11}(\mathbf{u}_2) + \mathbf{T}_{21}(\mathbf{u}_2) \\
\mathbf{R}_1^{22} &= \mathbf{T}_{22}(\mathbf{u}_1) \\
\mathbf{R}_1^{12} &= -\mathbf{S}_{21}(\mathbf{u}_1) - \mathbf{S}_{22}(\mathbf{u}_2) + 2\mathbf{T}_{12}(\mathbf{u}_1) + \mathbf{T}_{21}(\mathbf{u}_1) + \mathbf{T}_{22}(\mathbf{u}_2) \\
\mathbf{R}_2^{11} &= \mathbf{T}_{11}(\mathbf{u}_2) \\
\mathbf{R}_2^{22} &= -\mathbf{S}_{21}(\mathbf{u}_1) - \mathbf{S}_{22}(\mathbf{u}_2) + \mathbf{T}_{12}(\mathbf{u}_1) + 2\mathbf{T}_{22}(\mathbf{u}_2) \\
\mathbf{R}_2^{12} &= -\mathbf{S}_{12}(\mathbf{u}_2) - \mathbf{S}_{11}(\mathbf{u}_1) + 2\mathbf{T}_{21}(\mathbf{u}_2) + \mathbf{T}_{12}(\mathbf{u}_2) + \mathbf{T}_{11}(\mathbf{u}_1)
\end{aligned} \tag{7.4-23}$$

The **S** and **T** matrices are defined as

$$\mathbf{S}_{ij}(\mathbf{u}_k) = \int_{\Omega^e} \lambda \frac{\partial \mathbf{\Phi}}{\partial x_i} \mathbf{\Phi}^T \mathbf{u}_k \frac{\partial \mathbf{\Pi}^T}{\partial x_j}\, d\mathbf{x}$$

$$\mathbf{T}_{ij}(\mathbf{u}_k) = \int_{\Omega^e} \lambda \frac{\partial \mathbf{\Phi}}{\partial x_i} \frac{\partial \mathbf{\Phi}^T}{\partial x_j} \mathbf{u}_k \mathbf{\Pi}^T\, d\mathbf{x} \tag{7.4-24}$$

Finally, equations (20) and (21) can be expressed symbolically as

$$\mathbf{C}(\mathbf{u})\mathbf{U} + \mathbf{K}\mathbf{U} + \mathbf{R}(\mathbf{u})\tau = \mathbf{F} \qquad (7.4-25)$$

$$\mathbf{N}\tau - \mathbf{D}(\mathbf{u})\tau + \mathbf{C}^*(\mathbf{u})\tau - \mathbf{L}\mathbf{U} = 0 \qquad (7.4-26)$$

where the vector of unknowns is $\mathbf{U}^T = \{\mathbf{u}_1^T, \mathbf{u}_2^T, \mathbf{P}^T\}$. Equations (25) and (26) represent the finite element equations for a viscoelastic (Oldroyd B) fluid for the steady-state case; very little consideration has been given to the time–dependent case. The displayed matrix system is highly nonlinear and fully coupled. Equation (25) is recognized as the standard form for a Newtonian problem with the addition of the $\mathbf{R}(\mathbf{u})\sigma$ term which comes from the stress rate. Equation (26) is the finite element analog of the implicit constitutive equation and is seen to be an equation of the advection type for the extra–stress.

Before discussing solution procedures for the matrix system, some comments on the approximations used in (13) are required. In defining the approximations to the dependent variables all of the basis functions were taken to be different. From experience with mixed methods for Newtonian problems it is well known that the pressure approximation should be of lower order than the velocity approximation. For most applications in viscoelastic simulations this condition has been met by using quadratic functions for the velocity and linear approximations for the pressure. The choice of an appropriate approximation for the extra–stress is less clearly defined. As mentioned earlier, the problem formulation based on equations (7)–(9) has a restriction on the permissible extra–stress basis functions. A solvability condition for the resulting matrix problem requires that the extra–stresses be approximated to the same order as the velocity components [1]. In the present formulation there are no such restrictions and the choice of the interpolation functions used for stress components are quite arbitrary. Heuristic arguments were made for having the extra–stress and pressure be of the same order since they are both "stress–like" variables. However, many studies using both linear and quadratic stress approximations have been produced and in virtually all cases the mixed method has been limited to flows with small fluid elasticity. In the mid 1980s Marchal and Crochet [23,24] demonstrated that some improvement in behavior of a mixed method could be achieved by using higher–order approximations for the extra–stress. Their general criteria was that the stress approximation should be of sufficient order to represent velocity gradients. They achieved this approximation in practical computations by subdividing each velocity element into 3×3 or 4×4 bilinear subelements on which the extra–stresses were interpolated. This procedure, along with some upwinding methods, allowed computations to be made at large values of fluid elasticity. The benefits of higher–order stresses were also examined by Rao and Finlayson [25], where a cubic stress approximation was employed. Despite this work the problem of stress approximation is not understood nor resolved. Further comments will be made in Section 7.6.

7.4–4 Solution Methods

To discuss possible solution procedures for the discretized viscoelastic problem in (25) and (26), it is convenient to write the matrix system in a more compact form. Combining the incompressibility constraint with the momentum equation allows the following system to be derived

$$\mathbf{K_{uu}}(\mathbf{u})\mathbf{U} + \mathbf{K_{u\tau}}(\mathbf{u})\tau = \mathbf{F} \qquad (7.4-27a)$$

$$\mathbf{K_{\tau\tau}}(\mathbf{u})\tau + \mathbf{K_{\tau u}}\mathbf{U} = 0 \qquad (7.4-27b)$$

where the vector $\mathbf{U}$ now represents the velocity and pressure variables. The $\mathbf{K_{uu}}$ term contains the advection, diffusion, pressure gradient, and incompressibility terms from the momentum equation. The second equation in (27) is a rearranged version of the constitutive equation in (26).

As with any strongly coupled system of equations, there are two basic approaches to the solution of the system given in (27) — artificially decouple (split) the system and employ a cyclic solution process on the decoupled equations or directly solve the combined equation set. Though split equation algorithms are generally less effective than combined schemes, they have the potential for some significant computational benefits in the present system.

Consider the simplest cyclic Picard method available for (27), which is written here for cycle $n+1$ as

Step 1:

$$\mathbf{K_{uu}}(\mathbf{u}^n)\mathbf{U}^{n+1} = \mathbf{F} - \mathbf{K_{u\tau}}(\mathbf{u}^n)\tau^n \qquad (7.4-28a)$$

Step 2:

$$\mathbf{K_{\tau\tau}}(\mathbf{u}^{n+1})\tau^{n+1} = -\mathbf{K_{\tau u}}\mathbf{U}^{n+1} \qquad (7.4-28b)$$

At each step of (28a) a matrix solution is required. However, the matrices in (28a) are significantly smaller than the matrix associated with the combined system in (27). Reduction of the size of the matrix problem is especially important for the viscoelastic problem since the number of unknowns per nodal point is very large (*e.g.*, the number of unknowns in a viscoelastic flow can be more than double those of a Newtonian problem for the same finite element mesh). Unfortunately, little work has been done to improve the relatively poor iterative (convergence) performance of methods such as (28); virtually all viscoelastic problems to date have been solved using a combined equation method.

If the solution to the coupled matrix system is to be obtained by simultaneous solution of the equations in (26), then it is appropriate to rewrite the system in the operator form

$$\hat{\mathbf{K}}(\hat{\mathbf{U}})\hat{\mathbf{U}} = \hat{\mathbf{F}} \qquad (7.4-29)$$

where $\hat{\mathbf{U}}$ now contains all of the problem unknowns. The nonlinear problem shown in (19) may be solved using any of the fixed point iteration schemes. The Picard method

$$\hat{\mathbf{K}}(\hat{\mathbf{U}}^n)\hat{\mathbf{U}}^{n+1} = \hat{\mathbf{F}} \qquad (7.4-30)$$

was used by Coleman [21] for several viscoelastic flow problems. The most commonly used method, however, is Newton's method [9,26–29], which for (19) can be written as

$$\mathbf{J}(\hat{\mathbf{U}}^n)\left[\hat{\mathbf{U}}^{n+1} - \hat{\mathbf{U}}^n\right] = -\hat{\mathbf{K}}(\hat{\mathbf{U}}^n)\hat{\mathbf{U}}^n + \hat{\mathbf{F}} \qquad (7.4-31a)$$

where

$$\mathbf{J}(\hat{\mathbf{U}}^n) = \frac{\partial}{\partial \hat{\mathbf{U}}}\left[\hat{\mathbf{K}}(\hat{\mathbf{U}})\hat{\mathbf{U}} - \hat{\mathbf{F}}\right]\Big|_{\hat{\mathbf{U}}^n} \qquad (7.4-31b)$$

The Jacobian, $\mathbf{J}$, for the system given in (25) and (26) is very complex but can, in fact, be evaluated analytically. Newton's method converges rapidly and, in general, works very well for flows involving materials with low to moderate elasticity. For highly elastic fluids none of the above algorithms allow convergence, as will be discussed in Section 7.4–5.

There has been no discussion of solution methods for transient flows, mainly due to the lack of previous work in this area. It is certainly possible to extend the steady–state formulation of Section 7.3–2 into the time–dependent regime. Following the ideas developed for Newtonian problems [30], the resulting ordinary differential equation system could be integrated in time using an implicit method, such as the trapezoid rule. However, there are two basic difficulties in such an approach. First, since the matrix system for a viscoelastic formulation is so large, the repeated solution of an implicit integration algorithm may not be economically justifiable. Splitting the system during the integration procedure [as in (18)] could alleviate some of the computational burden but would increase concerns regarding stablility. The second concern is with the constitutive equation and the fact that the time scale for the material response can be significantly different than the characteristic time for the flow process [31]. Such a situation suggests that a single integration time scale may be inappropriate or uneconomical; the momentum and constitutive equations may require quite different integration procedures. A method developed by Gartling [32], which is similar to the work of Fortin and Fortin [33], addresses some of the questions regarding transient flows.

7.5 INTEGRAL FINITE ELEMENT MODEL

Finite element procedures for viscoelastic fluids of the integral type have generally received less attention than those for differential models. This can be attributed primarily to the difficulties in implementation of an integral constitutive equation. In this section we will outline the steps needed to use such a model and provide references to work that describes the details of several algorithms.

For consideration of a specific example the Maxwell fluid described by equation (7.2–22) will be used. The basic equations are rewritten here for reference

$$\frac{\partial u_i}{\partial x_i} = 0 \qquad (7.5-1)$$

$$\rho\left(\frac{\partial u_i}{\partial t} + u_j\frac{\partial u_i}{\partial x_j}\right) = -\frac{\partial P}{\partial x_j}\delta_{ij} + \frac{\partial}{\partial x_j}(\tau_{ij}) \qquad (7.5-2)$$

$$\tau_{ij} = \frac{\mu_0}{\lambda^2}\int_{-\infty}^{t} exp\left[-(t-t')/\lambda\right]\left[C_{ij}^{-1}(t') - \delta_{ij}\right]dt' \qquad (7.5-3)$$

Note that the stress decomposition procedure of equation (7.2–14) can also be used with an integral model; its use in the present case would lead to an integral form of the Oldroyd B fluid.

The balance laws, (7.4–22) and (7.4–23), can be discretized in the standard way, using a mixed (velocity, pressure) finite element procedure. The result of such a process is the following familiar set of matrix equations

$$\mathbf{Q}^T\mathbf{u} = 0 \qquad (7.5-4)$$

$$\mathbf{C}_i(\mathbf{u})\mathbf{u} + \mathbf{Q}_i\mathbf{P} = \mathbf{F}_i - \int_{\Omega^e} \frac{\partial \mathbf{\Phi}}{\partial x_j}\tau_{ij}d\mathbf{x} \qquad (7.5-5)$$

which are written here for the steady–state case. Note that in writing equation (5), $\mathbf{\Phi}$ is assumed to be the velocity basis (weighting) function; an integration by parts is responsible for the form of the stress integral. The arrangement of terms in (5) anticipates the use of an iterative solution procedure. In the first part of the iteration scheme equations (4) and (5) can be solved for the velocity and pressure given a known (or predicted) stress field. The task then is to develop an algorithm for the prediction of the extra–stress given a velocity field from (5).

From equation (3) it is apparent that to predict the extra–stress at a particular location, say, a nodal point or quadrature point, the strain history for that point must be evaluated over past time. This computation is usually done in three stages. In the first stage, the particle path (streamline for steady flow) through the point in question is constructed. Procedures for this construction range from the Lagrangian-like computation of displacement fields for each finite element [34] to the direct computation of the streamlines [35]. The construction of the trajectory allows the previous locations of the particle (*i.e.*, node or quadrature point) to be established. The second stage requires the evaluation of the strain measure (*e.g.*, C_{ij}^{-1}) along the particle trajectory. The final stage uses a numerical (Laguerre) quadrature to evaluate the history integral (3) and produce the extra–stress at a particular location.

The details of the particle tracking and strain evaluation depend directly on the particular type of finite element approximation used for the velocity field. A discussion of several different approaches may be found in the work of Viriyayuthakorn and Caswell [34], Dupont *et al.* [35] and Bernstein *et al.* [36]. Once a prediction of the stress field has been made the stress forcing function in (5) can be evaluated and a new velocity obtained. This cyclic procedure is continued until convergence of the dependent variables is achieved. Details regarding the implementation of the various iteration procedures may also be found in [34,35].

There are several drawbacks to the integral methods as currently implemented. The basic structure of the equations in (2)–(4) does not permit the easy use of higher-order iterative methods, such as Newton's method. Convergence is thus limited to a linear rate, which can be a severe computational burden. The particle tracking algorithms appear to be the weak point of the overall procedure. For recirculating flows the methods must be of sufficient accuracy to ensure the closure of streamlines. This is a rather difficult task when dealing with approximate numerical schemes and an incompressible fluid. Also, the extension of the particle tracking ideas to time-dependent flows would appear to be very difficult. Despite these disadvantages the work on integral constitutive equations should continue to increase, since these models are somewhat better at predicting realistic viscoelastic behavior than their differential counterparts.

7.6 UNRESOLVED PROBLEMS

7.6–1 General Comments

To conclude this brief treatment of finite element methods for viscoelastic flows, it is appropriate to outline some of the unresolved problems and areas of research in this field. Numerical difficulties in the simulation of Newtonian fluids are generally associated with the occurrence of large values of some relevant nondimensional parameter, *e.g.*, the Reynolds number, Peclet number or Rayleigh number. The situation is quite similar for the viscoelastic problem though the nondimensional parameters are not as familiar.

As noted in a previous section, there are two time scales involved in a viscoelastic flow — the time constant for the material (relaxation time, λ) and the characteristic time for the flow process, t_f. The ratio of these time parameters is called the Deborah number, $De = \lambda/t_f$. Very often the characteristic time for the flow is taken as the reciprocal of a typical (wall) shear rate, so that $De = \lambda\dot{\gamma}_w$. The nondimensionalization of simple constitutive equations, such as the Maxwell model, can lead to the definition of a second characteristic parameter. The Weissenberg number is typically defined as $W = \lambda U/L$ where U and L are a characteristic velocity and length scale for the flow. In many cases the Weissenberg and Deborah numbers can have the same definition; they also tend to be used interchangeably, which leads to some confusion in the literature.

The Deborah and Weissenberg numbers both indicate the relative importance of fluid elasticity for a given flow problem. High values of W (or De) indicate a material response that is very solid–like, while low values of the parameter represent "small" departures from normal viscous fluid behavior.

The continuing outstanding problem in the numerical simulation of viscoelastic fluids is generally termed the "high Weissenberg (or Deborah) number problem." In essence, this terminology refers to the observed behavior that, as the Weissenberg number is increased for a given problem, a critical value is reached (W_{cr}) beyond which the numerical algorithm fails. Failure in the present context normally consists of an initial degradation in the solution field (*e.g.*, spatial oscillations in the dependent variables, especially the stresses) followed by nonconvergence of the solution algorithm. The high Weissenberg limit is a universal phenomena in that it has been observed in all types of numerical procedures (finite element and finite difference) and for all types of constitutive equations. The critical value of the Weissenberg number is very sensitive to the particular flow geometry, numerical algorithm, constitutive equation, and computational mesh. Unfortunately, from an engineering analysis viewpoint the limiting value W_{cr} is usually small enough to make the obtainable solutions uninteresting (*i.e.*, they are perturbations to the Newtonian solution), while the important simulations at high Weissenberg number remain unsolved.

The failure of numerical methods to provide solutions for highly elastic fluids has spawned a great deal of research activity as evidenced by the numerous publications and several international workshops on the subject [37]. Though the basic causes of the problem have yet to be fully identified, a number of issues have been explored that contribute to the basic understanding of several parts of this complex problem. A summary of some of these topics is given below. For the reader interested in further details regarding the many facets of the "high Weissenberg number problem" the text by Crochet *et al.* [1], the recent reviews by Crochet [38] and Keunings [39], and recent issues of the *Journal of Non–Newtonian Fluid Mechanics* are recommended.

7.6–2 Choice of Constitutive Equation

Much of the early numerical work on viscoelastic flows was carried out using the Maxwell constitutive model. This was mainly due to the relative simplicity of the equation and the belief that it was prototypical of the more complex constitutive relations. Recent works by Phan Thien [40,42] and Phan Thien and Walsh [41] demonstrated that for some flows there are inherent instabilities in the Maxwell constitutive model that occur above a critical Weissenberg number. Such a proof demonstrates one of the subtle dangers of this class of problems — the use of approximate constitutive equations that are difficult or impossible to test analytically for non–physical behavior in complex flows. Many recent investigations have abandoned the Maxwell/Oldroyd family of fluid models in favor of more realistic (and complex) equations such as those due to Johnson–Segalman and Phan Thien–Tanner. There is of course no proof that the more complex models do not also contain built-in instabilities that may or may not be physically realistic. Despite the results shown in [33–35] most investigators do not feel that the high Weissenberg limit is due solely to the choice of the constitutive model, though it is certainly regarded as a contributing factor. It should also be noted that virtually all numerical work has considered only single relaxation time models, while it is clear that most constitutive models require multiple relaxation times to approximate realistic fluid behavior. The large computational burden for single relaxation time models has prevented more realistic relaxation spectra from being investigated.

7.6–3 Solution Uniqueness and Existence

The complexity of the equations describing viscoelastic flows precludes any proof of solution uniqueness or existence. Indeed, from the behavior of the simple Newtonian fluid it is to be expected that bifurcations and multiple solutions will exist for flows of these types of fluids. Various investigators [9,26,31] working with several different geometries and constitutive relations have demonstrated the occurrence of limit points (loss of solution with increasing Weissenberg number) and bifurcation points (multiple solutions) in numerical simulations. Typically such features of the flow are computed using a continuation (or imbedding) procedure and Newton's method. The question remains as to whether such behavior is an attribute of the physical model (differential constitutive equation) or is an artifact of the numerical discretization process. Recently, Keunings [43] and Debbaut and Crochet [44] have provided arguments for such phenomena being of a numerical nature in at least two particular simulations. Their findings, however, do not eliminate the concern over encountering true bifurcations or limit points under other circumstances.

Various experimental investigations of viscoelastic flows, such as [45], have demonstrated the occurrence of time–dependence and three–dimensional flow structure in relatively simple two-dimensional geometries. With the numerical emphasis on two-dimensional, steady simulations, this type of behavior is clearly not predictable. Issues such as these emphasize the difficulty in discriminating between behavior of the discretized equations, the continuum equations, and the physical system.

7.6–4 Numerical Algorithm Problems

A particularly disturbing behavior that has been observed in many numerical simulations is a pathology associated with mesh refinement studies. Typically, a solution on one mesh discretization can be obtained up to a critical Weissenberg number of, say, W_{cr}^{coarse}; a refined mesh solution will produce a solution up to W_{cr}^{fine}, where $W_{cr}^{fine} < W_{cr}^{coarse}$. Stated differently, the finite element procedure does not appear to converge as the mesh is refined — a fundamental property of a stable algorithm. Such behavior has been reported for several two-dimensional simulations [1,29] and studied in detail in a one-dimensional flow [27].

In [27] it was conjectured that this behavior was due to an improper choice of the approximating functions for the extra-stress. Such behavior may also stem from the use of an inappropriate solution algorithm or formulation for the constitutive equation. Note that the constitutive equation in a differential model is essentially of the hyperbolic type, which normally requires quite a different numerical treatment from the elliptic momentum equations [31]. As noted previously, the work of Marchal and Crochet [23,24] and others [25] has produced some support for the need for higher-order stress approximations. However, the key to success seems to be more closely tied to the use of upwinding techniques for the improved treatment of the hyperbolic constitutive equations.

Extensive work in this area by Dupret *et al.* [46] has led to some understanding of this particular aspect of the problem. Viscoelastic flows are particularly difficult to model due to: (a) the presence of normal stresses, which lead to very high stress gradients and thin boundary layers, (b) the occurrence of singularities and stress concentrations, and (c) the strong hyperbolic character of the constitutive equation. Large stress gradients and singularities place a high demand on mesh resolution for the stress variables. The hyperbolic nature of the constitutive equation requires that an appropriate numerical method be used for the equation set; upwind techniques or Petrov–Galerkin methods are one possible approach to this problem. When any of the above demands are not met, as is the case in most standard, mixed finite element methods, the discrete equation systems for the viscoelastic problem can produce improper evolutionary behavior. Dupret *et al.* [46] tracked this behavior by monitoring a characteristic tensor and correlating its loss of positive definiteness with the occurrence of numerical limit points in the solution. The combination of higher-order stress approximations and upwind techniques for the constitutive equation proved to stabilize the method, and allow solutions at very high Weissenberg numbers with appropriate mesh convergence behavior. However, upwinding and Petrov–Galarkin methods are not without their own problems. The addition of artificial diffusion with these methods must be carefully evaluated to ensure that a true solution is not masked/altered by the numerical technique.

At least two other approaches have been investigated to circumvent the numerical problems associated with viscoelastic flows. In the first, the basic partial differential equations are altered such that the elliptic nature of the momentum equations is emphasized and a Galerkin formulation will then produce an equation set that converges with mesh refinement. The so–called Explicitly Elliptic Momentum Equation (EEME) formulation by King *et al.* [47] still requires the use of upwind techniques for the constitutive equation. Simulations at reasonably high values of the Weissenberg number have been obtained, but the concern with upwinding is still present. A different approach to representation of the stress variables was taken by Fortin and Fortin [33] and Gartling [32]. In these studies, the stresses were defined on the elements by discontinuous interpolants. Fortin and Fortin [33] used an upwinding technique to treat the constitutive equation while Gartling [32] used an explicit time–dependent approach for the hyperbolic equations. The most significant difficulty with these methods was mesh resolution, especially with regard to singularities.

7.6–5 Equation Change of Type

The work of Joseph *et al.* [48] has recently demonstrated that the partial differential equations describing viscoelastic flows can undergo a formal change of type from elliptic to hyperbolic as the Weissenberg number increases. This is analogous to the transonic flow problem where local supercritical flow regions can be embedded in an otherwise subcritical flow. The implications of this work on present finite element methods, however, is not completely clear. The change of type is found to occur for the equations describing the fluid vorticity; the occurence of real characteristics in such an equation admits to the possibility of discontinuities in the vorticity. Unfortunately it is difficult to see what this type of vorticity behavior implies about solution algorithms that are based on continuous velocity and pressure fields, as are most of the current finite element methods. The occurrence of shear waves has been observed in several transient solutions of viscoelastic flows [32,49–51]. Standard velocity based finite element methods are capable of resolving such flows, as will be demonstrated in the next section. The question of change of type and the role of hyperbolicity in solution methods for these materials requires a great deal of further study.

7.7 NUMERICAL EXAMPLES

7.7–1 Introduction

The example computations in this section will be limited to a few problems that illustrate some of the non–Newtonian behavior of a viscoelastic fluid. A variety of simulations have been reported in the literature. However, as example problems they suffer from a common shortcoming in that the reported solutions are little different from those of a Newtonian fluid, mainly due to the "high Weissenberg limit" described above. One exception to the above comments is the work on die swell by Tanner [52], Crochet and Keunings [28,53,54], and others [55], which does show some departure from the Newtonian case for even slightly elastic fluids. The first problem to be considered here is a geometrically simple transient flow with a rather complex time–dependent response.

7.7–2 Time–Dependent Poiseuille Flow

Consider the problem of an initially quiescent viscoelastic fluid contained between infinite parallel plates. At time zero a specified, constant pressure gradient is imposed. The objective is to compute the time-dependent response of the fluid until it reaches a steady state (Poiseuille flow). A schematic of the problem is shown in Figure 7.7–1.

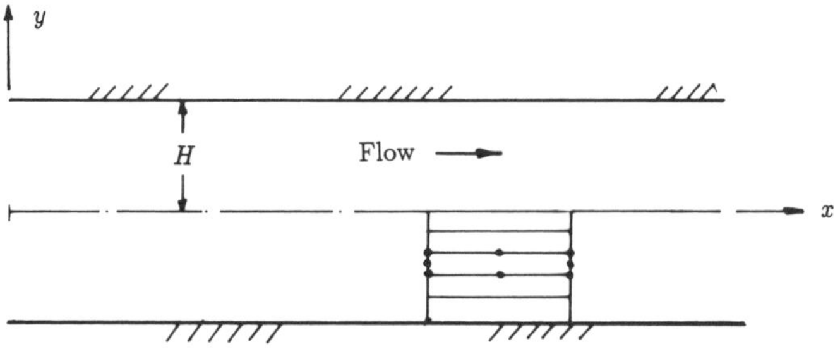

Figure 7.7–1: Schematic of Poiseuille flow.

The above problem was solved analytically by Rivlin [51] for both a Newtonian and an upper–convected Maxwell fluid. The numerical solutions presented here were obtained using the finite element scheme described in [32]; results are also reported for Newtonian and Maxwell fluids.

Shown in Figures 7.7–2 and 7.7–3 are representative time histories and velocity profiles for the response of a Newtonian fluid. The evolution from the rest state to a fully developed parabolic profile is seen to be a smooth, asymptotic process. Comparison of the numerical results with the analytical work of Rivlin shows excellent agreement (less than 1% difference).

The response of a viscoelastic fluid, such as a Maxwell material, is quite different. Figure 7.7–4 shows a typical time history for the Maxwell fluid. The damped oscillatory behavior is quite evident and represents a streamwise oscillation of the fluid. This is seen more clearly in Figures 7.7–5 and 7.7–6 where velocity profiles at various times are illustrated. Early in the start–up process the velocity field shows the propagation of a shear wave from the wall toward the centerline of the channel. The subsequent reflection of the wave causes the oscillatory behavior seen in Figure 7.7–4. At long times the velocity field approaches the expected steady-state profile. A comparison of this solution with the analytical result again shows excellent agreement. Figure 7.7–7 shows a series of shear stress profiles that illustrate the early time behavior of the shear wave as it propagates across the channel. The normal stress component has a similar behavior. Though the present example is quite simple, it does illustrate the complex behavior that can occur in fluids with memory.

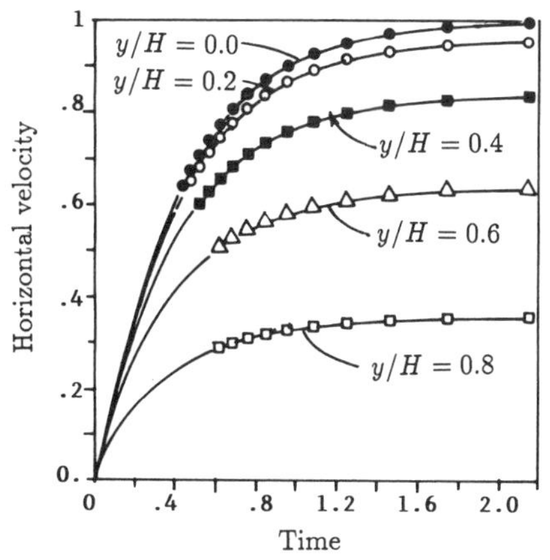

Figure 7.7–2: Velocity histories for Poiseuille flow, Newtonian fluid.

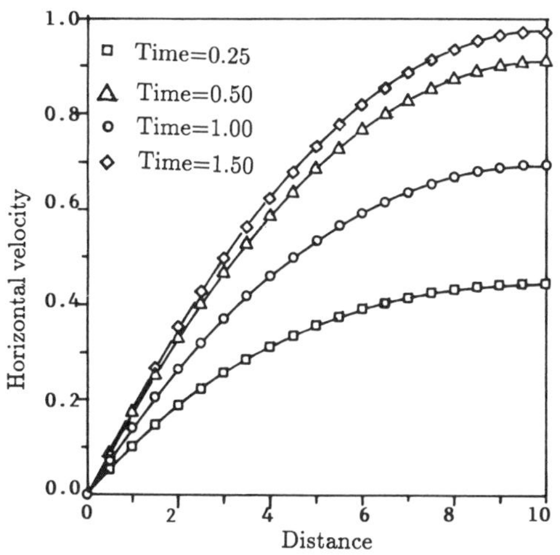

Figure 7.7–3: Velocity profiles for Poiseuille flow, Newtonian fluid.

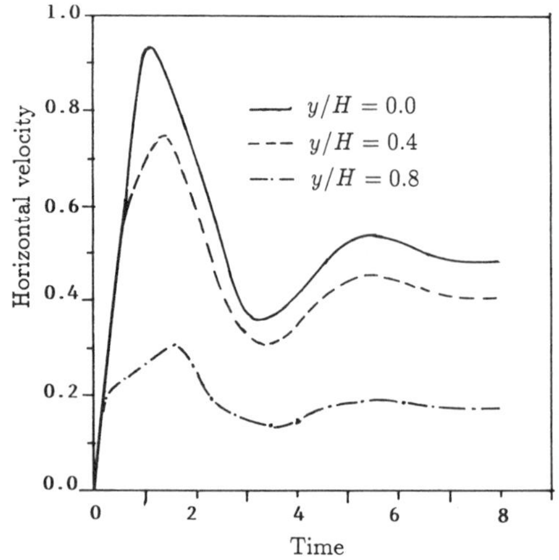

Figure 7.7–4: Velocity histories for Poiseuille flow, Maxwell fluid.

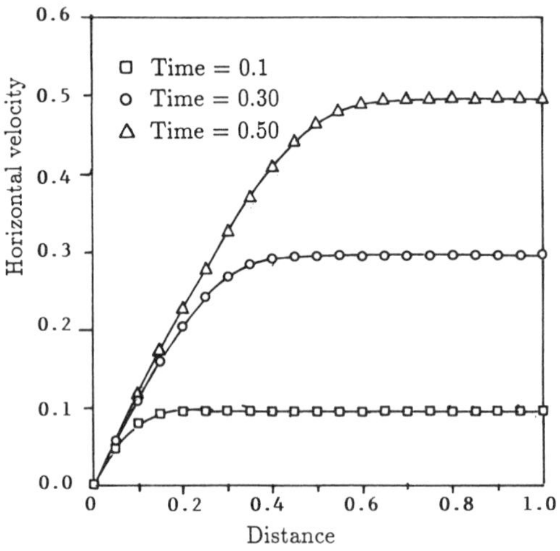

Figure 7.7–5: Velocity profiles for Poiseuille flow, Maxwell fluid.

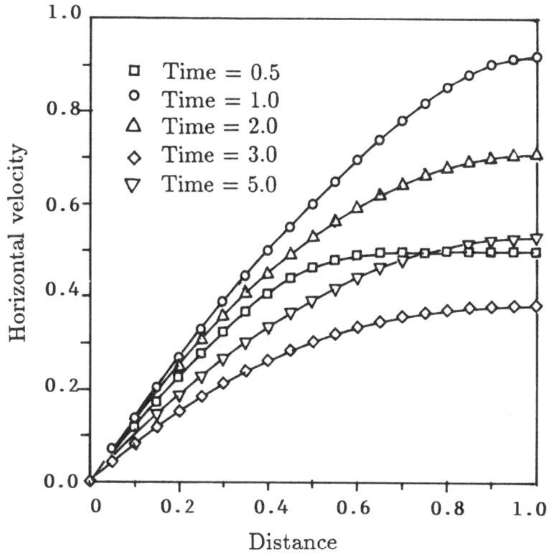

Figure 7.7–6: Velocity profiles for Poiseuille flow, Maxwell fluid.

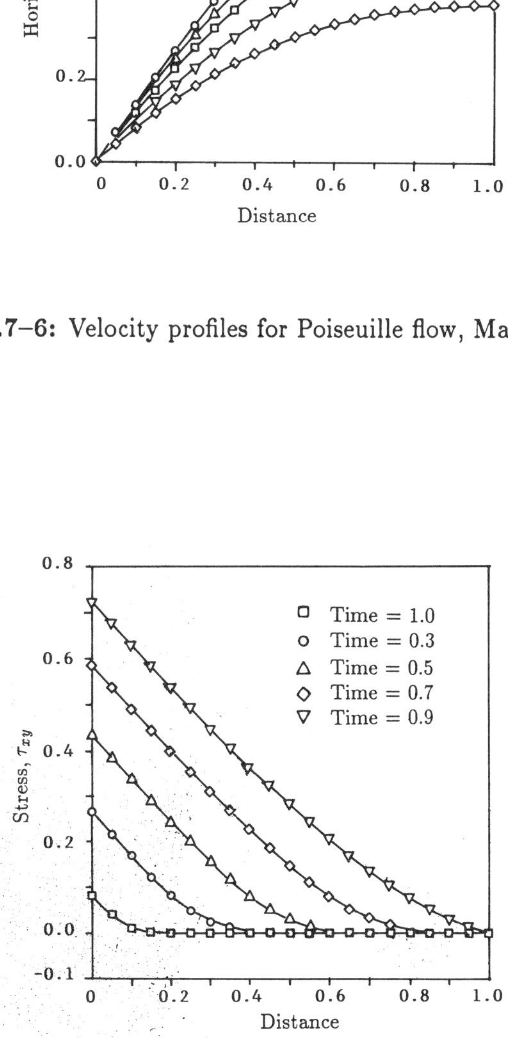

Figure 7.7–7: Shear stress profiles for Poiseuille flow, Maxwell fluid.

7.7–3 Four–to–One Contraction Problem

A standard test problem for the study of viscoelastic flow behavior is the entry flow in axisymmetric, four–to–one contraction. This geometry is popular since it is of technological importance for the polymer processing industry and also exhibits significant flow variation with changes in fluid elasticity. In particular, for several different polymers as the fluid elasticity is increased (*i.e.*, Weissenberg number is increased) the vortex structure in the largest entry tube undergoes a significant increase in length. The objective of many numerical simulations has been to predict this trend in vortex growth with a suitable constitutive model.

To illustrate typical results for the four–to–one contraction the method proposed by Rao and Finlayson [25,56] is employed. Figure 7.7–8 contains a schematic of the problem with the velocity boundary conditions shown; a typical quadrilateral mesh is shown in Figure 7.7–9. An Oldroyd–B constitutive model was used with a typical mixed method; a cubic approximation for the stress components was used in conjunction with quadratic velocities and a linear pressure. A Newton iteration method with continuation was used to solve the four–to–one contraction for a series of Weissenberg numbers. It was found that to achieve significant values of the Weissenberg number an upwind procedure for the constitutive equation was required. Full details of all the combinations of formulations and algorithms that were tried on this problem are available in [56].

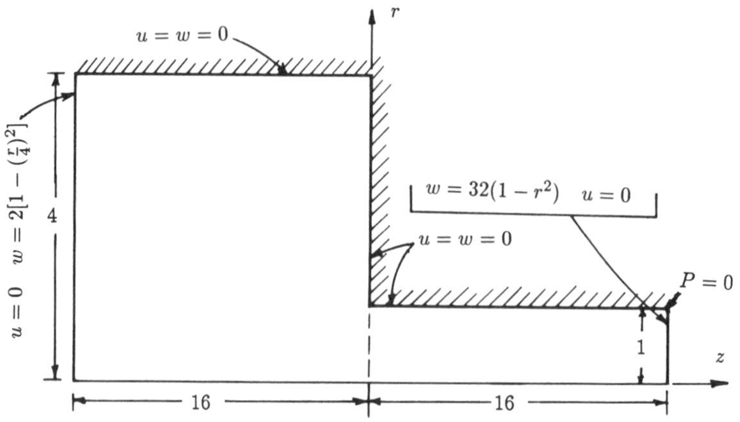

Figure 7.7–8: Schematic for axisymmetric 4:1 contraction problem.

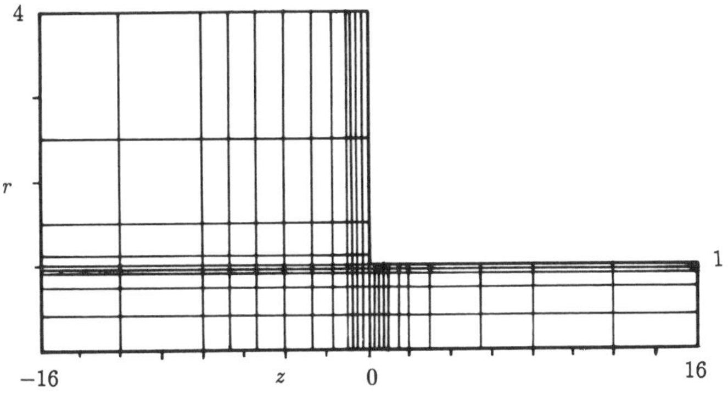

Figure 7.7–9: Quadrilateral mesh for 4:1 contraction. Stress approximation is cubic on each element.

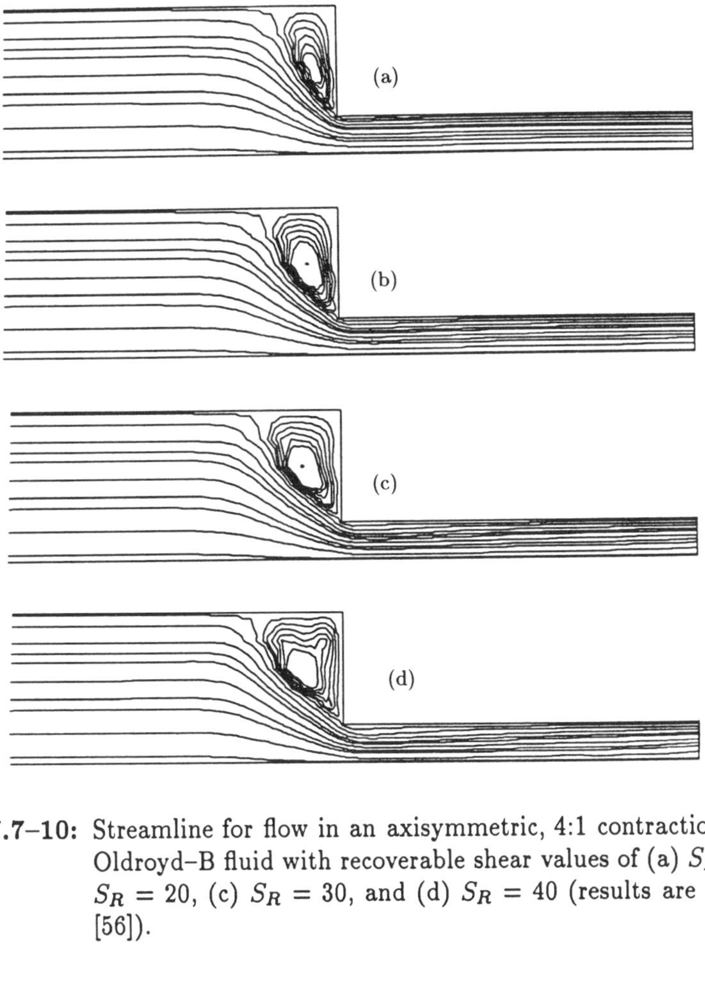

Figure 7.7–10: Streamline for flow in an axisymmetric, 4:1 contraction using an Oldroyd–B fluid with recoverable shear values of (a) $S_R = 10$, (b) $S_R = 20$, (c) $S_R = 30$, and (d) $S_R = 40$ (results are taken from [56]).

Computed results for the four–to–one contraction in the form of stream function contours are shown in Figure 7.7–10. The plots correspond to increasing values of the recoverable shear, S_R, which is directly proportional to the Weissenberg number. The simulations predict the correct trend in that the vortex length increases as the fluid elasticity increases. Quantitative comparisons with experimental data show that numerical simulations underpredict the vortex growth. This discrepancy is the result of using approximate constitutive models.

EXERCISE PROBLEMS

7.1 The general Oldroyd model is given by (see [1])

$$\tau_{ik} + \lambda \hat{\tau}_{ik} + \mu \tau_{jj} D_{ik} - \mu_1 \left(\tau_{ij} D_{jk} + \tau_{kj} D_{ji} \right) = 2\mu_0 \left(D_{ik} + \lambda_1 \hat{D}_{ik} - \mu_2 D_{ij} D_{jk} \right) \tag{1}$$

where μ_0 is a constant viscosity coefficient, and $(\lambda, \lambda_1, \mu, \mu_1, \mu_2)$ are material constants. Deduce the upper–convected, lower–convected, and corotational Maxwell models by assigning values to the material constants in the Oldroyd model.

7.2 The White–Metzner model [57] is defined by

$$\tau_{ik} + \lambda(I_2) \bar{\tau}_{ik} = 2\mu_0(I_2) D_{ik} \tag{2}$$

which describes viscoeleastic effects in flow problems that are dominated by the shear viscosity. Develop the finite element model governing viscoelastic flows in two–dimensional Cartesian geometries using the White–Metzner constitutive equation (see [55]).

7.3 Repeat Problem 3 for flows in axisymmetric two–dimensional geometries.

REFERENCES

1. M. J. Crochet, A. R. Davies, and K. Walters, *Numerical Simulation of Non–Newtonian Flow*, Elsevier, Amsterdam, The Netherlands (1984).

2. J. R. A. Pearson, *Mechanics of Polymer Processing*, Elsevier, London, U. K. (1985).

3. C. Truesdell and W. Noll, "The Non–Linear Field Theories of Mechanics," in *Encyclopedia of Physics, Vol III*, S. Flügge (ed.), Springer–Verlag, Berlin, Germany (1965).

4. M. J. Crochet and K. Walters, "Numerical Methods in Non-Newtonian Fluid Mechanics," *Annual Review of Fluid Mechanics*, **15**, 241–260 (1983).

5. A. C. Pipkin and R. I. Tanner, "A Survey of Theory and Experiment in Viscometric Flows of Viscoelastic Liquids," *Mechanics Today, Vol 1*, S. Nemat–Nasser (ed.), 262–321 (1972).

6. R. I. Tanner, "Recent Progress in Rheology," *Journal of Applied Mechanics*, **105**, 1181–1190 (1983).

7. M. J. Crochet and G. Pilate, "Numerical Study of the Flow of a Fluid of Second Grade in a Square Cavity," *Computers and Fluids*, **3**, 283–291 (1975).

8. K. R. Reddy and R. I. Tanner, "Finite Element Approach to Die-Swell Problems of Non–Newtonian Fluids," *Proceedings of the Sixth Australian Hydraulics and Fluid Mechanics Conference*, Adelaide, Australia, 431–434 (1977).

9. M. A. Mendelson, P. W. Yeh, R. A. Brown, and R. C. Armstrong, "Approximation Error in Finite Element Calculation of Viscoelastic Flow," *Journal of Non–Newtonian Fluid Mechanics*, **10**, 31–54 (1982).

10. M. W. Johnson and D. Segalman, "A Model for Viscoelastic Fluid Behavior Which Allows Non–Affine Deformation," *Journal of Non–Newtonian Fluid Mechanics*, **2**, 255–270 (1977).

11. N. Phan Thien and R. I. Tanner, "A New Constitutive Equation Derived From Network Theory," *Journal of Non–Newtonian Fluid Mechanics*, **2**, 353–365 (1977).

12. R. B. Bird, R. C. Armstrong, and O. Hassager, *Dynamics of Polymeric Liquids*, Vol. 1: Fluid Mechanics, Second Edition, John Wiley & Sons, New York (1971).

13. J. L. White and A. B. Metzner, "Development of Constitutive Equations for Polymeric Melts and Solutions," *Journal of Applied Polymer Science*, **7**, 1867–1889 (1963).

14. R. I. Tanner, *Engineering Rheology*, Clarendon Press, Oxford U. K. (1985).

15. C. J. S. Petrie, "Measures of Deformation and Convected Derivatives," *Journal of Non–Newtonian Fluid Mechanics*, **5**, 147–176 (1979).

16. L. E. Malvern, *Introduction to the Mechanics of a Continuous Medium*, Prentice–Hall, Englewood Cliffs, New Jersey (1969).

17. A. S. Lodge, *Elastic Liquids*, Academic Press, New York (1964).

18. Y. Shimazaki and E. G. Thompson, "Elasto Visco-Plastic Flow With Special Attention to Boundary Conditions," *International Journal for Numerical Methods in Engineering*, **17**, 97–112 (1981).

19. M. Kawahara and N. Takeuchi, "Mixed Finite Element Method for Analysis of Viscoelastic Fluid Flow," *Computers and Fluids*, **5**, 33–45 (1977).

20. M. J. Crochet and M. Bezy, "Numerical Solution for the Flow of Viscoelastic Fluids," *Journal of Non–Newtonian Fluid Mechanics*, **5**, 201–218 (1979).

21 C. J. Coleman, "A Finite Element Routine for Analysing Non-Newtonian Fluids, Part I: Basic Method and Preliminary Results," *Journal of Non–Newtonian Fluid Mechanics*, **7**, 289–301 (1980).

22. P. W. Chang, T. W. Patten, and B. A. Finlayson, "Collocation and Galerkin Finite Element Methods for Viscoelastic Fluid Flow, I. Description of Method and Problems with Fixed Geometries," *Computers and Fluids*, **7**, 267–283 (1979).

23. J. M. Marchal and M. J. Crochet, "Hermitian Finite Elements for Calculating Viscoelastic Flow," *Journal of Non-Newtonian Fluid Mechanics,* **20**, 187–207 (1986).

24. J. M. Marchal and M. J. Crochet, "A New Mixed Finite Element for Calculating Viscoelastic Flow," *Journal of Non-Newtonian Fluid Mechanics,* **26**, 77–114 (1987).

25. R. R. Rao and B. A. Finlayson, "Viscoelastic Flow Simulation Using Cubic Stress Finite Elements," *Journal of Non-Newtonian Fluid Mechanics,* **43**, 61–82 (1992).

26. P. W. Yeh, M. E. Kim, R. C. Armstrong, and R. A. Brown, "Multiple Solutions in the Calculation of Axisymmetric Contraction Flow of an Upper Convected Maxwell Fluid," *Journal of Non-Newtonian Fluid Mechanics*, **16**, 173–194 (1984).

27. D. K. Gartling, "One Dimensional Finite Element Solutions for a Maxwell Fluid," *Journal of Non-Newtonian Fluid Mechanics*, **17**, 203–231 (1985).

28. M. J. Crochet and R. Keunings, "Die Swell of a Maxwell Fluid: Numerical Predictions," *Journal of Non-Newtonian Fluid Mechanics*, **7**, 199–212 (1980).

29. R. Keunings and M. J. Crochet, "Numerical Simulation of the Flow of a Viscoelastic Fluid Through an Abrupt Contraction," *Journal of Non-Newtonian Fluid Mechanics*, **14** 279–299 (1984).

30. P. Gresho, R. Lee, R. Sani, and T. Stullich, "On the Time-Dependent FEM Solution of the Incompressible Navier-Stokes Equations in Two- and Three-Dimensions," in *Recent Advances in Numerical Methods in Fluids, Vol 1*, Pineridge Press, Swansea, U.K. (1980).

31. S. L. Josse and B. A. Finlayson, "Reflections on the Numerical Viscoelastic Flow Problem," *Journal of Non-Newtonian Fluid Mechanics*, **16**, 13–36 (1984).

32. D. K. Gartling, "A Finite Element Algorithm for Time-Dependent Viscoelastic Flows," Sandia National Laboratories Report, Albuquerque, New Mexico (1988).

33. M. Fortin and A. Fortin, "A New Approach for the FEM Simulation of Viscoelastic Flows," *Journal of Non-Newtonian Fluid Mechanics,* **32**, 295–310 (1989).

34. M. Viriyayuthakorn and B. Caswell, "Finite Element Simulation of Viscoelastic Flow," *Journal of Non-Newtonian Fluid Mechanics*, **6**, 245–267 (1980).

35. S. Dupont, J. M. Marchal, and M. J. Crochet, "Finite Element Simulation of Viscoelastic Fluids of the Integral Type," *Journal of Non-Newtonian Fluid Mechanics*, **17**, 157–183 (1985).

36. B. Bernstein, M. K. Kadivar, and D. S. Malkus, "Steady Flow of Memory Fluids with Finite Elements: Two Test Problems," *Computer Methods in Applied Mechanics and Engineering*, **27**, 279–302 (1981).

37. R. C. Armstrong, R. A Brown, and B. Caswell (eds.), "Papers From the Third International Workshop on Numerical Simulation of Viscoelastic Flows," *Journal of Non–Newtonian Fluid Mechanics*, **16** (1984).

38. M. J. Crochet, "Numerical Simulation of Viscoelastic Flow: A Review," *Rubber Reviews*, **62**, 426–455 (1989).

39. R. Keunings, "Progress and Challenges in Computational Rheology," *Rheologica Acta*, **29**, 556–570 (1990).

40. N. Phan Thien, "Coaxial–Disk Flow of an Oldroyd–B Fluid: Exact Solution and Stability," *Journal of Non–Newtonian Fluid Mechanics*, **13**, 325–340 (1983).

41. N. Phan Thien and W. Walsh, "Squeeze–Film Flow of an Oldroyd-B Fluid: Similarity Solution and Limiting Weissenberg Number," *Journal of Applied Mathematics and Physics (ZAMP)*, **35**, 747–759 (1984).

42. N. Phan Thien, "Cone–and–Plate Flow of the Odlroyd-B Fluid is Unstable," *Journal of Non–Newtonian Fluid Mechanics*, **17**, 37–44 (1985).

43. R. Keunings, "Mesh Refinement Analysis of the Flow of a Maxwell Fluid Through an Abrupt Contraction," *Proceedings of Fourth International Conference on Numerical Methods in Laminar and Turbulent Flow*, C. Taylor *et al.* (eds.), Pineridge Press, Swansea, U.K. (1985).

44. B. Debbaut and M. J. Crochet, "Further Results on the Flow of a Viscoelastic Fluid Through an Abrupt Contraction," *Journal of Non–Newtonian Fluid Mechanics*, **20**, 173–185 (1986).

45. J. V. Lawler, S. J. Muller, R. A. Brown, and R. C. Armstrong, "Laser Doppler Velocimetry, Measurements of Velocity Fields and Transitions in Viscoelastic Fluids," *Journal of Non–Newtonian Fluid Mechanics*, **20**, 51 (1986)

46. F. Dupret, J. M. Marchal, and M. J. Crochet, "On the Consequence of Discretization Errors in the Numerical Calculation of Viscoelastic Flow," *Journal of Non–Newtonian Fluid Mechnaics*, **18**, 173–186 (1985).

47. R. C. King, M. R. Apelian, R. C. Armstrong, and R. A. Brown, "Numerically Stable Finite Element Techniques for Viscoelastic Calculations in Smooth and Singular Geometries," *Journal of Non–Newtonian Fluid Mechanics*, **29**, 147–216 (1988).

48. D. D. Joseph, M. Renardy, and Jean–Claude Saut, "Hyperbolicity and Change of Type in the Flow of Viscoelastic Fluids," *Archives of Rational Mechanics and Analysis*, **87** 213–251 (1985).

49. N. Phan Thien, "Coaxial–Disk Flow and Flow About a Rotating Disk of a Maxwell Fluid," *Journal of Fluid Mechanics*, **128**, 427–442 (1983).

50. N. Phan Thien and R. I. Tanner, "Viscoelastic Squeeze–Film Flows — Maxwell Fluids," *Journal of Fluid Mechanics*, **129**, 265–281 (1983).

51. R. S. Rivlin, "Run–Up and Decay of Plane Poiseuille Flow," *Journal of Non–Newtonian Fluid Mechanics*, **14**, 203–217 (1984).

52. R. I. Tanner, "Extrudate Swell," in *Computational Analysis of Polymer Processing*, J. R. A. Pearson and S. M. Richardson (eds.), Elsevier, London, U. K., 63–91 (1983).

53. M. J. Crochet and R. Keunings, "On Numerical Die Swell," *Journal of Non–Newtonian Fluid Mechanics*, **10**, 85–94 (1982).

54. M. J. Crochet and R. Keunings, "Finite Element Analysis of Die–Swell of a Highly Elastic Fluid," *Journal of Non–Newtonian Fluid Mechanics*, **10**, 339–356 (1982).

55. J. N. Reddy and V. A. Padhye, "A Penalty Finite Element Model for Axisymmetric Flows of Non–Newtonian Fluids," *Numerical Methods in Partial Differential Equations*, **4**, 33–56 (1988).

56. R. R. Rao, "Adaptive Finite Element Analysis of Non–Newtonian Flow," Ph. D. Dissertation, Department of Chemical Engineering, University of Washington, Seattle, Washington (1990).

57. J. L. White and A. B. Metzner, "Development of Constitutive Equations for Polymeric Melts and Solutions," *Journal of Applied Polymer Science*, **7**, 1867–1889 (1963).

Chapter Eight

PRELUDE TO ADVANCED TOPICS

8.1 INTRODUCTION

This chapter provides a brief prelude to several advanced topics closely associated with incompressible flows and heat transfer. The complexities inherent to fluid dynamics and the extensive scope of the field continues to offer challenges for research, design, and analysis. The continued evolution of airplanes, missiles, cars, engines, heat exchangers, fans, turbines, and numerous other industrial components will require a detailed knowledge of the fluid and thermal fields. In addition, virtually every aspect of nature demonstrates some fundamental aspect of fluid mechanics. In the previous chapters, the essential aspects of finite element techniques for nonisothermal, viscous incompressible flows were demonstrated. Even within this relatively narrow class of problems a number of important topics had to be omitted. Here we will consider several extensions to the basic problem area. A primary focus will be on turbulent flows since they are of such importance to industrial problems. Brief introductions to simple chemically reactive flows and magnetohydrodynamic problems will also be presented. The last section of this chapter is devoted to some introductory ideas regarding the next technology improvement in computational mechanics, parallel processing.

8.2 TURBULENT FLOWS

8.2–1 Introduction

Turbulence is a highly complex physical phenomenon that is pervasive in flow problems of scientific and engineering concern. A simple, precise definition of turbulence is difficult though the phenomenon is often associated with the ideas of randomness, disorder, and chaos. Following Hinze [1], we say that turbulence is defined as an "irregular flow condition showing random variations with respect to both time and space coordinates with discernible statistical properties." Though a standard definition of turbulence may be elusive, a number of well-established characteristics of a turbulent flow can be listed. A turbulent flow is a:

- Highly nonlinear flow process
- Highly diffusive flow
- Three–dimensional flow
- Flow with multiple length and time scales
- Time–dependent (stochastic) phenomenon with identifiable statistical properties.

Turbulence is one of the unsolved problems in physics, especially in the sense that universally applicable mathematical models of the phenomenon are not available. Despite these theoretical difficulties, the demand for numerical computation of realistic (turbulent) flows remains high. Out interest here will therefore focus on the modeling and simulation of turbulence from an engineering point of view. This approach implies that the detailed resolution of a turbulent flow will be eliminated in favor of some type of averaged flow description. Turbulence effects will enter the flow description via a model that is typically based on a combination of theory and experiment.

In the following sections, the equations for the mean flow will be described followed by an outline of various types of turbulence models. Sections on finite element implementation and boundary conditions are also included.

8.2–2 Governing Equations

A majority of researchers accept the notion that, in principle, the Navier–Stokes equations are capable of fully describing a turbulent flow [2,3]. The natural question that follows this premise is if the Navier–Stokes equations are valid, why not solve them directly (via a numerical method) to obtain the needed turbulent solution? Recall from the brief introduction that turbulent flows are inherently unsteady and three–dimensional and contain a wide spectrum of length scales. To accurately simulate such a flow, the mesh resolution would have to be on the same order as the smallest length scale. Typically, the smallest eddies in a turbulent flow are $\approx O(0.001L)$ where L is the characteristic flow dimension. This translates into a three–dimensional grid with a minimum of $\approx 10^9$ mesh points. In addition, the flow must be resolved on the shortest relevant time scale, which is again associated with the smallest eddies. The computational demands of such a direct simulation of turbulence, even for simple geometries, is generally beyond the range of current computer resources. For industrially relevant problems, in complex geometries, the direct simulation approach is clearly not possible with present technologies.

The standard alternative to the direct numerical simulation approach involves the solution to some form of *averaged* Navier–Stokes equations. In most flow problems of interest it is the mean flow that is of most concern, with the turbulent fluctuations only being important in how they influence the mean flow evolution. By performing a suitable average on the instantaneous Navier-Stokes equations, a standard mean flow problem can be derived where the effects of the turbulence are relegated to a few terms that can be *modeled.* This approach forms the basis for most of the current computational work.

To outline this approach, let the instantaneous fluid velocity and pressure fields be expressed as the sum of a mean and fluctuating component. That is

$$u_i = U_i + u_i' \qquad (8.2-1)$$

$$p = P + P' \qquad (8.2-2)$$

where the upper case represents a mean quantity and the superscript prime denotes a fluctuation. Substituting these definitions into the incompressible, viscous flow equations of Chapter 4 and performing an ensemble average [1,3] produces the following

$$\frac{\partial U_i}{\partial x_i} = 0 \qquad (8.2-3)$$

$$\rho\left(\frac{\partial U_i}{\partial t} + U_j\frac{\partial U_i}{\partial x_j}\right) = -\frac{\partial P}{\partial x_i} + \frac{\partial}{\partial x_j}\left[\mu\left(\frac{\partial U_i}{\partial x_j} + \frac{\partial U_j}{\partial x_i}\right) - \rho\overline{u_i'u_j'}\right] + \rho g_i \qquad (8.2-4)$$

The above equations are written in Cartesian component form and hence the usual summation convention is implied; the horizontal overbar indicates an averaged quantity. Equations (3) and (4) describe the behavior of the mean fluid velocity and pressure fields and are very similar in form to the laminar equations discussed extensively in Chapter 4. In arriving at (3) and (4) an ensemble average was employed, though space and time averaging could also be utilized. The various averages can be related through the ergodic hypothesis. The details and subtleties of averaging are beyond this text and can be found in many standard references [3–5].

The extra term that appears in equation (4) is often termed the Reynolds stress and represents the effects of the turbulent velocity fluctuations on the mean flow. Note that these second–order moments form the nine components of a second–order tensor, though symmetry considerations reduce this to a total of six independent unknowns.

Since no additional equations were added to those appearing in (3) and (4), the appearance of the Reynolds stress unknowns leads to the well–known closure problem for turbulence. Through appropriate manipulation of the Navier–Stokes equations six equations for the Reynolds stress variables may be derived. Unfortunately, these derived equations contain additional unknowns in the form of their higher–order moments. Similarly, equations derived for the third–order moments contain fourth–order moments. This problem of cascading unknowns is known as the closure problem. In order to achieve closure to the boundary value problem, the equation derivation must be halted at some level and a model introduced for the remaining unknowns. A number of turbulence models will be discussed in the next section.

8.2–3 General Turbulence Models

The completion of the mean flow boundary value problem described by equations (3) and (4) requires that six additional equations for the Reynolds stresses be provided. This specification constitutes a turbulence model. There are an enormous variety of turbulence models, ranging in complexity from simple algebraic statements to descriptions involving multiple, nonlinear partial differential equations. Unfortunately, there is no universal method of classification for such models which adds greatly to the confusion within the field. Here we follow the classification scheme by Ferziger [6,7], which groups turbulence models according to the following labels:

- Correlations
- Integral methods

- One point closure
- Two point closure
- Large eddy simulation
- Direct numerical simulation

In this section we will briefly outline some of the salient features of the models in each category. A subsequent section focuses on the one point closure methods which are the most prominent in current numerical work.

Correlations

The general usefulness of models based on correlations is very limited as they are available for only a few standard problem geometries that have been extensively studied by experimental methods. Though this is a relatively primitive approach, it is still used in many industrial problems. Correlations are typically developed to predict friction factors and heat transfer coefficients for various types of channel flows. Correlations generally play no role in field equation simulations.

Integral Methods

Integral methods are useful when dealing with problems that have a parabolic nature, *e.g.*, boundary layer flows. Generally, this approach utilizes a standard mean field equation that has been integrated over one coordinate to reduce the mathematical complexity of the problem. Empirical information is easily incorporated with this approach and leads to a fast and simple numerical procedure. However, these models lack generality and are limited again to those types of flows that have been extensively studied. This type of model still finds use in industry for repetitive design comuptations, but has largely been abandoned by the research community. Reviews of this type of turbulence modeling can be found in [3,8,9].

One Point Closure

One point closure models are the most popular turbulence models, particularly for computational work. By one point closure it is meant that the correlations between the fluctuating velocity components, u_j' and u_i', always employ values taken at the same physical location. Despite this restriction, the one point methods encompass a wide variety of successful models. Additional discussion of one point closure models is provided in Section 8.2-4.

Two Point Closure

Two point closure models remove the restriction of the one point closure approach and explicitly introduce the effect of length scale. This approach is still in the development stage and has been limited in application to homogeneous and isotropic turbulence. An excellent introduction to this type of closure model can be found in References 2 and 10.

Large Eddy Simulation

As the name implies, Large Eddy Simulation (LES) or Subgrid Scale Modeling (SGM) attempts to simulate the large scale motions of the flow and only model the very small scale eddies. From theory and experiment it is known that the smallest scale motions tend to be universal in their behavior and thus LES would seem to be an ideal approach for a broadly applicable turbulence model. The drawback to LES is that it is still relatively expensive because grid refinement remains a primary consideration. A good introduction to LES can be found in the works of Rogallo and Moin [11] and Yoshizawa [12].

Direct Numerical Simulation

This type of turbulence model was discussed previously in Section 8.2–2.

8.2–4 One Point Closure Turbulence Models

As noted previously, the majority of computational work, especially for industrial applications, has relied on some form of the one point closure model. Even within this category there are a wide variety of possible models and various ways to catalog them. Shown in Figure 8.2–1 is a classification scheme that will be used in the following discussion. Of the two major branches shown in the figure, we will concentrate on the so–called eddy viscosity models. The Reynolds stress models (RSM), though generally more sophisticated than the eddy viscosity approach, lead to large systems of partial differential equations and a large number of empirical parameters. The computational burden of these models, coupled with a lack of certainty regarding the parameters, makes the RSM unattractive for present engineering applications.

Returning to the eddy viscosity models, it is important to note that this approach is based on one major assumption, the Boussinesq hypothesis. By analogy with the molecular diffusion of momentum, the Boussinesq hypothesis relates the turbulent momentum transport to the gradients of the mean velocity field. The Reynolds stresses in (4) are then expressed by

$$-\rho\overline{u_i'u_j'} = \mu_T\left(\frac{\partial U_i}{\partial x_j} + \frac{\partial U_j}{\partial x_i}\right) \qquad (8.2-5)$$

where μ_T is the eddy viscosity. Unlike the molecular viscosity, μ, which is a fluid property, the eddy viscosity is a local property of the flow. When the definition in equation (5) is substituted into the momentum equation then the equations for the mean flow become

$$\frac{\partial U_i}{\partial x_i} = 0 \qquad (8.2-6)$$

$$\rho\left(\frac{\partial U_i}{\partial t} + U_j\frac{\partial U_i}{\partial x_j}\right) = -\frac{\partial P}{\partial x_i} + \frac{\partial}{\partial x_j}\left[(\mu + \mu_T)\left(\frac{\partial U_i}{\partial x_j} + \frac{\partial U_j}{\partial x_i}\right)\right] + \rho g_i \qquad (8.2-7)$$

Once the form of the eddy viscosity is specified then the mean flow can be solved in the same manner as a laminar flow since the equations are the same except for an augmented viscosity.

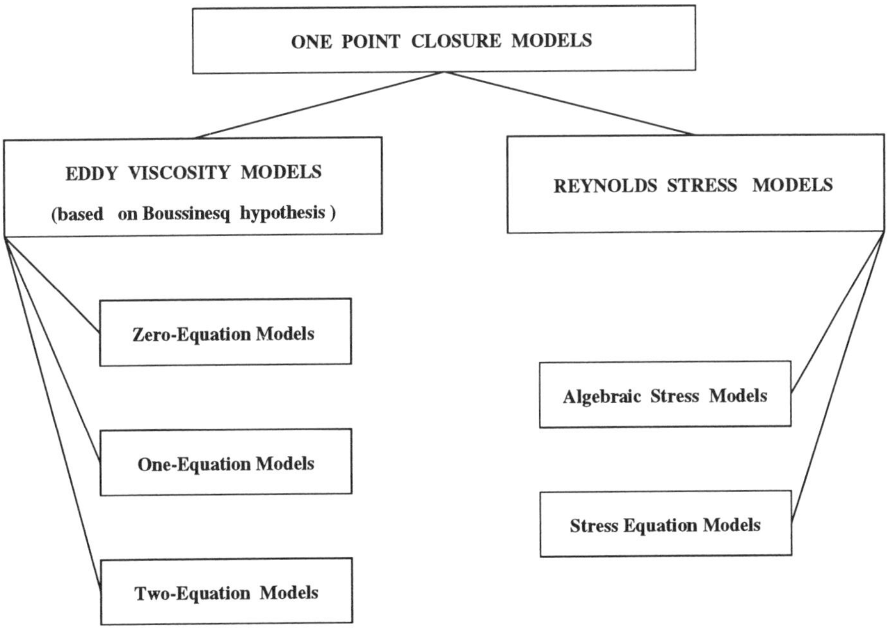

Figure 8.2–1: A classification scheme for one point closure models.

Though the turbulent flow problem has been reduced to a familiar set of partial differential equations, there remains the nontrivial task of specifying how the eddy viscosity varies with the flow field. Scaling arguments show that the eddy viscosity is proportional to a characteristic eddy velocity, u_e, and an eddy length, ℓ_e. That is

$$\mu_T \propto \rho u_e \ell_e \qquad (8.2-8)$$

In general, it is easier to specify the variation of u_e and ℓ_e for a turbulent flow, and most turbulence models use these parameters or some closely related variable. From Figure 8.2–1 the type of turbulence model being used is determined by the number of equations used to specify the variation of the variables in (8).

Zero Equation Models

These models calculate the eddy viscosity in (8) by an algebraic prescription of u_e and ℓ_e and therefore are the simplest of all the one point closure models. Most of these models are based on Prandtl's mixing length, which specifies ℓ_e to be the length scale across which turbulent mixing takes place. Prandtl gave the characteristic velocity u_e to be

$$u_e = \ell_e\left[\left(\frac{\partial U_i}{\partial x_j} + \frac{\partial U_j}{\partial x_i}\right)\frac{\partial U_i}{\partial x_j}\right]^{1/2} \tag{8.2-9}$$

which when substituted into (8) yields

$$\mu_T = \rho\ell_e^2\left[\left(\frac{\partial U_i}{\partial x_j} + \frac{\partial U_j}{\partial x_i}\right)\frac{\partial U_i}{\partial x_j}\right]^{1/2} \tag{8.2-10}$$

For a number of geometrically simple flows, the variation of the mixing length, ℓ_e, is well known and can be evaluated via simple formulas. Equation (10) then allows μ_T to be derived and the turbulence model completed. Flows that are amenable to such treatment include pipe and channel flows, jets, wakes, and boundary layers. A compilation of useful mixing length formulas can be found in references [13,14]. Though the above model is quite simple, good results can be obtained if the model is employed for the types of flows for which it is strictly intended.

One Equation Model

The limitation of the zero equation model to simple flows, and the need to simulate more complex geometries, has led naturally to the development of more sophisticated turbulence models. As a first step in this direction the algebraic specification of u_e can be replaced with a more generally applicable transport equation. Since the characteristic velocity u_e is proportional to the square root of the turbulent kinetic energy, k, then

$$u_e \propto k^{1/2} \tag{8.2-11}$$

and from equation (8)

$$\mu_T \propto \rho k^{1/2}\ell_e \tag{8.2-12}$$

A partial differential equation for k can be derived from the Navier-Stokes equation and is given by

$$\rho\left(\frac{\partial k}{\partial t} + U_j\frac{\partial k}{\partial x_j}\right) = \frac{\partial}{\partial x_j}\left(\frac{\mu_T}{\sigma_k}\frac{\partial k}{\partial x_j}\right) + \rho G - \rho\epsilon \tag{8.2-13}$$

where G is a generation term, ϵ is the turbulent dissipation, and σ_k is a constant.

Despite the apparent sophistication involving the evaluation of k, the model still relies on an algebraic specification of the mixing length. Like the zero equation model, variations in ℓ_e are problem dependent and are well characterized for only a few types of flows. The model is marginally better than the zero equation model but is not heavily used in numerical simulations.

Two Equation Model

A natural evolution of the one equation model involved the replacement of the algebraic relation for mixing length with a second transport equation. Dimensional arguments lead to the proportionality

$$\ell_e \propto \frac{k^{3/2}}{\epsilon} \qquad (8.2-14)$$

where ϵ is the viscous dissipation. Substituting (14) and (11) into equation (8) produces the proportionality

$$\mu_T \propto \rho \frac{k^2}{\epsilon}$$

or the Kolmogorov–Prandtl relation

$$\mu_T = C_\mu \rho \frac{k^2}{\epsilon} \qquad (8.2-15)$$

which relates the eddy viscosity directly to the turbulence variables, k and ϵ. In the two equation model, the turbulent kinetic energy is given by equation (13)

$$\rho\left(\frac{\partial k}{\partial t} + U_j \frac{\partial k}{\partial x_j}\right) = \frac{\partial}{\partial x_j}\left(\frac{\mu_T}{\sigma_k}\frac{\partial k}{\partial x_j}\right) + \rho G - \rho\epsilon \qquad (8.2-16)$$

and the turbulent dissipation is described by an equation of similar form

$$\rho\left(\frac{\partial \epsilon}{\partial t} + U_j \frac{\partial \epsilon}{\partial x_j}\right) = \frac{\partial}{\partial x_j}\left(\frac{\mu_T}{\sigma_\epsilon}\frac{\partial \epsilon}{\partial x_j}\right) + \rho\frac{\epsilon}{k}\left(c_1 G - c_2 \epsilon\right) \qquad (8.2-17)$$

where G is again a shear generation term and c_1 and c_2 are empirically derived constants as are σ_k and σ_g. Note that (16) and (17) are obtained from manipulation of the instantaneous momentum equation. However, to provide closure, certain terms, such as G, are modeled by relating their exact form to some function of the mean flow.

The two equation, $k-\epsilon$ model described by equations (16) and (17) can be used in conjunction with the mean flow equations and the definition of μ_T given by equation (15), to arrive at a continuum description of turbulent flow. The equation set is highly nonlinear, with a strong coupling between the various transport equations. However, this additional mathematical and computational complexity is offset by a substantial increase in realism and reliability of the turbulence predictions. The $k - \epsilon$ model is far from universal and has a number of weaknesses, though it remains one of the most heavily used methods for flow simulation. A good discussion of this model can be found in Reference 15.

8.2–5 Turbulent Heat Transfer Models

The previous sections have been limited to consideration of isothermal, turbulent flows though the extension to nonisothermal flows is of considerable engineering importance. Most of the ideas and models for isothermal flows have a direct counterpart when the energy equation is included in the formulation. Here we will outline some of the most important concepts for the case of forced convection heat transfer; the theory for strongly coupled, buoyancy driven flows is somewhat less developed.

Extending the assumption given in equations (1) and (2), the instantaneous temperature can be written as

$$\theta = \Theta + \theta' \qquad (8.2-18)$$

where Θ is the average or mean temperature and the primed quantity is the fluctuation. When this definition and equation (1) are substituted into the instantaneous energy equation and ensemble averaged, the result is an equation of the form

$$\rho C\left(\frac{\partial\Theta}{\partial t} + U_j\frac{\partial\Theta}{\partial x_j}\right)) = \frac{\partial}{\partial x_j}\left(k\frac{\partial\Theta}{\partial x_j}\right) - \rho C\overline{u_j'\theta'} \qquad (8.2-19)$$

The averaging process has introduced the three new variables $\overline{u_j'\theta'}$ which are the components of the Reynolds heat flux vector. These quantities can be modeled in a variety of ways, all of which are analogous to the methods used for the Reynolds stresses. The current status of turbulent heat transfer modeling is reviewed by Launder [16]. Considering only one point closure models, the Reynolds heat flux can be related to the mean temperature field through a Boussinesq type hypothesis. That is, let

$$-\rho C\overline{u_j'\theta'} = \frac{k_T}{C}\frac{\partial\Theta}{\partial x_j} = \frac{\mu_T}{\mathrm{Pr}_T}\frac{\partial\Theta}{\partial x_j} \qquad (8.2-20)$$

which can be used in (19) to produce

$$\rho C\left(\frac{\partial\Theta}{\partial t} + U_j\frac{\partial\Theta}{\partial x_j}\right) = \frac{\partial}{\partial x_j}\left[\left(k + \frac{\mu_T}{\mathrm{Pr}_T}\right)\frac{\partial\Theta}{\partial x_j}\right] \qquad (8.2-21)$$

where k_T is the eddy conductivity and Pr_T is the turbulent Prandtl number. Once the form of the eddy conductivity or the turbulent Prandtl number is specified, then the energy equation in (21) can be solved in conjunction with the flow equations.

Methods for specifying the eddy conductivity or turbulent Prandtl number parallel the ideas for the eddy viscosity specification. Of all the possible approaches, one of the most popular is the specification of Pr_T as an empirical relation. Reynolds [17] describes approximately thirty different formulas for Pr_T, all of which have a limited applicability. Clearly, the modeling of turbulent heat transfer processes remains an active area of research, though its progress is directly related to progress made in turbulent flow modeling.

8.2–6 Finite Element Modeling of Turbulence

Once a turbulence model has been selected for a particular application, it is important to consider how a numerical technique, such as the finite element method, would be used to produce useful flow simulations. In previous chapters, various solution methods for the mean flow equations such as (6), (7), and (21) were described in detail and need not be reconsidered here. However, it is appropriate to outline some of the added complications that must be addressed when a turbulence model is added to the mean flow system. To limit the discussion, only the zero equation and two equation, $k - \epsilon$, forms of the one point closure models will be considered. These two approaches are the ones most commonly encountered in practical applications.

Zero Equation Model

The finite element implementation of a zero equation turbulence model is straightforward since the eddy viscosity is prescribed by a simple algebraic relation, such as equation (10). Since the eddy viscosity varies over the flow domain, the methods introduced in Chapters 4 and 6 for modeling variable coefficients may be utilized. From equation (10) it is apparent that the turbulent viscosity depends on velocity gradients, which is directly analogous to the behavior of inelastic, non–Newtonian fluids. For the best accuracy, the velocity gradients should be evaluated at element integration points. The dependence of μ_T on the mean field velocity produces another source of nonlinearity in the numerical problem, though this is not of major significance for most iterative solution algorithms.

The most difficult aspect of these models comes from the specification of the mixing length over the flow domain. These empirical formulas will vary from region to region as different types of flows (*e.g.*, boundary layers, shear layers, jets) are encountered. Also, many of the ℓ_e functions depend on distance from a boundary, a dependence that may be awkward to specify easily and uniquely in a complex geometry. Despite these inconveniences the simplicity of the model makes it a popular and useful choice for modeling.

Two Equation Model

The finite element implementation of a $k - \epsilon$ model is also conceptually straightforward since the turbulent transport equations (16) and (17) are of the familiar advection–diffusion type discussed in Chapter 5. A weak form of the $k - \epsilon$ system can be readily constructed, and when coupled with finite element interpolations for the unknowns, a standard set of coupled ordinary differential equations can be defined. The discretized forms of (16) and (17) will be highly nonlinear and strongly coupled to the mean flow equations. It is obvious that the size of the computational problem for a two equation turbulent simulation has increased substantially over its laminar counterpart. Not so obvious is the fact that the numerical difficulties have also increased significantly.

The major problems that occur with the $k - \epsilon$ model come from two sources: the nonlinear behavior of the partial differential equations and boundary conditions for the turbulent transport and mean field equations. It has been observed by a number of investigators that the dissipation equation may cause instabilities in the flow simulation that lead to a poor or nonconvergence of the numerical solution algorithm. Another manifestation of these stability problems is the prediction of negative values for both k and ϵ [15,18–21]. This nonphysical occurrence is sometimes attributed to the inaccurate modeling of the source terms for k and ϵ. The practical solution, in many cases, involves a clipping procedure in which negative values are replace by small positive values [15].

The sensitive nature of the $k-\epsilon$ model leads to some practical recommendations for solution methods in turbulent flows. Generally, it has been found advantageous to employ Picard iteration methods as opposed to the higher–order Newton schemes. Though the convergence rate is reduced in this case, the overall stability of the algoritm is improved. Relaxation factors greater than 0.5 have also been observed to substantially enhance the solution process. Also related to the solution stability is the grading of the computational mesh. Computations that have abrupt changes in mesh density have been reported to show poor behavior in the k and ϵ variables and in some cases can lead to divergence of the solution process. In monitoring the convergence of a turbulent flow simulation it is noteworthy that the velocity and pressure fields converge first; convergence of the k and ϵ fields is very slow and may not be monotonic.

Many of the above experiences were encountered in the solution of the submarine flow problems described in Chapter 5. Details of the $k - \epsilon$ model and its behavior during the solution process are available in reference [22].

The topic of boundary conditions for the $k - \epsilon$ model and the modeling of turbulence close to solid boundaries is quite involved and will only be summarized here. The $k-\epsilon$ model described in equations (16) and (17) is known as a high Reynolds number model, since its derivation was predicated on a single length scale associated with an unbounded, fully developed flow. Close to solid boundaries, the turbulent nature of the flow changes to a low Reynolds number model, where the small eddies associated with dissipation become important. Of course, very near the wall a thin viscous sublayer exists in which a turbulence model is no longer appropriate. From a computational viewpoint these near–wall regions are extremely challenging, since not only does the type of turbulence model vary with distance from the boundary, but all of the flow variables experience their greatest variation in this region. In most cases, attempts to solve the mean flow and $k - \epsilon$ equations all the way to the wall would be prohibitively expensive, especially for complex three–dimensional geometries.

An alternative to the above approach was developed at Imperial College [23,24] and relies on the use of universal models or wall laws to empirically estimate the flow behavior in the viscous sublayer and the buffer layer. The development of such a model allows the computational domain to be displaced from the wall and located in a fully turbulent region where the $k - \epsilon$ model is valid. Boundary conditions for the mean flow variables, k and ϵ, at the edge of the computational domain are provided by the wall laws. A wall law is typically a "universal" velocity profile that is derived from a combination of theory and experimental observation; standard profiles for k and ϵ near a boundary can also be derived. These parameterized profiles provide values of the mean field shear stress, k and ϵ that can be directly applied as boundary conditions to the computational domain. A good description of wall laws and modeling is available in [15,24]. The major disadvantage of this approach is that the computational domain must be offset from the wall by a distance that depends on the wall law and the flow field. The location of the domain may therefore have to be altered during or after the solution process. Despite this difficulty, the wall law approach is the only practical way to model turbulent flows in the vicinity of a boundary.

An improvement to the wall law formulation for finite element applications was developed in conjunction with the FIDAP code [25]. A standard representation of a universal velocity profile (wall law) consists of a linear profile for the viscous sublayer, a log–linear profile in the buffer region, and a logarithmic profile in the fully developed turbulent region. It is possible to develop an interpolation or shape function that represents this composite profile and therefore construct a special finite element that represents the wall law region of a turbulent flow. Though the problem with proper location of the edge of the special wall element remains, the development of such an element solves many other problems involving mesh generation and the specification of boundary conditions on complex surfaces. The precursor to this work, and some of the first work on specialized elements for fluid mechanics, can be found in References 19 and 26.

Though the primary discussion has been related to boundary conditions at solid walls, it must be remembered that the $k - \epsilon$ equations require boundary values to be specified on all computational boundaries. Especially troublesome in this regard are k and ϵ values for inlet boundaries, since to a large extent these values will set the turbulence level for the entire flow. Haroutunian [15] provided a method for consistently generating these values.

8.3 CHEMICALLY REACTING SYSTEMS

8.3–1 Introduction

The inclusion of chemical reactions in a fluid or thermal problem adds considerably to the difficulty of the numerical simulation. However, such effects are central to many important technology areas. The majority of these applications involve combustion or other gas phase processes that are described by compressible flows and are therefore outside the scope of this book. Chemical reactions in incompressible flows are often associated with gelation or polmerization reactions; some low–speed gas flows, such as chemical vapor deposition processing, are frequently modeled as incompressible. Thermal diffusion problems coupled with chemical kinetics are important for thermal ignition studies in energetic materials (*e.g.*, pyrotechnics and explosives) and in the simulation of material decomposition or ablation. In the following sections we will outline the general form at the continuum equations that describe a chemically reactive system, make these equations specific for a few particular applications, and comment on some of the difficulties associated with numerical computation.

8.3–2 Governing Equations

For a chemically reactive fluid, the boundary value problem described in Chapter 1 is altered by the fact that the incompressible fluid must now be considered a mixture of various chemical species. Conservation principles must now be invoked for the individual constituents of the fluid as well as for the fluid mixture as a whole. The overall mass, momentum, and energy balances for the mixture remain as given in Chapter 1. That is

$$\frac{\partial u_i}{\partial x_i} = 0 \tag{8.3 – 1}$$

$$\rho\left(\frac{\partial u_i}{\partial t} + u_j\frac{\partial u_i}{\partial x_j}\right) = -\frac{\partial P}{\partial x_i} + \frac{\partial}{\partial x_j}\left[\mu\left(\frac{\partial u_i}{\partial x_j} + \frac{\partial u_j}{\partial x_i}\right)\right] + \rho g_i \tag{8.3 – 2}$$

$$\rho C\frac{\partial T}{\partial t} + \rho C u_j\frac{\partial T}{\partial x_j} = \frac{\partial}{\partial x_i}\left(k\frac{\partial T}{\partial x_i}\right) + Q_R \tag{8.3 – 3}$$

In equations (1)–(3) the thermophysical properties will normally be strong functions of the chemical composition of the fluid. Assume that the fluid is composed of I individual species and that these species may react according to J reactions. The primary effect of the reaction process on the flow is to alter the internal energy of the system; this change in energy provides the volume source term Q_R in equation (3).

A mass conservation equation for each of the I species is required and these have the form

$$\frac{\partial[N_i]}{\partial t} + u_j\frac{\partial[N_i]}{\partial x_j} = \frac{\partial}{\partial x_j}\left(D_{jk}\frac{\partial[N_i]}{\partial x_k}\right) + R_i \quad \text{for } i = 1, 2, \ldots, I \tag{8.3 – 4}$$

where $[N_i]$ is the concentration variable (mole fraction) for species i, D_{jk} is the diffusion tensor and R_i is the chemical reaction rate. Note that there is no summation on i in Equation (4). In writing (4) an incompressible flow was assumed and the diffusion process was assumed to follow Fick's law. More general species conservation equations can be derived that account for other types of diffusion processes [27,28], but these need not be considered here.

To describe the details of the chemical reaction, the stoichiometry, reaction kinetics, and material property behavior must be specified. For a material with I species and J reactions, the stoichiometry is provided by

$$\sum_{i=1}^{I}\nu'_{ij}\mathcal{M}_i \rightarrow \sum_{i=1}^{I}\nu''_{ij}\mathcal{M}_i \quad \text{for } \; j = 1, 2, \ldots, J \tag{8.3 – 5}$$

where ν'_{ij} and ν''_{ij} are stoichiometric coefficients (usually integer values) and $\mathcal{M}_i$ is the chemical symbol for the ith species. Generally, these expressions are given as reversible reactions; however, they are treated here as irreversible, and the reversed reactions are specified as additional reaction steps. To accommodate expressions for global reactions, the stoichiometric coefficients are allowed to be non–integer.

For each step of the reaction, a reaction rate, r_j, is defined in the form:

$$r_j = k_j(T) \prod_{i=1}^{I} [N_i]^{\mu_{ij}} \quad \text{for} \quad j = 1, 2, \ldots, J \tag{8.3-6}$$

where μ_{ij} are the concentration exponents (usually $\mu_{ij} = \nu'_{ij}$ in kinetic theory, but here they are treated independently) and $\prod$ indicates the product operation. Typically, the expressions for the kinetic coefficients, $k_j(T)$, are given in an Arrhenius form

$$k_j(T) = T^{\beta_j} A_j \; exp(-E_j/RT) \tag{8.3-7}$$

where β_j is the coefficient for a steric factor, A_j is the pre–exponential factor, E_j is the activation energy, and the universal gas constant is R. It is convenient to define $\nu_{ij} = (\nu''_{ij} - \nu'_{ij})$ and thus the rate of change of the species (neglecting diffusion) is given as

$$R_i = \sum_{j=1}^{J} \nu_{ij} r_j \quad \text{for} \quad i = 1, 2, \ldots, I \tag{8.3-8}$$

The chemical reaction process is coupled directly to the energy equation (3) by the volumetric source term

$$Q_r = \sum_{j=1}^{J} q_j r_j \tag{8.3-9}$$

where q_j represents the known endothermic or exothermic energy release for reaction step j.

The material properties for the mixture are usually represented as mole fraction weighted averages of the I constituents. That is

$$\begin{aligned} (\rho C) = (\rho C)_{mix} &= \sum_{i=1}^{I} [N_i](\rho C)_i \\ \mu = (\mu)_{mix} &= \sum_{i=1}^{I} [N_i](\mu)_i \\ k = (k)_{mix} &= \sum_{i=1}^{I} [N_i](k)_i \end{aligned} \tag{8.3-10}$$

where the constituent properties could still be functions of the temperature or other variable.

The equations described above include a wide spectrum of multistep chemical reactions. For many flow or heat transfer problems such generality is not required and the chemical process can be simplified. For a polymerization or curing reaction, or in fact any binary mixture problem, the set of mass conservation equations for the species reduces to a single equation

$$\frac{\partial \alpha}{\partial t} + u_j \frac{\partial \alpha}{\partial x_j} = \frac{\partial}{\partial x_j}\left(D \frac{\partial \alpha}{\partial x_j}\right) + R \qquad (8.3-11)$$

where α is now integrated as a mass fraction or extent of reaction variable that varies from 0 to 1. Equation (11) describes the simple process of a material being transformed from species A to species B with an accompanying change in energy. The reaction ratio R in (11) would typically be of the Arrhenius type. In a fluid system the heat release from the reaction could cause a buoyancy–induced flow and thus couple the flow problem to (11). Such a problem was described in Chapter 5 in the context of an epoxy solidification example.

In the case of a thermal conduction problem, the elimination of the fluid velocity allows additional simplification of the species equations by elimination of the advection term in (4). Also, for most solids, the species diffusion is negligibly small and the conservation equations in (4) can be reduced to a set of ordinary differential equations

$$\frac{d[N_i]}{dt} = R_i = \sum_{j=1}^{J} u_{ij} v_j \qquad (8.3-12)$$

The equations in (12) must be solved in conjuction with the standard heat conduction equation and the appropriate definitions for the heat source and kinetic parameters. Studies of reactive systems of this type have been performed by Baer *et al.* [29], McGuire and Tarver [30], and others.

8.3–3 Finite Element Modeling of Chemical Reactions

The development of a finite element model for chemically reactive flows follows the usual prescription. The continuity, momentum, and energy equations for the mixture are standard and were considered in Chapters 4 and 5. The conservation equations for the species are of the advection-diffusion type and can therefore be treated by the same methods as used for the energy equation. Recall that the auxiliary equations defined in Chapter 1 could be used directly as species equations. The discretized form of the species equations provides an additional set of I matrix equations that can, in principle, be added to the matrix system for the basic nonisothermal flow problem.

Though the development of the reactive flow model apears to be straightforward, the solution of the resulting equations is extremely difficult. First, from the standpoint of problem size, the addition of multistep chemistry leads to a very rapid growth in the number of degrees of freedom considered in any given problem. However, problem size alone is not the most significant difficulty. The equation set is highly coupled and very nonlinear, making the convergence by standard solution methods unpredictable. More importantly, the length and time scales for the chemical processes tend to be significantly different (smaller) than the scales for the fluid and thermal response. This disparity in scales makes the use of fully coupled solution procedures impractical. In mathematical terms, the equation system becomes very stiff. The most effective solution algorithms allow the flow problem to run on its natural time scale, while the chemistry solution is decoupled and run as a subcycled process on a shorter time scale. Obviously, some of the rigorous coupling between variables is destroyed by this technique and variable coefficients must be lagged or predicted and corrected. Despite the drawbacks, this type of approach is the most widely used for problems with "difficult" chemistry.

Chemical reactions also tend to produce large gradients in the dependent variables over small distances. Since the location, extent, and movement of such reaction zones is not known *a priori*, excessively refined meshes are often employed to ensure spatial accuracy. Again, this leads to a large problem size and excessive computer resources. A more efficient solution to the length scale problem is the use of adaptive meshes. An example of this technique for a reactive problem can be found in reference [29].

8.4 ELECTROMAGNETIC FIELDS

8.4–1 Introduction

Problems involving the coupling of electromagnetic (EM) fields with fluid and thermal transport have a broad spectrum of applications ranging from astrophysics to manufacturing. Here we will limit the discussion to incompressible fluids that are good electrical conductors. This eliminates many of the interesting problems involving plasmas, and concentrates the applications in the area of liquid metal flows. In many of these problems an applied or induced magnetic field provides an additional body force to the fluid which results in a convective motion. For high current applications resistive heating may also be important as a volumetric energy source.

In the following section an outline of the field equations for electromagnetics is given along with their coupling to the nonisothermal, viscous flow problem. Subsequent sections show how the EM problem is redefined in terms of potential functions that are more suitable for finite element approximation. A final section describes some of the numerical issues associated with EM field simulation. A good introduction to coupled fluid–EM problems is available in [31]; general EM field theory is available in texts such as the one by Jackson [32]. Finite element models for EM applications are well covered in the texts by Jin [33], Sadiku [34], and Binns *et al.* [35].

8.4–2 Maxwell's Equations

The appropriate mathematical description of electromagnetic phenomena in a conducting material region, Ω_C, is given by Maxwell's equations. In rational MKSA notation these equations are expressed as (see [31,32])

$$\nabla \times \mathbf{E} = -\frac{\partial \mathbf{B}}{\partial t} \tag{8.4-1}$$

$$\nabla \times \mathbf{H} = \mathbf{J} + \frac{\partial \mathbf{D}}{\partial t} \tag{8.4-2}$$

$$\nabla \cdot \mathbf{B} = 0 \tag{8.4-3}$$

$$\nabla \cdot \mathbf{D} = \rho \tag{8.4-4}$$

where the field variables are the electric field intensity, **E**, the magnetic field intensity, **H**, the magnetic flux density, **B**, the electric flux (displacement) density, **D**, the conduction current density, **J**, and the source charge density, ρ. A continuity condition on the current density is also defined by

$$\nabla \cdot \mathbf{J} = \frac{\partial \rho}{\partial t} \tag{8.4-5}$$

To complete the formulation, the constitutive relations for the material are required. The fluxes are related to the field variables by

$$\mathbf{D} = \epsilon \cdot \mathbf{E} \tag{8.4-6}$$

$$\mathbf{B} = \mu \cdot \mathbf{H} \quad \text{or} \quad \mathbf{H} = \nu \cdot \mathbf{B} \tag{8.4-7}$$

$$\mathbf{J} = \sigma \cdot \mathbf{E} + \sigma \cdot (\mathbf{u} \times \mathbf{B}) \tag{8.4-8}$$

where ϵ is the permittivity of the material, μ is the permeability, ν is the reluctivity, and σ is the conductivity. In equation (8) the second term gives the current induced by the motion of a conductor (*e.g.*, fluid) with velocity **u** in the presence of a magnetic field. For many problems of interest this term is identically zero or relatively small. Note also that in general the material properties are tensorial in nature and may be field dependent and anisotropic. The above equations have been written for the case of a conducting medium; the system can be simplified for the case of free space.

The coupling of electromagnetic fields with a fluid or thermal problem occurs through the dependence of material properties on EM field quantities and the production of EM–induced body forces and volumetric energy production. The Lorentz body force in a conductor due to the presence of electric currents and magnetic fields is given by

$$\mathbf{F}_B = \mathbf{J} \times \mathbf{B} \tag{8.4-9}$$

where in the general case the current is defined by equation (8). The energy generation or Joule heating in a conductor is described by

$$Q_J = \mathbf{J} \cdot \mathbf{E} \tag{8.4-10}$$

which takes on a more familiar form if the simplified (**u**=0) form of (8) is used to produce

$$Q_J = \frac{\mathbf{J} \cdot \mathbf{J}}{\sigma} \qquad (8.4-11)$$

The above forces and heat source occur in the fluid momentum and energy equations, respectively.

Quasi–Static Approximation

For good conductors, the conduction current, **J**, is large compared to the displacement current, **D**, for most frequencies of interest. Neglecting the displacement current allows Ampere's law (2) to be simplified and Coulomb's law (4) to be omitted. Also, the continuity relation is simplified since the divergence of $\nabla \times \mathbf{H}$ is zero by an identity. The omission of the displacement current is termed a quasi–static approximation since the propagation of electromagnetic waves is precluded. Under this assumption Maxwell's equations for a conducting region become

$$\nabla \times \mathbf{E} = -\frac{\partial \mathbf{B}}{\partial t} \qquad (8.4-12)$$

$$\nabla \times \mathbf{H} = \mathbf{J} \qquad (8.4-13)$$

$$\nabla \cdot \mathbf{B} = 0 \qquad (8.4-14)$$

The continuity condition is

$$\nabla \cdot \mathbf{J} = 0 \qquad (8.4-15)$$

the required constitutive relations are

$$\mathbf{B} = \mu \cdot \mathbf{H} \quad \text{or} \quad \mathbf{H} = \nu \cdot \mathbf{B} \qquad (8.4-16)$$

$$\mathbf{J} = \sigma \cdot \mathbf{E} + \sigma \cdot (\mathbf{u} \times \mathbf{B}) \quad \text{or} \quad \mathbf{J} = \sigma \cdot \mathbf{E} \qquad (8.4-17)$$

8.4–3 Electromagnetic Potentials

For many static and quasi–static applications it is usual to introduce a set of potential functions to represent the electric and magnetic field variables and reduce the number of partial differential equations requiring solution. Two basic systems of potentials may be considered: (a) the electric scalar potential, V, and a magnetic vector potential, **A**, and (b) the electric vector potential, **T**, and the scalar magnetic potential, ψ. The **T**–ψ formulation is of limited value for general analyses since there are significant difficulties in representing multiply–connected domains. Though the **A**–V formulation generally leads to a larger number of differential equations, it is preferred for numerical computation due to its complete generality.

From Gauss' relation $\nabla \cdot \mathbf{B} = 0$, it follows that $\mathbf{B}$ is derivable from a vector potential. By definition, then

$$\mathbf{B} = \nabla \times \mathbf{A} \tag{8.4-18}$$

where $\mathbf{A}$ is the magnetic vector potential. In addition, from Faraday's law (12), one has

$$\nabla \times \mathbf{E} = -\frac{\partial \mathbf{B}}{\partial t} = -\frac{\partial(\nabla \times \mathbf{A})}{\partial t} \tag{8.4-19a}$$

Rearranging the above equation, we obtain

$$\nabla \times \left(\mathbf{E} + \frac{\partial \mathbf{A}}{\partial t}\right) = 0 \tag{8.4-19b}$$

For a scalar V, the vector identity, $\nabla \times \nabla V = 0$ holds and allows the definition

$$-\nabla V \equiv \mathbf{E} + \frac{\partial \mathbf{A}}{\partial t}$$

or

$$\mathbf{E} = -\nabla V - \frac{\partial \mathbf{A}}{\partial t} \tag{8.4-20}$$

The scalar function V is known as the *electric potential.* The definitions in (18) and (20) may be used in the appropriate forms of Ampere's law and the current continuity equation to produce the needed field equations for $\mathbf{A}$ and V in conductors and in free–space regions. Using equations (16)–(18) and (20) in (13), one obtains

$$\nabla \times [\nu \cdot (\nabla \times \mathbf{A})] = -\sigma \cdot \frac{\partial \mathbf{A}}{\partial t} - \sigma \cdot \nabla V \tag{8.4-21}$$

Also, in the conduction regions the electric field is described by the continuity equation (15) which is rewritten in terms of $\mathbf{A}$ and V by the use of equations (17b) and (20). That is

$$\nabla \cdot \left(-\sigma \cdot \nabla V - \sigma \cdot \frac{\partial \mathbf{A}}{\partial t}\right) = 0 \tag{8.4-22}$$

Equations (21)–(22) describe the general, quasi–static electromagnetic field problem for free space and conductors in terms of the magnetic vector potential, $\mathbf{A}$, and electric scalar potential, V. These equations are appropriate for solution by the finite element method.

8.4–4 Finite Element Models for EM Fields

The potential equations in (21) and (22) represent the three components of the vector magnetic potential and the scalar electric potential. Using the standard method of weighted residual techniques these equations can be converted to appropriate weak forms and utilized with a finite element approximation. Details of such a derivation are available in [33,35]. The result of this process is a large set of coupled, nonlinear ordinary differential equations. For AC currents a harmonic approximation can be used to eliminate the time dependence in favor of an equation system that is complex. Time–dependent DC problems can be integrated using standard methods for diffusion equations. Steady–state simulations would normally be solved via an iterative algorithm such as Picard's method or Newton's method. At the conclusion of an EM simulation, currents, forces, and volume heat sources can be derived from the potential fields for use in the flow model. It is usual for the EM and flow problem to be solved sequentially since the coupling usually is not too strong. Exceptions, however, do occur, in which case the fully coupled problem represents a significant computational challenge. Examples of strongly coupled applications include metal melting (or remelting) by high current vacuum, arc processing, and induction hardening of metal surfaces.

Though many aspects of the finite element models of EM problems seem very standard there are a few areas of continuing difficulty. The use of the magnetic vector potential leads to a problem of uniqueness in the computed solution. Because of the definition of $\mathbf{A}$, the curl of $\mathbf{A}$ (*i.e.*, $\nabla \times \mathbf{A}$) is uniquely defined, but $\mathbf{A}$ itself is only defined up to the gradient of an arbitrary scalar function. Typically, this arbitrariness in $\mathbf{A}$ is resolved by defining the divergence of $\mathbf{A}$. The incorporation of $\nabla \cdot \mathbf{A}$ as a constraint, termed gauging, may be accomplished in any of several ways. The Coulomb gauge is one particular choice that has found extensive use in numerical simulations. In this case, the potential is made unique by enforcing

$$\nabla \cdot \mathbf{A} = 0 \tag{8.4-23}$$

Other choices for the gauge condition, such as the Lorentz gauge, select a nonhomogeneous form for equation (23). Note that in some cases of reduced dimensionality the explicit use of (23) is not required, since $\mathbf{A}$ is automatically divergence free. The actual implementation of (23) in a numerical method may take any of several forms, including modification of the field equations, penalty methods, projection methods, and the construction of divergence free basis functions. Like the analogous divergence free velocity constraint for incompressible flow, each of the above gauge methods incurs a computational penalty either in terms of additional work or the modification of the equation system to a less desirable form.

8.5 PARALLEL PROCESSING

8.5–1 Introduction

Parallel processing is a relatively new facet of computing, though it promises to have a major impact on computational mechanics. Most of the current computing systems are based on the sequential processing, von Neumann architecture, in which instructions are executed in sequence, one at a time, on a single data element. For computational algorithms that have sequential character, this type of processing is optimal. However, most computational problems are not strictly sequential, but have processes that are independent of each other, and could therefore be executed simultaneously. For these parallel algorithms, sequential processing is obviously not optimal. In the past, increases in computer performance have been primarily achieved through increases in speed of the central processing unit (CPU). However, as the rate of increase in CPU speed slows down, and costs for complex cpu's increase, computer developers have turned to alternative architectures to achieve high computing performance. A logical choice is to use multiple processors and exploit the parallelism inherent in most numerical algorithms. This hardware development has lead to the rapidly growing field of parallel computing.

8.5–2 Terminology

Classification

The standard terminology often used for classifying parallel computers is the scheme proposed by Flynn [36]. Even though this scheme does not include many new developments in classification [37], it still forms a good basis. The four classes in the scheme proposed by Flynn are

1. Single Instruction Single Data (SISD)
2. Single Instruction Multiple Data (SIMD)
3. Multiple Instruction Single Data (MISD)
4. Multiple Instruction Multiple Data (MIMD).

An instruction stream is the sequence of instructions executed by the processing element (PE) and the data stream is the sequence of data on which instructions are performed.

The traditional sequential computers (Micro- and Mini computers) fall under the class of the SISD system. They have a single processing element executing instructions in sequence on a single stream of data. The SIMD systems have many processing elements. They execute the same set of instructions in each of these processing elements but on different data streams. Vector computers fall under this class. On the other hand, MISD computers execute a different set of instructions in each processing element but on the same data stream. These systems are useful in application areas like signal processing and are often termed pipeline processors. The fourth category, MIMD machines, is the most general type among the parallel processing systems. They have many processing elements executing their own set of instructions on different data streams.

Parallel processing systems can be further classified based on the following characteristics. Most of the parallel machines of interest in computational mechanics come under the MIMD or SIMD class of computers. The classification discussed here is relevant primarily to these systems.

Granularity of the processing elements

Granularity of the system can be either fine or coarse, where a coarse grain system has fewer processors than the fine grain system. A coarse grain system typically operates with a few (say, 100) powerful processing elements, while a fine grain system would usually has many (say, 1000) ordinary processing elements. The boundary between fine and coarse grain computers is not rigidly defined.

Topology of interconnections

The topology of interconnection is the manner in which the individual processing elements are connected. In terms of hardware, the connections may be arranged as a ring, tree, pipeline, or a hypercube. With the advent of new systems with a virtual communication facility, the programmer can assume that each processor is connected to every other processor. The system will take care of the data communication through the intermediate processors based on the actual physical topology. Efficiency of such a system will be high if the actual layout and virtual layout are identical. Figure 8.5–1 shows the commonly used interconnect topologies.

Distribution of control across the processing elements

The control of the system may be with a single processor, which directs the system, as is often the case in SIMD machines. Also, in MIMD machines, even though a master(or main) processor is identified, each processor operates on its own, independently. However, interprocessor communication demands synchronous operation.

Memory access

Each processing element can have its own local memory and/or share the global memory of the system. Both types of architectures have advantages and disadvantages. In a local memory system, synchronous communication becomes vital and deadlocks/delays in communication may cause drastic deterioration of the system efficiency. In a shared memory system, processing elements compete with each other in addressing the memory locations. Bottlenecks in communication remain one of the major problems in a shared memory system. A detailed discussion of these characterstics is available in References 36–39.

Languages

Special programming languages are required to effectively program an algorithm for a parallel machine. A parallel programming language should, at a minimum, offer the following features:

1. Commands to identify the parallel and sequential parts of the code.
2. Tools to configure the system: specification of the jobs allocated to each processing element and the communication route.

Currently parallel versions of FORTRAN and C are popular for general scientific programming. For systems based on transputers [39], the language of choice is Occam, which was developed specifically for this type of processor [39,40]. However, these systems do support parallel FORTRAN and C.

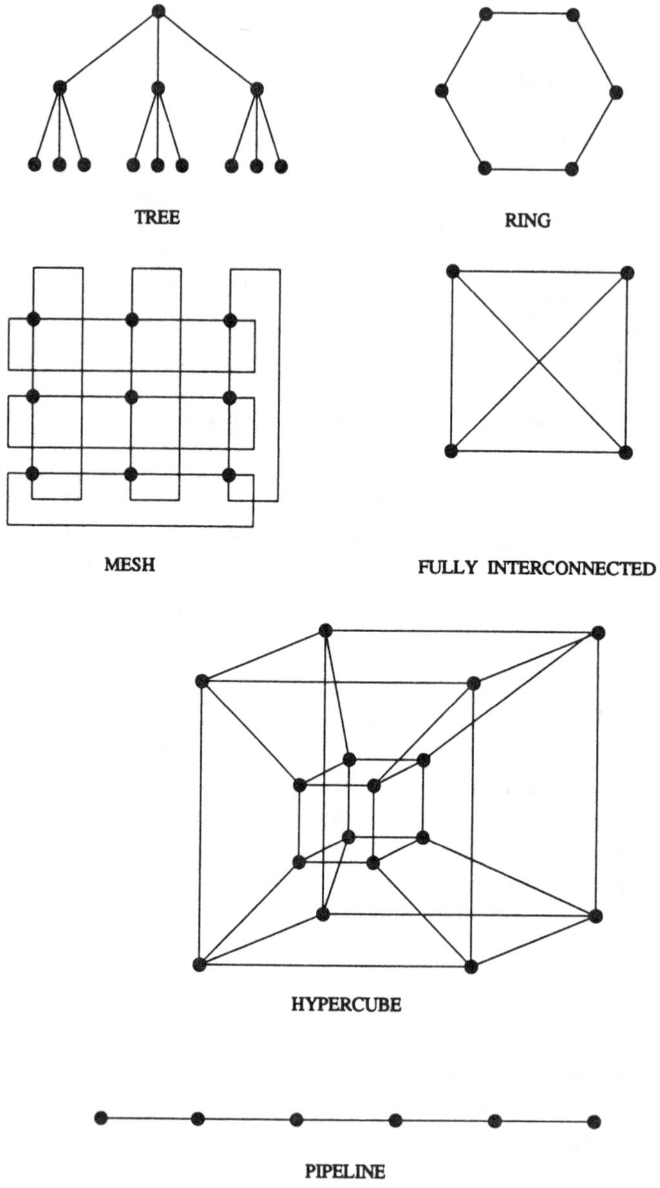

Figure 8.5–1: Topology of interconnections.

Generally a parallel program consists of two separate programs. One having the instructions to be executed on different processors and other specifying the configuration. For instance, a parallel code written in Occam will have instructions to be executed on the master transputer in one file, while the configuration and instructions to be executed on the slave transputers are located in a separate file. This may vary depending on the supporting software available in the system.

Rating

The performance of a parallel code is generally estimated its parallel efficiency, which will indicate how much is gained by solving the problem in parallel. The best way to estimate this measure is to compare it with a sequential version of the same code. This can be done in two different ways [41].

Algorithmic efficiency

This is defined as the the ratio of CPU time taken for running the parallel code on N_p processing elements to the time taken when the code is run on a single processing element. The sequential code used here is essentially a parallel code run on a single processor. Hence the sequential code will also have the delays in communication reflected in its timing. This will indicate the parallel efficiency of the algorithm rather than the problem. The formula for algorithmic efficiency is

$$\eta_{alg} = \frac{t_{par}}{t_{seq1} N_p} 100 \qquad (8.5-1)$$

where t_{par} is the time taken to execute the program on a parallel machine with N_p processors and t_{seq1} is the time taken to run the same code on a single processor.

Actual/Beneficial Efficiency

Another measure, the actual or beneficial efficiency is defined as the ratio of CPU time taken for running the parallel code on N_p processing elements to the time taken by the fastest sequential code run on one processing element (t_{seq2}). This efficiency will reflect the parallel efficiency of the problem, or in other words it will reflect both the level of parallelism in the problem under consideration and the efficiency of the parallel algorithm used. It indicates the actual benefit derived by resorting to parallel processing. It is given by the following equation:

$$\eta_{act} = \frac{t_{par}}{t_{seq2} N_p} 100 \qquad (8.5-2)$$

The actual efficiency will be always be less than or equal to the algorithmic efficiency. If these efficiencies are less than 50%, then it is not usually worth considering the case. Sometimes a speed–up ratio is also used to rate a program. This is just the ratio of time taken by the parallel program to the time taken by the sequential program. The speed–up ratio should be greater than one. The speed–up factor can be calculated by multiplying the efficiency by $N_p/100$.

8.5–3 FEM and Parallel Processing

The finite element method extends itself naturally to parallel computing. In most cases, evaluation of matrices of an element is independent of other elements.

Hence, generation of element matrices can be carried out simultaneously. This feature renders an enviable advantage to FEM. However, the solution process requires a direct or an indirect assembly of the element matrices. Due to this, the interprocessor communication becomes crucial at this stage. Even in sequential processing, the solver takes most of the computing time. The steps preceding the solver are highly parallel, hence it deserves little attention. The efficiency of the finite element code directly depends on the efficiency of the parallel solver used. The next subsection is devoted to the discussion of the solvers in the context of parallel processing.

A typical finite element program consists of the following logical units:

1. Preprocessor, in which the data and parameters of the problem are read.

2. Processor 1, in which element matrices and vectors are generated, assembled, and boundary conditions of the problem are implemented.

3. Processor 2, where the resulting system of global equations are solved for the nodal values of the unknowns.

4. Post–processor, where secondary quantities of interest are computed from the nodal values.

For unsteady–state problems, the steps in processor and postprocessor are repeated for each time step till the final time is reached. For nonlinear problems, steps in processor unit are repeated until convergence is achieved. Note that most computations in processor 1 can be parallelized; solution of equations can be parallelized only when an iterative method of solution is used. The solution of equations is the most time–consuming part of the analysis.

Suppose that there are N_p processors in a parallel machine. Often the parallel machine identifies one of them as the main or master processor, and the remaining ones as slave processors. The master processor reads the data from the input file/keyboard. Typically, systems have the master processor connected to host and that alone enjoys the facility of having I/O processes. Some of the recently developed systems permit access to I/O operations from all the processors. In the former case, data has to be distributed to all the processors from the master processor. Efficiency of the data distribution depends on the topology of interconnection of the processors. The type of topology selected depends on the problem being solved. The so–called 'tree topology' of one level, where the master processor is connected directly to the slave processors, can be used for many finite element problems with an element–by–element solver. A fully interconnected topology will be ideal for any category of problems, as any processor in the array of processors can contact any other processor.

The more recent parallel processing systems come with a virtual communication facility, which is done through electronic circuits and software. In this case, the user can assume that each processor has access to every other processor in the network or the topology of his/her choice. But the efficiency of the communication depends on how closely the virtual layout chosen resembles the actual layout and wiring of the processing elements. This is critical as the interprocessor communication plays a significant role in determining the parallel efficiency.

In finite element analyses, it is advantageous to distribute the data in such a way that all the processors have an equal number of equations. Distribution is an important factor in developing an efficient parallel code. It is preferable to have elements with closely placed globally numbered nodes in one processor. The use of higher-order elements is advantageous in parallel processing as it will result in a finite element mesh with few elements. Also, the numerical integration does not add any burden to the communication load. It is preferable to have more computation than communication. This will result in a greater parallel efficiency.

Assembly of element matrices will involve massive interprocessor communication, thereby affecting the efficiency of the code. Hence, it is preferable to choose an element-by-element (EBE) data structure of the finite element method, which will not only avoid the assembly process but also be an efficient solver (the word efficiency in this section is associated only with parallel efficiency). Solution of equations is the least parallelizable step in the analysis. Hence, it plays a crucial role in determining the efficiency of the parallel code developed. This is one of the important reasons for using explicit schemes of solution (*e.g.*, Gauss–Siedel iteration), where solution to a system of equations is not required. The limitations on the magnitude of the time step used will not matter if the problem is solved in a parallel machine with thousands of processing elements. However, certain classes of physical problems cannot be solved accurately with explicit schemes, and therefore, it is essential to have efficient parallel solvers.

8.5–4 Solvers

The solvers used can be iterative or direct. Direct solvers obtain the solution to the system of simultaneous equations in a fixed number of steps. On the other hand, iterative solvers do not converge to the solution in a fixed number of iterations. The number of iterations required depends on the mathematical structure of the system of equations being solved. Nonlinearity of the problem and the diagonal dominance of the system of equations affect the convergence rate. In some cases, it may not even be possible to obtain a solution using an iterative solver—especially, without a preconditioning of the coefficient matrix. The major advantage of most iterative solvers is that they are amenable to the element-by-element (EBE) solution approach. The system of equations is solved without direct assembly of the element matrices. These algorithms have very high parallel efficiency when compared with the direct solvers.

Most of the direct solvers use Gauss elimination scheme as their base algorithm. This algorithm consists of two steps, the forward elimination and the backward substitution. The forward elimination can be programmed in parallel with a good parallel efficiency. However, the backward substitution is sequential in nature, and this will decrease the parallel efficiency of the solver. Also, the forward elimination will be less efficient for banded systems.

The Gauss–Jordan solver is a less efficient solver in a sequential mode, and it is rarely used in sequential programming. If N is the number of unknowns, Gauss elimination performs $O(N^3/3)$ multiplications and the same number of additions during the elimination operation (multiplication and division and addition and subtraction are taken as identical operations). The back substitution scheme involves $O(N^2/2)$ multiplications and additions. On the otherhand, the Gauss–Jordan solver performs $O(N^3/2)$ multiplications and additions during the elimination step, and there is no back substitution. Since there are more operations in the Gauss–Jordan method than the Gauss elimination scheme, it is not preferred; but the issues take different priorities in parallel processing. The Gauss–Jordan algorithm is devoid of the back substitution step, which is highly sequential. This is a major advantage. Also more computation is performed for a unit of communication in the Gauss–Jordan method than the Gauss elimination scheme. This will result in an improved parallel efficiency [42]. For a complete system, Gauss–Jordan will break even with Gauss elimination, even in terms of CPU time, in few processors. However, for a banded system, a greater number of processors will be required to achieve it.

Among the many iterative solvers, like the Jacobi iteration method and the Gauss–Seidel method, the Conjugate Gradient Method (CGM) is very efficient. In principle, it will yield convergent solution in less than N iterations, where N is the number of unknowns; but, in reality, this is not so due to the round–off errors. It normally takes more than N iterations, hence restarting procedures have to be used. Even then, it is faster than the Gauss–Seidel iteration method by an order of magnitude.

The conjugate gradient method is often used for minimizing a function. It uses the gradient information of the function to minimize it, without evaluating the function itself. This method was originally developed for solving systems of simultaneous equations. The original method can solve only symmetric, positive–definite systems. Using an appropriate preconditioning matrix, other systems can also be solved. Details of the conjugate gradient method algorithm with preconditioning can be found in Reference 43.

In developing a parallel algorithm for iterative schemes, two key features are exploited. Algorithms like the conjugate gradient method involve matrix–vector product and vector dot product. These steps can be easily parallelized by expressing the product of a matrix with a vector as the sum of the products of their components. The key idea is

$$[A]\{x\} = \left[\sum_{i=1}^{N_{elem}} [A_i]\right] \{x\} \tag{8.5 – 3}$$

$$[A]\{x\} = \sum_{i=1}^{N_p} \left[\sum_{j=M_{i1}}^{M_{i2}} [A_j]\right] \{x\} \tag{8.5 – 4}$$

where N_p is the number of processors and N_{elem} is the number of elements. The values of M_{i1} and M_{i2} can be calculated in different ways. Only the matrix $[A]$ is expressed as a sum and not the load vector. Hence the load vector generated in the processors has to be added to get the final load vector. This involves communication. If the load vector is also expressed as a sum, the process will be even more complicated. Also it will be computationally intensive with a heavy communication load. This can be readily seen from the following equations:

$$A \cdot x = (A_1 + A_2)(x_1 + x_2) = A_1x_1 + A_1x_2 + A_2x_1 + A_2x_2 \tag{8.5 – 5}$$

$$A \cdot x = \left[\sum_{i=1}^{N} A_i\right]\left[\sum_{j=1}^{N} x_j\right] = \sum_{i=1}^{N}\sum_{j=1}^{N} A_i x_j \tag{8.5-6}$$

If there are N values of A and x, each value of A is associated with all the N components of x. This places a requirement on the processors to know all the values, and there are more number multiplications as well. Hence, it is not advisable to express the vectors involved in the matrix–vector product as a sum of its element contributions.

The other product frequently encountered is a vector dot product. Consider two vectors $\{x\}$ and $\{y\}$ of dimension N. The dot product can be written as

$$\{x\} \cdot \{y\} = \sum_{i=1}^{N} x_i y_i \tag{8.5-7}$$

$$\{x\} \cdot \{y\} = \sum_{i=1}^{N_p} \left[\sum_{j=M_{i3}}^{M_{i4}} x_j y_j\right] \tag{8.5-8}$$

The values of M_{i3} and M_{i4} are calculated from the element connectivity of the elements present in the i–th processor. Care should be taken to make sure that a particular node is identified with only one processor in calculating the dot product, even though it may be associated with many processors. This is due to the fact that the vectors are assembled and they are global. Depending upon the particular algorithm under consideration, the methodology can be restructured. More details on parallel algorithms are available in Reference 41.

8.5–5 Summary

The future of computing will be influenced to a great extent by the developments in parallel processing. Especially, computational fluid dynamics has a lot to benefit from parallel processing. Currently, most of the parallel algorithms are the parallel versions of existing sequential algorithms. It will be more useful to conceive and develop algorithms which are fundamentally parallel. Other than CFD, problems like artificial vision, signal processing, neural networks, optimization, and finite element analyses will be benefited with the advances in parallel computing. The scope is much broader than what is mentioned here.

REFERENCES

1. J. O. Hinze, *Turbulence,* McGraw–Hill, New York (1975).

2. D. C. Leslie, *Developments in the Theory of Turbulence,* Oxford University Press, Oxford (1984).

3. P. Bradshaw (ed.), *Turbulence,* Second Edition, Springer–Verlag, Berlin (1978).

4. G. K. Batchelor, *The Theory of Homogeneous Turbulence,* Cambridge University Press, Cambridge, U.K. (1953).

5. A. S. Monin and A. M. Yaglom, *Statistical Fluid Mechanics,* MIT Press, Cambridge, Massachusetts (1973).

6. J. H. Ferziger, "Simulation of Incompressible Turbulent Flows," *Journal of Computational Physics,* **69**, 1–48 (1987).

7. J. M. Ferziger, "Higher Level Simulations of Turbulent Flows," in *Computational Methods for Turbulent, Transonic and Viscous Flows,* J. A. Essers (ed.), 93–182, Hemisphere Publishing, Washington, D.C. (1983).

8. P. T. Harsha, "Kinetic Energy Methods," in *Handbook of Turbulence,* W. Frost and T. Moulden (ed.), 187–236, Plenum Press, New York (1977).

9. J. D. Murphy, "Turbulence Modeling," in *Encyclopedia of Fluid Mechanics,* N. Chermisinoff (Ed.), **6**, 1131–1151, Gulf Publishing, Houston, Texas (1988).

10. M. Lesieur, *Turbulence in Fluids,* Martinius Nijhoff, Dordrecht, The Netherlands (1987).

11. R. S. Rogallo and P. Moin, "Numerical Simulation of Turbulent Flows," in *Annual Review of Fluid Mechanics,* **16**, 99–137 (1984).

12. A. Yoshizawa, "Large Eddy Simulation of Turbulent Flows," in *Encyclopedia of Fluid Mechanics,* N. Chermisinoff (ed.), **6**, 1277–1297, Gulf Publishing, Houston, Texas (1968).

13. W. Rodi, "Turbulence Models and Their Applications to Hydraulics," IAHR Report, Delft, The Netherlands (1984).

14. T. Cebeci and A. M. O. Smith, "Analysis of Turbulent Boundary Layers," *Applied Mathematics and Mechanics,* Academic Press, New York (1974).

15. V. Haroutunian, "Turbulent Flows with FIDAP," Fluid Dynamics International Seminar Notes, Evanston, Illinois (1988).

16. B. Launder, "On the Computation of Convective Heat Transfer in Complex Turbulent Flows," *Journal of Heat Transfer,* **110**, 1112–1128 (1988).

17. A. J. Reynolds, "The Prediction of Turbulent Prandtl and Schmidt Numbers," *International Journal of Heat and Mass Transfer,* **18**, 1055–1069 (1975).

18. M. Nallasamy, "Turbulence Models and Their Applications to the Prediction of Internal Flows: A Review," *Computers and Fluids,* **15**, 151–194(1987).

19. R. M. Smith, "On Finite Element Calculations of Turbulent Flow Using the $k-\epsilon$ Model," *International Journal for Numerical Methods in Fluids,* **4**, 303–319 (1984).

20. R. M. Smith, "A Practical Method of Two Equation Modelling Using Finite Elements," *International Journal for Numerical Methods in Fluids,* **4**, 321–336 (1984).

21. A. G. Hutton, R. M. Smith and S. Hickmott, "The Computation of Turbulent Flows of Industrial Complexity by Finite Element Methods-Progress and Prospects," in *Finite Elements in Fluids,* Vol. 7, R. H. Gallagher *et al.* (eds.), John Wiley & Sons, New York (1988).

22. R. C. Givler, D. K. Gartling, M. S. Engleman, and V. Haroutunian, "Navier–Stokes Simulations of Flow Past Three–Dimensional Submarine Models," *Computer Methods in Applied Mechanics and Engineering,* **87**, 175–200 (1991).

23. B. E. Launder and D. B. Spalding, *Lectures in Mathematical Models of Turbulence,* Academic Press, London, U.K. (1972).

24. B. E. Launder and D. B. Spalding, "The Numerical Computation of Turbulent Flow," *Computer Methods in Applied Mechanics and Engineering,* **23**, 249–270 (1974).

25. FIDAP Manual, Version 6.0, Fluid Dynamics International, Evanston, Illinois (1991).

26. A. G. Hutton, "Finite Element Boundary Techniques for Improved Performance in Computing Navier–Stokes and Related Heat Transfer Problems," in *Finite Elements in Fluids,* Vol. 4, R. H. Gallagher *et al.* (eds.) John Wiley & Sons, New York (1982).

27. J. O. Hirschfelder, C. F. Curtiss, and R. B. Bird, *Molecular Theory of Gases and Liquids,* John Wiley & Sons, New York (1954).

28. E. S. Oran and J. P. Boris, *Numerical Simulation of Reactive Flow,* Elsevier, New York (1987).

29. M. R. Baer, R. E. Benner, R. J. Gross, and J. W. Nunziato, "Modeling and Computation of Deflagration–to–Detonation Transition (DDT) in Reactive Granular Materials," *Lectures in Applied Mathematics,* **24**, American Mathematical Society, Providence, Rhode Island (1986).

30. R. R. McGuire and C. M. Tarver, "Chemical Decomposition Models for the Thermal Explosion of Confined HMX, TATB, RDX and TNT Explosives," *Seventh International Symposium on Detonation,* NSWC MP (1981).

31. W. F. Hughes and F.J. Young, *The Electromagnetodynamics of Fluids,* John Wiley & Sons, New York (1966).

32. J. D. Jackson, *Classical Electrodynamics,* Second Edition, John Wiley & Sons, New York (1975).

33. J. Jin, *The Finite Element Method in Electromagnetics,* John Wiley & Sons, New York (1993).

34. M. N. O. Sadiku, *Numerical Techniques in Electromagnetics,* CRC Press, Boca Raton, Florida (1992).

35. K. J. Binns, P. J. Lawrence, and C. W. Trowbridge, *The Analytical and Numerical Solution of Electric and Magnetic Fields,* John Wiley & Sons, New York (1992).

36. M. J. Flynn, "Very High Speed Computing Systems," *Proceedings of IEEE*, **54**, 1901–1909 (1966).

37. B. Davidson, "A Parallel Processing Tutorial," *IEEE Antennas and Propagation Society Magazine*,**32**, 6–19 (1990).

38. M. R. Hord, *Parallel Supercomputing in SIMD Architecture*, CRC Press, Boca Raton, Florida (1990).

39. A. Carling, *Parallel Processing: the Transputer and Occam*, Sigma Press, London, U.K. (1992).

40. J. Galletly, *Occam 2*, Krieger Publishing, Melbourne, Florida (1990).

41. T. L. Freeman, *Parallel Numerical Algorithms*, Prentice Hall, Englewood Cliffs, New Jersey (1992).

42. J. J. Modi, *Parallel Algorithms and Matrix Computations*, Oxford University Press, Oxford, U.K. (1988).

43. G. Gambolati, "Fast Solution to Finite Element Flow Equations by Newton Iteration and Modified Conjugate Gradient Method," *International Journal for Numerical Methods in Engineering*, **15**, 661–675 (1980).

44. R. Narasimha and S. M. Deshpande (eds.), *Surveys in Fluid Mechanics*, Indian Academy of Sciences, Bangalore, India (1982).

Appendix A

COMPUTER PROGRAM *HEATFLOW*

A–1 INTRODUCTION

The computer program HEATFLOW is a finite element analysis program for the solution of two–dimensional heat transfer and viscous, incompressible fluid flow problems. The program is a modification of the computer program FEM2DV2 from the book by Reddy [1]. The program is discussed here to aid the readers with the basic computer implementation aspects of the finite element method. The program is educational in nature and it does not contain powerful pre– or post–processors found in commercial finite element software.

A–2 HEAT TRANSFER AND RELATED PROBLEMS

The heat transfer part of the program is based on a slightly general form of equation (2.2–1) on page 26 of this book:

$$-\left[\frac{\partial}{\partial x}\left(a_{11}\frac{\partial u}{\partial x}\right)+\frac{\partial}{\partial y}\left(a_{22}\frac{\partial u}{\partial y}\right)\right]+a_{00}u=f \quad \text{in } \Omega \qquad (A-1)$$

with the boundary conditions [*c.f.*, equation (2.2–5)]

$$u=\hat{u}(s) \quad \text{on } \Gamma_u \qquad (A-2a)$$

$$\left(a_{11}\frac{\partial u}{\partial x}n_x+a_{22}\frac{\partial u}{\partial y}n_y\right)+\beta(u-u_\infty)=q \quad \text{on } \Gamma_q \qquad (A-2b)$$

where Γ_u and Γ_q are disjoint portions of the boundary Γ such that $\Gamma=\Gamma_u\cup\Gamma_q$. For a heat transfer problem, a_{11} and a_{22} denote conductivities in the x and y directions, respectively, $f(x,y)$ denotes the known internal heat generation per unit volume, (n_x,n_y) denote the direction cosines of the unit normal vector on the boundary, β denotes the convective heat transfer coefficient, and u_∞ is the ambient temperature.

Of course, equation (1) arises in a variety of other fields and the program HEATFLOW can be used to analyze them as long as the governing equation of the problem solved is a special case of equation (1). For example, slow flows of inviscid fluids are often described in terms of either the *velocity potential* ϕ

$$u\equiv-\frac{\partial\phi}{\partial x},\quad v\equiv-\frac{\partial\phi}{\partial y} \qquad (A-3)$$

or the *stream function* ψ

$$v \equiv -\frac{\partial \psi}{\partial x}, \quad u \equiv \frac{\partial \psi}{\partial y} \qquad (A-4)$$

Here (u, v) denote the velocity components. Thus, the program HEATFLOW can be used to solve the inviscid flow problems in terms of the velocity potential or the stream function, and then post–compute the velocity field as per equations (3) and (4).

A–3 VISCOUS INCOMPRESSIBLE FLOWS

The finite element model of viscous, incompressible flows used in the HEATFLOW program is the *reduced integration penalty model* discussed in Section 4.3–3. The flows considered here have no inertia and are therefore termed Stokes flows. The governing equations of the model are a combination of the continuity equation (4.3–1) and momentum equations (4.3–2):

$$-\frac{\partial}{\partial x}(2\mu\frac{\partial u}{\partial x}) - \frac{\partial}{\partial y}[\mu(\frac{\partial u}{\partial y} + \frac{\partial v}{\partial x})] - \gamma\frac{\partial}{\partial x}(\frac{\partial u}{\partial x} + \frac{\partial v}{\partial y}) = 0 \qquad (A-5a)$$

$$-\frac{\partial}{\partial x}[\mu(\frac{\partial u}{\partial y} + \frac{\partial v}{\partial x})] - \frac{\partial}{\partial y}(2\mu\frac{\partial v}{\partial y}) - \gamma\frac{\partial}{\partial y}(\frac{\partial u}{\partial x} + \frac{\partial v}{\partial y}) = 0 \qquad (A-5b)$$

with the boundary conditions

$$u = \hat{u} \text{ on } \Gamma_u \quad \text{or} \quad \mathcal{T}_1 = \hat{\mathcal{T}}_1 \quad \text{on} \quad \Gamma_{\mathcal{T}_1} \qquad (A-6a)$$

$$v = \hat{v} \text{ on } \Gamma_v \quad \text{or} \quad \mathcal{T}_2 = \hat{\mathcal{T}}_2 \quad \text{on} \quad \Gamma_{\mathcal{T}_2} \qquad (A-6b)$$

where Γ_u and $\Gamma_{\mathcal{T}_1}$, and Γ_v and $\Gamma_{\mathcal{T}_2}$ are disjoint portions of the boundary Γ such that $\Gamma = \Gamma_u \cup \Gamma_{\mathcal{T}_1}$ and $\Gamma = \Gamma_v \cup \Gamma_{\mathcal{T}_2}$. The boundary stress components $\mathcal{T}_1$ and $\mathcal{T}_2$ are given by

$$\mathcal{T}_1 = (2\mu\frac{\partial u}{\partial x} - P)n_x + \mu(\frac{\partial u}{\partial y} + \frac{\partial v}{\partial x})n_y \qquad (A-7a)$$

$$\mathcal{T}_2 = \mu(\frac{\partial u}{\partial y} + \frac{\partial v}{\partial x})n_x + (2\mu\frac{\partial v}{\partial y} - P)n_y \qquad (A-7b)$$

In equations (6a) and (6b) γ denotes the *penalty parameter*, and in equations (7a) and (7b) P denotes the pressure.

A–4 DESCRIPTION OF THE INPUT DATA

The following table describes the input data required for the solution of heat transfer and related problems and viscous fluid flow problems using program

HEATFLOW. Sample input data files are included. For additional examples, the reader may consult the book by Reddy [1, Chapter 13].

Table A.1: Description of the input variables to the program HEATFLOW.

```
* Data Card 1:
        TITLE   - Title of the problem being solved (80 characters)
* Data Card 2:
        ITYPE   - Problem type
                  ITYPE = 0, Heat transfer (and like) problems
                  ITYPE = 1, Viscous incompressible flow problems
        IGRAD   - Indicator for computing the gradient of the solution
                  IGRAD = 0, No postprocessing is required
                  IGRAD > 0, Postprocessing is required
                  When ITYPE=0 and IGRAD=1, the gradient is computed
                  as in Eq. (A.3);  for ITYPE = 0 and IGRAD > 1, the
                  gradient is computed as in Eq. (A.4).
        ITEM    - Indicator for transient analysis:
                  ITEM  = 0, Steady--state analysis is required
                  ITEM  > 0, Transient analysis is required
* Data Card 3:
        IELTYP  - Element type used in the analysis:
                  IELTYP= 0, Triangular elements
                  IELTYP> 0, Quadrilateral elements
        NPE     - Nodes per element:
                  NPE   = 3, Linear triangle (IELTYP=0)
                  NPE   = 4, Linear quadrilateral (IELTYP>0)
                  NPE   = 6, Quadratic triangle (IELTYP=0)
                  NPE   = 8 or 9, Quadratic quadrilateral (IELTYP>0)
        MESH    - Indicator for mesh generation by the program:
                  MESH  = 0, Mesh is not generated by the program
                  MESH  = 1, Mesh is generated by the program for
                             rectangular domains (by MSH2DR)
                  MESH  > 1, Mesh is generated by the program for
                             general domains (by MSH2DG)
        NPRNT   - Indicator for printing certain output:
                  NPRNT = 0, Not print array NOD, element matrices,
                             or global matrices
                  NPRNT = 1, Print array NOD and Element 1 matrices:
                             [ELK] and {ELF}
                  NPRNT = 2, Print array NOD and assembled matrices,
                             [GLK] and {GLF}
                  NPRNT > 2, Combination of NPRNT=1 and NPRNT=2

* Data Card 4:   SKIP the card if MESH.EQ.1
        NEM     - Number of elements in the mesh when the user inputs
                  the mesh or the mesh is generated by MSH2DG
        NNM     - Number of  nodes   in the mesh when the user inputs
                  the mesh or the mesh is generated by MSH2DG
```

```
* Data Card 5:  SKIP the card if  MESH.NE.0; otherwise, the card is
                read in a loop on the number of elements (N=1, NEM)
       NOD(N,I)-Connectivity for the N-th element (I=1,NPE)
* Data Card 6:  SKIP the card if  MESH.NE.0
       GLXY(I,J)-Global x and y coordinates of I-th global node in
                 the mesh (J=1, x-coordinate, J=2, y-coordinate)
                 Loops on I and J are: ((J=1,2), I=1,NNM); the NNM
                 pairs of (x,y)--coordinates are read sequentially

_____ The next FOUR data cards are read in subroutine  MSH2DG ______
* * * * *  SKIP Cards 7, 8, 9,  and  10  unless MESH.GT.1  * * * * *

* Data Card 7:
       NRECL    - Number of line records to be read in the mesh

* Data Card 8: Read the following variables NRECL times
       NOD1     - First global node number of the line segment
       NODL     - Last global node number of the line segment
       NODINC   - Node increment on the line
       X1       - The global x-coordinate of NOD1
       Y1       - The global y-coordinate of NOD1
       XL       - The global x-coordinate of NODL
       YL       - The global y-coordinate of NODL
       RATIO    - The ratio of the first element length to the last
                  element length
* Data Card 9:
       NRECEL   - Number of rows of elements to be read in the mesh
* Data Card 10: Read the following variables NRECEL times
       NEL1     - First element number of the row
       NELL     - Last element  number of the row
       IELINC   - Increment of element number in the row
       NODINC   - Increment of global node number in the row
       NPE      - Number of nodes in each element
       NODE(I)  - Connectivity array of the first element in the row
                  (I=1,NPE)
* Data Card 11:  SKIP the card if  MESH.NE.1
       NX       - Number of elements in the x-direction
       NY       - Number of elements in the y-direction
* Data Card 12:  SKIP the card if  MESH.NE.1
       X0       - The x-coordinate of global node 1
       DX(I)    - The x-dimension of the I-th element  (I=1,NX)
* Data Card 13:  SKIP the card if  MESH.NE.1
       Y0       - The y-coordinate of global node 1
       DY(I)    - The y-dimension of the I-th element  (I=1,NY)
* Data Card 14:
       NSPV     - The number of specified primary variables
* Data Card 15:  SKIP the card if NSPV.EQ.0
       ISPV(I,J)-Node number and LOCAL degree of freedom number of
                 the I-th specified primary variable:
                 ISPV(I,1)=Node number; ISPV(I,2)=Local DOF number
                 The loops on I and J are: ((J=1,2),I=1,NSPV)
```

```
* Data Card 16:  SKIP the card if NSPV.EQ.0
        VSPV(I) - Specified value of the I-th primary variable
                  (I=1,NSPV)
* Data Card 17:
        NSSV    - Number of (nonzero) specified secondary variables
* Data Card 18:  SKIP the card if NSSV.EQ.0
        ISSV(I,J)-Node number and LOCAL degree of freedom number of
                  the I-th specified secondary variable:
                  ISSV(I,1)=Node number; ISSV(I,2)=Local DOF number
                  The loops on I and J are: ((J=1,2),I=1,NSSV)
* Data Card 19:  SKIP the card if NSSV.EQ.0
        VSSV(I) - Specified value of the I-th secondary variable
                  (I=1,NSSV)
--------------------------------------------------------------------
Data Cards 20 - 26 are for HEAT TRANSFER PROBLEMS (ITYPE = 0) only

* Data Card 20:  SKIP the card if ITYPE.NE.0
        A10      |
        A1X      |  Coefficients of the differential equation
        A1Y      |  a11 = A10 +A1X*X + A1Y*Y
* Data Card 21:  SKIP the card if ITYPE.NE.0
        A20      |
        A2X      |  Coefficients of the differential equation
        A2Y      |  a22 = A20 +A2X*X + A2Y*Y
* Data Card 22:  SKIP the card if ITYPE.NE.0
        A00      -  Coefficient of the differential equation
* Data Card 23:  SKIP the card if ITYPE.NE.0
        ICONV    -  Indicator for convection boundary conditions
                    ICONV = 0, No convection boundary conditions
                    ICONV > 0, Convection boundary conditions present
* Data Card 24:  SKIP the card if ITYPE.NE.0 or ICONV.EQ.0
        NBE      -  Number elements with convection
* Data Card 25:  SKIP the card if ITYPE.NE.0 or ICONV.EQ.0
                 The following cards are read for each I, I=1,NBE
        IBN(I)   -  I-th element number with convection
        BETA(I)  -  Film coefficient for convection on I-th element
        TINF(I)  -  Ambient temperature of the I-th element
* Data Card 26:  SKIP the card if ITYPE.NE.0 or ICONV.EQ.0
        INOD(I,J)-  Local node numbers of the side with convection
                    (J=1,2; for quadratic elements, give end nodes)
                    Loops on I and J are: ((J=1,2), I=1,NBE)
--------------------------------------------------------------------
_____ Data Card 27 is for VISCOUS FLUID FLOWS (ITYPE = 1) only _____

* Data Card 27:  SKIP the card if ITYPE.EQ.0
        AMU     - Viscosity of the fluid
        PENLTY  - Penalty parameter used
--------------------------------------------------------------------
__________ Remaining data cards are for ALL problem types __________
```

```
* Data Card 28:
        F0     |
        FX     | Coefficients to define the source term:
        FY     |  f = F0 + FX*x + FY*y
-------------------------------------------------------------------
__________ Cards 29 thru 32 are for TRANSIENT ANALYSIS only ________

* Data Card 29:  SKIP the card if ITEM.EQ.0
        C0     | Coefficients defining the temporal parts of the
        CX     | differential equations:
        CY     | CT = C0 + CX*x + CY*y
* Data Card 30:  SKIP the card if ITEM.EQ.0
        NTIME  - Number of time steps for the transient solution
        NSTP   - Time step number at which the source is removed
        INTVL  - Time step interval at which to print the solution
        INTIAL - Indicator for nature of initial conditions:
                 INTIAL=0, Zero initial conditions are used
                 INTIAL>0, Non-zero initial conditions are used
* Data Card 31:  SKIP the card if ITEM.EQ.0
        DT     - Time step used for the transient solution
        ALFA   - Parameter in the alfa-family of time approximation
                 used:
                 ALFA=0,   The forward difference scheme (C.S.)@
                 ALFA=0.5, The Crank-Nicolson scheme (stable)
                 ALFA=2/3, The Galerkin scheme (stable)
                 ALFA=1,   The backward difference scheme (stable)
                @C.S.=Conditionally Stable; For all schemes with
                 ALFA < 0.5, the time step DT is restricted to:
                 DT < 2/[MAXEGN*(1-2*ALFA)], where MAXEGN is the
                 maximum eigenvalue of the discrete problem.
        EPSLN  - A small parameter to check if the solution has
                 reached a steady state
* Data Card 32:  SKIP the card if ITEM or INTIAL.EQ.0, or NEIGN.NE.0
        GLU(I) - Vector of initial values of the primary variables
                 (I=1,NEQ, NEQ=Number of nodal values in the mesh)
```

Example 1: Convective heat transfer in a square region (see Figure A–1) is solved using 2×2 mesh of nine–node rectangular elements. The conductivities are taken to be $a_{11} = a_{22} = 10$ W/(m °C). The input and an edited output from program HEATFLOW is given below.

```
Example A-1: CONVECTIVE heat transfer in a square region
 0   1   0   0                                 ITYPE,IGRAD,ITEM,NEIGN
 2   9   1   0                                 IEL, NPE, MESH, NPRNT
 2   2
 0.0  0.5   0.5
 0.0  0.5   0.5
 9                                              NSPV, ISPV, VSPV
```

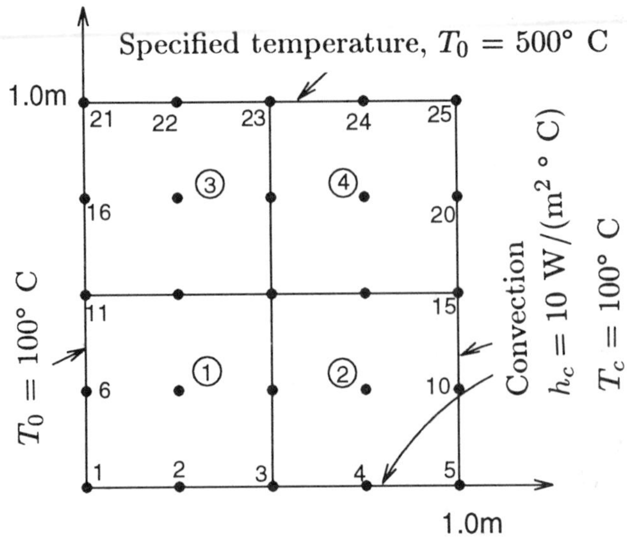

Figure A–1: Domain, boundary conditions, and mesh used for the convective heat transfer problem in Example 1.

```
1 1  6 1  11 1  16 1  21 1  22 1  23 1  24 1   25 1
100.0 100.0  100.0  100.0  500.0  500.0  500.0  500.0  500.0
0                                                NSSV
10.0  0.0  0.0                                   A10, A1X, A1Y
10.0  0.0  0.0                                   A20, A2X, A2Y
 0.0                                             A00
1                                                ICONV
4
1 10.0 100.0   2 10.0 100.0   2 10.0 100.0   4 10.0 100.0
1 2   1 2    2 3  2 3
0.0      0.0  0.0                                F0,  FX,  FY
```

```
Example A-1: CONVECTIVE heat transfer in a square region
________________________________________________________________

        OUTPUT  FROM  PROGRAM  *HEATFLOW*  BY  J. N. REDDY
________________________________________________________________

             ANALYSIS  OF  AN   HEAT  TRANSFER   PROBLEM

  COEFFICIENTS OF THE DIFFERENTIAL EQUATION:

     Coefficient, A10 .......................=  0.1000E+02
     Coefficient, A1X .......................=  0.0000E+00
     Coefficient, A1Y .......................=  0.0000E+00
     Coefficient, A20 .......................=  0.1000E+02
     Coefficient, A2X .......................=  0.0000E+00
     Coefficient, A2Y .......................=  0.0000E+00
     Coefficient, A00 .......................=  0.0000E+00
```

```
CONVECTIVE HEAT TRANSFER DATA:

   Number of elements with convection, NBE .=   4
   Elements, their LOCAL nodes and convective
   heat transfer data:
   Ele. No.     End Nodes         Film Coeff.      T-Infinity

       1          1    2        0.10000E+02       0.10000E+03
       2          1    2        0.10000E+02       0.10000E+03
       2          2    3        0.10000E+02       0.10000E+03
       4          2    3        0.10000E+02       0.10000E+03

FINITE ELEMENT MESH INFORMATION:

   Element type: 0 = Triangle; >0 = Quad.)..=   2
   Number of nodes per element, NPE ........=   9
   No. of primary deg. of freedom/node, NDF =   1
   Number of elements in the mesh, NEM .....=   4
   Number of nodes in the mesh, NNM ........=  25
   Number of equations to be solved, NEQ ...=  25
   Half bandwidth of the matrix GLK, NHBW ..=  13
   Mesh subdivisions, NX and NY ............=   2    2
   No. of specified PRIMARY variables, NSPV =   9
```

Node	x-coord.	y-coord.	Speci. primary & secondary variables (0, unspecified; >0, specified) Primary DOF	Secondary DOF
1	0.0000E+00	0.0000E+00	1	0
2	0.2500E+00	0.0000E+00	0	0
3	0.5000E+00	0.0000E+00	0	0
4	0.7500E+00	0.0000E+00	0	0
5	0.1000E+01	0.0000E+00	0	0
6	0.0000E+00	0.2500E+00	1	0
7	0.2500E+00	0.2500E+00	0	0
8	0.5000E+00	0.2500E+00	0	0
9	0.7500E+00	0.2500E+00	0	0
10	0.1000E+01	0.2500E+00	0	0
11	0.0000E+00	0.5000E+00	1	0
12	0.2500E+00	0.5000E+00	0	0
13	0.5000E+00	0.5000E+00	0	0
14	0.7500E+00	0.5000E+00	0	0
15	0.1000E+01	0.5000E+00	0	0
16	0.0000E+00	0.7500E+00	1	0
17	0.2500E+00	0.7500E+00	0	0
18	0.5000E+00	0.7500E+00	0	0
19	0.7500E+00	0.7500E+00	0	0
20	0.1000E+01	0.7500E+00	0	0
21	0.0000E+00	0.1000E+01	1	0
22	0.2500E+00	0.1000E+01	1	0
23	0.5000E+00	0.1000E+01	1	0
24	0.7500E+00	0.1000E+01	1	0
25	0.1000E+01	0.1000E+01	1	0

S O L U T I O N :

Node	x-coord.	y-coord.	Temperature
1	0.00000E+00	0.00000E+00	0.10000E+03
2	0.25000E+00	0.00000E+00	0.11380E+03
3	0.50000E+00	0.00000E+00	0.13273E+03
4	0.75000E+00	0.00000E+00	0.13350E+03
5	0.10000E+01	0.00000E+00	0.11697E+03
6	0.00000E+00	0.25000E+00	0.10000E+03
7	0.25000E+00	0.25000E+00	0.14268E+03
8	0.50000E+00	0.25000E+00	0.17135E+03
9	0.75000E+00	0.25000E+00	0.17561E+03
10	0.10000E+01	0.25000E+00	0.15241E+03
11	0.00000E+00	0.50000E+00	0.10000E+03
12	0.25000E+00	0.50000E+00	0.18453E+03
13	0.50000E+00	0.50000E+00	0.24111E+03
14	0.75000E+00	0.50000E+00	0.24392E+03
15	0.10000E+01	0.50000E+00	0.21538E+03
16	0.00000E+00	0.75000E+00	0.10000E+03
17	0.25000E+00	0.75000E+00	0.29562E+03
18	0.50000E+00	0.75000E+00	0.34509E+03
19	0.75000E+00	0.75000E+00	0.35267E+03
20	0.10000E+01	0.75000E+00	0.30641E+03
21	0.00000E+00	0.10000E+01	0.50000E+03
22	0.25000E+00	0.10000E+01	0.50000E+03
23	0.50000E+00	0.10000E+01	0.50000E+03
24	0.75000E+00	0.10000E+01	0.50000E+03
25	0.10000E+01	0.10000E+01	0.50000E+03

Example 2: Flow of an incompressible viscous fluid in a slider bearing (see Figure A–2) is analyzed using a mesh of six nine–node elements. The mesh used is to illustrate the input data, and it is not representative of the mesh required to obtain an accurate solution. The input data and an edited output from program HEATFLOW is given below.

```
Example A-2: Flow of a viscous fluid in a slider bearing
  1   1   0                                            ITYPE,ISTRS,ITEM
  1   9   2   0                                        IELTYP,NPE,MESH,NPRNT
  6  35                                                NEM, NNM
  5                                                    NRECL
  1   7   1   0.0  0.0     0.36  0.0     1.0           NOD1,NODL,NODINC,...
  8  14   1   0.0  5.0E-5  0.36  5.0E-5  1.0
 15  21   1   0.0  1.0E-4  0.36  1.0E-4  1.0
 22  28   1   0.0  3.5E-4  0.36  1.5E-4  1.0
 29  35   1   0.0  8.0E-4  0.36  4.0E-4  1.0
```

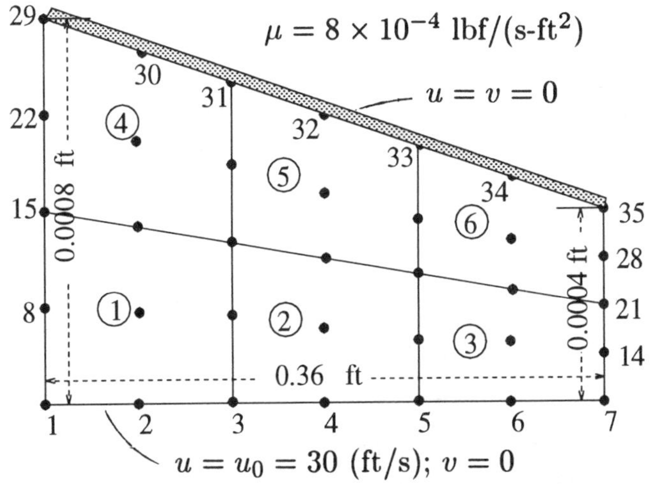

Figure A–2: Domain and mesh used for the flow of a viscous incompressible fluid in a slider bearing (Example 2).

```
2                                                            NRECEL
1   3   1   2   9    1   3  17  15   2  10  16   8   9      NEL1,NELL,..
4   6   1   2   9   15  17  31  29  16  24  30  22  23
28                                              NSPV
1  1  1  2   2  1   2  2   3  1  3  2   4  1  4  2  5  1  5  2
6  1  6  2   7  1   7  2  29  1 29  2  30  1 30  2  31 1  31 2
32 1  32 2  33  1  33 2  34  1 34  2  35  1 35  2  ISPV(I,J)
30.0   0.0   30.0   0.0   30.0   0.0  30.0   0.0  30.0   0.0
30.0   0.0   30.0   0.0    0.0   0.0   0.0   0.0   0.0   0.0
 0.0   0.0    0.0   0.0    0.0   0.0   0.0   0.0     VSPV(I)
0                                               NSSV
8.0E-4  8.0E12                                  AMU, PENLTY
0.0   0.0   0.0                                 F0, FX, FY
```

```
Example A-2: Flow of a viscous fluid in a slider bearing
________________________________________________________________

        OUTPUT  FROM  PROGRAM  *HEATFLOW*  BY  J. N. REDDY
________________________________________________________________

           A  VISCOUS INCOMPRESSIBLE FLOW IS ANALYZED

   PARAMETERS OF THE FLUID FLOW PROBLEM:

      Viscosity of the fluid, AMU ............=  0.8000E-03
      Penalty parameter, PENLTY ..............=  0.8000E+13

   FINITE ELEMENT MESH INFORMATION:

      Element type: 0 = Triangle; >0 = Quad.)..=   1
      Number of nodes per element, NPE .......=   9
```

```
No. of primary deg. of freedom/node, NDF =    2
Number of elements in the mesh, NEM .....=    6
Number of nodes in the mesh, NNM ........=   35
Number of equations to be solved, NEQ ...=   70
Half bandwidth of the matrix GLK, NHBW ..=   34
No. of specified PRIMARY variables, NSPV =   28
```

Node	x-coord.	y-coord.	Speci. primary & secondary variables (0, unspecified; >0, specified) Primary DOF		Secondary DOF	
1	0.0000E+00	0.0000E+00	1	2	0	0
2	0.6000E-01	0.0000E+00	1	2	0	0
3	0.1200E+00	0.0000E+00	1	2	0	0
4	0.1800E+00	0.0000E+00	1	2	0	0
5	0.2400E+00	0.0000E+00	1	2	0	0
6	0.3000E+00	0.0000E+00	1	2	0	0
7	0.3600E+00	0.0000E+00	1	2	0	0
8	0.0000E+00	0.5000E-04	0	0	0	0
9	0.6000E-01	0.5000E-04	0	0	0	0
10	0.1200E+00	0.5000E-04	0	0	0	0
11	0.1800E+00	0.5000E-04	0	0	0	0
12	0.2400E+00	0.5000E-04	0	0	0	0
13	0.3000E+00	0.5000E-04	0	0	0	0
14	0.3600E+00	0.5000E-04	0	0	0	0
15	0.0000E+00	0.1000E-03	0	0	0	0
16	0.6000E-01	0.1000E-03	0	0	0	0
17	0.1200E+00	0.1000E-03	0	0	0	0
18	0.1800E+00	0.1000E-03	0	0	0	0
19	0.2400E+00	0.1000E-03	0	0	0	0
20	0.3000E+00	0.1000E-03	0	0	0	0
21	0.3600E+00	0.1000E-03	0	0	0	0
22	0.0000E+00	0.3500E-03	0	0	0	0
23	0.6000E-01	0.3167E-03	0	0	0	0
24	0.1200E+00	0.2833E-03	0	0	0	0
25	0.1800E+00	0.2500E-03	0	0	0	0
26	0.2400E+00	0.2167E-03	0	0	0	0
27	0.3000E+00	0.1833E-03	0	0	0	0
28	0.3600E+00	0.1500E-03	0	0	0	0
29	0.0000E+00	0.8000E-03	1	2	0	0
30	0.6000E-01	0.7333E-03	1	2	0	0
31	0.1200E+00	0.6667E-03	1	2	0	0
32	0.1800E+00	0.6000E-03	1	2	0	0
33	0.2400E+00	0.5333E-03	1	2	0	0
34	0.3000E+00	0.4667E-03	1	2	0	0
35	0.3600E+00	0.4000E-03	1	2	0	0

S O L U T I O N :

Node	x-coord.	y-coord.	Velocity, u	Velocity, v
1	0.00000E+00	0.00000E+00	0.30000E+02	0.00000E+00
2	0.60000E-01	0.00000E+00	0.30000E+02	0.00000E+00
3	0.12000E+00	0.00000E+00	0.30000E+02	0.00000E+00
4	0.18000E+00	0.00000E+00	0.30000E+02	0.00000E+00
5	0.24000E+00	0.00000E+00	0.30000E+02	0.00000E+00
6	0.30000E+00	0.00000E+00	0.30000E+02	0.00000E+00
7	0.36000E+00	0.00000E+00	0.30000E+02	0.00000E+00
8	0.00000E+00	0.50000E-04	0.26008E+02	-0.45304E-02
9	0.60000E-01	0.50000E-04	0.25983E+02	0.22683E-02
10	0.12000E+00	0.50000E-04	0.26085E+02	-0.46442E-02
11	0.18000E+00	0.50000E-04	0.26308E+02	0.21300E-02
12	0.24000E+00	0.50000E-04	0.26982E+02	-0.50555E-02
13	0.30000E+00	0.50000E-04	0.28270E+02	0.14714E-02
14	0.36000E+00	0.50000E-04	0.31533E+02	-0.68447E-02
15	0.00000E+00	0.10000E-03	0.22292E+02	0.10406E-01
16	0.60000E-01	0.10000E-03	0.22259E+02	-0.51970E-02
17	0.12000E+00	0.10000E-03	0.22421E+02	0.10016E-01
18	0.18000E+00	0.10000E-03	0.22847E+02	-0.57272E-02
19	0.24000E+00	0.10000E-03	0.23933E+02	0.86438E-02
20	0.30000E+00	0.10000E-03	0.26237E+02	-0.81796E-02
21	0.36000E+00	0.10000E-03	0.31469E+02	0.26246E-02
22	0.00000E+00	0.35000E-03	0.87401E+01	0.14297E-01
23	0.60000E-01	0.31667E-03	0.10255E+02	-0.80107E-02
24	0.12000E+00	0.28333E-03	0.12079E+02	0.12563E-01
25	0.18000E+00	0.25000E-03	0.14350E+02	-0.99138E-02
26	0.24000E+00	0.21667E-03	0.17197E+02	0.97258E-02
27	0.30000E+00	0.18333E-03	0.20956E+02	-0.13382E-01
28	0.36000E+00	0.15000E-03	0.25982E+02	0.38986E-02
29	0.00000E+00	0.80000E-03	0.00000E+00	0.00000E+00
30	0.60000E-01	0.73333E-03	0.00000E+00	0.00000E+00
31	0.12000E+00	0.66667E-03	0.00000E+00	0.00000E+00
32	0.18000E+00	0.60000E-03	0.00000E+00	0.00000E+00
33	0.24000E+00	0.53333E-03	0.00000E+00	0.00000E+00
34	0.30000E+00	0.46667E-03	0.00000E+00	0.00000E+00
35	0.36000E+00	0.40000E-03	0.00000E+00	0.00000E+00

x-coord.	y-coord.	sigma-x	sigma-y	sigma-xy	pressure
0.2536E-01	0.2113E-04	-0.2325E+04	-0.2325E+04	-0.6464E+02	0.2325E+04
0.2536E-01	0.7887E-04	-0.2324E+04	-0.2324E+04	-0.5930E+02	0.2324E+04
0.9464E-01	0.2113E-04	-0.8666E+04	-0.8666E+04	-0.6391E+02	0.8666E+04
0.9464E-01	0.7887E-04	-0.8666E+04	-0.8666E+04	-0.5883E+02	0.8666E+04
0.1454E+00	0.2113E-04	-0.1311E+05	-0.1311E+05	-0.6235E+02	0.1311E+05
0.1454E+00	0.7887E-04	-0.1311E+05	-0.1311E+05	-0.5731E+02	0.1311E+05
0.2146E+00	0.2113E-04	-0.1771E+05	-0.1771E+05	-0.5387E+02	0.1771E+05
0.2146E+00	0.7887E-04	-0.1771E+05	-0.1771E+05	-0.5184E+02	0.1771E+05
0.2654E+00	0.2113E-04	-0.1858E+05	-0.1858E+05	-0.4341E+02	0.1858E+05
0.2654E+00	0.7887E-04	-0.1859E+05	-0.1859E+05	-0.4380E+02	0.1859E+05
0.3346E+00	0.2113E-04	-0.9608E+04	-0.9608E+04	-0.2520E+00	0.9608E+04
0.3346E+00	0.7887E-04	-0.9607E+04	-0.9607E+04	-0.1734E+02	0.9607E+04
0.2536E-01	0.1753E-03	-0.2323E+04	-0.2323E+04	-0.4786E+02	0.2323E+04
0.2536E-01	0.5632E-03	-0.2323E+04	-0.2323E+04	-0.1607E+02	0.2323E+04
0.9464E-01	0.1590E-03	-0.8666E+04	-0.8666E+04	-0.4855E+02	0.8666E+04
0.9464E-01	0.5025E-03	-0.8667E+04	-0.8667E+04	-0.2187E+02	0.8667E+04
0.1454E+00	0.1471E-03	-0.1311E+05	-0.1311E+05	-0.4832E+02	0.1311E+05
0.1454E+00	0.4580E-03	-0.1311E+05	-0.1311E+05	-0.2752E+02	0.1311E+05
0.2146E+00	0.1309E-03	-0.1771E+05	-0.1771E+05	-0.4730E+02	0.1771E+05
0.2146E+00	0.3973E-03	-0.1771E+05	-0.1771E+05	-0.3830E+02	0.1771E+05
0.2654E+00	0.1190E-03	-0.1859E+05	-0.1859E+05	-0.4532E+02	0.1859E+05
0.2654E+00	0.3529E-03	-0.1859E+05	-0.1859E+05	-0.4934E+02	0.1859E+05
0.3346E+00	0.1027E-03	-0.9605E+04	-0.9605E+04	-0.6196E+02	0.9605E+04
0.3346E+00	0.2922E-03	-0.9606E+04	-0.9606E+04	-0.7193E+02	0.9606E+04

A–5 SOURCE LISTINGS OF SELECTIVE SUBROUTINES

The computer implementation of the finite element formulations presented in Chapters 2 and 3 can be found in Chapter 13 of [1]. A complete listing of the FORTRAN source of FEM2DV2 is also included in [1]. Listings of subroutines ELKMFR, SHPRCT, and TEMPORAL from program HEATFLOW, which illustrate the computation of element matrices for rectangular (quadrilateral) elements, are presented here. The complete FORTRAN source of program HEATFLOW (or FEM2DV2 [1]) may be obtained for a small charge from the first author.

REFERENCE

1. J. N. Reddy, *An Introduction to the Finite Element Method*, Second Edition, McGraw–Hill, New York (1993).

```
      SUBROUTINE ELKMFR(NCOUNT,NPE,NN,ITYPE,ITEM,INTIAL)
C     ________________________________________________________________
C
C     Element  calculations based on  linear and quadratic rectangular
C     elements and  isoparametric  formulation are carried out for the
C     heat transfer and penalty model of fluid flow.
C     ________________________________________________________________
C
      IMPLICIT REAL*8(A-H,O-Z)
      COMMON/STF/ELF(18),ELK(18,18),ELM(18,18),ELXY(9,2),ELU(18),A1,A2
      COMMON/PST/A10,A1X,A1Y,A20,A2X,A2Y,A00,C0,CX,CY,F0,FX,FY,
     1          AMU,PENLTY
      COMMON/SHP/SF(9),GDSF(2,9),SFH(16)
      COMMON/PNT/IPDF,IPDR,NIPF,NIPR
      DIMENSION  GAUSPT(5,5),GAUSWT(5,5)
      COMMON/IO/IN,ITT
C
      DATA GAUSPT/5*0.0D0, -0.57735027D0, 0.57735027D0, 3*0.0D0,
     2  -0.77459667D0, 0.0D0, 0.77459667D0, 2*0.0D0, -0.86113631D0,
     3  -0.33998104D0, 0.33998104D0, 0.86113631D0, 0.0D0, -0.90617984D0,
     4  -0.53846931D0,0.0D0,0.53846931D0,0.90617984D0/
C
      DATA GAUSWT/2.0D0, 4*0.0D0, 2*1.0D0, 3*0.0D0, 0.55555555D0,
     2   0.88888888D0, 0.55555555D0, 2*0.0D0, 0.34785485D0,
     3 2*0.65214515D0, 0.34785485D0, 0.0D0, 0.23692688D0,
     4   0.47862867D0, 0.56888888D0, 0.47862867D0, 0.23692688D0/
C
      NDF = NN/NPE
      NET=NPE
C
C     Initialize the arrays
C
      DO 120 I = 1,NN
      ELF(I)  = 0.0
      DO 120 J = 1,NN
      IF(ITEM.NE.0) THEN
          ELM(I,J)= 0.0
      ENDIF
  120 ELK(I,J)= 0.0
C
C     Do-loops on numerical (Gauss) integration begin here. Subroutine
C     SHPRCT (SHaPe functions for ReCTangular elements) is called here
C
      DO 200 NI = 1,IPDF
      DO 200 NJ = 1,IPDF
      XI  = GAUSPT(NI,IPDF)
      ETA = GAUSPT(NJ,IPDF)
      CALL SHPRCT (NPE,XI,ETA,DET,ELXY,NDF,ITYPE)
      CNST = DET*GAUSWT(NI,IPDF)*GAUSWT(NJ,IPDF)
      X=0.0
      Y=0.0
      DO 140 I=1,NPE
      X=X+ELXY(I,1)*SF(I)
  140 Y=Y+ELXY(I,2)*SF(I)
C
```

```
      SOURCE=F0+FX*X+FY*Y
      IF(ITEM.NE.0) THEN
          CT=C0+CX*X+CY*Y
      ENDIF
      IF(ITYPE.LE.0) THEN
          A11=A10+A1X*X+A1Y*Y
          A22=A20+A2X*X+A2Y*Y
      ENDIF
C
      II=1
      DO 180 I=1,NET
      JJ=1
      DO 160 J=1,NET
      S00=SF(I)*SF(J)*CNST
      S11=GDSF(1,I)*GDSF(1,J)*CNST
      S22=GDSF(2,I)*GDSF(2,J)*CNST
      S12=GDSF(1,I)*GDSF(2,J)*CNST
      S21=GDSF(2,I)*GDSF(1,J)*CNST
      IF(ITYPE.EQ.0) THEN
C
C     Heat transfer and like problems (i.e. single DOF problems):________
C
          ELK(I,J) = ELK(I,J) + A11*S11 + A22*S22 + A00*S00
          IF(ITEM.NE.0) THEN
              ELM(I,J) = ELM(I,J) + CT*S00
          ENDIF
      ELSE
C
C     Viscous incompressible fluids:_________________________________
C     Compute coefficients associated with viscous terms (full integ.)
C
          ELK(II,JJ)     = ELK(II,JJ)      + AMU*(2.0*S11 + S22)
          ELK(II+1,JJ)   = ELK(II+1,JJ)    + AMU*S12
          ELK(II,JJ+1)   = ELK(II,JJ+1)    + AMU*S21
          ELK(II+1,JJ+1)= ELK(II+1,JJ+1)  + AMU*(S11 + 2.0*S22)
          IF(ITEM.NE.0) THEN
              ELM(II,JJ)     = ELM(II,JJ)      + CT*S00
              ELM(II+1,JJ+1)= ELM(II+1,JJ+1)  + CT*S00
          ENDIF
      ENDIF
  160 JJ = NDF*J+1
C
C     Source of the form fx = F0 + FX*X + FY*Y is assumed
C
      L=(I-1)*NDF+1
      ELF(L) = ELF(L)+CNST*SF(I)*SOURCE
  180 II = NDF*I+1
  200 CONTINUE
C
      IF(ITYPE.GT.0) THEN
C
C     Use reduced integration to evaluate coefficients associated with
C     penalty terms for flows of viscous incompressible fluids.
C
```

```
            DO 280 NI=1,IPDR
            DO 280 NJ=1,IPDR
            XI  = GAUSPT(NI,IPDR)
            ETA = GAUSPT(NJ,IPDR)
            CALL SHPRCT (NPE,XI,ETA,DET,ELXY,NDF,ITYPE)
            CNST=DET*GAUSWT(NI,IPDR)*GAUSWT(NJ,IPDR)
C
            II=1
            DO 260 I=1,NPE
            JJ = 1
            DO 240 J=1,NPE
            S11=GDSF(1,I)*GDSF(1,J)*CNST
            S22=GDSF(2,I)*GDSF(2,J)*CNST
            S12=GDSF(1,I)*GDSF(2,J)*CNST
            S21=GDSF(2,I)*GDSF(1,J)*CNST
            ELK(II,JJ)     = ELK(II,JJ)     + PENLTY*S11
            ELK(II+1,JJ)   = ELK(II+1,JJ)   + PENLTY*S21
            ELK(II,JJ+1)   = ELK(II,JJ+1)   + PENLTY*S12
            ELK(II+1,JJ+1)= ELK(II+1,JJ+1) + PENLTY*S22
  240 JJ=NDF*J+1
  260 II=NDF*I+1
  280 CONTINUE
      ENDIF
C
      IF(ITEM.NE.0) THEN
C
C     Compute the coefficient matrices of the final algebraic equations
C     (i.e., after time approximation) in the transient analysis:_______
C
          CALL TEMPORAL(NCOUNT,INTIAL,ITEM,NN)
      ENDIF
      RETURN
      END
      SUBROUTINE SHPRCT(NPE,XI,ETA,DET,ELXY,NDF,ITYPE)
C     ________________________________________________________________
C
C     The subroutine evaluates the interpolation functions (SF(I)) and
C     their derivatives with respect to global coordinates (GDSF(I,J))
C     for Lagrange linear & quadratic rectangular elements, using  the
C     isoparametric formulation. The subroutine also evaluates Hermite
C     interpolation functions and their  global derivatives  using the
C     subparametric formulation.
C
C     SF(I).......Interpolation function for node I of the element
C     DSF(J,I).....Derivative of SF(I) with respect to XI if J=1 and
C                 and ETA if J=2
C     GDSF(J,I)....Derivative of SF(I) with respect to  X if J=1 and
C                 and  Y  if J=2
C     XNODE(I,J)...J-TH (J=1,2) Coordinate of node I of the element
C     NP(I).......Array of element nodes (used to define SF and DSF)
C     GJ(I,J)......Jacobian matrix
C     GJINV(I,J)...Inverse of the jacobian matrix
C     ________________________________________________________________
```

```
C
      IMPLICIT REAL*8 (A-H,O-Z)
      DIMENSION ELXY(9,2),XNODE(9,2),NP(9),DSF(2,9),GJ(2,2),GJINV(2,2)
      COMMON/SHP/SF(9),GDSF(2,9)
      COMMON/IO/IN,ITT
      DATA XNODE/-1.0D0, 2*1.0D0, -1.0D0, 0.0D0, 1.0D0, 0.0D0, -1.0D0,
     *        0.0D0, 2*-1.0D0, 2*1.0D0, -1.0D0, 0.0D0, 1.0D0, 2*0.0D0/
      DATA NP/1,2,3,4,5,7,6,8,9/
C
      FNC(A,B) = A*B
      IF(NPE.EQ.4) THEN
C
C     LINEAR Lagrange interpolation functions for FOUR-NODE element
C
          DO 10 I = 1, NPE
          XP  = XNODE(I,1)
          YP  = XNODE(I,2)
          XI0 = 1.0+XI*XP
          ETA0=1.0+ETA*YP
          SF(I)    = 0.25*FNC(XI0,ETA0)
          DSF(1,I)= 0.25*FNC(XP,ETA0)
   10     DSF(2,I)= 0.25*FNC(YP,XI0)
      ELSE
          IF(NPE.EQ.8) THEN
C
C     QUADRATIC Lagrange interpolation functions for EIGHT-NODE element
C
              DO 20 I = 1, NPE
              NI   = NP(I)
              XP   = XNODE(NI,1)
              YP   = XNODE(NI,2)
              XI0  = 1.0+XI*XP
              ETA0 = 1.0+ETA*YP
              XI1  = 1.0-XI*XI
              ETA1 = 1.0-ETA*ETA
              IF(I.LE.4) THEN
                  SF(NI)     = 0.25*FNC(XI0,ETA0)*(XI*XP+ETA*YP-1.0)
                  DSF(1,NI)  = 0.25*FNC(ETA0,XP)*(2.0*XI*XP+ETA*YP)
                  DSF(2,NI)  = 0.25*FNC(XI0,YP)*(2.0*ETA*YP+XI*XP)
              ELSE
                  IF(I.LE.6) THEN
                      SF(NI)     = 0.5*FNC(XI1,ETA0)
                      DSF(1,NI)  = -FNC(XI,ETA0)
                      DSF(2,NI)  = 0.5*FNC(YP,XI1)
                  ELSE
                      SF(NI)     = 0.5*FNC(ETA1,XI0)
                      DSF(1,NI)  = 0.5*FNC(XP,ETA1)
                      DSF(2,NI)  = -FNC(ETA,XI0)
                  ENDIF
              ENDIF
   20         CONTINUE
          ELSE
C
C     QUADRATIC Lagrange interpolation functions for NINE-NODE element
C
```

```
            DO 30 I=1,NPE
            NI   = NP(I)
            XP   = XNODE(NI,1)
            YP   = XNODE(NI,2)
            XI0  = 1.0+XI*XP
            ETA0 = 1.0+ETA*YP
            XI1  = 1.0-XI*XI
            ETA1 = 1.0-ETA*ETA
            XI2  = XP*XI
            ETA2 = YP*ETA
            IF(I .LE. 4) THEN
                 SF(NI)     = 0.25*FNC(XI0,ETA0)*XI2*ETA2
                 DSF(1,NI)= 0.25*XP*FNC(ETA2,ETA0)*(1.0+2.0*XI2)
                 DSF(2,NI)= 0.25*YP*FNC(XI2,XI0)*(1.0+2.0*ETA2)
            ELSE
                 IF(I .LE. 6) THEN
                     SF(NI)      = 0.5*FNC(XI1,ETA0)*ETA2
                     DSF(1,NI)  = -XI*FNC(ETA2,ETA0)
                     DSF(2,NI)  = 0.5*FNC(XI1,YP)*(1.0+2.0*ETA2)
                 ELSE
                     IF(I .LE. 8) THEN
                         SF(NI)      = 0.5*FNC(ETA1,XI0)*XI2
                         DSF(2,NI)  = -ETA*FNC(XI2,XI0)
                         DSF(1,NI)  = 0.5*FNC(ETA1,XP)*(1.0+2.0*XI2)
                     ELSE
                         SF(NI)      = FNC(XI1,ETA1)
                         DSF(1,NI)  = -2.0*XI*ETA1
                         DSF(2,NI)  = -2.0*ETA*XI1
                     ENDIF
                 ENDIF
            ENDIF
   30       CONTINUE
        ENDIF
    ENDIF
C
C     Compute the Jacobian matrix [GJ] and its inverse [GJINV]
C
      DO 40 I = 1,2
      DO 40 J = 1,2
      GJ(I,J)   = 0.0
      DO 40 K = 1,NPE
   40 GJ(I,J)   = GJ(I,J) + DSF(I,K)*ELXY(K,J)
C
      DET = GJ(1,1)*GJ(2,2)-GJ(1,2)*GJ(2,1)
      GJINV(1,1) = GJ(2,2)/DET
      GJINV(2,2) = GJ(1,1)/DET
      GJINV(1,2) = -GJ(1,2)/DET
      GJINV(2,1) = -GJ(2,1)/DET
C
C     Compute the derivatives of the interpolation functions with
C     respect to the global coordinates (x,y): [GDSF]
C
      DO 50 I   = 1,2
      DO 50 J   = 1,NPE
      GDSF(I,J) = 0.0
```

```
      DO 50 K  = 1, 2
   50 GDSF(I,J) = GDSF(I,J) + GJINV(I,K)*DSF(K,J)
      RETURN
      END


      SUBROUTINE TEMPORAL(NCOUNT,INTIAL,ITEM,NN)
C    ________________________________________________________________
C
C    The  subroutine computes  the algebraic equations associated with
C    the  parabolic and hyperbolic differential equations bu using the
C    alfa-family and Newmark family of approximations, respectively.
C    ________________________________________________________________
C
      IMPLICIT REAL*8(A-H,O-Z)
      COMMON/STF/ELF(27),ELK(27,27),ELM(27,27),ELXY(9,2),ELU(27),
     1            ELV(27),ELA(27),A1,A2,A3,A4,A5
C
      IF(ITEM.EQ.1) THEN
C
C     The alfa-family of time approximation for parabolic equations
C
          DO 20 I=1,NN
          SUM=0.0
          DO 10 J=1,NN
          SUM=SUM+(ELM(I,J)-A2*ELK(I,J))*ELU(J)
   10     ELK(I,J)=ELM(I,J)+A1*ELK(I,J)
   20     ELF(I)=(A1+A2)*ELF(I)+SUM
      ELSE
C
C     The Newmark intergration scheme for hyperbolic equations
C
          IF(NCOUNT.EQ.1 .AND. INTIAL.NE.0) THEN
              DO 40 I = 1,NN
              ELF(I)  = 0.0
              DO 40 J = 1,NN
              ELF(I)  = ELF(I)-ELK(I,J)*ELU(J)
   40         ELK(I,J)= ELM(I,J)
          ELSE
              DO 70 I = 1,NN
              SUM     = 0.0
              DO 60 J = 1,NN
              SUM     = SUM+ELM(I,J)*(A3*ELU(J)+A4*ELV(J)+A5*ELA(J))
   60         ELK(I,J)= ELK(I,J)+A3*ELM(I,J)
   70         ELF(I)  = ELF(I)+SUM
          ENDIF
      ENDIF
      RETURN
      END
```

Appendix B

FIXED POINT METHODS AND CONTRACTION MAPPINGS

B–1 FIXED POINT THEOREM

The discussion in this section will focus on the use of fixed point iteration schemes for the solution of the system of nonlinear equations of the form

$$\mathcal{R}(\mathbf{x}) = 0 \qquad (B-1)$$

or its fixed point form

$$\mathbf{x} = \mathcal{G}(\mathbf{x}) \equiv \mathbf{x} - \mathcal{R}(\mathbf{x}) \qquad (B-2)$$

where $\mathbf{x}$ represents a vector of unknowns. Solutions of equation (2) are called *fixed points* of the mapping $\mathcal{G}$, because they are unchanged by the operation of $\mathcal{G}$. The form of (2) immediately suggests the following iteration scheme [1]:

$$\mathbf{x}^{n+1} = \mathcal{G}(\mathbf{x}^n) \qquad (B-3)$$

where $n = 0, 1, \cdots$. Equation (3) is the method of successive substitutions, known as the Picard iteration method. The convergence of the method in (3) is guaranteed by the *Banach Fixed Point Theorem*, which gives the sufficient conditions for the existence and uniqueness of solutions (see Reddy [1], pp. 202–208). Before we present the theorem, it is necessary to define a contraction mapping.

An operator $\mathcal{G}$ is termed a *contraction mapping* if there exists a positive number λ, $0 < \lambda < 1$, such that

$$\| \mathcal{G}(\mathbf{x}) - \mathcal{G}(\mathbf{y}) \| \leq \lambda \| \mathbf{x} - \mathbf{y} \| \qquad (B-4)$$

holds for all $\mathbf{x}, \mathbf{y}$ in the closed ball $\mathcal{N} = \{\mathbf{x} : \| \mathbf{x} - \mathbf{x}_0 \| \leq r\}$ of a normed space $\mathcal{V}$, with center at $\mathbf{x}_0$ and radius r. Here λ denotes the contraction factor. The notation $\| \cdot \|$ is an appropriate norm for the operator in $\mathcal{N}$ (see [1]). It is clear from equation (4) that a contraction mapping "contracts distances": the distance between the images $\mathcal{G}(\mathbf{x})$ and $\mathcal{G}(\mathbf{y})$ is smaller by a scale factor of λ than the distance between the elements $\mathbf{x}$ and $\mathbf{y}$.

Theorem: Let $\mathcal{G}(\mathbf{x})$ be a contraction mapping on a closed subset $\mathcal{N}$ of a closed normed space (or a Banach space) $\mathcal{V}$, with a contraction factor λ and let $\mathbf{x}_0$ be such that

$$\frac{1}{1-\lambda} \parallel \mathcal{G}(\mathbf{x}_0) - \mathbf{x}_0 \parallel = r_0 \leq r$$

Then

1. The sequence defined by $\mathbf{x}^{n+1} = \mathcal{G}(\mathbf{x}^n)$ converges to a point $\mathbf{x}^*$ in $\mathcal{N}$.
2. $\mathbf{x}^*$ is a fixed point of the operator $\mathcal{G}$.
3. $\mathbf{x}^*$ is the unique fixed point of $\mathcal{G}$ in $\mathcal{N}$.

The proof of this theorem and a detailed background on nonlinear solution methods is available in [1,2].

Based on the above theorem, a further result can be obtained if a few restrictions are placed on $\mathcal{G}$. Assume that (3) has a solution $\mathbf{x}^*$ and that $\mathcal{G}$ has a continuous Jacobian that satisfies

$$\parallel \mathcal{G}'(\mathbf{x}) \parallel_\infty \leq \lambda < 1 \qquad (B-5)$$

where $\parallel \cdot \parallel_\infty$ denotes the *sup–norm* ("sup" stands for supremum). Then it can be shown that for any initial estimate $\mathbf{x}^0$ satisfying

$$\parallel \mathbf{x}^0 - \mathbf{x}^* \parallel_\infty \leq r \qquad (B-6)$$

the iterative scheme in equation (3) will converge to the unique solution $\mathbf{x}^*$. The conditions (5) and (6) basically state that the norm of the Jacobian of the iteration operator must be less than unity and the initial estimate for the solution must be within the ball of contraction, if the fixed point scheme is to converge. The quantity r is also referred to as the radius of convergence of the iteration scheme.

Generally, it is quite difficult to prove that a given operator $\mathcal{G}$ is a contraction mapping, especially for the systems generated by finite element applications. It is somewhat easier to use (5) and estimate the norm of $\mathcal{G}'$ to test for a convergent process. However, the real benefit from the ideas of fixed point operators is derived from the iteration scheme in (3) and its generalization,

$$\mathbf{x}^{n+1} = \mathbf{x}^n - \alpha \mathbf{A}(\mathbf{x}^n)\mathcal{F}(\mathbf{x}^n) \qquad (B-7)$$

where α is a constant and $\mathbf{A}$ is a nonsingular matrix function of $\mathbf{x}$. In the following, a few of the basic iteration schemes will be defined based on the definition of $\mathbf{A}$.

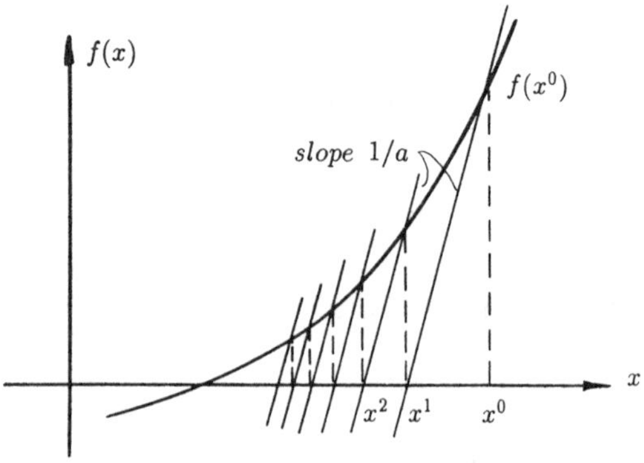

Figure B–1. Illustration of chord method for one degree–of–freedom problem.

B–2 CHORD METHOD

The simplest choice for **A** is a constant which defines the parallel chord method. For a one degree–of–freedom problem, this becomes

$$x^{n+1} = x^n - af(x^n)$$

which is shown graphically in Figure B–1. In the multi–dimensional case, the line with slope $\frac{1}{a}$ is replaced by a hyperplane. The behavior of this method depends strongly on the form of **A**. One possibility is to set $\mathbf{A} = \alpha\mathbf{I}$ (**I** is the identity matrix) in which case each component of **x** is corrected in proportion to α. If $\mathbf{A} = \alpha^T\mathbf{I}$ where α is a vector, then each component of **x** can be corrected individually. The problem remains, however, as to how to select α.

B–3 MODIFIED NEWTON'S METHOD

Extending the idea that **A** is in some sense an inverse "slope", it is appropriate to relate **A** to the derivative of $\mathcal{F}$. Therefore, let

$$\mathbf{A} = \left[\frac{\partial \mathcal{F}}{\partial \mathbf{x}}\right]^{-1} = \mathbf{J}^{-1}$$

in which case (B–7) becomes

$$\mathbf{x}^{n+1} = \mathbf{x}^n - \alpha\mathbf{J}^{-1}(\mathbf{x}^0)\mathcal{F}(\mathbf{x}^n)$$

where the Jacobian is evaluated at $\mathbf{x}^0$. For the one degree–of–freedom case, this becomes ($\alpha = 1$)

$$x^{n+1} = x^n - \frac{1}{f'(x^0)} f(x^n)$$

which is illustrated in Figure B–2. This is recognized as a modified Newton's method since the Jacobian is not updated at each iteration.

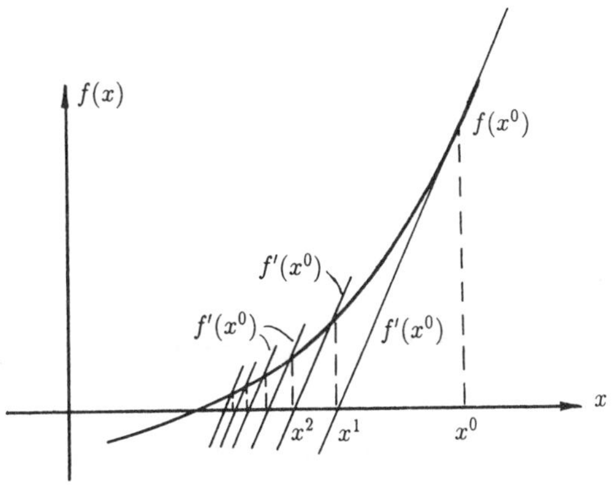

Figure B–2: Illustration of modified Newton's method.

B–4 NEWTON'S METHOD

The modified Newton's scheme suggests a rational manner in which to select **A**. However, its convergence is little better than the chord method due to the use of a constant Jacobian. Allowing the Jacobian to change at each iteration produces the standard Newton (or Newton–Raphson) procedure

$$\mathbf{x}^{n+1} = \mathbf{x}^n - \alpha \mathbf{J}^{-1}(\mathbf{x}^n)\mathcal{F}(\mathbf{x}^n)$$

This is shown graphically in Figure B–3 for a one–component system. Note that if $\alpha \neq 1$ a damped Newton scheme is produced.

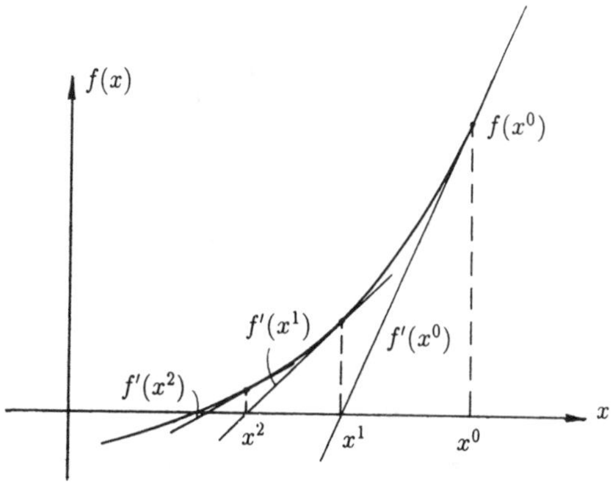

Figure B–3. Illustration of Newton's method.

B–5 DESCENT METHODS

Descent methods reinterpret the general algorithm in (7) as

$$\mathbf{x}^{n+1} = \mathbf{x}^n - \alpha \mathbf{D}^n \qquad (B-8)$$

where $\mathbf{D}$ is a vector that determines the direction of the correction and α controls the magnitude of the correction. In addition, rather than directly finding the solution to (1), descent methods work by trying to minimize the function

$$R(\mathbf{x}) = \mathcal{F}^T(\mathbf{x})\mathcal{F}(\mathbf{x}) \qquad (B-9)$$

The simplest descent methods are the univariant or relaxation schemes in which $\mathbf{D}$ is selected so that only one component of $\mathbf{x}$ is corrected at each iteration. A better approach involves the gradients of R, such as the method of steepest descent. In this case

$$\mathbf{D}^n = \mathbf{g}^n = \nabla R(\mathbf{x}^n) \qquad (B-10)$$

The method of steepest descent has a linear rate of convergence.

Finally, the conjugate gradient method is a sequential method in which the correction vectors are computed from the recursion relation

$$\mathbf{D}^n = \nabla R(\mathbf{x}^n) + \beta^{n-1}\mathbf{D}^{n-1} \qquad (B-11)$$

Here the new correction vector is a combination of the gradient at $\mathbf{x}^n$ plus the old correction direction. The parameter β is defined to insure that the new and old correction vectors are properly orthogonal (conjugate).

Full details for all of the above algorithms and the general concepts behind nonlinear solution methods can be found in References 3–5.

REFERENCES

1. J. N. Reddy, *Applied Functional Analysis and Variational Methods in Engineering*, McGraw–Hill, New York (1986); reprinted by Krieger Publishers, Melbourne, Florida (1991).

2. M. R. Hestenes, *Optimization Theory: The Finite Dimensional Case*, John Wiley & Sons, New York (1975).

3. E. B. Becker, G. F. Carey, and J. T. Oden, *Finite Elements, An Introduction*, Vol. I, Prentice–Hall, Englewood Cliffs, New Jersey (1981).

4. J. M. Ortega and W. C. Rheinboldt, *Iterative Solution of Nonlinear Equations in Several Variables*, Academic Press, New York (1970).

5. C. Johnson, *Numerial Solution of Partial Differential Equations by the Finite Element Method*, Cambridge University Press, Cambridge, U.K. (1987).

SUBJECT INDEX